# Robotics, Automation and Control

Edited by

Pavla Pecherková, Miroslav Flídr
and Jindřich Duník

**Robotics, Automation and Control**

Edited by Pavla Pecherková, Miroslav Flídr and Jindřich Duník

This is a reprint of the book published in 2008 by In-Teh

**Republished by InTech**

Janeza Trdine 9, 51000 Rijeka, Croatia



**Technical Editor** Teodora Smiljanic

**Cover Designe**r Dino Smrekar

Additional hard copies can be obtained from orders@intechopen.com

Robotics, Automation and Control,
Edited by Pavla Pecherková, Miroslav Flídr and Jindřich Duník
p. cm.
ISBN 978-953-7619-18-3

For EU product safety concerns:
IN TECH d.o.o., Prolaz Marije Krucifikse Kozulić 3, 51000 Rijeka, Croatia,
info@intechopen.com or visit our website at intechopen.com.

# Preface

This book was conceived as a gathering place of new ideas from academia, industry, research and practice in the fields of robotics, automation and control. The aim of the book was to point out interactions among the various fields of interests in spite of diversity and narrow specializations which prevail in the current research.

We believe that the resulting collection of papers fulfills the aim of the book. The book presents twenty four chapters in total. The scope of the topics presented in the individual chapters ranges from classical control and estimation problems to the latest artificial intelligence techniques. Moreover, whenever possible and appropriate, the proposed solutions and theories are applied to real-world problems.

The common denominator of all included chapters appears to be a synergy of various specializations. This synergy yields deeper understanding of the treated problems. Each new approach applied to a particular problem, may enrich and inspire improvements of already established approaches to the problem.

We would like to express our gratitude to the whole team who made this book possible. We hope that this book will provide new ideas and stimulation for your research.

October 2008

Editors

**Pavla Pecherková**
**Miroslav Flídr**
**Jindřich Duník**

# Contents

# 1

# Multi-Domain Modelling and Control in Mechatronics: the Case of Common Rail Injection Systems

Paolo Lino and Bruno Maione
*Dipartimento di Elettrotecnica ed Elettronica, Politecnico di Bari*
*Via Re David 200, 70125 Bari,*
*Italy*

## 1. Introduction

The optimal design of a mechatronic system calls for the proper dimensioning of mechanical, electronic and embedded control subsystems (Dieterle, 2005; Isermann, 1996; Isermann, 2008). According to the current approach, the design problem is decomposed into several sub-problems, which are faced separately, thus leading to a sub-optimal solution. Usually, the mechanical part and the control system are considered independently of each others: the former is designed first, then the latter is synthesized for the already existing physical system. This approach does not exploit many potential advantages of an integrated design process, which are lost in the separate points of view of different engineering domains. The physical properties and the dynamical behaviour of parts, in which energy conversion plays a central role, are not determined by the choices of the control engineers and therefore are of little concern to them. Their primary interests, indeed, are signal processing and information management, computer power requirements, choice of sensors and sensor locations, and so on. So it can happen that poorly designed mechanical parts do never lead to good performances, even in presence of advanced controllers. On the other hand, a poor knowledge of how controllers can directly influence and balance for defects or weaknesses in mechanical components does not help in achieving quality and good performances of the whole process.

Significant improvements to overall system performances can be achieved by early combining the physical system design and the control system development (Isermann, 1996b; Stobart et al., 1999; Youcef-Toumi, 1996). Nevertheless, some obstacles have to be overcome, as this process requires the knowledge of interactions of the basic components and sub-systems for different operating conditions. To this end, a deep analysis considering the system as a whole and its transient behaviour seems necessary. In this framework, simulation represents an essential tool for designing and optimizing mechatronic systems. In fact, it can help in integrating the steps involved in the whole design process, giving tools to evaluate the effect of changes in the mechanical and the control subsystems, even at early stages. Available or suitably built models may be exploited for the geometric optimization of components, the design and test of control systems, and the characterization of new systems.

Since models are application oriented, none of them has absolute validity. Models that differ for complexity and accuracy can be defined to take into account the main physical phenomena at various accuracy levels (Bertram et al., 2003; Dellino et al., 2007b; Ollero et al., 2006). Mathematical modelling in a control framework requires to trade off between accuracy in representing the dynamical behaviour of the most significant variables and the need of reducing the complexity of controller structure and design process. Namely, if all engineering aspects are taken into account, the control design becomes very messy. On the other hand, using virtual prototyping techniques allows characterizing system dynamics, evaluate and validate the effects of operative conditions and design parameters, which is appropriate for mechanical design (Ferretti et al., 2004); nevertheless, despite of good prediction capabilities, models obtained in such a way are completely useless for designing a control law, as they are not in the form of mathematical equations. Instead, from the control engineer point of view, the use of detailed modelling tools allows the safe and reliable evaluation of the control systems.

It is clear that an appropriate modelling and simulation approach cannot be fitted into the limitations of one formalism at time, particularly in the early stages of the design process. Hence, it is necessary a combination of different methodologies in a multi-formalism approach to modelling supported by an appropriate simulation environment (van Amerongen, 2003; van Amerongen & Breedveld, 2003; Smith, 1999). The use of different domain-specific tools and software packages allows to take advantage of the knowledge from different expertise fields and the power of the specific design environment.

In this chapter, we consider the opportunity of integrating different models, at different level of details, and different design tools, to optimize the design of the mechanical and control systems as a whole. The effectiveness of the approach is illustrated by means of two practical case studies, involving both diesel and CNG injection systems for internal combustion engines, which represent a benchmark for the evaluation of performances of the approach. As a virtual environment for design integration, we choose AMESim (Advanced Modelling Environment for Simulation): a simulation tool, which is oriented to lumped parameter modelling of physical elements, interconnected by ports enlightening the energy exchanges between element and element and between an element and its environment (IMAGINE S.A., 2007). AMESim, indeed, is capable of describing physical phenomena with great precision and details and of accurately predicting the system dynamics. In a first step, we used this tool to obtain virtual prototypes of the injection systems, as similar as possible to the actual final hardware. Then, with reference to these prototypes, we also determined reduced order models in form of transfer function and/or state space representations, more suitable for analytical (or empirical) tuning of the pressure controllers. Using virtual prototypes in these early design stages enabled the evaluation of the influence of the geometrical/physical alternatives on the reduced models used for the controller tuning. Then, based on these reduced models, the controller settings were designed and adjusted in accordance with the early stages of the mechanical design process. Finally, the detailed physical/geometric models of the mechanical parts, created by the AMESim package, were exported ad used as a module in a simulation program, which enabled the evaluation of the controllers performances in the closed-loop system. In other words, the detailed simulation models surrogated for a real hardware. Experimental and simulation proved the validity of the proposed approach.

## 2. Steps in the multi-domain design approach

An integrated design approach gives more degrees of freedom for the optimization of both the mechanical and its control system than the classical approach. In particular, the improvement of the design process could be obtained by considering the following aspects: iteration of the design steps, use of different specific-domains interacting tools for design, application of optimization algorithms supported by appropriate models (Dellino et al., 2007a). The use of different domain-specific tools allows one to take advantage of the knowledge of engineers from different expertise fields and the power of the specific design environment. The interaction during the design process can be realized by using automatic optimization tools and a proper management of communication between different software environments, without the need of the expertise intervention. Instead, the expertise opinion takes place during the analysis phase of performances. The resulting integrated design process could consist in the following steps (Fig. 1):

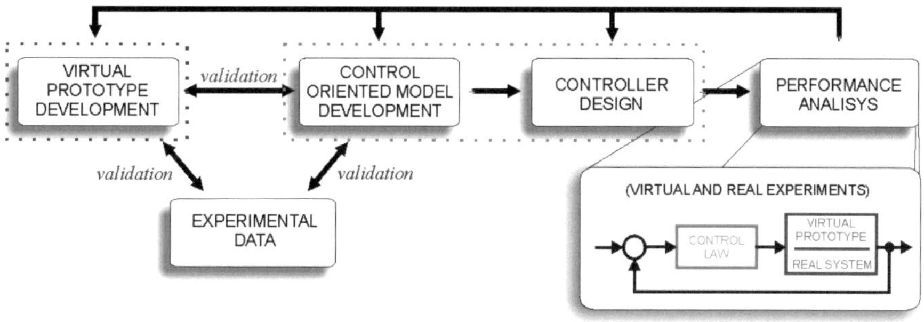


Fig. 1. Integrated design approach for mechatronic systems development.

- Development of a virtual prototype of the considered system using a domain-specific tool (e.g. AMESim, Modelica, etc.) and analysis of the system performances.
- Eventually, realization of a real prototype of the system. Alternatively, a virtual prototype of an existing process can be built and these first two steps have to be swapped.
- Validation of the virtual prototype by comparing simulation results and real data. At the end of this step, the virtual prototype could be assumed as a reliable model of the real system.
- Derivation of a simplified control-oriented analytical model of the real system (white box or black box models). Solving equation of such analytical models is made easier by employing specific software packages devoted to the solution of differential equations (e.g. MATLAB/Simulink).
- Validation of the analytical model against the virtual prototype: this step can be considerably simplified by simulation of different operating conditions.
- Design of control algorithms based on the analytical model parameters. Complex and versatile algorithms are available in computational tools like MATLAB/Simulink to design and simulate control systems. Nevertheless, the construction of accurate models in the same environment could be a complex and stressful process if a deep knowledge of the system under study is not achieved.
- Evaluation of performances of the control laws on the virtual prototype. The use of the virtual prototype allows to perform safer, less expensive, and more reliable tests than

using the real system. In this chapter, the AMESim-Simulink interface allows to integrate AMESim models within the Simulink environment, taking advantage of peculiarities of both software packages.

- The final step consists in evaluating the control algorithm performances on the real system.

The described process could be suitably reiterated to optimize the system and the controller design by using automatic optimization tools. In the next Sections, two case studies involving the common rail injection systems for both CNG and diesel engines are considered to show the feasibility of the described design approach.

## 3. Integrated design of a compressed natural gas injection system

We consider a system composed of the following elements (Fig. 2): a fuel tank, storing high pressure gas, a mechanical pressure reducer, a solenoid valve and the fuel metering system, consisting of a common rail and four electro-injectors. Two different configurations were compared for implementation, with different arrangements of the solenoid valve affecting system performances (i.e. cascade connection, Fig. 2(a), and parallel connection, Fig. 2(b), respectively). Detailed AMESim models were developed for each of them, providing critical information for the final choice. Few details illustrate the injection operation for both layouts.

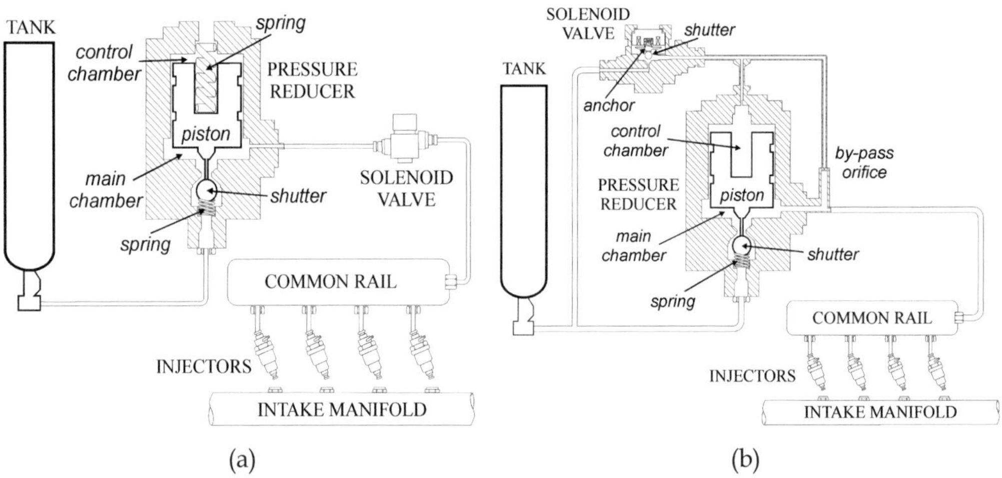


Fig. 2. Block schemes of the common rail CNG injection systems; (a) cascade connection of solenoid valve; (b) parallel connection of solenoid valve.

With reference to Fig. 2(a), the pressure reducer receives fuel from the tank at a pressure in the range between 200 and 20 bars and reduces it to a value of about 10 bar. Then the solenoid valve suitably regulates the gas flow towards the common rail to control pressure level and to damp oscillations due to injections. Finally, the electronically controlled injectors send the gas to the intake manifold for obtaining the proper fuel air mixture. The injection flow only depends on rail pressure and injection timings, which are precisely driven by the Electronic Control Unit (ECU). The variable inflow section of the pressure reducer is varied by the axial displacement of a spherical shutter coupled with a moving piston. Piston and shutter dynamics are affected by the applied forces: gas pressure in a main chamber acts on the piston lower surface pushing it at the top, and elastic force of a preloaded spring holden in a control chamber pushes it down and causes the shutter to open. The spring preload value sets the desired equilibrium reducer pressure: if the pressure exceeds the reference value the shutter closes and the gas inflow reduces, preventing a further pressure rise; on the contrary, if the pressure decreases, the piston moves down and the shutter opens, letting more fuel to enter and causing the pressure to go up in the reducer chamber (see Maione et al., 2004, for details).

As for the second configuration (Fig. 2(b)), the fuel from the pressure reducer directly flows towards the rail, and the solenoid valve regulates the intake flow in a secondary circuit including the control chamber. The role of the force applied by the preloaded spring of control chamber is now played by the pressure force in the secondary circuit, which can be controlled by suitably driving the solenoid valve. When the solenoid valve is energized, the fuel enters the control chamber, causing the pressure on the upper surface of the piston to build up. As a consequence, the piston is pushed down with the shutter, letting more fuel to enter in the main chamber, where the pressure increases. On the contrary, when the solenoid valve is non-energized, the pressure on the upper side of the piston decreases, making the piston to raise and the main chamber shutter to close under the action of a preloaded spring (see Lino et al., 2008, for details).

On the basis of a deep analysis performed on AMESim virtual prototypes the second configuration was chosen as a final solution, because it has advantages in terms of performances and efficiency. To sum up, it guarantees faster transients as the fuel can reach the common rail at a higher pressure. Moreover, leakages involving the pressure reducer due to the allowance between cylinder and piston are reduced by the lesser pressure gradient between the lower and upper piston surfaces. Finally, allowing intermediate positions of the shutter in the pressure reducer permits a more accurate control of the intake flow from the tank and a remarkable reduction of the pressure oscillations due to control operations. A detailed description of the AMESim model of the system according the final layout is in the following (Fig. 3a).

### 3.1 Virtual prototype of the compressed natural gas injection system

By assumption, the pressures distribution within the control chamber, the common rail and the injectors is uniform, and the elastic deformations of solid parts due to pressure changes are negligible. The pipes are considered as incompressible ducts with friction and a non uniform pressure distribution. Temperature variations are taken into account, affecting the pressure dynamics in each subcomponent. Besides, only heat exchanges through pipes are considered, by properly computing a thermal exchange coefficient. The tank pressure plays the role of a maintenance input, and it is modelled by a constant pneumatic pressure source. To simplify the AMESim model construction some supercomponents have been suitably created, collecting elements within a single one.

The main components for modelling the pressure reducer are the *Mass block with stiction and coulomb friction and end stops*, which computes the piston and the shutter dynamics through the Newton's second law of motion, a *Pneumatic ball poppet with conical seat*, two *Pneumatic piston*, and an *Elastic contact* modelling the contact between the piston and the shutter.

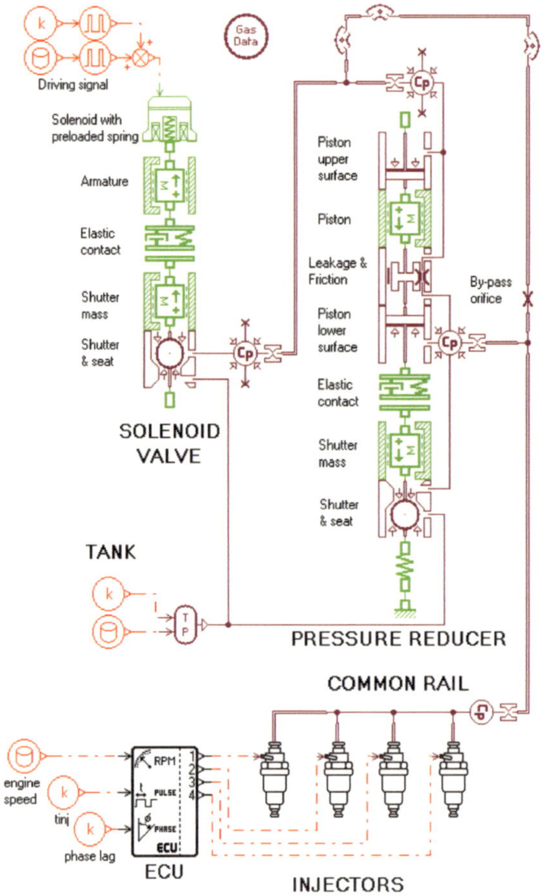


(a)

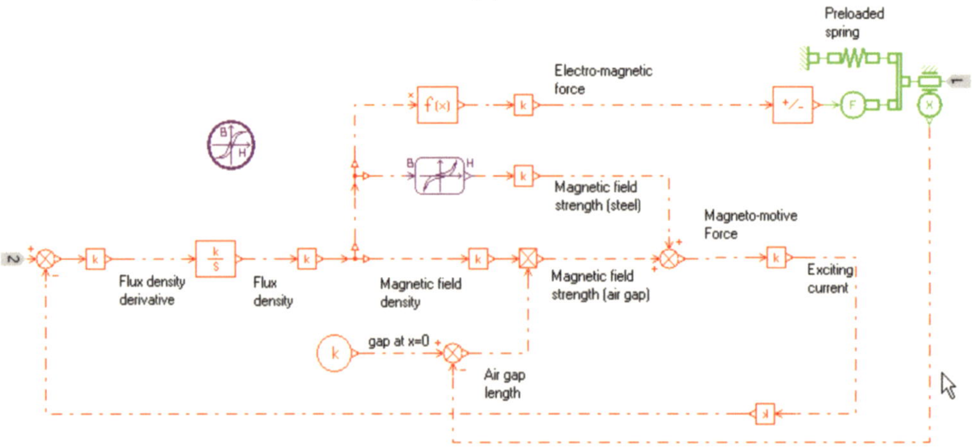


(b)

Fig. 3. (a) AMESim model of the CNG injection system; (b) solenoid with preloaded spring supercomponent.

The *Pneumatic piston* components compute the pressure forces acting upon the upper and lower piston surfaces. The viscous friction and leakage due to contact between piston and cylinder are taken into account through the *Pneumatic leakage and viscous friction* component, by specifying the length of contact, the piston diameter and the clearance. Finally, a *Variable volume with pneumatic chamber* is used to compute the pressure dynamics as a function of temperature $T$ and intake and outtake flows $\dot{m}_{in}$, $\dot{m}_{out}$, as well as of volume changes due to mechanical part motions, according to the following equation:

$$\frac{dp}{dt} = \frac{RT}{V}\left(\dot{m}_{in} - \dot{m}_{out} + \rho\frac{dV}{dt}\right) \tag{1}$$

where $p$ is the fuel pressure, $\rho$ the fuel density and $V$ the taken up volume. The same component is used to model the common rail by neglecting the volume changes. Both pressure and viscous stresses contribute to drag forces acting on a body immersed in a moving fluid. In particular, the total drag acting on a body is the sum of two components: the pressure or form drag, due to pressure gradient, and the skin friction or viscous drag, i.e. *Drag force* = *Form drag* + *Skin friction drag*. By introducing a drag coefficient $C_D$ depending on Reynolds number, the drag can be expressed in terms of the relative speed $v$ (Streeter et al., 1998):

$$Drag = C_D\rho A v^2/2 \tag{2}$$

Moving shutters connecting two different control volumes are subject to both form drag and skin friction drag. The former one is properly computed by AMESim algorithms for a variety of shutters, considering different poppet and seat shapes. As for the latter, it is computed as a linear function of the fluid speed by the factor of proportionality. It can be obtained by noting that for a spherical body it holds *Form Drag* = $2\pi D\mu v$ (Streeter et al., 1998), being $\mu$ the absolute viscosity and $D$ the shutter diameter. The moving anchor in the solenoid valve experiences a viscous drag depending on the body shape. The skin friction drag can be computed using eq. (2) by considering the appropriate value of $C_D$. Since, by hypothesis, the anchor moves within a fluid with uniform pressure distribution, the form drag is neglected.

The continuity and momentum equations are used to compute pressures and flows through pipes so as to take into account wave propagation effects. In case of long pipes with friction, a system of nonlinear partial differential equations is obtained, which is implemented in the *Distributive wave equation submodel of pneumatic pipe* component from the pneumatic library. This is the case of pipes connecting pressure reducer and common rail. The continuity and momentum equations can be expressed as follows (Streeter et al., 1998):

$$\frac{\partial\rho}{\partial t} + \rho\frac{\partial v}{\partial x} = 0 \tag{3}$$

$$\frac{\partial v}{\partial t} + \frac{\alpha^2}{\rho}\frac{\partial\rho}{\partial x} + \frac{f}{2d}v|v| = 0 \tag{4}$$

where $\alpha$ is the sound speed in the gas, $d$ is the pipe internal diameter, $f$ is the D'Arcy friction coefficient depending on the Reynolds number. AMESim numerically solves the above equations by discretization.

For short pipes, the *Compressibility + friction submodel of pneumatic pipe* is used, allowing to compute the flow according the following equation:

$$q = \sqrt{\frac{2d\Delta p}{L\rho f}} \tag{5}$$

where $\Delta p$ is the pressure drop along the pipe of length $L$. The pipes connecting common rail and injectors are modelled in such a way.
Heat transfer exchanges are accounted for by the above mentioned AMESim components, provided that a heat transfer coefficient is properly specified. For a cylindrical pipe of length $L$ consisting of a homogeneous material with constant thermal conductivity $k$ and having an inner and outer convective fluid flow, the thermal flow $Q$ is given by (Zucrow& Hoffman, 1976):

$$Q = \frac{2\pi k L \Delta T}{\ln r_o / r_i} \tag{6}$$

where $\Delta T$ is the temperature gradient between the internal and external surfaces, and $r_o$ and $r_i$ are the external and internal radiuses, respectively. With reference to the outside surface of the pipe, the heat-transfer coefficient $U_o$ is:

$$U_o = \frac{k}{r_o \ln r_o / r_i} \tag{7}$$

The AMESim model for the solenoid valve is composed of the following elements: a solenoid with preloaded spring, two moving masses with end stops subject to viscous friction and representing the magnet anchor and the shutter respectively, and a component representing the elastic contact between the anchor and the shutter. The intake section depends on the axial displacement of the shutter over the conical seat and is computed within the *Pneumatic ball poppet with conical seat* component, which also evaluates the drags acting on the shutter. The solenoid valve is driven by a peak-hold modulated voltage. The resulting current consists of a peak phase followed by a variable duration hold phase. The valve opening time is regulated by varying the ratio between the hold phase duration and signal period, namely the control signal *duty cycle*. This signal is reconstructed by using a *Data from ASCII file signal source* that drives a *Pulse Width Modulation* component.
To compute the magnetic force applied to the anchor, a supercomponent *Solenoid with preloaded spring* in Fig. 3a modelling the magnetic circuit has been suitably built, as described in the following (Fig. 3b). The magnetic flux within the whole magnetic circuit is given by the Faraday law:

$$\dot{\varphi} = \left(e_{ev} - \mathrm{Re}_v\, i_{ev}\right)/n \tag{8}$$

where $\varphi$ is the magnetic flux, $R$ the $n$ turns winding resistance, $e_{ev}$ the applied voltage and $i_{ev}$ the circuit current. Flux leakage and eddy-currents have been neglected. The magnetomotive-force *MMF* able to produce the magnetic flux has to compensate the magnetic tension drop along the magnetic and the air gap paths. Even though most of the circuit reluctance is applied to the air gap, nonlinear properties of the magnet, due to saturation and hysteresis, sensibly affect the system behaviour. The following equation holds:

$$MMF = MMF_s + MMF_a = H_s l_s + H_a l_a \tag{9}$$

where $H$ is the magnetic field strength and $l$ is the magnetic path length, within the magnet and the gap respectively. The air gap length depends on the actual position of the anchor. The magnetic induction within the magnet is a nonlinear function of $H$. It is assumed that the magnetic flux cross section is constant along the circuit, yielding:

$$B = \varphi / A_m = f(H_s) = \mu_0 H_a \tag{10}$$

where $A_m$ is the air gap cross section and $\mu_0$ is the magnetic permeability of air. The $B$-$H$ curve is the hysteresis curve of the magnetic material. Arranging the previous equations yields to $\varphi$, $B$ and $H$. The resulting magnetic force and circuit current are:

$$F_{ev} = A_m B^2 / \mu_0 \tag{11}$$

$$i_{ev} = MMF / n \tag{12}$$

The force computed by the previous equation is applied to the mass component representing the anchor, so that the force balance can be properly handled by AMESim.
The injectors are solenoid valves driven by the ECU in dependence of engine speed and load. The whole injection cycle takes place in a 720° interval with a 180° delay between each injection. A supercomponent including the same elements as for the solenoid valve has been built to model the electro-injectors. The command signal generation is demanded to the ECU component, which provides a square signal driving each injector and depending on the current engine speed, injector timings and pulse phase angle.

### 3.2 Controller design for a compressed natural gas injection system

In designing an effective control strategy for the injection pressure it is necessary to satisfy physical and technical constraints. In this framework, model predictive control (MPC) techniques are a valuable choice, as they have shown good robustness in presence of large parametric variations and model uncertainties in industrial processes applications. They predict the output from a process model and then impress a control action able to drive the system to a reference trajectory (Rossiter, 2003). A 2nd order state space analytical model of the plant (Lino et al., 2008) is used to derive a predictive control law for the injection pressure regulation. The model trades off between accuracy in representing the dynamical behaviour of the most significant variables and the need of reducing the computational effort and complexity of controller structure and development. The design steps are summarized in the following. Firstly, the model is linearized at different equilibrium points, in dependence of the working conditions set by the driver power request, speed and load. From the linearized models it is possible to derive a discrete transfer function representation by using a backward difference method. Finally, a discrete Generalised Predictive Contorl (GPC) law suitable for the implementation in the ECU is derived from the discrete linear models equations.
By considering the *duty cycle* of the signal driving the solenoid valve and the rail pressure as the input $u$ and output $y$ respectively, a family of ARX models can be obtained, according the above mentioned design steps (Lino et al., 2008):

$$\left(1 - a_1 z^{-1}\right) y(t) = \left(b_0 z^{-1} - b_1 z^{-2}\right) u(t) \tag{13}$$

where $z^{-1}$ is the shift operator and $a_1$, $b_0$, $b_1$ are constant parameters. The $j$-step optimal predictor of a system described by eq. (13) is (Rossiter, 2003):

$$\hat{y}(t+j\mid t)=G_j\Delta u(t+j-1)+F_j y(t) \tag{14}$$

where $G_j$ and $F_j$ are polynomials in $q^{-1}$, and $\Delta$ is the discrete derivative operator. Let $\mathbf{r}$ be the vector of elements $y(t+j)$, $j=1, ..., N$, depending on known values at time $t$. Then eq. (14) can be expressed in the matrix form $\hat{\mathbf{y}}=\mathbf{G}\tilde{\mathbf{u}}+\mathbf{r}$, being $\tilde{\mathbf{u}}=[\Delta u(t),\ldots,\Delta u(t+N-1)]^T$, and $\mathbf{G}$ a lower triangular $N\times N$ matrix (Rossiter, 2003).
If the vector $\mathbf{w}$ is the sequence of future reference-values, a cost function taking into account the future errors can be introduced:

$$J=E\left\{(\mathbf{G}\tilde{\mathbf{u}}+\mathbf{r}-\mathbf{w})^T(\mathbf{G}\tilde{\mathbf{u}}+\mathbf{r}-\mathbf{w})+\lambda\tilde{\mathbf{u}}^T\tilde{\mathbf{u}}\right\} \tag{15}$$

where $\lambda$ is a sequence of weights on future control actions. The minimization of $J$ with respect of $\tilde{\mathbf{u}}$ gives the optimal control law for the prediction horizon $N$:

$$\tilde{\mathbf{u}}=\left(\mathbf{G}^T\mathbf{G}+\lambda I\right)^{-1}\mathbf{G}^T(\mathbf{w}-\mathbf{r}) \tag{16}$$

At each step, the first computed control action is applied and then the optimization process is repeated after updating all vectors. It can be shown (Lino et al., 2008) that the resulting control law for the case study becomes:

$$\Delta u(t)=k_1 w(t)+\left(k_2+k_3 q^{-1}\right)y(t)+k_4\Delta u(t-1) \tag{17}$$

where $[k_1, k_2, k_3, k_4]$ depends on $N$.

## 4. The common rail injection system of diesel engines

The main elements of the common rail diesel injection system in Fig. 4 are a low pressure circuit, including the fuel tank and a low pressure pump, a high pressure pump with a delivery valve, a common rail and the electro-injectors (Stumpp & Ricco, 1996). Few details illustrate the injection operation. The low pressure pump sends the fuel coming from the tank to the high pressure pump. Hence the pump pressure raises, and when it exceeds a given threshold, the delivery valve opens, allowing the fuel to reach the common rail, which supplies the electro-injectors.
The common rail hosts an electro-hydraulic valve driven by the Electronic Control Unit (ECU), which drains the amount of fuel necessary to set the fuel pressure to a reference value. The valve driving signal is a square current with a variable duty cycle (i.e. the ratio between the length of "on" and the "off" phases), which in fact makes the valve to be partially opened and regulates the rail pressure. The high pressure pump is of reciprocating type with a radial piston driven by the eccentric profile of a camshaft. It is connected by a small orifice to the low pressure circuit and by a delivery valve with a conical seat to the high pressure circuit. When the piston of the pump is at the lower dead centre, the intake orifice is open, and allows the fuel to fill the cylinder, while the downstream delivery valve is closed by the forces acting on it. Then, the closure of the intake orifice, due to the camshaft rotation, leads to the compression of the fuel inside the pump chamber.

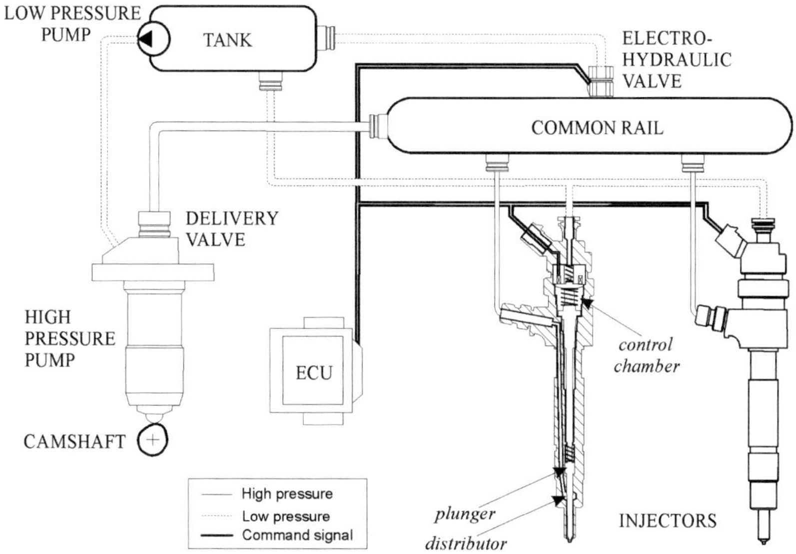


Fig. 4. Block schemes of the common rail diesel injection systems.

When the resultant of valve and pump pressures overcomes a threshold fixed by the spring preload and its stiffness, the shutter of the delivery valve opens and the fuel flows from the pump to the delivery valve and then to the common rail.

As the flow sustained by the high pressure pump is discontinuous, a pressure drop occurs in the rail due to injections when no intake flow is sustained, while the pressure rises when the delivery valve is open and injectors closed. Thus, to reduce the rail pressure oscillations, the regulator acts only during a specific camshaft angular interval (activation window in the following), and its action is synchronized with the pump motion.

The main elements of an electro-injector for diesel engines are a distributor and a control chamber. The control chamber is connected to the rail and to a low pressure volume, and both its inlet and outlet sections are regulated by an electro-hydraulic valve. The distributor includes the feeding pipes and a plunger pushed by a spring against the injection orifices. The plunger axial position depends on the balance of forces acting upon its surfaces, i.e. the control chamber pressure force, the spring force pulling it in closed position, and the distributor pressure force, in the opposite direction. During normal operations the valve electro-magnetic circuit is off and the control chamber is fed by the high pressure fuel coming from the common rail. When the electro-magnetic circuit is excited, the control chamber intake orifice closes while the outtake orifice opens, causing a pressure drop; after a short transient, the plunger reaches the top position disclosing the injection orifices, allowing the injection of the fuel in the cylinders. The Energizing Time ($E_T$) depends on the fuel amount to be injected. When the electro-magnetic circuit is off the control chamber is filled in, so that the plunger is pulled back by the preloaded spring towards the closed position. In the system under study, the whole injection cycle takes place in a complete camshaft revolution and consists of two injections starting every 180 degrees of rotation.

In the described system, the pressure regulation aims at supplying the engine precisely with the specific amount of fluid and the proper air/fuel mixture demanded by its speed and load.

### 4.1 Virtual prototype of the common rail diesel injection system

To build the AMESim model for the common rail diesel injection system, similar assumptions than the previous case are made concerning pressure distributions within lumped volumes like common rail, high pressure pump and injectors control volumes. Diversely, temperature does not affect pressure dynamics. Drag forces acting on moving shutters are computed as previously described. Further, the low pressure pump delivers fuel towards the high pressure pump at a constant pressure, so it is considered as an infinite volume source. On the other hand, because of the isobaric expansion during injection, the cylinders' pressure is slightly variable within a range that can be determined experimentally. For this reason, the cylinders are considered as infinite volumes of constant, albeit uncertain, pressure. Since most of the relevant components have been previously described in dept, a brief discussion introduces those used to assemble the virtual prototype of the diesel injection system shown in Fig. 5a. The high pressure pump model is composed of two subsystems, the former representing the pump dynamics, the latter describing the delivery valve behaviour. In particular, the *Cam and cam follower* block is used to represent the cam profile and its rotary motion, which affects the piston axial displacement. The *Spool with annular orifice* models the orifice connecting the low pressure circuit to high pressure pump; its section varies according to the piston displacement. The piston inertia is neglected in this model. The leakage due to contact between piston and cylinder is taken into account through the *Viscous frictions and leakages* component, by specifying the length of contact, the piston diameter and the clearance. Finally, a *Hydraulic volume with compressibility* is used to compute the pressure dynamics as a function of intake and outtake flows $q_{in}$ and $q_{out}$, as well as of volume changes $dm/dt$ due to mechanical part motions, according to the following equation (IMAGINE S.A., 2007):

$$\frac{dp}{dt} = \frac{K_f}{V}\left( q_{in} - q_{out} - \frac{dm}{dt} \right) \tag{18}$$

where $p$ is the fuel pressure, $V$ the instantaneous volume of liquid and $K_f$ is the fuel bulk modulus of elasticity. The intake and outtake flows come from the energy conservation law. The same component is used to model the delivery valve internal volume, the control chamber and the distributor volumes inside the electro-injectors, and the common rail.

The components included in the delivery valve model are a *Mass block with stiction and coulomb friction and end stops*, which computes the shutter dynamics, a *Poppet with sharp edge seat*, a *Hydraulic volume with compressibility*, and a *Piston with spring* representing the force applied by the preloaded spring on the delivery valve shutter.

To model pipes within the diesel common rail injection system two situations are considered, i.e. short pipes and long pipes, both accounting for friction, fuel compressibility and expansion of pipes due to high pressures. Short pipes are modelled by the *compressibility + friction hydraulic pipe* sub-model, which uses an effective bulk modulus $K_B$ to take into account both compressibility of the fluid and expansion of the pipe wall with pressure; the effective bulk modulus depends on the wall thickness and Young's modulus for the wall material. The equation describing the pressure dynamics at the mid-point is:

$$\frac{\partial p}{\partial t} + \frac{K_B}{A}\frac{\partial q}{\partial x} = 0 \tag{19}$$

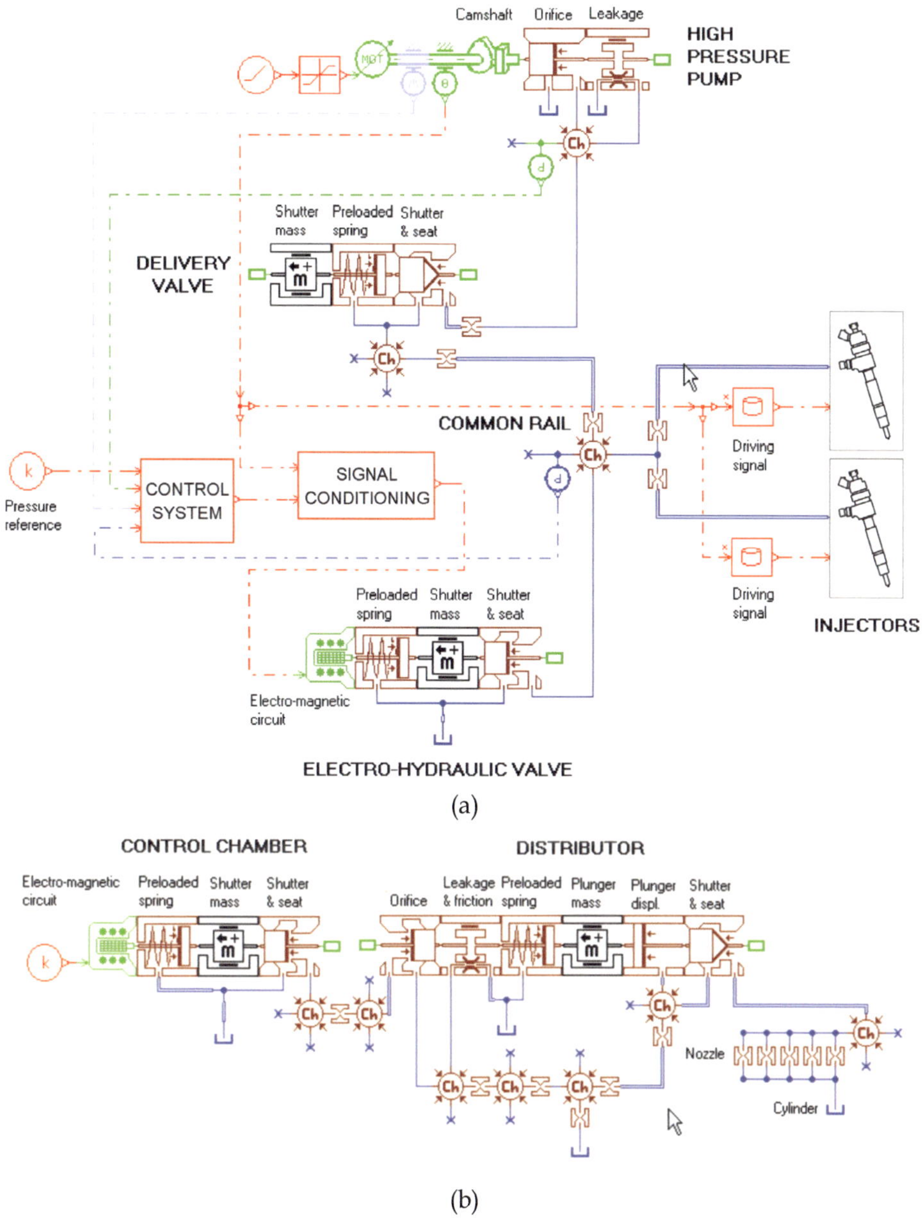


Fig. 5. (a) AMESim model of the common rail diesel injection system; (b) Injector supercomponent.

where $A$ is the pressure dependent cross sectional area of pipe. Pipe friction is computed using a friction factor based on the Reynolds number and relative roughness (IMAGINE SA, 2007). The resulting flow is calculated by means of Eq. (5). This model is used for pipes connecting common rail and injectors. The long pipe connecting delivery valve to common rail is modelled by using the *Simple wave equation hydraulic pipe*, which is based on the continuity equation (19) and momentum equation, giving for uncompressible fluids:

$$\frac{\partial q}{\partial t} - \frac{A}{\rho}\frac{\partial p}{\partial x} + v\frac{\partial q}{\partial x} + \frac{f}{2dA}q|q| = 0 \tag{20}$$

being ρ the fuel density, $d$ the pipe internal diameter, $v$ the mean flow speed and $f$ the friction factor.

The electro-hydraulic valve model includes: a *Mass block with stiction and coulomb friction and end stops* representing the shutter dynamics; a *Piston with spring*; the supercomponent *Electro-magnetic circuit*, which is obtained similarly to the *Solenoid with preloaded spring* supercomponent and converts the controller signal into a force applied to the shutter; a *Spool with annular orifice* modelling the shutter.

Finally, a supercomponent has been used to model the electro-injectors (Fig. 5b), which is a slightly modified version of the block available within the AMESim library. In particular, it consists of two sub-models representing the control chamber and the distributor, respectively. The former sub-model is equal to the electro-hydraulic valve model. The latter includes a *Mass block with stiction and coulomb friction and end stops* and a *Poppet with conical seat* for the plunger, a *Piston with spring*, a *Piston* computing volume changes due to plunger motion, and a *Viscous frictions and leakages* component to take into account flows between the control chamber and the distributor.

### 4.2 Controller design of the diesel common rail

As previously stated, to develop both an appropriate control strategy and an effective controller tuning a simplified model of the diesel common rail injection system is necessary. In fact, considering too many details and a high number of adjustable parameters make the design of the control law quite difficult. Hence, to this aim, a lumped parameter nonlinear model is considered (Lino et al., 2007), which is suitable for control purposes and can be adapted to different injection systems with the same architecture. The model is validated by simulation, using the package AMESim. The model is expressed in state space form, where the pump pressure $p_p$ and the common rail pressure $p_r$ are the state variables, while the camshaft angular position θ and speed $\omega_{rpm}$, the regulator exciting signal $u$ and the injectors driving signal $E_T$ are the inputs. Assuming, without loss of generality, that no reversal flows occur, the state space representation is (Lino et al., 2007):

$$\begin{aligned}\dot{p}_p &= \eta_a\left(p_p, p_r, \omega_{rpm}, \theta\right)\\ \dot{p}_r &= f_1\left(p_p, p_r\right) + f_2\left(p_r\right)\cdot u + \eta_b\left(p_r, E_T\right)\end{aligned} \tag{21}$$

where $f_1$ and $f_2$ are known functions of pressures, which are accessible for measurement and control. Functions $\eta_a$ and $\eta_b$ also depend on parameters which are uncertain (i.e. camshaft angular position and cylinders pressure) and not available for control purpose. The aim of the control action $u$ is to take $p_p$ and $p_r$ close to the constant set-points $P_p$ and $P_r$ . Hence, by defining $e_p = p_p - P_p$ and $e_r = p_r - P_r$, equation (21) become:

$$\begin{aligned}\dot{e}_p &= \eta_a\left(e_p, e_r, \omega_{rpm}, \theta\right)\\ \dot{e}_r &= f_1\left(e_p, e_r\right) + f_2\left(e_r\right)\cdot u + \eta_b\left(e_r, E_{T12}\right)\end{aligned} \tag{22}$$

Given a system described by equations (22), it is possible to design a sliding mode control law that can effectively cope with system nonlinearities and uncertainties (Khalil, 2002). The aim of the sliding mode approach is to design a control law $u$ able to take the system trajectory on a sliding surface $s = e_r - \psi(e_p) = 0$ and, as soon as the trajectory lies on this surface, $u$ must also make $e_p = 0$. In (22), $e_r$ plays the role of the control input; therefore the control can be achieved by solving a stabilization problem for $\dot{e}_p = \eta_a(e_p, e_r, \omega_{rpm}, \theta)$. A control law $u$ is designed to bring $s$ to 0 in finite time and to make $s$ maintain this value for all future time. Since by (22)

$$\dot{s} = f_1(e_p, e_r) - \frac{\partial \psi}{\partial e_p}\eta_a(e_p, e_r, \omega_{rpm}, \theta) + f_2(e_r)\cdot u + \eta_b(e_r, E_{T12}) \tag{23}$$

the control law $u$ can be designed to cancel $f_1(e_p, e_r)$ on the right-hand side of (23):

$$u = \frac{1}{f_2(e_r)}\left[-f_1(e_p, e_r) + \varepsilon\right] \tag{24}$$

where $\varepsilon$ must be chosen to compensate the other nonlinear terms in (23). If the absolute value of the remaining term in (23) is bounded by a positive function $\sigma(e_p, e_r) \geq 0$, it is possible to design $\varepsilon$ to force $s$ toward the sliding surface $s = 0$. More precisely, the sliding surface is attractive if $\varepsilon$ is given by:

$$\begin{aligned} &\varepsilon = -\beta(e_p, e_r)\mathrm{sgn}(s) \\ &\beta(e_p, e_r) \geq \sigma(e_p, e_r) + \beta_0 \end{aligned} \tag{25}$$

with $\beta_0 > 0$ coping with uncertainties (Lino et al., 2007). This ensures that $s\dot{s} \leq 0$, so that all the trajectories starting off the sliding surface reach it in finite time and those on the surface cannot leave it. The sliding surface is chosen as $s = e_r + ke_p = 0$, where $k$ is an appropriate constant representing the sliding surface slope. In particular, $e_p \to 0$ only if $k \to \infty$. With a finite $k$, $e_p$, and consequently $e_r$, are finite and can only be made smaller than a certain value. However, to avoid saturation of the control valve, $k$ cannot be chosen too high. Finally, the sliding surface is made attractive for the error trajectory by a proper choice of $\beta(e_p, e_r)$. To compensate the rail pressure drop $\Delta p_r$ caused by the injection occurring within the angular interval [180°, 360°] during regulator inactivity, a compensation term is introduced in the pressure reference which is derived from the injection flow equation. Thus, the new pressure reference becomes $P_r - \Delta p_r$ (Lino et al., 2007).

## 5. Simulation and experimental results

### 5.1 The CNG injection system

To assess the effectiveness of the AMESim model in predicting the system behaviour, a comparison of simulation and experimental results has been performed. Since, for safety reasons, air is used as test fluid, the experimental setup includes a compressor, providing air at a constant input pressure and substituting the fuel tank. The injection system is equipped with four injectors sending the air to a discharging manifold. Moreover, a PC system with a National Instrument acquisition board is used to generate the engine speed and load signals, and a programmable MF3 development master box takes the role of ECU driving the injectors and the control valve.

Figure 6 refers to a typical transient operating condition, and compares experimental and simulation results. With a constant 40 bar input pressure, the system behaviour for a constant $t_j$ = 3ms injectors opening time interval, while varying engine speed and solenoid valve driving signal has been evaluated. The engine speed is composed of ramp profiles (6c), while the *duty cycle* changes abruptly within the interval [2%, 12%] (Fig. 6d). Figures 6a and 6b show that the resulting dynamics is in accordance with the expected behaviour. A maximum error of 10% confirms the model validity.

After the validation process, the AMESim virtual prototype was used to evaluate the GPC controller performances in simulation by employing the AMESim-Simulink interface, which enabled us to export AMESim models within the Simulink environment. The interaction between the two environments operates in a *Normal* mode or a *Co-simulation* mode. As for the former, a compiled S-function containing the AMESim model is generated and included in the Simulink block scheme, and then integrated by the Simulink solver. As for the latter, which is the case considered in this chapter, AMESim and Simulink cooperate by integrating the relevant portions of models.

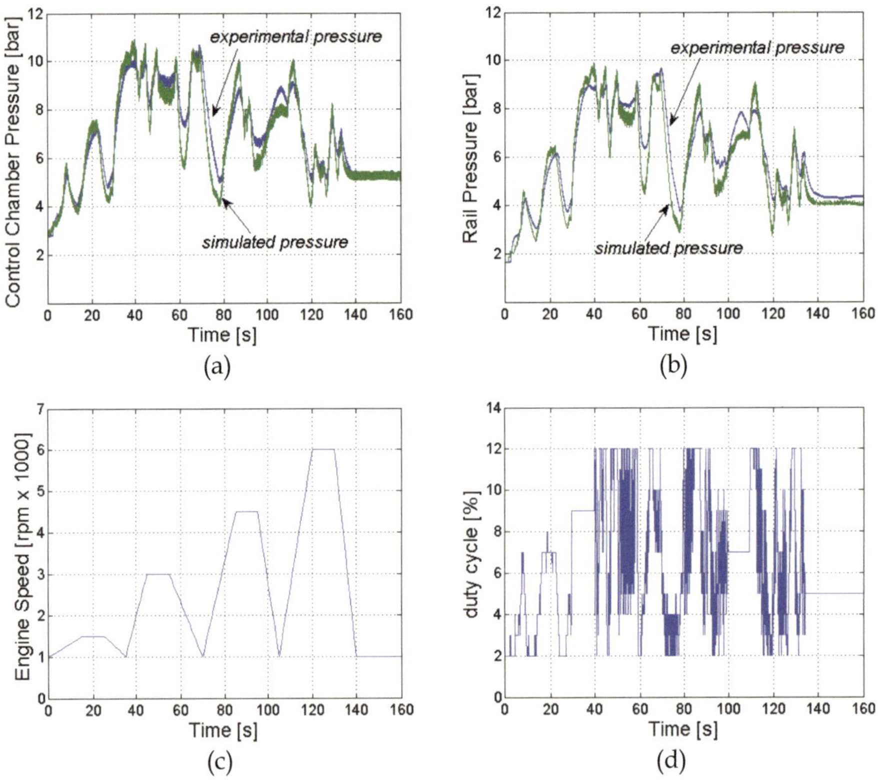


Fig. 6. Simulation and experimental results when varying *duty cycle* and engine speed, with a constant $t_j$ = 3ms; (a) control chamber pressure; (b) common rail pressure; (c) engine speed; (d) control signal *duty cycle*.

The GPC controller was tuned referring to models linearized at the starting equilibrium point, according to design steps of Section 3.2. The test considered ramp variations of the engine speed and load, for the system controlled by a GPC with a $N$ = 5 (0.5s) prediction horizon. The input air pressure from the compressor was always 30bar. The rail pressure reference was read from a static map depending on the working condition and had a sort of ramp profile as well. The final design step consisted in the application of the GPC control law to the real system.

In Fig. 7, the engine speed accelerates from 1100rpm to 1800rpm and then decelerates to 1100rpm, within a 20s time interval (Fig. 7b). The control action applied to the real system guarantees a good reference tracking, provided that its slope does not exceed a certain value (Fig. 7a, time intervals [0, 14] and [22, 40]). Starting from time 14s, the request of a quick pressure reduction causes the control action to close the valve completely (Fig. 7c) by imposing a *duty cycle* equal to 0. Thanks to injections, the rail pressure (Fig. 7a) decreases to the final 5bar reference value, with a time constant depending on the system geometry; the maximum error amplitude cannot be reduced due to the actuation variable saturation. Fig. 7d shows the injectors' exciting time during the experiment. It is worth to note that simulation and experimental results are in good accordance, supporting the proposed approach.

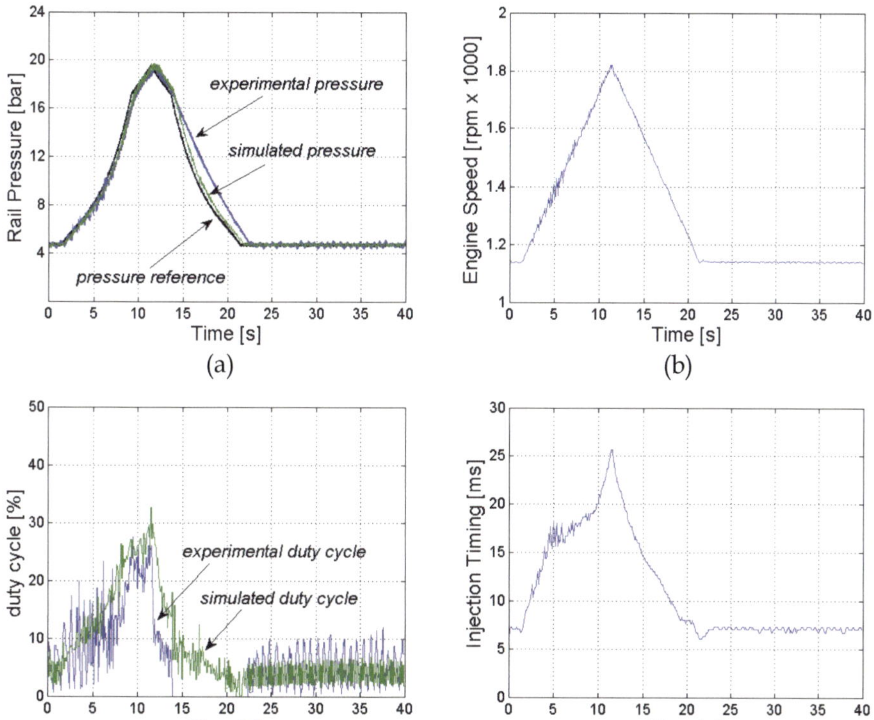


Fig. 7. Model and real system responses for speed and load ramp variations and a 30bar input pressure, when controlled by a GPC with N = 5; (a) common rail pressure; (b) engine speed (c) *duty cycle*; (d) injectors exciting time interval.

### 5.2 The diesel injection system

The state space model used for designing the pressure controller has been implemented and simulated in the Matlab/Simulink® environment. To assess its capability of predicting the rail pressure dynamics for each injection cycle, simulation results have been compared both with experimental data obtained on a Common Rail injection system (Dinoi, 2002) and with those provided by the AMESim software for fluid dynamic simulation. Modelling and simulation within this prototyping environment are used for verifying alternative designs and parameterizations. For the sake of brevity, experimental results are not shown in this chapter (see Lino et al., 2007 for details).

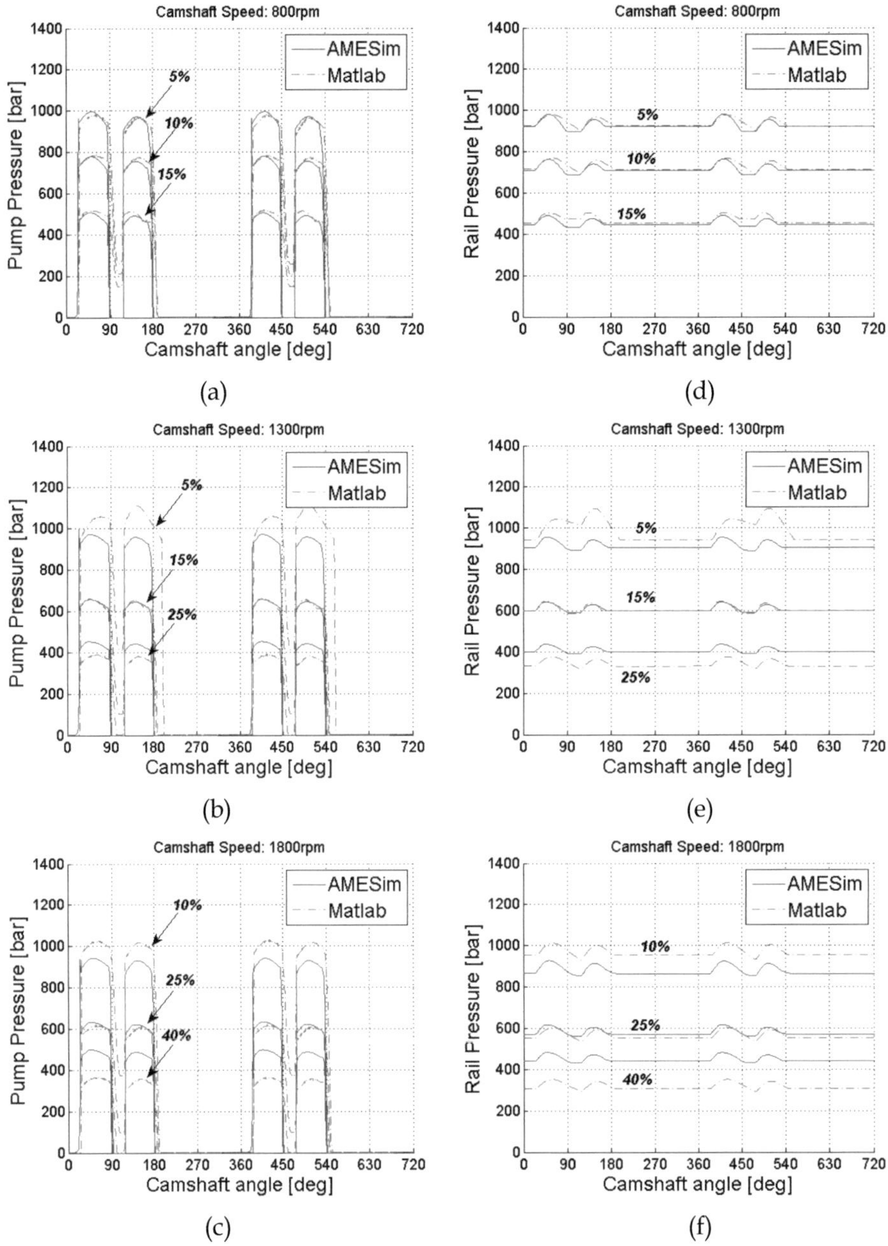


Fig. 8. Comparison of AMESim and Matlab simulated pump and rail pressures, by varying the solenoid valve driving signal and for different camshaft speeds: (a), (b), (c) pump pressure; (d), (e), (f) rail pressure.

Fig. 8 compares Matlab and AMESim simulations referred to two complete camshaft revolutions. This figure represents pump and rail pressures, for 800, 1300 and 1800 rpm camshaft speeds respectively, and different values of the electro-hydraulic valve duty-cycle. According to Figure 8, the pump pressure increases because of the piston motion, until the delivery valve opens. From this moment on, the pressure decreases because of the outflow towards the rail. Subsequently, the pressure increases again because of the camshaft profile. The rail pressure is constant during the angular interval in which the delivery valve is closed. The opening of the delivery valve causes a pressure increase, which is immediately compensated by the intervention of the regulation valve. We can conclude that the pressure dynamics are well modelled, both in amplitude and in timing. The difference in the steady values is due to the approximation introduced in the state space model for the electro-hydraulic valve.

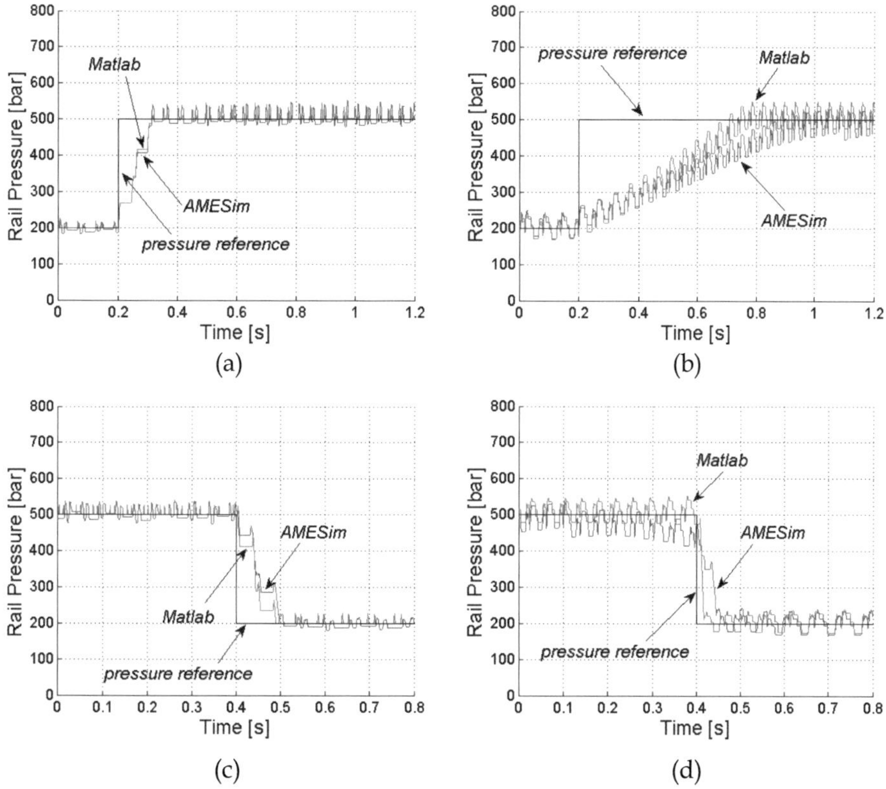


Fig. 9. AMESim and Matlab rail pressure dynamics with reference step variations and ramp camshaft speed variations; (a) step and ramp increments in absence of injections; (b) step and ramp increments in presence of injections; (c) step and ramp decrements in absence of injections; (d) step and ramp decrements in presence of injections.

To test the sliding mode controller tracking and disturbance rejection capabilities we have extensively simulated different operating conditions by using AMESim software. To check the effectiveness of the approach AMESim simulations with Matlab simulations of the state space model have been compared. Significant results are discussed in the following.

First of all, it is considered a reference pressure step variation, by varying the camshaft speed, without injections. In Fig. 9a, a 300 bar set-point variation occurs at time 0.2, starting from 200bar up to 500bar. The initial 1000rpm camshaft speed increases to 1600rpm following a ramp profile, within a 0.5s time interval starting at time 0.2s. The injectors are kept completely closed, so that the controller copes with the pressure disturbances due to the pump motion. Moreover, the control action, computed at the beginning of each injection cycle, is applied during the valve activation window. The rail pressure is properly taken close to the set-point without any overshoot, but it suffers of undesirable oscillations, as the control action is held constant during the whole camshaft revolution. Figure 9b considers analogous operating conditions, but in presence of injections. The injectors' driving signal acts to keep the angular opening interval constant, regardless of the camshaft speed. The sliding mode controller is still able to maintain the rail pressure close to the reference value, with a good rejection of the disturbance due to the injection flow. The high pressure oscillations, due to the injectors' operation, cannot be removed as the control valve acts only during the activation window. Finally, the rising transient is slower than the previous case, as a fraction of the fuel delivered by the pump is sent to cylinders. In both cases, Matlab simulations are in good accordance with those performed within the AMESim environment, showing the feasibility of the derivation of the control law from the reduced order model.
In Figures 9c and 9d, a pressure reference step variation occurs at time 0.4, while the speed decreases from 1600rpm to 1000rpm, within a 0.5s time interval starting at time 0.2s, following a ramp profile. In the first case (Fig. 9c), the injectors are kept completely closed. In the second case (Fig. 9d), the injectors opening time is proportional to the camshaft speed. It is possible to note that, within the time interval 0.2-0.4s, the control action is not able to maintain the rail pressure at set-point, as decreasing the pump speed reduces fuel supply for each camshaft revolution. Even in these working conditions the comparison of Matlab and AMESim results confirms the proposed approach.

## 7. Conclusion

In this chapter, we presented a procedure for integrating different models and tools for a reliable design, optimization and analysis of a mechatronic system as a whole, encompassing the real process and the control system. The effectiveness of the methodology has been illustrated by introducing two practical case studies involving the CNG injection system and the common rail diesel injection system for internal combustion engines. The design process included the possibility of analyzing different candidate configurations carried out with the help of virtual prototypes developed in the AMESim environment, design and performance evaluation of controllers designed on simpler models of the plants by employing the virtual prototypes, validation of the control laws on the real systems. Simulation and experimental results proved the validity of the approach.

## 8. References

van Amerongen, J. (2003). Mechatronic design. *Mechatronics*, Vol. 13, pp. 1045-1066, ISSN 0957-4158.

van Amerongen, J. & Breedveld, P. (2003). Modelling of physical systems for the design and control of mechatronic systems. *Annual Reviews in Control*, Vol. 27, pp. 87–117, ISSN 1367-5788.

Bertram, T., Bekes, F., Greul, R., Hanke, O., Ha, C., Hilgert, J., Hiller, M., Ottgen, O., Opgen-Rhein, P., Torlo, M. & Ward, D. (2003). Modelling and simulation for mechatronic design in automotive systems. *Control Engineering Practice*, Vol. 11, pp. 179–190, ISSN 0967-0661.

Dellino, G., Lino, P., Meloni, C. & Rizzo, A. (2007a). Enhanced evolutionary algorithms for multidisciplinary design optimization: a control engineering perspective, In: *Hybrid Evolutionary Algorithms*, Grosan, C., Abraham, A., Ishibuchi H. (Ed.), 39-76, Springer Verlag, ISBN 978-3-540-73296-9, Berlin, Germany.

Dellino, G., Lino, P., Meloni, C. & Rizzo, A. (2007b). Kriging metamodel management in the design optimization of a CNG injection system. To appear on *Mathematics and Computers in Simulation*, ISSN 0378-4754.

Dieterle, W. (2005). Mechatronic systems: automotive applications and modern design methodologies. *Annual Reviews in Control*, Vol. 29, pp. 273-277, ISSN 1367-5788.

Dinoi, A. (2002). *Control issues in common rail diesel injection systems* (In Italian), Master Theses, Bari, Italy.

Ferretti, G., Magnani, G. & Rocco P. (2004). Virtual prototyping of mechatronic systems. *Annual Reviews in Control*, Vol. 28, No 2, pp. 193-206, ISSN 1367-5788.

IMAGINE S.A. (2007). *AMESim Reference Manual rev7*, Roanne.

Isermann, R. (1996a). Modeling and design methodology for mechatronic systems. *IEEE Transactions on Mechatronics*, Vol. 1, No 1, pp. 16-28, ISSN 1083-4435.

Isermann, R. (1996b). On the design and control of mechatronic systems - a survey. *IEEE Transactions on Industrial Electronics*, Vol. 43, No. 1, pp. 4-15, ISSN 0278-0046.

Isermann, R. (2008). Mechatronic systems - Innovative products with embedded control. *Control Engineering Practice*, Vol. 16, pp. 14–29, ISSN 0967-0661.

Khalil, H. K., (2002). *Nonlinear Systems*, Prentice Hall, ISBN 978-0130673893, Upper Saddle River.

Lino, P., Maione, B. & Rizzo, A. (2007). Nonlinear modelling and control of a common rail injection system for diesel engines. *Applied Mathematical Modelling*, Vol. 31, No 9, pp. 1770-1784, ISSN 0307-904X.

Lino, P., Maione, B. & Amorese, C. (2008). Modeling and predictive control of a new injection system for compressed natural gas engines. *Control Engineering Practice*, Vol. 16, No 10, pp. 1216-1230, ISSN 0967-0661.

Maione, B., Lino, P., DeMatthaeis, S., Amorese, C., Manodoro, D. & Ricco, R. (2004). Modeling and control of a compressed natural gas injection system. *WSEAS Transactions on Systems*, Vol. 3, No 5, pp. 2164-2169, ISSN 1109-2777.

Ollero, A., Boverie, S., Goodall, R., Sasiadek, J., Erbe, H. & Zuehlke, D. (2006). Mechatronics, robotics and components for automation and control. *Annual Reviews in Control*, Vol. 30, pp. 41–54, ISSN 1367-5788.

Rossiter, J.A. (2003). *Model-Based Predictive Control: a Practical Approach*, CRC Press, ISBN 978-0849312915, New York.

Smith, M. H. (1999). Towards a more efficient approach to automotive embedded control system development, *Proceedings of IEEE CACSD Conference*, pp. 219-224, ISBN 0-7803-5500-8, Hawaii, USA, August 1999.

Stobart, R., May, A., Challen, B. & Morel, T. (1999). New tools for engine control system development. *Annual Reviews in Control*, Vol. 23, pp. 109–116, ISSN 1367-5788.

Streeter, V., Wylie, K. & Bedford, E. (1998). *Fluid Mechanics*, McGraw-Hill, ISBN 978-0070625372, New York.

Stumpp, G. & Ricco, M. (1996). Common rail - An attractive fuel injection system for passenger car DI diesel engine. *SAE Technical Paper 960870.*

Youcef-Toumi, K. (1996) Modeling, design, and control integration: a necessary step in mechatronics. *IEEE Transactions on Mechatronics*, Vol. 1, No. 1, pp. 29-38, ISSN 1083-4435.

Zucrow, M. & Hoffman J. (1976). *Gas Dynamics*, John Wiley & Sons, ISBN 978-0471984405, New York.

# 2

# Time-Frequency Representation of Signals Using Kalman Filter

Jindřich Liška and Eduard Janeček
*Department of Cybernetics, University of West Bohemia in Pilsen,*
*Univerzitní 8, 306 14 Pilsen,*
*Czech Republic*

## 1. Introduction

Data analysis is a necessary part of practical applications as well as of pure research. The problem of data analysis is of great interest in engineering areas such as signal processing, speech recognition, vibration analysis, time series modelling, etc. Sensing of physical effects transformed to time series of voltage represents measured reality for us. However, the knowledge of time characteristics of measured data is often quite insufficient for description of real signal properties. Therefore, frequency analysis, as an alternative signal description approach, has been engaged in signal processing methods. Unfortunately, data from real applications are mostly non-stationary and represent non-linear processes in general. This fact has leaded to introduction of joint time-frequency representation of measured signals.

Former time-frequency methods dealing with non-stationary data analysis, such as the short-time Fourier transform, repeatedly use a block of data processing with the assumption that frequency characteristics are time-invariant and/or a process is stationary within the usage of the data block. The resolution of such methods is then given by the Heisenberg-Gabor uncertainty principle. This limitation is mainly addressed by Fourier pairs and by the block processing.

In this chapter, the definition of instantaneous frequency introduces a new approach to acquire a proper time-frequency representation of a signal. The description of limitations which have to be taken into account in order to achieve meaningful results is also given there.

The section that deals with the instantaneous frequency phenomenon is followed by the part describing the construction of the signal model. A signal is modelled as an output of a system that consists of sum of auto-regressive (AR), linear time-invariant (LTI), second-order subsystems. In fact, such a model corresponds to a system of resonators formed in parallel. The reason for such model selection is simplicity in matrix implementation as well as preservation of a physical meaning of decomposed signal components. The signal model is developed in state space and there is also demonstrated how to select adequately all system matrices.

The estimation of particular signal components is then obtained through the adaptive Kalman filtering which is based on the previously defined state space model. The Kalman filter recursively estimates time-varying signal components in a complex form. The complex form is required in order to compute the instantaneous frequency and amplitude of the

signal components. Furthermore, the algorithm recursive form facilitates implementations in signal processing devices. Initial parameters of the Kalman filter are obtained from frequency spectrum characteristics and from estimation of signal spectral density through the adaptive algorithm by matching the auto-regressive prediction model and the signal.
To illustrate performance of the proposed method, experimental results are presented in order to show comparisons of common time-frequency techniques and the Kalman filter method. Moreover, results from analysis of example signals (chirp signal, step frequency change, etc.) are added. The contribution of this method mainly consists in improvement of the time-frequency resolution as discussed in the last section, which concludes the whole chapter.

## 2. Instantaneous frequency and the complex signal

In mechanics, the frequency of vibration motion is defined as the number of oscillations per time period. In the course of one oscillation, the body deflects from the equilibrium, goes through the extremes and the oscillation ends again in the equilibrium position. The simple harmonic sine wave is often used for representation of such a vibration motion. The harmonic motion actually represents projection of circle body move with uniform velocity on a circle diameter.
However, in many applications the motion velocity and therefore also the oscillation frequency changes in time. Signals with these properties are often referred to as nonstationary and their important characteristic is the dependence of frequency on time, therefore the notion of instantaneous frequency. In other words, instantaneous frequency is the time-varying parameter which describes the location of the signal's spectral peak in time. It can be interpreted as the frequency of a sine wave which locally fits the analyzed signal, but physically it has meaning only for monocomponent signals, where there is only one frequency or a narrow range of frequencies varying as a function of time (Boashash, 1992a).
Let's assume the simple harmonic motion in following form:

$$s_r(t) = a\cos(\omega t + \theta), \quad \omega = 2\pi f \tag{1}$$

where $a$ is amplitude, $\omega$ is the angular frequency, $\theta$ is a phase constant and the argument of the cosine function is the instantaneous phase $\phi(t)$ (namely $\omega t + \theta$). In the case of frequency changing the instantaneous phase $\phi(t)$ is then the integral of frequency in time and the signal form should be rewritten as

$$s_r(t) = a\cos(\int_0^t \omega(t)dt + \theta) \tag{2}$$

Considering the monocomponent signal, instantaneous frequency $\omega(t)$ is defined, with respect to (2), as a derivation of phase $\phi(t)$

$$\omega(t) = \frac{d\phi(t)}{dt} = 2\pi f(t) \tag{3}$$

In most cases, there is no way how to determine the instantaneous phase directly from the real signal. One of the tricks how to obtain the unknown phase is the introduction of complex signal $z(t)$ which somehow corresponds to the real signal. As described in (Hahn, 1996) or in (Huang, 1998), the Hilbert transform is the elegant solution for generating of complex signal from the real one. The Hilbert transform of a real signal $s_r(t)$ of the continuous variable $t$ is

$$s_i(t) = \frac{1}{\pi} P \int_{-\infty}^{\infty} \frac{s_r(\eta)}{\eta - t} d\eta \,, \tag{4}$$

where $P$ indicates the Cauchy Principle Value integral. The complex signal $z(t)$

$$z(t) = s_r(t) + j \cdot s_i(t) = a(t)e^{j\varphi(t)} \,, \tag{5}$$

whose imaginary part is the Hilbert transform $s_i(t)$ of the real part $s_r(t)$ is then called the analytical signal and its spectrum is composed only of the positive frequencies of the real signal $s_r(t)$.

From the complex signal, an instantaneous frequency and amplitude can be obtained for every value of $t$. Following (Hahn, 1996) the instantaneous amplitude is simply defined as

$$a(t) = \sqrt{s_r(t)^2 + s_i(t)^2} \tag{6}$$

and similarly the instantaneous phase as

$$\varphi(t) = \arctan \frac{s_i(t)}{s_r(t)} \,. \tag{7}$$

The instantaneous frequency is then equal to

$$\omega(t) = \frac{d}{dt}(\arctan \frac{s_i(t)}{s_r(t)}) = \frac{s_r(t)\dot{s}_i(t) - s_i(t)\dot{s}_r(t)}{s_r(t)^2 + s_i(t)^2} \,. \tag{8}$$

Even with the Hilbert transform, there is still considerable controversy in defining the instantaneous frequency as mentioned also in (Boashash, 1992a). Applying the Hilbert transform directly to a multicomponent signal provides values of $a(t)$ and $\omega(t)$ which are unusable for describing the signal. The idea of instantaneous frequency and amplitude does not make sense when a signal consists of multiple components at different frequencies. It leads (Huang, 1998) to introduce a so called Empirical Mode Decomposition method to decompose the signal into monocomponent functions (Intrinsic Mode Functions). In this work, another method for signal decomposition is introduced.

## 3. Complex signal component model

Let's consider the multicomponent real signal $s_r(t)$

$$s_r(t) = \sum_{n=1}^{N} s_r^{(n)}(t) + \rho(t) \tag{9}$$

which consists from noise $\rho(t)$ representing any undesirable components and from $N$ single component nonstationary signals described by envelopes $a$ and frequencies $\omega$.

$$s_r^{(n)}(t) = a_n \cdot \cos(\omega_n \cdot t) \tag{10}$$

The first derivation of the signal component is

$$\dot{s}_r^{(n)}(t) = -a_n \cdot \omega_n \cdot \sin(\omega_n \cdot t) \tag{11}$$

and the second one is described as follows

$$\ddot{s}_r^{(n)}(t) = -a_n \cdot \omega_n^{\ 2} \cos(\omega_n \cdot t) = -\omega_n^{\ 2} \cdot s_r^{(n)}(t) \tag{12}$$

On the basis of equation (12) the state space model of signal component $s_r^{(n)}(t)$ can be derived as second-order ($n = 2$) model of auto-regressive (AR), linear time-invariant (LTI) system.

Let's assume an AR state space model with state *x(t)*

$$\dot{x}(t) = A \cdot x(t) \tag{13}$$

and system's model output *y(t)*

$$y(t) = C \cdot x(t) \tag{14}$$

where $A$ is the state matrix and $C$ is the output matrix. The state vector *x(t)* consists, in our case, of two internal states: real state part $x_r(t)$ and imaginary part $x_i(t)$.

With respect to the former introduced notation of the signal components and considering the analytical signal, we can choose the state vector components as $x_r^{(n)} = \cos(\omega_n t)$ and $x_i^{(n)} = \sin(\omega_n t)$. This choice takes into account the Hilbert transform and the orthogonal character of both components. Hence, the corresponding state space model is represented by the following state equation

$$\begin{bmatrix} \dot{x}_r^{(n)}(t) \\ \dot{x}_i^{(n)}(t) \end{bmatrix} = \begin{bmatrix} 0 & -\omega_n \\ \omega_n & 0 \end{bmatrix} \cdot \begin{bmatrix} x_r^{(n)}(t) \\ x_i^{(n)}(t) \end{bmatrix} \tag{15}$$

and the model's output is

$$y_n(t) = \begin{bmatrix} 1 & 0 \end{bmatrix} \cdot \begin{bmatrix} x_r^{(n)}(t) \\ x_i^{(n)}(t) \end{bmatrix}. \tag{16}$$

In this case the model output $y_n(t)$ represents the signal component $s_r^{(n)}(t)$. The state matrix $A$ of the model is a 2D rotation matrix whose eigenvalues are pure imaginary numbers. The trajectory in state space is a circle and the model represents in fact an undamped resonator (oscillator) with natural frequency in $\omega_n$. The solution of the state equation consist only from the homogenous part (there is no model input) and is described by the following form

$$x_r^{(n)}(t) = e^{A(t-t_0)} x_r^{(n)}(t_0). \tag{17}$$

Computing state transition matrix $e^{At}$ and using discretization step of $\Delta t = h$, the discrete state space representation is acquired as

$$\begin{bmatrix} x_r^{(n)}(k+1) \\ x_i^{(n)}(k+1) \end{bmatrix} = \begin{bmatrix} \cos(h\cdot\omega_n) & -\sin(h\cdot\omega_n) \\ \sin(h\cdot\omega_n) & \cos(h\cdot\omega_n) \end{bmatrix} \cdot \begin{bmatrix} x_r^{(n)}(k) \\ x_i^{(n)}(k) \end{bmatrix} + \Gamma(k)\cdot\xi(k)\,. \tag{18}$$

and the output equation is

$$y_n(k) = \begin{bmatrix} 1 & 0 \end{bmatrix} \cdot \begin{bmatrix} x_r^{(n)}(k) \\ x_i^{(n)}(k) \end{bmatrix}. \tag{19}$$

The generalized output equation for all signal components is then

$$y(k) = C\cdot x(k) + \Delta\cdot\eta(k)\,. \tag{20}$$

In previous equations, $\xi(k)$ is the state noise and $\eta(k)$ is the output noise of state model. Both noise vectors, $\xi(k)$ and $\eta(k)$, are zero-centred with identity covariance matrices. The specific features of the noises are characterized by the covariance matrix $\Gamma$ and value $\Delta$.
This derived resonator model forms together with Kalman filtering approach an estimator of analytical signal. The estimation of the first model state is a real part (sine function) of the signal component and estimation of the second state is an imaginary part (cosine function).

### 3.1 Selection of matrices $\Gamma$ and value $\Delta$

The choice of the proper matrix $\Gamma$ and value $\Delta$ is the important part in definition of the signal component model. These two parameters decide what amount of energy of the original signal will be assigned to the actual signal component.
We have already defined the description of the complex signal in equation (5) which represents the general output of the component model. Let's assume that the amplitude $a$ of the n<sup>th</sup> complex signal component $z_n(k)$ is not constant but it changes from sample $k$ to sample $k+1$ of deviation $\varepsilon(k)$:

$$z_n(k+1) = (a(k) + \varepsilon(k))\cdot e^{j\varphi(k+1)}\,. \tag{21}$$

It means that the amplitude in sample $k+1$ is described in recursive form by the following equation

$$a(k+1) = a(k) + \varepsilon(k)\,,\ \text{where}\ \varepsilon(k) \sim N(0,\sigma_\varepsilon^{\ 2})\,. \tag{22}$$

The time changing phasor with its corresponding amplitude $a(k)$, phase $\varphi(k)$ and deviation $\varepsilon(k)$ is displayed in Fig.1. The real trajectory of the complex signal mode is marked by black curve and phasor trajectory in case of the constant amplitude in time $k$ is signed as grey-dashed circle. The recursive form of the amplitude progression model is shown in the left picture in Fig. 1. Each phasor amplitude differs from the previous one (green dotted line) of deviation $\varepsilon$ (red line).
The right picture of the Fig. 1. shows in detail the progression of the phasor in samples $k$ and $k+1$. The characteristics of the $\gamma_1$ and $\gamma_2$ parameters are derived in the following text.
The state in the sample $k+1$ in dependence on the state in sample $k$ is described, as above defined, in the following form

$$x(k+1) = A\cdot x(k) + \Gamma(k)\cdot\xi(k)\,,\ \text{where}\ \xi(k) \sim N(0,1) \tag{23}$$

Equation (23) describes model, where the complex signal $x(k)$ is firstly rotated by the matrix $A$ and the sum with $\varepsilon(k)$ produces a new state vector $x(k+1)$. Let's assume equation (23) in a new form, where vector $x(k)$ is firstly summed with $\varepsilon(k)$ and as late as then rotated by matrix $A$:

$$x(k+1) = A \cdot \left(x(k) + \overline{\Gamma}(k) \cdot \xi(k)\right) \tag{24}$$

This form is used in order to reveal the relation between $\varepsilon$ and $\Gamma$ and that is why the new vector $\overline{\Gamma}$ is introduced into the state equation.

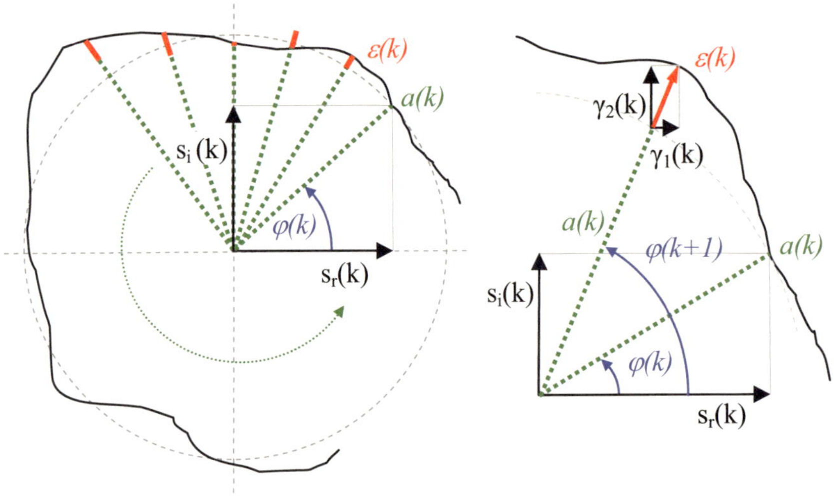


Fig. 1. The time changing phasor with its corresponding amplitude $a(k)$, phase $\varphi(k)$ and deviation $\varepsilon(k)$.

Equation (24) can be also rewritten in quadratic form (in vector domain, the transposed vectors and matrices are used)

$$x^T(k+1)x(k+1) = \left(A \cdot x(k) + A \cdot \overline{\Gamma}(k) \cdot \xi(k)\right)^T \cdot \left(A \cdot x(k) + A \cdot \overline{\Gamma}(k) \cdot \xi(k)\right) \tag{25}$$

The converted form of the equation

$$\begin{aligned} x^T(k+1) \cdot x(k+1) &= x(k)^T \cdot A^T \cdot A \cdot x(k) + x(k)^T \cdot A^T \cdot A \cdot \overline{\Gamma}(k) \cdot \xi(k) + \\ &+ \xi(k) \cdot \overline{\Gamma}(k)^T \cdot A^T \cdot A \cdot x(k) + \overline{\Gamma}(k)^T \cdot A^T \cdot A \cdot \overline{\Gamma}(k) \cdot \xi(k)^2 \end{aligned} \tag{26}$$

The quadratic form equality is valid also for the recursive amplitude equation

$$a(k+1)^2 = \left(a(k) + \varepsilon(k)\right)^2 = a(k)^2 + 2a(k)\varepsilon(k) + \varepsilon(k)^2 \tag{27}$$

Using (27) and the following knowledge of the state product

$$a(k)^2 = x^T(k)x(k) \tag{28}$$

the equation (26) can be rewritten into (30) where product of system matrices $A^T \cdot A$ is substituted by the identity matrix $I$ as follows

$$A^T \cdot A = I \Rightarrow x(k)^T \cdot A^T \cdot A \cdot x(k) = x(k)^T \cdot x(k); \quad \overline{\Gamma}(k)^T \cdot A^T \cdot A \cdot \overline{\Gamma}(k) = \overline{\Gamma}(k)^T \cdot \overline{\Gamma}(k) \tag{29}$$

The result is following

$$\begin{aligned} &\underbrace{x^T(k+1) \cdot x(k+1)}_{a(k+1)^2} = \\ &= \underbrace{x(k)^T \cdot x(k)}_{a(k)^2} + x(k)^T \cdot \overline{\Gamma}(k) \cdot \xi(k) + \xi(k) \cdot \overline{\Gamma}(k)^T \cdot x(k) + \overline{\Gamma}(k)^T \cdot \overline{\Gamma}(k) \cdot \xi(k)^2 \end{aligned} \tag{30}$$

The main reason of this derivative description is the need to obtain a relation between $\varepsilon$ and $\overline{\Gamma}$. Application of mean value operator on (30) results in the equation where only the required variables are presented:

$$\begin{aligned} &E\left(a(k)^2\right) + \underbrace{E\left(2a(k)\varepsilon(k)\right)}_{0} + E\left(\varepsilon(k)^2\right) = \\ &= E\left(a(k)^2\right) + \underbrace{E\left(x(k)^T \cdot \overline{\Gamma}(k) \cdot \xi(k)\right)}_{0} + \underbrace{E\left(\xi(k) \cdot \overline{\Gamma}(k)^T \cdot x(k)\right)}_{0} + E\left(\overline{\Gamma}(k)^T \cdot \overline{\Gamma}(k) \cdot \xi(k)^2\right) \end{aligned} \tag{31}$$

and resulting equation which connects $\varepsilon$ and $\overline{\Gamma}$ is

$$\sigma_\varepsilon^{\ 2} = \overline{\Gamma}(k)^T \cdot \overline{\Gamma}(k) \tag{32}$$

The $\overline{\Gamma}$ vector consists of two components (because of second order model) and the result from multiplying of the $\overline{\Gamma}$ vectors is the sum of the square vector components

$$\overline{\Gamma}(k) = \begin{bmatrix} \overline{\gamma}_1(k) \\ \overline{\gamma}_2(k) \end{bmatrix} \Rightarrow \overline{\Gamma}(k)^T \cdot \overline{\Gamma}(k) = \overline{\gamma}_1(k)^2 + \overline{\gamma}_2(k)^2 \tag{33}$$

Using (32) the first condition for selection of $\overline{\Gamma}$ is obtained as

$$\sigma_\varepsilon^{\ 2} = \overline{\gamma}_1(k)^2 + \overline{\gamma}_2(k)^2 . \tag{34}$$

The second condition can be obtained from the presumption that the phasor of the state *x(k)* and of the deviation $\varepsilon(k)$ have the same normal line. Going from the estimation of the state components $x_r(k)$ and $x_i(k)$ the ratio between them should be the same as the ration between $\overline{\gamma}_1(k)$ and $\overline{\gamma}_2(k)$ (see(35)). To be precise, there is a product of vector $\overline{\Gamma}$ and stochastic variable $\xi$ in the ratio equation but because of the same stochastic value in sample *k* the variable $\xi$ is then cancelled out from the ratio (therefore the $\gamma_1$ and $\gamma_2$ coordinates of $\varepsilon$ in Fig. 1. are also used without variable $\xi$)

$$\frac{\overline{\gamma}_2(k) \cdot \xi(k)}{\overline{\gamma}_1(k) \cdot \xi(k)} = \frac{x_i(k)}{x_r(k)} \tag{35}$$

The converted form of the ratio equation

$$\overline{\gamma}_2(k) = \frac{x_i(k)}{x_r(k)} \cdot \overline{\gamma}_1(k) \tag{36}$$

is then substituted into (34) and results in formula for the first parameter of vector $\overline{\Gamma}(k)$

$$\bar{\gamma}_1(k)=\sqrt{\frac{\sigma_{\varepsilon}^{\ 2}\cdot x_r(k)^2}{x_r(k)^2+x_i(k)^2}} \tag{37}$$

and for the second one

$$\bar{\gamma}_2(k)=\sqrt{\frac{\sigma_{\varepsilon}^{\ 2}\cdot x_i(k)^2}{x_r(k)^2+x_i(k)^2}} \tag{38}$$

These equations rewritten for the original model definition with the vector $\Gamma(k)$ have the following form

$$\Gamma(k)=\begin{bmatrix}\gamma_1(k)\\ \gamma_2(k)\end{bmatrix}=A\cdot\begin{bmatrix}x_r(k)\cdot\sqrt{\dfrac{\sigma_{\varepsilon}^{\ 2}}{x_r(k)^2+x_i(k)^2}}\\ x_i(k)\cdot\sqrt{\dfrac{\sigma_{\varepsilon}^{\ 2}}{x_r(k)^2+x_i(k)^2}}\end{bmatrix}. \tag{39}$$

This formula describes the characteristics of the amplitude deviation between samples $k$ and $k+1$ namely in the state model in (18) which is then together with (19) used as a model for Kalman estimation. In Kalman filter computation of $\Gamma(k)$ an estimate $\mu(k)$ of state $x(k)$ is used (see(46)). Selection of value $\Delta$ in the output equation (20) is the next step of defining the state model. Output $y(k)$ of $n$ component model is formed only from the real parts of the components. This is acquired through the vector C and the product of $C\cdot x(t)$. The value $\Delta$ makes it possible to include the measurement noise ($\Delta\cdot\eta(k)$) into the output equation. In fact, $\Delta$ is the variance of additive error of the measuring chain. For example in case where the sensor accuracy class is defined in statistical sense so that the absolute additive error is 95% quantile in limits of product of accuracy class$\times$measuring range, the 95% quantile represents the range of $0\pm2\sigma$ where $\sigma$ is the standard deviation of additive error. In this case the following form of the measurement error variance can be accepted

$$\Delta=\left(\frac{\text{accuracy class}\times\text{measuring range}}{2}\right)^2. \tag{40}$$

## 4. Use of Kalman filter for estimation of the signal components

In this work, an adaptive Kalman filter based approach is used for estimation of the analyzed signal. Signal is modelled as sum of resonators (signal components) and it is required that the estimated components are complex functions because of efficient computation of the instantaneous frequency.

### 4.1 Discrete Kalman filter

A discrete-time Kalman filter realizes a statistical estimation of the internal states of noisy linear system and it is able to reject uncorrelated measurement noise – a property shared by all Kalman filters. Let's assume a system with more components as mentioned above. Then the state matrix consists of following blocks:

$$A_n = \begin{bmatrix} \cos(h\cdot\omega_n) & \sin(h\cdot\omega_n) \\ -\sin(h\cdot\omega_n) & \cos(h\cdot\omega_n) \end{bmatrix};. \tag{41}$$

and the state noise vector blocks are defined as in (39). Then in state-variable representation, the description of the whole system, which is characterized by the sum of resonators, is given by the following matrices:

$$A = \underbrace{\begin{bmatrix} A_1 & 0 & 0 & \dots & 0 \\ 0 & A_2 & 0 & \dots & 0 \\ 0 & 0 & \ddots & & \vdots \\ \vdots & \vdots & & \ddots & 0 \\ 0 & 0 & \dots & 0 & A_n \end{bmatrix}}_{2n\times 2n};. \tag{42}$$

the output matrix is following

$$C = \underbrace{[1 \quad 0 \quad 1 \quad 0 \quad \dots \quad 1 \quad 0]}_{1\times 2n};. \tag{43}$$

and state and measurement noise is characterized by the following parameters

$$\Gamma = \underbrace{\begin{bmatrix} \Gamma_1 \\ \Gamma_2 \\ \vdots \\ \Gamma_n \end{bmatrix}}_{2n\times 1}; \quad \underbrace{\Delta}_{1\times 1}. \tag{44}$$

Commonly, the Kalman estimation includes two steps - prediction and correction phase. Let's assume that the state estimate $\mu(0)$ is known with an error variance $P(0)$. A priori value of the state at instant *k+1* can be obtained as

$$\mu(k+1) = A\cdot\mu(k) \tag{45}$$

The measured value $y(k)$ is then used to update the state at instant $k$. The additive correction of the a priori estimated state at *k+1* is according to (Vaseghi, 2000) proportional to the difference between the a priori output at instant $k$ defined as $C\cdot\mu(k)$ and the measured $y(k)$:

$$\mu(k+1) = A\cdot\mu(k) + K(k)\cdot\left(y(k) - \left(C\cdot\mu(k)\right)\right) \tag{46}$$

where $K(k)$ is the Kalman gain which guarantees the minimal variance of the error $x(k)-\mu(k)$.

Also, at each step the variance $P(k+1)$ of the error of $\mu(k+1)$ is calculated (see (Vaseghi, 2000)):

$$P(k+1) = A \cdot P(k) \cdot A^T + \Gamma \cdot \Gamma^T - K(k) \cdot \left(C \cdot P(k) \cdot A^T + \Delta \cdot \Gamma^T\right) \tag{47}$$

This variance matrix is then used for calculation of Kalman gain in the next step of the recursive calculation (correction phase):

$$K(k) = \left(A \cdot P(k) \cdot C^T + \Gamma \cdot \Delta^T\right) \cdot \left(C \cdot P(k) \cdot C^T + \Delta \cdot \Delta^T\right)^{-1} \tag{48}$$

### 4.2 Estimation of initial parameters

The initial parameters for Kalman filter are obtained from the estimation of the signal spectrum. In principle, there are many ways how to fix the parameters, let's present two of them. Generally, there is need to define which modes, or more precisely which frequencies $\omega_n$, should be estimated and what is the variance of the amplitude on these frequencies.

The first way how to acquire the initial parameters is based on the short-time Fourier transform (STFT) algorithm. The half length of the STFT window can be used as an order of the modelled system (number of modelled resonators) and the corresponding frequencies are then the frequencies of the modelled signal components. The variation of the amplitude in the STFT frequency bands serves then as the estimation of the variation $\sigma_\varepsilon^2$ of $\varepsilon$ for each resonator. This approach shows the Kalman filter method based on the short-time Fourier frequency analysis.

The second alternative is based on the estimation of spectral density by fitting an AR prediction model to the signal. The used estimation algorithm is known as Burg's method (Marple, 1987), which fits an AR linear prediction filter model of a specified order to the input signal by minimizing the arithmetic mean of the forward and backward prediction errors. The spectral density is then computed from the frequency response of the prediction filter. The AR filter parameters are constrained to satisfy the Levinson-Durbin recursion. The initial Kalman filter parameters (frequencies of the resonators) are then obtained as local maxima of estimated spectral density which are greater than a predefined level. These values indicate significant frequencies in spectral density and determine the order of the model.

The Fig. 2. shows the schema of the whole method. The estimated parameters from the input signal form the system model which is then used in the estimation phase of the algorithm.

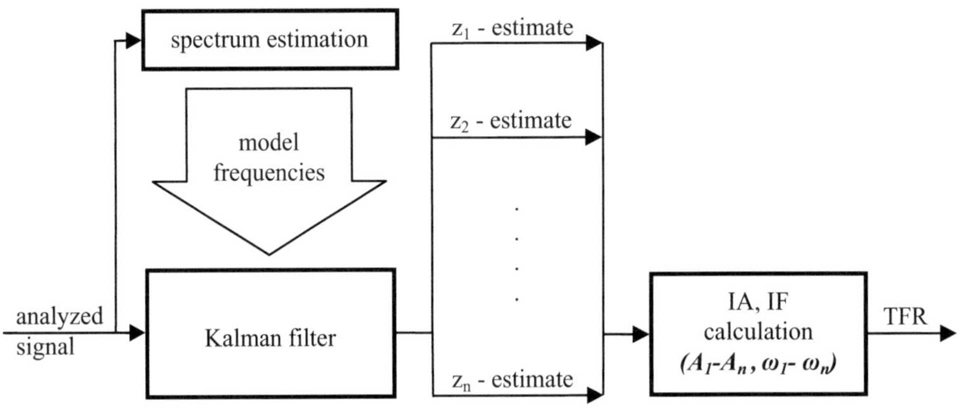


Fig. 2. Schema of the method for time-frequency decomposition of the signal using Kalman filter

Kalman filter estimates the states of the modelled system in form of complex signal. It means that the imaginary parts of components are estimated simultaneously with the estimation of real signal components. This procedure substitute the Hilbert transform in generation of analytical signal and is performed recursively.
Each component in complex form serves then for calculation of the instantaneous amplitude and instantaneous frequency (see (6) and (8)). The output of the method is the time-frequency representation of the analyzed signal. The representation consists of amplitude curves with changing frequency in contrast to other time-frequency methods whose TF representation is computed in bands that is why the resolution in time and frequency is limited.

## 5. Examples

In this section, some examples of the time-frequency analysis using Kalman filter method are shown. The most important indicators in comparison with other analysis methods are the sharpness of the estimated frequency curves and the adaptability to the changes in the component amplitudes. These attributes are deciding for the next signal processing in the class of problems which are solved by authors.
Let's have a signal with three harmonic components, with $1kHz$ sampling rate and the total analyzed length of the signal is 1 second ($N = 1000$ points). Signal is formed by sinus functions with oscillation frequencies $f_1 = 10Hz$, $f_2 = 30Hz$ and $f_3 = 50Hz$. The amplitude for all three components is set to 10, the second component is set to zero for the first 0.5 seconds. The output noise with mean $m_\eta = 0$ and variance $\sigma_\eta^2 = 1$ was added to the simulated signal. The initial parameters for Kalman filter were obtained through Burg's AR linear prediction filter of order 10 and the level for local maxima in the Burg's spectral density was determined as $\max > 1$. The initial conditions of Kalman estimator were set up in the following way: $\mu(0) = [1\ldots1]^T$, $P(0) = 10^6 \cdot I$, $\Delta = 1$, where $\dim(\mu) = n \times 1$ and $\dim(P) = 2n \times 2n$.
The time-frequency representation of the instantaneous frequency and amplitude is shown in Fig. 3 (left). The result shows that the adaptation of one of the model component has an influence also on the instantaneous frequency estimates of other components which is expressed in the oscillation between 0.5 and 0.6 second (the overall oscillation of the instantaneous frequency is the effect of the output noise).
The adaptation rate is shown in Fig. 3 (right), where the comparison of estimate convergence of the component at $f_2 = 30Hz$ using Hilbert approach and Kalman filter is performed. The Hilbert transform was applied artificially to the component itself, not to the analyzed signal (in order to compare the adaptation). The reason is the disadvantage of the Hilbert transform that requires the pre-processing of the signal through some signal decomposition method.
The above introduced Kalman filter method uses for decomposition of the signal into its components, as already mentioned, the model of sum of resonators and thus simultaneously decomposes the signal and estimates the time progression of its components.

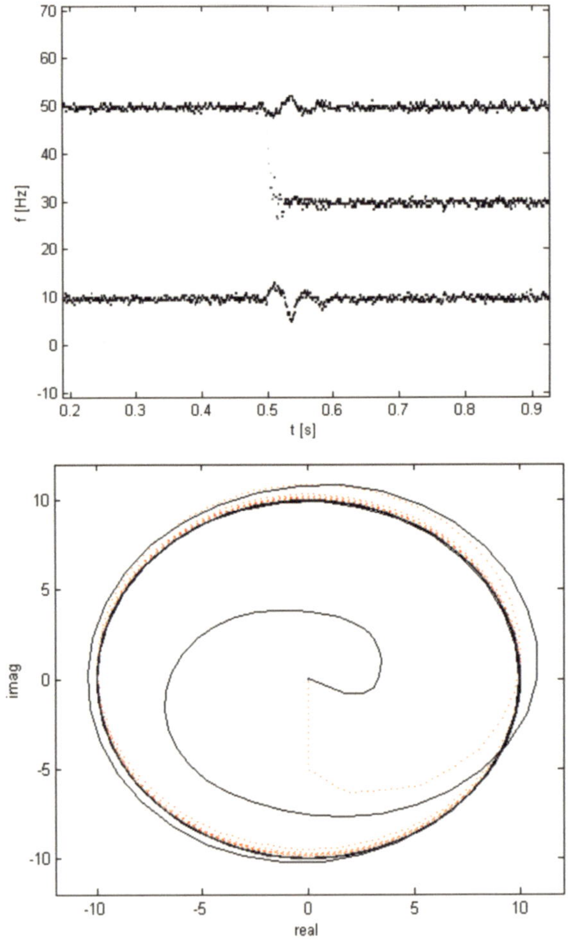


Fig. 3. Simulation signal with three components (left). Complex signal (right) obtained through Hilbert transform (dotted line) and through Kalman estimation (solid line).

The algorithm based on Kalman estimation is also illustrated on another signal example (see Fig. 4). It consists from signal components, where the first one is the stationary harmonic signal with constant frequency. This signal part is summed in time with the concave parabolic chirp signal where its frequency changes from 0.45 to 0.1 of normalized frequency. Both components exist in time between $t = 100$ and $t = 900$ . The initial conditions of Kalman estimator were set up as mentioned in previous example and initial frequencies of the model were obtained through Burg's AR linear prediction filter of order 25 (number of estimated frequencies was *n=10* and thus order of the estimated system was 20).

The Kalman filter method is compared with other typical time-frequency methods and the results are shown in Fig. 4. For comparison, the short-time Fourier transform - STFT, wavelet transform – WT (using Morlet wavelet) and smoothed pseudo Wigner-Ville distribution – SPWVD, were used. The output of Kalman estimation in time-frequency domain has comparable resolution as the output of the SPWVD. The other methods have relative wide frequency bands containing the signal energy.

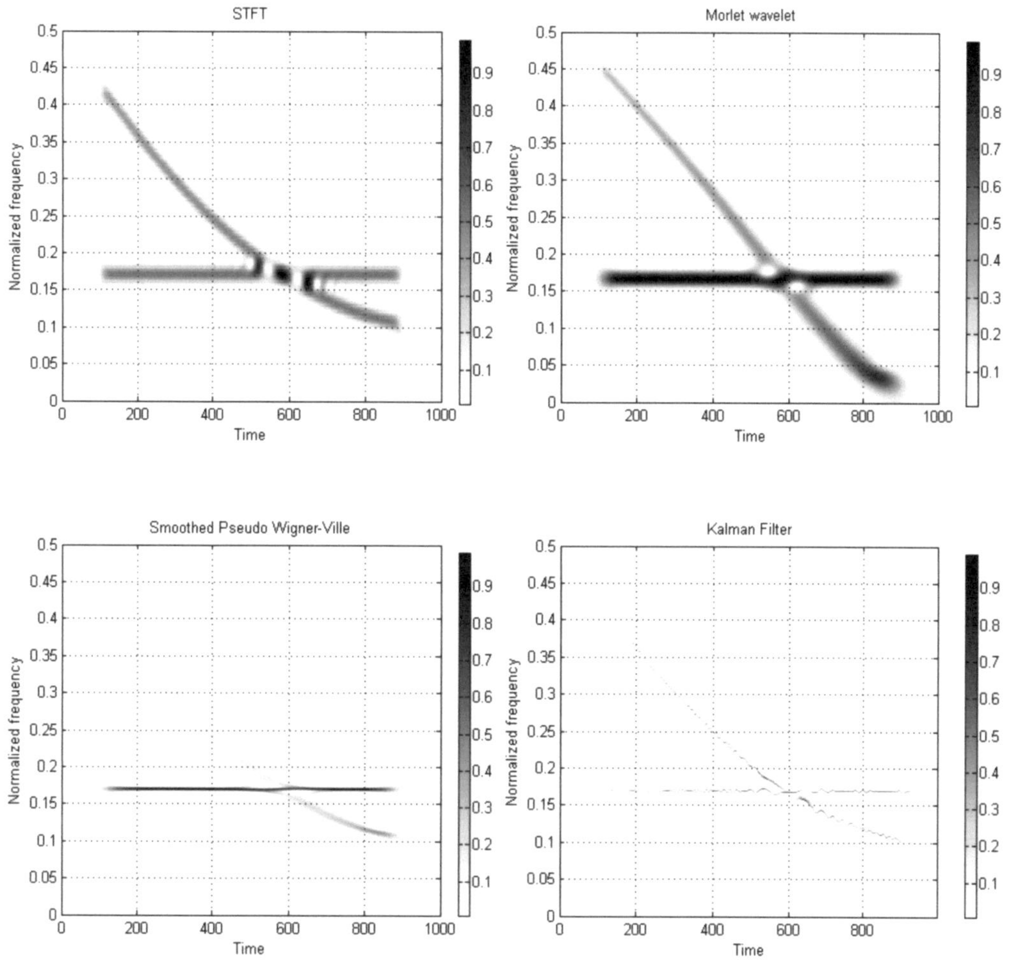


Fig. 4. Estimation of instantaneous frequency and amplitude of harmonic and chirp signal components.

Similarly, the result of methods comparison in Fig. 5. shows the Kalman filter method and SPWVD as more detailed time-frequency approaches. The test signal consists, in this case, of four harmonic components and the accuracy of the methods to identify the frequency and time of the component origin or end is tested. The signal begins again in time $t = 100$ and ends in $t = 900$. The frequency changes in $t = 300$ and $t = 600$, whereas there are two components simultaneously between these times. The ability of methods to distinguish between these two component frequencies is visible. SPWVD and Kalman filter representations display these two simultaneous components separately. On the contrary STFT and WT representations contain time-frequency artefacts between components and that's why the identifying of each component separately using these methods could be in such cases really difficult.

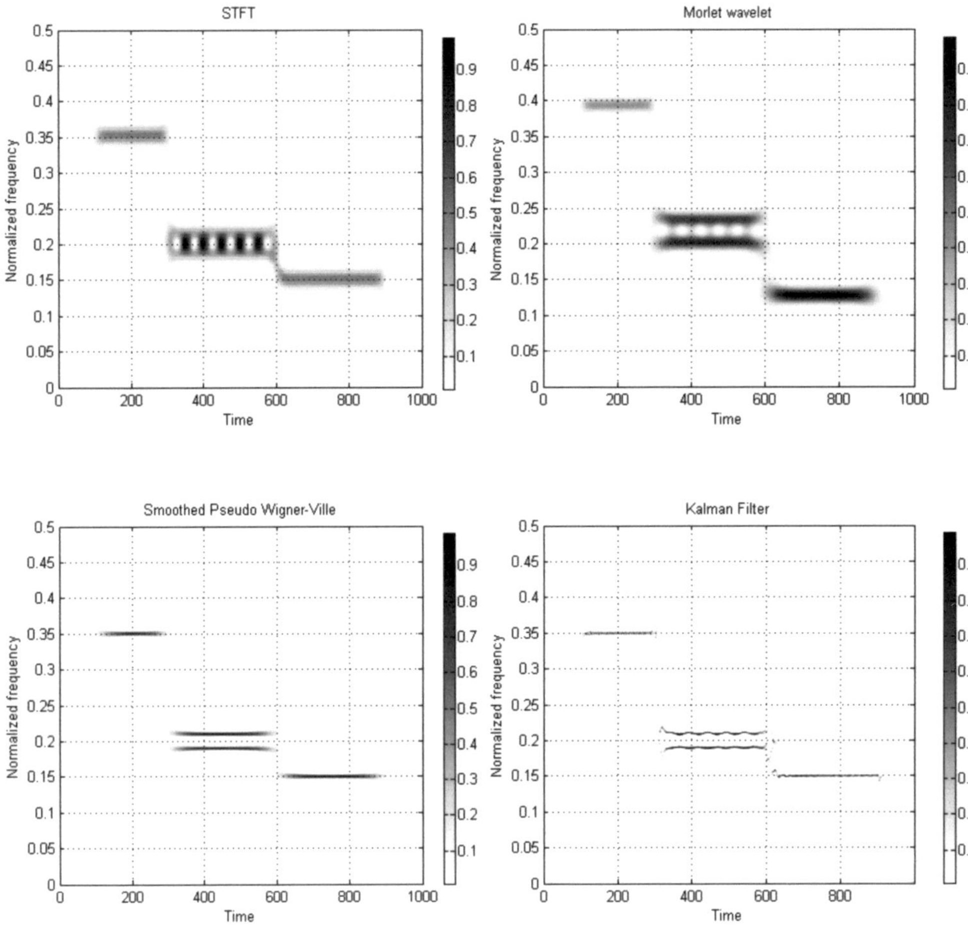


Fig. 5. Estimation of instantaneous frequency and amplitude of harmonic components of the signal.

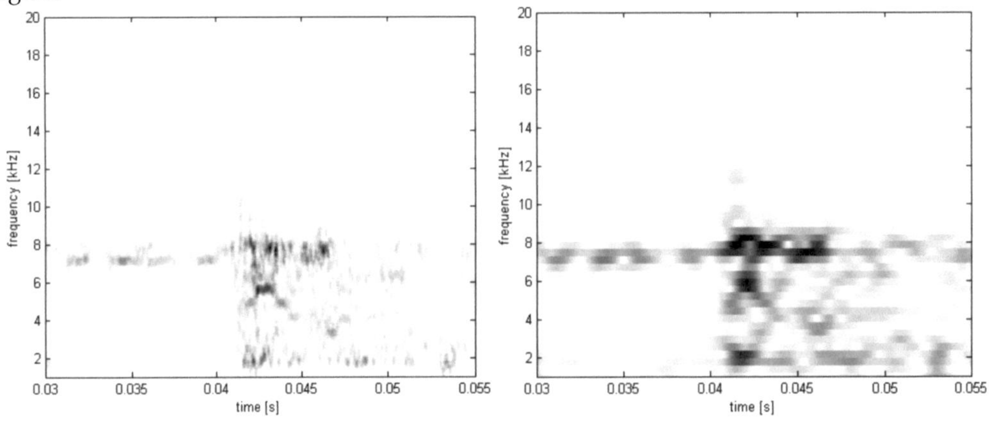


Fig. 6. Estimation of instantaneous frequency and amplitude of loosened part impact signal: Kalman filter method (left) and STFT (right).

The last example is the transform of the acoustic signal from the real equipment where the nonstationary event took place. This event is measured result of loosed part impact in primary circuit of nuclear power plant. Signal was measured with 100 kHz sampling rate. For comparison, in Fig. 6, the time-frequency-amplitude responses of STFT (right) and of the Kalman method (left) are compared. Kalman filter signal model was initialized with frequencies from the frequency analysis with window of length $2^8$ samples. The same window length with 75% overlapping was used for analysis by means of STFT (Fig. 6 right). An event (impact) which occurs at time 0.041 seconds is well visible in both spectrograms (see the frequency band 1 - 12 kHz). The advantage of the Kalman version of spectrogram is its better resolution in time and frequency than in STFT spectrogram (which is used by authors for improvement of impact location).

## 6. Conclusion and perspective

The new approach for time-frequency signal analysis and for representation of instantaneous frequency and amplitude has been introduced in this chapter. The procedure is based on the Kalman estimation and shares its advantages regarding the suppression of measurement noise. In this method the Kalman filter serves for dissociation of signal into modes with well defined instantaneous frequency. The analyzed signal is modelled as sum of second-order subsystems (resonators) whereby the Kalman filter decompose the system by estimating these subsystems. Simultaneously with the signal decomposition the time progression of signal components in complex form is evaluated. This procedure utilizes the adaptive feature of the Kalman filter and it is done recursively for each sample.

In cases where the short-time Fourier transform cannot offer sufficient resolution in time-frequency domain, there can be taken the advantage of this method despite of higher computational severity. Also the experimental results show that the resolution of the introduced method is equal or higher than in other usual time-frequency techniques.

In vibro-diagnostic methods, where time-frequency information is used for location of non-stationary events, the sharpness of the introduced method is helpful for the improvement of the non-stationary event location. The next tasks of the Kalman filter method development are the systematization of the results from different actually solved problems in loose part monitoring and finding the new branches where the method could be effectively apllied.

## 8. References

Boashash, B. (1992a). Estimating and Interpreting the Instantaneous Frequency of a Signal – Part 1: Fundamentals, *Proceedings of the IEEE*, Vol. 80, No. 4, pp. 520-538, April 1992, ISSN 0018-9219

Boashash, B. (1992b). Estimating and Interpreting the Instantaneous Frequency of a Signal – Part 2: Algorithms and Applications, *Proceedings of the IEEE*, vol. 80, no. 4, pp. 540-568, April 1992, ISSN 0018-9219

Cohen, L. (1995). *Time Frequency Analysis: Theory and Applications*, Prentice Hall PTR., ISBN 978-0135945322, New Jersey

Fairman, F.W. (1998). *Linear Control Theory: The State Space Approach*, John Wiley& Sons, ISBN 0471974897, Toronto

Hahn, S.L. (1996). *Hilbert Transforms in Signal Processing*, Artech House Publishers, ISBN 978-089006886, Boston

Huang, N.E., et al. (1998). The empirical mode decomposition and the Hilbert spectrum for nonlinear and non-stationary time series analysis, *Proceedings: Mathematical, Physical and Engineering Sciences*, Vol. 454, No. 1971, pp. 903-995, March 1998, ISSN 1364-5021

Marple, S.L. (1987). *Digital Spectral Analysis with Applications*, Prentice Hall Signal Processing Series, Prentice Hall, Englewood Cliffs, ISBN 978-0132141499, New Jersey.

Vaseghi, S.V. (2000). *Advanced Digital Signal Processing and Noise Reduction*, John Wiley & Sons Ltd, ISBN 0471626929, New York

3

# Discrete-Event Dynamic Systems Modelling Distributed Multi-Agent Control of Intermodal Container Terminals

Guido Maione
*Technical University of Bari*
*Department of Environmental Engineering and Sustainable Development (DEESD)*
*Viale del Turismo, 8 – I74100 – Taranto*
*Italy*

## 1. Introduction

Maritime intermodal container terminals (CTs) are complex hub systems in which multiple transport modes receive and distribute freight to various destinations. They can be considered as interchange places in a transport system network where seaways, railways, and motorways intersect.

Freight is usually organized, transported, stacked, handled and delivered in standard units of a typical container, which is called TEU (Twenty Equivalent Unit), 20 feet long, 8 feet wide, and 8 feet high. TEUs easily fit to ships, trains and trucks that are built and work for this kind of cargo units. Usually, ships travelling on long routes (e.g. from Taiwan to a Mediterranean Sea port) are called vessel or 'mother' ships, and ships offering service on short distances in a local area (e.g. from Turkey to Mediterranean ports in other countries) are called feeder ships. Vessel and feeder ships differ in size and container capacity.

A maritime CT is usually managed to offer three main services: a railway/road 'export cycle', when containers arrive to the terminal by trains/trucks and depart on vessel ships; a railway/road 'import cycle', when containers arrive on vessel ships and depart by trains/trucks; a 'transshipment cycle', when containers arrive on vessel (or feeder) ships and depart on feeder (or vessel) ships. These activities cause different concurrent and competition processes for the available resources. The aim is to achieve efficiency in flows of TEUs and in information distribution to guarantee fast operations and low costs.

Many problems were investigated and many of them have connections and cannot be solved separately. The main focused problems are: berth allocation of arriving ships; loading and unloading of ships (crane assignment, stowage planning); transfer of containers from ships to storage area and *viceversa* (quayside operation); stacking operations (yardside operation); transfer to/from other transport modes (landside operation); workforce scheduling. In a word, managing a CT requires:

- planning, monitoring and controlling all internal procedures for handling TEUs;
- optimizing handling equipment, human operators, and information and communication technologies (control software and hardware, PDAs, wireless sensors

and actuators, etc.); optimization must take into account connections and relations between humans and hardware resources.

Then, an intelligent and reactive control system is required to meet terminal specifications, independently of disturbances and parameter variations. As a matter of fact, robustness is important to guarantee quick reaction to different phenomena perturbing normal (or steady-state) operating conditions. Perturbations may come from: increase of ship traffic volumes or urgent demands for service; infrastructure development (reduction/expansion of berthing or stacking spaces, changes in yard organization, acquisition of new resources, etc.); changes in routing of transport vehicles or traffic congestion inside the CT; faults and malfunctions or sudden lack of hardware resources.

Complexity of the issues involved justifies trends towards distributing the computing resources and decision controllers while aiming at robustness, and hence motivates heterarchical control by means of a *Multi-Agent System* (MAS). In a MAS, indeed, information, decision, and control are physically and logically distributed across agents. This chapter reports some preliminary results and ideas on how to model MAS controlling intermodal CTs, by using the DEVS-Scheme (Zeigler, 2000). The accurate model could be used to develop a simulation platform, which can be useful to test different control strategies to be applied in real cases. For example, the CT in Taranto (Italy) can be considered as a real system to be simulated in different operating scenarios.

The organization of this chapter is as follows. Section 2 discusses about literature contributions to modelling, simulation and control of CTs. Motivations are explained to use the methodology of *Discrete-Event Dynamic Systems* to represent the MAS architecture for simulation and control of complex CTs. Section 3 describes the typical services and processes to be guaranteed in an intermodal container terminal. Then, the basic components of the proposed MAS are presented, and their roles and relations are generally specified. Section 4 specifies how agents are modelled as atomic discrete-event dynamic systems and focuses on the interactions between agents, when containers are downloaded from ships to the terminal yard area. Section 5 gives some ideas and details about the plans of a simulation platform to test efficiency and robustness of the proposed MAS architecture. Section 6 overviews the benefits of the approach and enlightens the open issues.

## 2. Literature overview: motivation of DEVS modelling for MAS simulation and control

Planning processes and scheduling resources in a maritime CT pose very complex modelling and control problems to the scientific community. In particular, flexible and powerful modelling and simulation tools are necessary to represent intermodal hub systems made up of many different infrastructures and services. Modelling, simulation and control of CTs is a relatively new field of research. Several literature contributions developed models and simulation tools, but no standard exists to be applied to the different real terminals and scenarios. In the absence of standard tools, several research studies are based on discrete-event simulation techniques (Vis & de Koster, 2003), on mathematical models or empirical studies (Crainic et al., 1993; Peterkofsky & Daganzo, 1990; Gambardella et al., 1998).

Analytical models, based on different approaches, have been proposed as tool for the simulation of terminals useful to define the optimal design and layout, organization, management policies and control. A thorough literature review on modelling approaches is given in (Steenken et al., 2004). As regards control issues, determining the best management and control policies is a challenging problem (Mastrolilli et al., 1998). Literature highlights two main classes of approaches: microscopic and macroscopic modelling approaches (Cantarella et al., 2006).

Microscopic modelling approaches are generally based on discrete-event system simulation that may include Petri Nets (Degano & Di Febbraro, 2001; Fischer & Kemper, 2000; Liu & Ioannou, 2002), object-oriented approaches (Bielli et al., 2006; Yun & Choi, 1999), and queuing networks theory approaches (Legato & Mazza, 2001). Even if it requires high computational effort, microscopic simulation allows the explicit modelling of each activity within the terminal as well as of the whole system by considering the single containers as entities. Then, it is possible to estimate performance as consequence of different system design and/or management scenarios.

Macroscopic modelling is suitable for terminal system analysis. It assumes continuous containers flow along the whole sequence of activities. This representation is useful and appropriate for supporting strategic decisions, system design, terminal layout, handling equipment investments. The most frequent problems in this class are berth planning, marshalling strategies, space allocation and system layout, handling equipment capacity and technologies. A network-based approach is presented in (Kozan, 2000) for optimising the terminal efficiency by using a linear programming method. In (de Luca et al., 2005) a macroscopic model is based on a space-time domain.

In the context of intelligent control of transport systems, this research contribution proposes a rigorous approach based on the *Discrete EVent System* (DEVS) specification technique (Zeigler et al., 2000) to completely and unambiguously characterize a MAS for controlling an intermodal container terminal. Namely, no standard exists for modelling and simulating complex MAS controlling CTs, but a generic and unambiguous framework is needed to describe the discrete and asynchronous actions of agents, and to guarantee modular analysis and a feasible computational burden. These requirements are easily enforced by the DEVS formalism. Connecting agents, each modelled as an atomic DEVS, makes the whole MAS represented as a DEVS, as the formalism is closed under coupling. Moreover, DEVS theory provides a strong mathematical foundation and models can be easily translated in a simulation environment.

The DEVS approach is then useful to develop a simulation platform to test the MAS efficiency in controlling the CT activities. In particular, both static and dynamically adapted decision strategies can be tested for the agents defining the MAS. Then, performance is measured in terms of commonly used indices (ship service time, throughput or lateness of containers, resource utilization, etc.) and other indices of the MAS efficiency (number of requests in negotiation, waiting time before decision, etc.). Performance can be evaluated both in steady-state operating conditions and in perturbed conditions, when disturbances or parameter variations occur.

Autonomous agents play as atomic DEVS dynamic systems. They exchange messages one with another to negotiate services in a common environment. Agents are local controllers, represented by software modules devoted to a hardware component or a specific function (like mediation between different agents). They behave autonomously and concurrently, by communicating each other to negotiate tasks, so exchanging messages and information to 'buy' or 'sell' services. Usually, there is no hierarchy between agents. In this context, cooperation may be explicitly designed or implicitly achieved through adaptation of the agents' decision mechanism, by using feedback of their effects. The general recognized

benefits of a MAS architecture are reducing the programming complexity, guaranteeing scalability and re-configurability, obtaining an intelligent and reactive control software.
The DEVS technique is fully compatible with the heterarchical design principles, it leads to MAS where all information and control functions are distributed across agents, and it is also suitable for a rigorous theoretical analysis of structural properties of the MAS. Moreover, the DEVS formalism is an interesting alternative to other recently proposed tools for MAS specification, e.g. the Unified Modeling Language (Huhns & Stephens, 2001) and Petri Nets (Lin & Norrie, 2001). This formalism is suitable to develop useful models both for discrete-event simulation and for implementation of the software controllers for the considered system.
As in MAS for manufacturing control (Heragu et al., 2002; Shen & Norrie, 1999), agents may use decision algorithms emulating micro-economic environments. Each 'buyer' agent uses a fictitious currency to buy services from other 'seller' agents which, on their turn, use pricing strategies. Sellers and buyers have to reach an equilibrium between conflicting objectives, i.e. to maximize profit and to minimize costs, respectively. Recently developed analytical models of negotiation processes (Hsieh, 2004), underline the need of a systematical analysis and validation method for distributed networks of autonomous control entities. Other researches have focused on the experimental and detailed validation of MAS on distributed simulation platforms (Schattenberg & Uhrmacher, 2001; Logan & Theodoropoulos, 2001).
This work proposes a DEVS model of a MAS architecture for controlling an intermodal container terminal system. In particular, a detailed DEVS model of the interactions, occurring between the agents concurrently operating during the critical downloading process of containers from a ship, is developed.
The author aims at giving a contribution to define a complete DEVS model to develop a detailed simulation platform for testing and comparing the proposed MAS with other centralized or distributed control architectures for intermodal container terminals. The simulation model could be used to design and test alternative system layouts and different control policies, to be used in standard or perturbed operating conditions.

## 3. Multi-agent system framework for intermodal container terminals

To define the autonomous agents operating and interacting in an intermodal container terminal environment, the main processes executed in the terminal area have to be examined to represent the most significant and critical aspects. These processes are associated to the offered services, syntethically described as follows.

### 3.1 Import, export and transshipment cycles in an intermodal container terminal

As recalled before, an intermodal terminal usually offers three different kinds of services interconnecting different transport modes:

1. an *import* cycle, when containers arrive on a vessel or feeder ship and depart by trains or trucks, corresponding to a transition from sea to railway or road modes;
2. an *export* cycle, when containers arrive by trains or trucks and depart on a vessel ship, for a transition from railway or road to sea mode;
3. a *transshipment* cycle, when containers arrive and depart by ship: cargo is moved from vessel to feeder ships to reach close destinations, or, *viceversa*, from feeder to vessel ships to reach far ports.

The possible flows of containers are synthetically described in Fig. 1, where TRS, IMP and EXP represent transhipment, import and export cycles, respectively, and RA and RO symbolize a railway or road transport mode. Note that full containers can be imported, exported or transshipped. Obviously, imported TEUs are only downloaded from ships, exported TEUs are only loaded on ships, while transshipped TEUs occur both processes. Empty containers are downloaded from feeder ships or arrive on trains or trucks, then they are loaded on vessel ships. So, these TEUs are transhipped or exported.

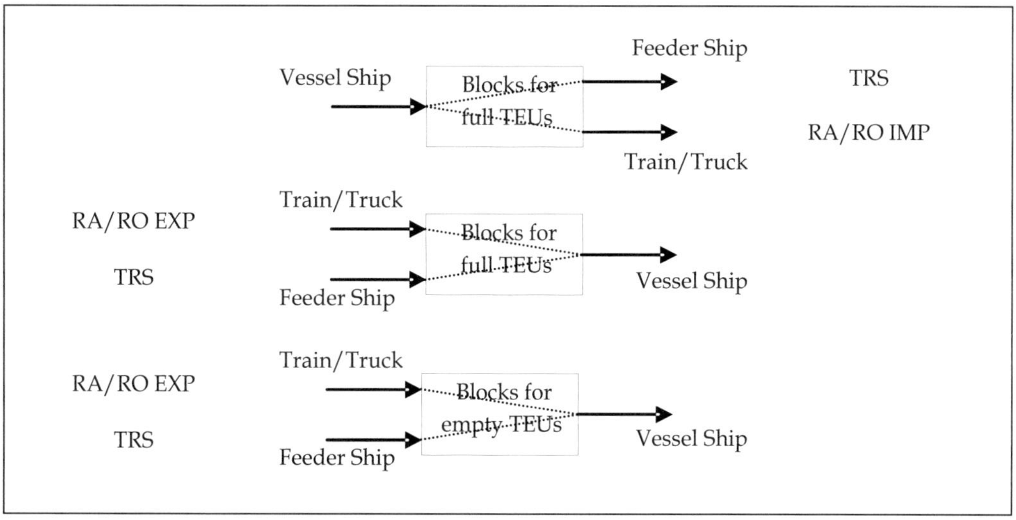


Fig. 1. Flows of containers in an intermodal container terminal

To execute these cycles, the companies managing container terminals provide activities like: loading/downloading containers on/from ships; delivering/picking containers to/from trucks or trains; stacking and keeping containers in dedicated areas, called blocks, in which the terminal yard is organized; transferring containers from ship/train/truck to yard blocks and backwards; inspecting containers for customs, safety and other requirements; consolidating, i.e. redistributing containers between blocks to allow fast retrieval.

Several dedicated or shared resources are used to execute the above processes: cranes (quay cranes, yard cranes, railway cranes, jolly mobile cranes); internal transport vehicles (trailers, automatically guided vehicles or AGVs); reach stackers for handling containers between trailers and trucks, and for dangerous or inspected containers; side loaders for handling empty containers; other special areas and infrastructures like quays and berths, lanes for internal transport, terminal gates, railway tracks; skilled human operators.

For example, the layout of the *Taranto Container Terminal* (TCT, see official website at *http://www.tct-it.com/*), located in the Taranto city harbour area in the south-east of Italy and currently managed by a private company, is organized as sketched in Fig. 2. A quay receives ships, yard stacking blocks are used for keeping full or empty TEUs. A gate (G) let trucks enter or exit the terminal, a railway connection allows trains to enter or exit. Special blocks are for parking of trailers (PARK), fuel station (FU), customs (EX), inspections, and other functions. A control centre (CC) is used to follow the processes. Stacking blocks are divided between blocks for full containers (from 11 to 46), blocks for empty containers (M's) and blocks for dangerous containers (DG). High-level planning operators schedule and monitor all activities, while low-level quay and yard specialized operators execute the planned activities. Most of activities in TCT are for transshipment services.

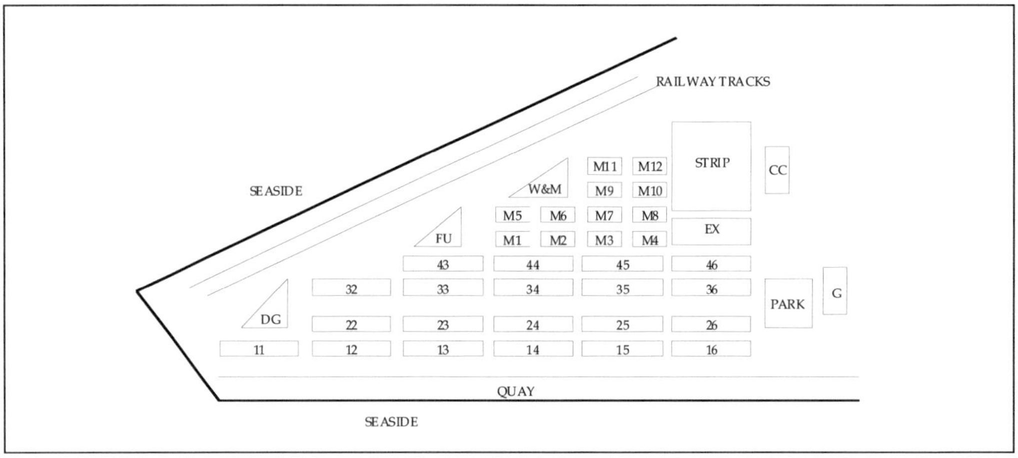


Fig. 2. Taranto Container Terminal organization

A road import cycle may be divided into three steps. Firstly, quay cranes download containers from a berthed ship to trailers, which transfer cargo into blocks. Here, yard cranes pick-up containers from trailers and stack them in assigned positions. Secondly, containers stay in the blocks for a certain time, while waiting for their destination; they are eventually relocated by yard cranes in a more proper position, according to a consolidation procedure which may use trailers to move containers between blocks. Thirdly and after consolidation, containers are loaded from blocks to trucks, which exit from the terminal gate. Similarly, in a railway import cycle, yard cranes pick-up containers from blocks and load them on trailers moving from yard to the railway connection, where special railway cranes pick-up containers to put them on departing trains.

In a road/railway export cycle, the sequence goes in the opposite direction: from the terminal gate or railway connection to blocks and then to vessel ships. If arriving on trucks, containers are transferred to yard blocks where they are picked up by yard cranes; then, they are stacked and consolidated; finally, they are moved by trailers to the quayside where quay cranes load them on ships. If containers arrive on trains, railway cranes load them on trailers for the transfer to yard blocks.

In a transshipment cycle, when a vessel ship arrives, containers are downloaded from ships to trailers using quay cranes, transferred to blocks using trailers, picked-up and stacked by yard cranes. After a consolidation and/or a delay in their position, containers are picked-up by trailers and transferred back to the quay area, where they are loaded on a feeder ship. The opposite occurs if a feeder arrives after a short trip and a vessel departs for a long one.

Hereinafter, the modelling focus will be on the first step of the import and transshipment cycles, when a downloading process from a ship is executed. Namely, this is considered as a critical phase for the terminal efficiency. But similar models and observations can be obtained for the subsequent steps and also for the export cycles.

### 3.2 Classification of agents in the MAS

Now, a MAS is specified modelling the negotiations occurring in an intermodal container terminal which is mainly devoted to transshipment cycles. The main considered agents are:

- the *Container Agent* (CA): each CA is an autonomous entity in charge of controlling the flow of a single container unit or a group of containers;
- the *Quay crane Agent* (QA): it is an autonomous controller for a (set of) quay crane(s) with the same performance characteristics, or with the same physical possibility to reach and serve ship bays;
- the *Trailer Agent* (TA): it is an entity associated to a (set of) trailer(s) with the same performances, or with the same reachable portions of quay or yard spaces;
- the *Yard crane Agent* (YA): it manages the control of a (set of) yard crane(s) guaranteeing the same performances, or associated to the same yard blocks;
- the *Railway crane Agent* (RA): it is a software controller of a (set of) railway crane(s), used to receive containers from trailers and to load them on trains, or to deliver containers from trains to trailers;
- the *Truck Agent* (KA): it follows the operations executed by a (set of) truck(s), entering or leaving the terminal by the gate.

If related to a set of containers, a CA represents units physically stowed in a ship bay when considering the downloading process or units stacked in yard blocks when considering the loading process.

In import processes, the CA should identify the most suitable quay crane to download containers from ship, then the most suitable trailers to transport containers to their assigned yard blocks, and finally the most proper yard cranes to pick-up and stack containers in their assigned block positions. All these choices result from negotiations between the CA and several QAs, TAs, and YAs. The CA has also responsibilities in export processes: it selects the yard cranes to pick-up containers from blocks, trailers to transport them to the quay area, and quay cranes to load them into their assigned bay-row-tier location in the ship. In this case, the CA negotiates with YAs, TAs, and finally YAs.

During consolidation, containers are sometimes moved around and relocated to new positions in yard blocks, such that subsequent export operations are optimized or made easier. These moves are the consequence of other negotiations of CAs with YAs and TAs.

The decisions taken by a CA are based upon real-time updated information received from agents of the alternative available cranes and trailers.

The global control of the activities in the terminal emerges from the behaviour of concurrently operating agents. The dynamical interaction between agents has to be analysed to specify the desired global system behaviour. For instance, the precedent observations lead us to examine interactions between a CA and several QAs, TAs, and YAs for downloading, transferring, and stacking containers, or for picking, transferring, and loading containers. Interactions exist also between a CA and YAs, TAs, and RAs/KAs when railway/road transport modes are considered.

The interaction between agents is usually based on a negotiation mechanism, which is typically organized in the following steps. *Announcement*: an agent starts a bid and requires availability to other agents for a service. *Offer*: the agent requests data to the agents which declared availability. These data regard the offered service the queried agents can guarantee. *Reward*: the agent selects the best offer between the collected replies from the queried agents, and sends a rewarding output message. *Confirmation*: the agent waits for a confirmation message from the rewarded agent, after which it acquires the negotiated service. If confirmation is delayed or does not arrive, then the agent selects another offer in the rank or starts the bid over again.

In this work, the focus is on the interactions between a CA with QAs, TAs and YAs for downloading containers from ships to yard blocks during transhipment cycles. The reason is twofold: transhipment is the main and more complex service in intermodal container terminals (e.g., see the TCT case), and the downloading part has critical effects on terminal efficiency. The developments may be easily extended to other negotiations between agents for the loading part of transshipment cycle, export and import cycles or other processes.
The agents' negotiations and decisions are only limited by:

- constraints of terminal spaces and resources (e.g., the limited number of quay cranes that may physically serve a fixed ship bay, considering that quay cranes are sequentially lined up and move along fixed tracks; the limited number of yard cranes serving blocks);
- fixed working schedules for downloading/loading processes (they often establish the number of containers moved for each ship bay, their exact location in the ship hold or cover, the sequence of handling moves, sometimes even the preferred handling quay crane).

Agents' decisions are fulfilled by human operators devoted to the associated resources (i.e. crane operators, trailer/truck drivers). The network of interacting agents may appear and behave as a unique distributed supervisor for the physical terminal system.

## 4. Discrete-event systems modelling of agents' dynamics

Now, each agent in the previously identified classes is described as an atomic DEVS (Zeigler, 2000). Namely, all agents interact by transmitting outputs and receiving inputs, which are all considered as event messages. Events are instantaneous, then timed activities are defined by a start-event and a stop-event.
For each agent, internal events are triggered by internal mechanisms, external input events (i.e. inputs) are determined by exogenous entities, for example other agents, and external output events (i.e. outputs) are generated and directed to other entities.
These external or internal events change the agent state. Namely, an agent stays in a state until either it receives an input X or the time scheduled by a time advance function *ta* elapses (this time specifies the time before the occurrence of an internal event I). In the first case, an external transition function $\delta_{ext}$ determines the state next to the occurrence of the received input; in the second case, an internal transition function $\delta_{int}$ gives the state next to the occurrence of the internal event. An output function $\lambda$ is used to generate the reactions of the agent, each output being indicated with a symbol Y.
It is important to note that the DEVS formalism makes a difference between the *total state*, **q**, and the *sequential state*, **s**. This latter refers to the transition mechanism due to internal events. It is based on the current value of the so-called *status*, i.e. the condition in which an agent stays between two consecutive events during negotiations, and other characteristic information *inf* peculiar to the considered agent:

$$\mathbf{s} = (status, inf) \tag{1}$$

The total state is composed by **s**, the time *e* elapsed since the last transition, and some additional information like the decision logic **DL** currently used by the agent to rank and choose the offers received by other agents during negotiation:

$$\mathbf{q} = (\mathbf{s}, e, \mathbf{DL}) \tag{2}$$

For a CA, **s** may include information on: the current container position; the quay cranes available for negotiating downloading/loading operations; the trailers available for negotiating the transport from quay to yard or *viceversa;* the yard cranes available for negotiating pick-up and stacking operations (from trailer to a block position or backwards); the time scheduled in current state before the next internal event, if no external input event occurs.
For QAs, YAs, TAs, RAs, KAs, the sequential state may include the time prospected before the next internal event and the queued requests coming from CAs for availability, for data about offered service, for the confirmation of assigned service, etc..
To summarize, each agent can be represented as an atomic DEVS in the following way:

$$A = < \mathbf{X}, \mathbf{Y}, \mathbf{S}, \delta_{int}, \delta_{ext}, \lambda, ta > \quad (3)$$

where **X** is the set of input events, **Y** is the set of output events, **S** is the set of sequential states, $\delta_{int}$: $\mathbf{S} \rightarrow \mathbf{S}$ is the internal transition function, $\delta_{ext}$: $\mathbf{Q}\times\mathbf{X} \rightarrow \mathbf{S}$ is the external transition function, $\mathbf{Q} = \{\mathbf{q} = (\mathbf{s}, e, \mathbf{DL}) \mid \mathbf{s}\in\mathbf{S}, 0 \le e \le ta(\mathbf{s})\}$ is the set of total states, $\lambda$: $\mathbf{S} \rightarrow \mathbf{Y}$ is the output function, $ta$: $\mathbf{S} \rightarrow \Re_0^+$ is the time advance function, with $\Re_0^+$ set of positive real numbers with 0 included.
Here, the *status*-transitions triggered by events are examined since the *status* is considered as the main component of the sequential state. This assumption let the dynamics of each agent be described by *status*-transitions.
The same DEVS methodology can be applied to represent interactions between agents in the loading part of a transhipment cycle, when containers are moved from yard blocks to ships, or in import or export cycles.

### 4.1 Dynamics of interactions between agents in a downloading process

The MAS can be regarded as 'container-driven' because each container follows a sequence of handling operations executed by terminal resources. The container is the main entity flowing in the system and flows of containers are controlled by agents.
To download containers from a ship bay, transport and stack them into a yard block, each CA follows three phases in negotiation. Firstly, it interacts with QAs to choose the quay crane for downloading. Secondly, with TAs to select the trailers for moving the containers from the quay to the yard area. Thirdly, with YAs to determine the yard cranes for stacking.
Then, it is assumed that the CA firstly communicates exclusively with QAs, then with TAs only, finally with YAs. The negotiation is based on data about the offered services: a QA gives the estimated time to wait before the associated quay crane can start downloading containers, and the estimated time to execute the operation; a TA the estimated time to wait a trailer, and the estimated time for the transport task; a YA the estimated time for the yard crane to be ready close to the block, and the estimated time for the stacking task.
Note that, for the loading part of the transshipment cycle from yard to ship, the CA will sequentially interact with YAs, TAs, and QAs.

#### 4.1.1 Interactions between a CA and QAs

For $t<t_{C0}$ let the CA, say $C$, associated with a generic container be in a quiescent *status* (QUIESC) and let it begin its activity at $t_{C0}$ (input event $X_{C0}$). Then $C$ spends the time interval $[t_{C0}, t_{C1}]$ to send outputs $Y_{C01}, Y_{C02},\ldots, Y_{C0q}$ at instants $t_{01}>t_{C0}, t_{02},\ldots, t_{0q}=t_{C1}$. These messages request the availability to all the $q$ alternative QAs of quay cranes that can serve the container. The sequence of requests cannot be interrupted by any internal or external occurrence. For sake of simplicity, instead of modelling a sequence of $q$ different and subsequent *status*-values, REQQAV represents the whole duration of the activity and it is assumed that C makes transition at $t_{C1}$ (internal event $I_{C1}$).

In $[t_{C1}, t_{C2}]$ agent C waits for answers (WAIQAV) from QAs. Namely, the request C transmits to each QA may queue up with similar ones sent by other CAs. Next transition occurs at $t_{C2}$ when either C receives all the answers from the queried QAs ($X_{C1}$), or a specified time-out of WAIQAV expires before C receives all the answers. In case it receives no reply within the time-out ($I_{C2}$), C returns to REQQAV and repeats the request procedure. In case of time-out expiration and some replies received ($I_{C3}$), C considers only the received answers to proceed. The repeated lack of valid replies may occur for system congestion, for crane failures or communication faults, or for other unpredictable circumstances. In all cases permanent waits or deadlocks may occur. To avoid further congestion and improve system fault-tolerance, time-outs are used and C is allowed to repeat the cycle REQQAV-WAIQAV only a finite number of times, after which C is replaced by another software agent.

If all or some replies are received before the time-out expiration, C starts requesting service to the $g \leq q$ available QAs at $t_{C2}$. In $[t_{C2}, t_{C3}]$ C requests information to these QAs by sending them outputs $Y_{C11}$, $Y_{C12}$, ..., $Y_{C1g}$ at instants $t_{11}>t_{C2}$, $t_{12}$,..., $t_{1g}=t_{C3}$. As the sequence of requests cannot be interrupted, REQQSE *status* is referred for the whole activity: at $t_{C3}$ agent C is assumed to make transition ($I_{C4}$).

Then, agent C spends $[t_{C3}, t_{C4}]$ waiting for offers from the available QAs (WAIQOF), as the request C transmits to each QA may queue up with those sent by other CAs. Next transition occurs at $t_{C4}$ when either C receives all the answers from the queried QAs ($X_{C2}$) or a time-out of WAIQOF expires. In case no reply is received within the time-out ($I_{C5}$), C returns to REQQSE and repeats the procedure. In case of time-out expiration and some replies are received ($I_{C6}$), C considers only the received offers to select the crane. Again, to avoid congestion, C repeats the cycle REQQSE-WAIQOF a finite number of times, then it is discharged.

Once received the offers from QAs, C utilizes $[t_{C4}, t_{C5}]$ to take a decision for selecting the quay crane (TAKQDE). At $t_{C5}$ the decision algorithm ends ($I_{C7}$), after selecting a QA and building a rank of all the offers from QAs.

Subsequently, C reserves the chosen crane by transmitting a booking message ($Y_{C2}$) to the corresponding QA. So C takes $[t_{C5}, t_{C6}]$ for communicating the choice to the 'winner' QA (COMCHQ). At $t_{C6}$ the communication ends ($I_{C8}$). Now, the selected QA has to send a rejection, if there is a conflict with another CA, or a booking confirmation ($X_{C5}$). Hence, C uses $[t_{C6}, t_{C7}]$ to wait for a confirmation from the selected QA (WAIQCO). The confirmation is necessary because the availability of the cranes can be modified by actions of CAs other than C during the decision interval, and the selected crane can be no longer available. If C receives a rejection ($X_{C3}$), or does not receive any reply within a time-out ($I_{C9}$), it returns to COMCHQ, and sends a new request of confirmation to the second QA in the decision rank. If C has no other alternative destinations and the rejection ($X_{C4}$) or the time-out ($I_{C10}$) occurs, it returns to REQQAV and repeats the negotiation. Note that WAIQAV, WAIQOF and WAIQCO cannot lead to indefinite circular waits (deadlocks), thanks to the time-out mechanism.

At $t_{C7}$, after receiving a confirmation ($X_{C5}$) from the selected QA, *C* makes a transition to issue a downloading command (DWNLDG). It takes the interval [$t_{C7}$, $t_{C8}$] to issue the command $Y_{C3}$ for the quay crane downloading the container.

#### 4.1.2 Other interactions

When, at time $t_{C8}$, the downloading command is complete ($I_{C11}$), *C* starts the second negotiation-phase with TAs for a trailer to carry the container to its assigned block.
This new negotiation follows the same procedure as in the interaction with QAs. Agent *C* requests and waits for availability, requests information and waits for offers about trailer service, evaluates and ranks the offers to take a decision, communicates the choice to the best offering TA, and waits for a confirmation/rejection. This can be noted from Fig. 3, where the second part of the *status*-transition graph has the same repeated structure as in the first part, and the *status*-values have the same meaning. Also, time-outs are used to limit waiting and the repeated requests. This fact guarantees the model is modular, which is very important for simulation and control purposes.
When a confirmation is received, *C* makes a transition to issue a transport command (TRANSP), and the container is loaded on the vehicle associated to the selected TA.
Then, *C* starts the third negotiation-phase with YAs for a yard crane to stack the container in an assigned position inside a specified block. Again, due to the modularity of the approach, the sequence of allowed *status*-values follows the same protocol, and, finally, if a confirmation is received, *C* makes a transition to issue a stacking command (STCKNG). When the command is complete, *C* gets back to QUIESC.
From $t_{C24}$ to the beginning of the next negotiation cycle (if any) for downloading, consolidating or loading another container, *C* stops making decisions, receiving and sending messages, and remains quiescent. The associated container is downloaded, transported and stacked in a block where it waits for the next destination (a new block or a ship bay): all these processes do not involve agent activities. Only when they are over, *C* is ready to start a new negotiation for the same container. If faults occur to the selected cranes or trailer, *C* remains in QUIESC and there is no need to restart negotiations with QAs, TAs, and YAs. Terminal operators manage the repair process and when the normal operating conditions are restored, the container can be handled by the selected resources.
Fig. 3 depicts the complex interaction dynamics previously described for the negotiation between a CA with QAs, TAs, and YAs. Circles represent the CA *status*-values, and the period spent in each *status* is indicated aside of it. Arrows represent the (internal or input) events, labelling the arrows themselves (the event time is indicated below the event symbol). The outputs, directly associated with *status*-values, are encapsulated in the circles. The output function simply defines outputs for the allowed *status*-values. The time advance function gives the residual time $ta(\mathbf{s})$ in state $\mathbf{s}$ before the scheduled occurrence of next internal invent. For instance, at the time $t^*$ of entering a waiting *status*, the related time-out fixes the maximum time to wait $T_w$, such that $ta(\mathbf{s}) = t^*+T_w-t$, where $t$ is the current time.
To synthesize, one can use a 'macro-status' for each of the negotiation phases and depict the diagram of Fig. 4. Macro-status *QA-neg* represents all the diagram corresponding to the first negotiation-phase between the CA and QAs. In the same way, *TA-neg* and *YA-neg* aggregate the parts of diagram in Fig. 3, which are related to the second and third negotiation-phases of the CA with TAs and YAs, respectively.

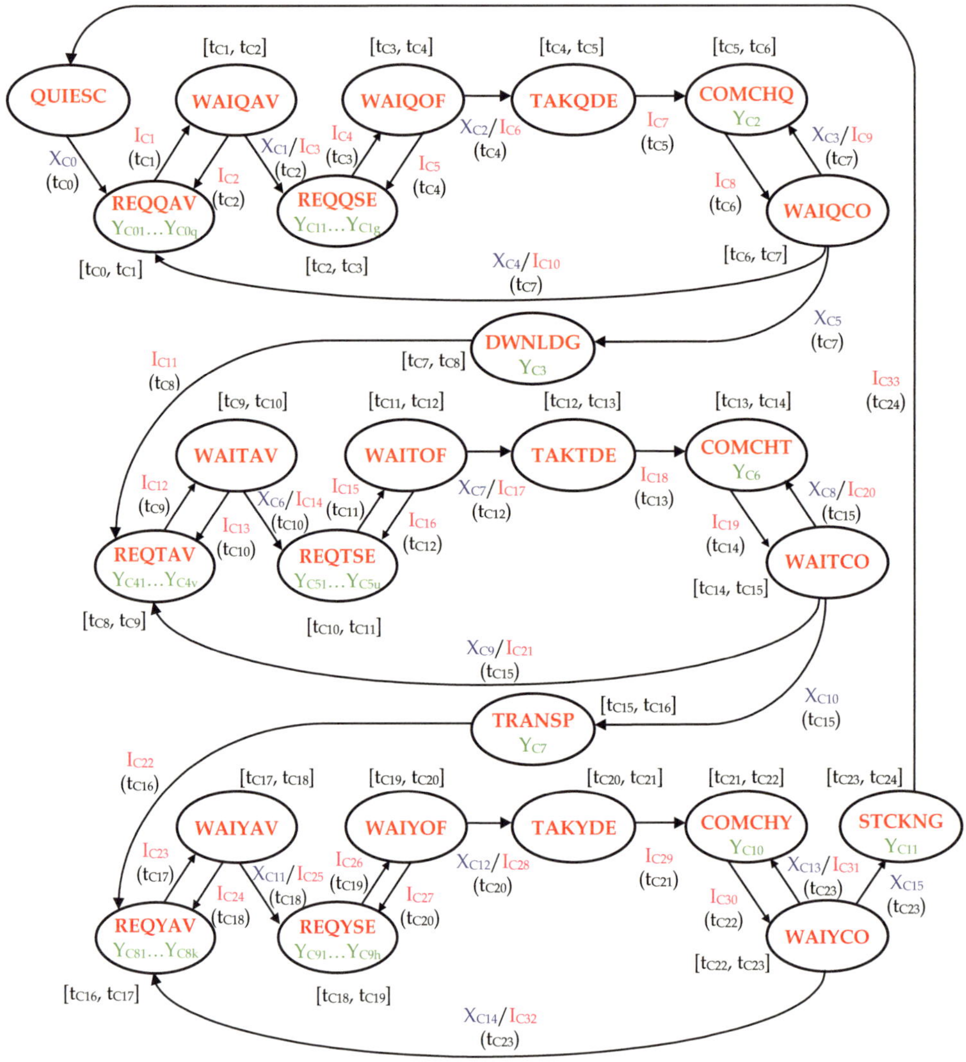


Fig. 3. Downloading in a transshipment cycle: negotiation of a CA with QAs, TAs, and Yas

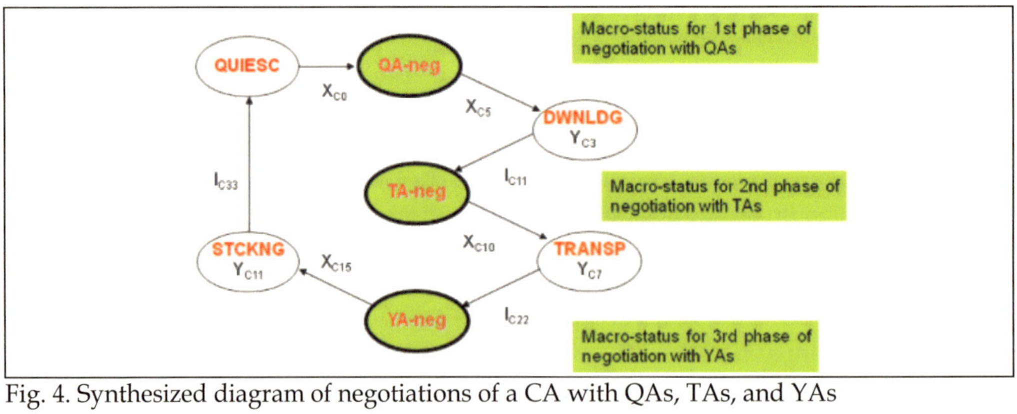


Fig. 4. Synthesized diagram of negotiations of a CA with QAs, TAs, and YAs

### 4.1.3 Remarks about negotiation

As Fig. 3 shows, the CA may receive confirmation from a QA, TA or YA after several successive loops COMCHQ-WAIQCO, COMCHT-WAITCO or COMCHY-WAIYCO.

Time-outs can bring the CA back to REQQAV from WAIQAV if no availability signal comes from QAs, or from WAIQCO if all selected QAs rejected the selection. The CA can also go back to REQTAV from WAITAV if no availability signal comes from TAs, or from WAITCO after rejection from all selected TAs. Finally, the CA comes back to REQYAV from WAIYAV if no availability signal comes from YAs, or from WAIYCO after rejection from all selected YAs.

Other time-outs rule the loops between WAIQOF and REQQSE, between WAITOF and REQTSE, and between WAIYOF and REQYSE.

One could merge the requests for availability and for information to get a more compact representation. But this would imply a lower effectiveness in reducing CA waiting times and in preventing deadlocks. Namely, the detailed model separates and reduces the effects of delays and losses of messages due to communication faults or to the unavailability of cranes and trailers, when particular conditions occur (faults, overloading conditions, etc.). Moreover, the number of *status*-loops and the consequent delays in decision are reduced if the described detailed model is adopted.

### 4.2 Specification of the DEVS model of a container agent

On the basis of the negotiation mechanism described in section 4.1 and by Fig. 3, the components of the DEVS model of a CA are identified. Table 1 reports and explains all the admissible *status*-values.

The sequential state $\mathbf{s}$ = (*C-status*, *C-inf*) of a container agent is defined as function of the current *status*-value, with *C-status* belonging to the set of all admissible values:

$$\begin{aligned}C\text{-}status \in \{&\text{QUIESC, REQQAV, WAIQAV, REQQSE, WAIQOF, TAKQDE, COMCHQ,}\\ &\text{WAIQCO, DWNLDG, REQTAV, WAITAV, REQTSE, WAITOF, TAKTDE, COMCHT,}\\ &\text{WAITCO, TRANSP, REQYAV, WAIYAV, REQYSE, WAIYOF, TAKYDE, COMCHY,}\\ &\text{WAIYCO, STCKNG}\}\end{aligned} \tag{4}$$

and as function of other information *C-inf* about the CA, which is defined as follows:

$$C\text{-}inf = (p, AQ, AT, AY, ta(\mathbf{s})) \tag{5}$$

which, in turn, depends on the current position $p$ of the container (on ship, picked by quay crane, on trailer, picked by yard crane, in the yard block); the set $AQ$ of alternative quay cranes available for the currently negotiated downloading operations, or the set $AT$ of trailers available for the currently negotiated transport tasks, or the set $AY$ of yard cranes available for the currently negotiated stacking tasks; the time $ta(\mathbf{s})$ scheduled in current state $\mathbf{s}$ before the next internal event occurrence.

A detailed description of inputs, outputs and internal events is reported in Tables 2, 3 and 4, respectively, for sake of clarity. It can be easily derived from the interactions between a CA and QAs, TAs, and YAs, as explained in section 4.1 and indicated in Fig. 3.

Most inputs received and outputs sent by a CA obviously correspond to outputs coming from and inputs to other agents, respectively. Then, these messages will concur to define the DEVS atomic models of the associated agents. More specifically, the agents interacting with a CA will change their *status*, then their sequential state, on the basis of the inputs received by the CA. Moreover, they will generate outputs by using their specific output functions to answer the CA during the different negotiation-phases.

| *Phase of negotiation* | *Status* | *Activity Description* |
|---|---|---|
| | QUIESC | Agent quiescent (inactive) |
| First phase | REQQAV | Request availability to all alternative QAs |
| | WAIQAV | Wait for availability from QAs |
| | REQQSE | Request service to available QAs |
| | WAIQOF | Wait for offers from available QAs |
| | TAKQDE | Rank QAs and take decision for the best QA |
| | COMCHQ | Communicate choice to selected QA |
| | WAIQCO | Wait confirmation/rejection from selected QA |
| | DWNLDG | Command selected QA to download container |
| Second phase | REQTAV | Request availability to all alternative TAs |
| | WAITAV | Wait for availability from TAs |
| | REQTSE | Request service to available TAs |
| | WAITOF | Wait for offers from available TAs |
| | TAKTDE | Rank TAs and take decision for the best TA |
| | COMCHT | Communicate choice to selected TA |
| | WAITCO | Wait confirmation/rejection from selected TA |
| | TRANSP | Command selected TA to transport container |
| Third phase | REQYAV | Request availability to all alternative YAs |
| | WAIYAV | Wait for availability from YAs |
| | REQYSE | Request service to available YAs |
| | WAIYOF | Wait for offers from available YAs |
| | TAKYDE | Rank YAs and take decision for the best YA |
| | COMCHY | Communicate choice to selected YA |
| | WAIYCO | Wait confirmation/rejection from selected YA |
| | STCKNG | Command selected YA to stack container |

Table 1. *Status*-values for a Container Agent

Finally, note how the successive inputs, outputs and internal events of the CA in the three subsequent negotiation-phases repeat with the same role, which gives the DEVS model a structure that can be easily implemented in simulation.

DEVS models can be also specified for the other agents, by following the same methodology. In particular, the sequential state of each QA, TA or YA will include information about the queued requests coming from different CAs competing for the same controlled quay crane, trailer or yard crane.

| *Phase of negotiation* | *Input* | *Time* | *Event Description* |
|---|---|---|---|
| | $X_{C0}$ | $t_{C0}$ | Start of negotiation activity for a new operation |
| First phase | $X_{C1}$ | $t_{C2}$ | Last of *q* replies for availability received from QAs |
| | $X_{C2}$ | $t_{C4}$ | Last of *g* replies for offer received from QAs |
| | $X_{C3}$ | $t_{C7}$ | Rejection & other QAs in the CA rank |
| | $X_{C4}$ | $t_{C7}$ | Rejection & no other QA in the CA rank |
| | $X_{C5}$ | $t_{C7}$ | Confirmation from selected QA |
| Second phase | $X_{C6}$ | $t_{C10}$ | Last of *v* replies for availability received from TAs |
| | $X_{C7}$ | $t_{C12}$ | Last of *u* replies for offer received from TAs |
| | $X_{C8}$ | $t_{C15}$ | Rejection & other TAs in the CA rank |
| | $X_{C9}$ | $t_{C15}$ | Rejection & no other TA in the CA rank |
| | $X_{C10}$ | $t_{C15}$ | Confirmation from selected TA |
| Third phase | $X_{C11}$ | $t_{C18}$ | Last of *k* replies for availability received from YAs |
| | $X_{C12}$ | $t_{C20}$ | Last of *h* replies for offer received from YAs |
| | $X_{C13}$ | $t_{C23}$ | Rejection & other YAs in the CA rank |
| | $X_{C14}$ | $t_{C23}$ | Rejection & no other YA in the CA rank |
| | $X_{C15}$ | $t_{C23}$ | Confirmation from selected YA |

Table 2. Input events received by a Container Agent in negotiation with QAs, TAs, and Yas

| *Phase of negotiation* | *Output* | *Time* | *Event Description* |
|---|---|---|---|
| First phase | $Y_{C01}, Y_{C02}, \ldots, Y_{C0q}$ | $t_{01}>t_{C0}, t_{02}, \ldots, t_{0q}=t_{C1}$ | Requests of availability to *q* QAs |
| | $Y_{C11}, Y_{C12}, \ldots, Y_{C1g}$ | $t_{11}>t_{C2}, t_{12}, \ldots, t_{1g}=t_{C3}$ | Requests of information to *g* available QAs |
| | $Y_{C2}$ | $t_{C6}$ | Choice communication to the selected QA |
| | $Y_{C3}$ | $t_{C8}$ | Command for downloading container |
| Second phase | $Y_{C41}, Y_{C42}, \ldots, Y_{C4v}$ | $t_{41}>t_{C8}, t_{42}, \ldots, t_{4v}=t_{C9}$ | Request of availability to *v* TAs |
| | $Y_{C51}, Y_{C52}, \ldots, Y_{C5u}$ | $t_{51}>t_{C10}, t_{52}, \ldots, t_{5u}=t_{C11}$ | Requests of information to *u* available TAs |
| | $Y_{C6}$ | $t_{C14}$ | Choice communication to the selected TA |
| | $Y_{C7}$ | $t_{C16}$ | Command for transporting container |
| Third phase | $Y_{C81}, Y_{C82}, \ldots, Y_{C8k}$ | $t_{81}>t_{C16}, t_{82}, \ldots, t_{8k}=t_{C17}$ | Request of availability to *k* YAs |
| | $Y_{C91}, Y_{C92}, \ldots, Y_{C9h}$ | $t_{91}>t_{C18}, t_{92}, \ldots, t_{9h}=t_{C19}$ | Requests of information to *h* available YAs |
| | $Y_{C10}$ | $t_{C22}$ | Choice communication to the selected YA |
| | $Y_{C11}$ | $t_{C24}$ | Command for stacking container |

Table 3. Output events sent by a Container Agent in negotiation with QAs, TAs, and Yas

| *Phase of negotiation* | *Internal Event* | *Time* | *Event Description* |
|---|---|---|---|
| First phase | $I_{C1}$ | $t_{C1}$ | End of request for availability of QAs |
| | $I_{C2}$ | $t_{C2}$ | Time-out & no availability signal received from QAs |
| | $I_{C3}$ | $t_{C2}$ | Time-out & $g$ availability signals received from QAs |
| | $I_{C4}$ | $t_{C3}$ | End of request for offered service from available QAs |
| | $I_{C5}$ | $t_{C4}$ | Time-out & no offer received from the $g$ available QAs |
| | $I_{C6}$ | $t_{C4}$ | Time-out & $oq \leq g$ offers received from available QAs |
| | $I_{C7}$ | $t_{C5}$ | End of decision for choosing quay crane (agent) |
| | $I_{C8}$ | $t_{C6}$ | End of choice communication to the selected QA |
| | $I_{C9}$ | $t_{C7}$ | Time-out & no confirmation received from the selected QA & ranked offers are available from other QAs |
| | $I_{C10}$ | $t_{C7}$ | Time-out & no confirmation received from the selected QA & no ranked offers are available from other QAs |
| Transition to 2nd phase | $I_{C11}$ | $t_{C8}$ | End of downloading command |
| Second phase | $I_{C12}$ | $t_{C9}$ | End of request for availability of TAs |
| | $I_{C13}$ | $t_{C10}$ | Time-out & no availability signal received from TAs |
| | $I_{C14}$ | $t_{C10}$ | Time-out & $u$ availability signals received from TAs |
| | $I_{C15}$ | $t_{C11}$ | End of request for offered service from available TAs |
| | $I_{C16}$ | $t_{C12}$ | Time-out & no offer received from the $u$ available TAs |
| | $I_{C17}$ | $t_{C12}$ | Time-out & $ot \leq u$ offers received from available TAs |
| | $I_{C18}$ | $t_{C13}$ | End of decision for choosing trailer (agent) |
| | $I_{C19}$ | $t_{C14}$ | End of choice communication to the selected TA |
| | $I_{C20}$ | $t_{C15}$ | Time-out & no confirmation received from the selected TA & ranked offers are available from other TAs |
| | $I_{C21}$ | $t_{C15}$ | Time-out & no confirmation received from the selected TA & no ranked offers are available from other TAs |
| Transition to 3rd phase | $I_{C22}$ | $t_{C16}$ | End of transport command |
| Third phase | $I_{C23}$ | $t_{C17}$ | End of request for availability of YAs |
| | $I_{C24}$ | $t_{C18}$ | Time-out & no availability signal received from YAs |
| | $I_{C25}$ | $t_{C18}$ | Time-out & $h$ availability signals received from YAs |
| | $I_{C26}$ | $t_{C19}$ | End of request for offered service from available YAs |
| | $I_{C27}$ | $t_{C20}$ | Time-out & no offer received from the $h$ available YAs |
| | $I_{C28}$ | $t_{C20}$ | Time-out & $oy \leq h$ offers received from available YAs |
| | $I_{C29}$ | $t_{C21}$ | End of decision for choosing yard crane (agent) |
| | $I_{C30}$ | $t_{C22}$ | End of choice communication to the selected YA |
| | $I_{C31}$ | $t_{C23}$ | Time-out & no confirmation received from the selected YA & ranked offers are available from other YAs |
| | $I_{C32}$ | $t_{C23}$ | Time-out & no confirmation received from the selected YA & no ranked offers are available from other YAs |
| Transition to QUIESC | $I_{C33}$ | $t_{C24}$ | End of stacking command |

Table 4. Internal events in a Container Agent in negotiation with QAs, TAs, and YAs

Namely, let *Q*, *T* and *Y* be three of such agents. Then, each agent will be characterized by a queue $\mathbf{c}_1$ for the availability requests associated to one of the messages: $Y_{C0i}$ $(1 \le i \le q)$ arriving to *Q*, $Y_{C4i}$ $(1 \le i \le v)$ arriving to *T*, $Y_{C8i}$ $(1 \le i \le k)$ arriving to *Y*; a queue $\mathbf{c}_2$ for the information requests associated to one of the messages: $Y_{C1i}$ $(1 \le i \le g)$ arriving to *Q*, $Y_{C5i}$ $(1 \le i \le u)$ arriving to *T*, $Y_{C9i}$ $(1 \le i \le h)$ arriving to *Y*; a queue $\mathbf{c}_3$ for the confirmation requests associated to messages: $Y_{C2}$ arriving to *Q*, $Y_{C6}$ arriving to *T*, $Y_{C10}$ arriving to *Y*. Moreover, the state of *Q*, *T*, or *Y* will be defined by the current number of containers served by the associated quay crane, trailer or yard crane.

## 5. Ideas for simulation and evaluation of efficiency and robustness of the MAS control

The DEVS atomic models can be integrated to obtain a complete networked system, which can be used as a platform for simulating the MAS architecture for controlling an intermodal container terminal. E.g. the TCT in Taranto can be used as a test-bed.

In this context, it is possible to simulate not only the dynamics of terminal activities, the flow of containers, and the utilization of terminal resources (cranes, trailers, human operators, etc.), but also the efficiency of the MAS and its agents (flow of event messages, *status* transitions, waiting loops, etc.).

Then, two types of performance indices can be defined. Namely, it is possible to measure the following conventional indices: the total number of handled (imported, exported, transshipped) containers; the average throughput, during downloading (from ship to yard) or loading (from yard to ship) processes; the average lateness of containers in the terminal; the utilization of resources; the ship turn-around time, i.e. the average time required to serve a ship for downloading and loading containers. Moreover, it is possible to measure the behaviour of the MAS and the efficiency of the agents' decision policies by means of: the average number of requests for each negotiation; the number of repeated negotiation loops of *status*-values before a final decision is taken by a CA, expressed in percentile terms with respect to the total number of operations executed by every CA. The lower is this index, the better is the agent capability to obtain a service at the first request. The higher is the value of the index, the higher is the lack of feasible replies due to congestion of the other agents or of the communication system.

More specifically, the terminal performance measures can be evaluated both in steady-state operating conditions and in perturbed conditions. Perturbations may arise from: hardware faults or malfunctions; abrupt increase/decrease of containers to be handled due to changes in maritime traffic volumes; sudden increase/reduction of yard space; traffic congestion of trailers; congestion, delays, message losses in the communications between agents.

For example, the private company managing TCT usually plans and controls the activities to serve one ship at a day. This is due to the fact that ship arrivals are known and scheduled in advance with ship agencies. But the company itself has recently foreseen a traffic increase in the following years, due to expected cargo movements coming from China and eastern countries to the Mediterranean Sea. Then, it is quite reasonable to think about working days in which more than one ship is berthed and served at the same time. In this case, at least two ships berthed to the quay would give a big perturbation to the required operations and terminal efficiency. Fig. 5 shows a snapshot of a discrete-event simulation made in this condition (2 ships in quay in the schematic view of the terminal), by using a conventional centralized control architecture, based on the current policies used in the terminal. The performance obtained were much lower than in standard conditions.

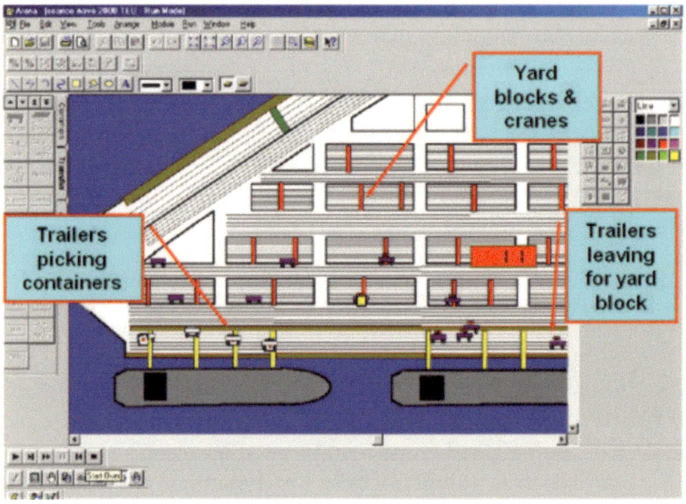


Fig. 5. Simulation snapshot of TCT in perturbed conditions - centralized control

Then, it is important to use distributed MAS control architectures and to measure robustness of agents' decision laws, to see how they dynamically react to disturbances and parameter variations, and eventually to adapt them. The adaptation aims to make the autonomous agents learn the most appropriate decision laws in all terminal conditions.

In this sense, the decision logic **DL** of a CA could be partly constant to encapsulate the most reliable strategies, and partly adaptable according to a learning algorithm. The constant part is constituted by a set of *R* heuristic decision rules, each related to a different evaluation parameter provided by QAs, TAs, and YAs. Some decision rules may be more effective in perturbed conditions (anomalies, faults, congestions), whereas a tradeoff between different rules may be more appropriate in other cases. Therefore, weights assigned to the rules represent the adjustable part of **DL** = $\{\alpha_1, \alpha_2, \dots, \alpha_R\}$, where each $\alpha_j$ ($j = 1, \dots, R$) specifies the factor weighting the role of the *j*-th heuristic in the global decision criterion. Then, an evolutionary algorithm can be used to adapt factors. In this way, in any operating condition, the worst performance of an agent should never be significantly lower than the performance of the worst decision rule.

To conclude, the discrete-event simulation platform also allows the comparison of alternative types of control architectures which can be defined by:

- a static MAS in which CAs use static logics based on heuristic decision parameters (estimated delivery time of the requested task, distance of cranes or trailers);
- a dynamic MAS in which CAs take decisions by fuzzy weighted combinations of heuristic criteria; the weights are adapted by an evolutionary algorithm;
- other distributed control architectures.

## 6. Conclusion

A MAS architecture was proposed for controlling operations in intermodal container terminal systems. The autonomous agents are represented as atomic DEVS components. The interactions between agents are modelled according to the DEVS formalism to represent negotiations for tasks when downloading containers from ships to yard stacking blocks. The

developed model can be easily extended to describe other processes, like loading containers from yard to ships, redistributing containers in the yard, import or export cycles.
An accurate DEVS model of the MAS can be used in a detailed simulation environment of a real system (the TCT in Taranto), which allows the measurement of standard terminal performance indices and of the efficiency of the MAS in real scenarios.
Moreover, open issues are testing and comparing static MAS and dynamically adapted MAS, if for example evolutionary adaptation mechanisms are used, especially with reference to different operating scenarios and to possible perturbations with respect to steady-state operating conditions.

## 7. References

Bielli, M.; Boulmakoul, A. & Rida, M. (2006). Object oriented model for container terminal distributed simulation. *European Journal of Operational Research*, Vol. 175, No. 3, pp. 1731-1751, ISSN 0377-2217

Cantarella, G. E.; Cartenì, A. & de Luca, S. (2006). A comparison of macroscopic and microscopic approaches for simulating container terminal operations, *Proceedings of the EWGT2006 International Joint Conferences*, pp. 556-558, ISBN 88-901798-2-1, Bari, Italy, 27-29 September 2006, 01Media, Molfetta (Bari), Italy

Crainic, T. G.; Gendreau, M. & Dejax, P. (1993). Dynamic and stochastic models for the allocation of empty containers. *Operations Research*, Vol. 41, No. 1, pp. 102-126, ISSN 0030-364X

de Luca, S.; Cantarella, G. E. & Cartenì, A. (2005). A macroscopic model of a container terminal based on diachronic networks, *Second Workshop on the Schedule-Based Approach in Dynamic Transit Modelling*, Ischia, Naples, Italy, 29-30 May 2005

Degano, C. & Di Febbraro, A. (2001). Modelling Automated Material Handling in Intermodal Terminals, *Proceedings 2001 IEEE/ASME International Conference on Advanced Intelligent Mechatronics (AIM'01)*, Vol. 2, pp. 1023-1028, ISBN 0-7803-6737-5, Como, Italy, 8-12 July 2001, IEEE, Piscataway, NJ, USA

Fischer, M. & Kemper, P. (2000). Modeling and Analysis of a Freight Terminal with Stochastic Petri Nets, *Proceedings of the 9th IFAC Int. Symp. Control in Transportation Systems*, Vol. 2, pp. 195-200, ISBN-13 978-0-08-043552-7, Braunschweig, Germany, 13-15 June 2000, Eds. Schnieder, E. & Becker, U., Elsevier-Pergamon, Oxford, UK

Gambardella, L. M.; Rizzoli, A. E. & Zaffalon, M. (1998). Simulation and planning of an intermodal container terminal. *Simulation, Special Issue on Harbour and Maritime Simulation*, Vol. 71, No. 2, pp. 107-116, ISSN 0037-5497

Heragu, S. S.; Graves, R. J.; Kim, B.-I. & Onge, A. St. (2002). Intelligent Agent Based Framework for Manufacturing Systems Control. *IEEE Trans. Systems, Man, and Cybernetics - Part A: Systems and Humans*, Vol. 32, No. 5, pp. 560-573, ISSN 1083-4427

Hsieh, F.-S. (2004). Model and control holonic manufacturing systems based on fusion of contract nets and Petri nets. *Automatica*, Vol. 40, No. 1, pp. 51-57, ISSN 0005-1098

Huhns, M. N. & Stephens, L. M. (2001). Automating supply chains. *IEEE Internet Computing*, Vol. 5, No. 4, pp. 90-93, ISSN 1089-7801

Kozan, E. (2000). Optimising container transfers at multimodal terminals. *Mathematical and Computer Modelling*, Vol. 31, No. 10-12, pp. 235-243, ISSN 0895-7177

Legato, P. & Mazza, R. M. (2001). Berth planning and resources optimisation at a container terminal via discrete event simulation. *European Journal of Operational Research*, Vol. 133, No. 3, pp. 537-547, ISSN 0377-2217

Lin, F. & Norrie, D. H. (2001). Schema-based conversation modeling for agent-oriented manufacturing systems. *Computers in Industry*, Vol. 46, No. 3, pp. 259-274, ISSN 0166-3615

Liu, C. I. & Ioannou, P. A. (2002). Petri Net Modeling and Analysis of Automated Container Terminal Using Automated Guided Vehicle Systems. *Transportation Research Record*, No. 1782, pp. 73-83, ISSN 0361-1981

Logan, B. & Theodoropoulos, G. (2001). The distributed simulation of multi-agent systems. *Proceedings of the IEEE*, Vol. 89, No. 2, pp. 174-185, ISSN 0018-9219

Mastrolilli, M.; Fornara, N.; Gambardella, L. M.; Rizzoli, A. E. & Zaffalon, M. (1998). Simulation for policy evaluation, planning and decision support in an intermodal container terminal, *Proceedings Int. Workshop "Modeling and Simulation within a Maritime Environment"*, pp. 33-38, ISBN 1-565-55132-X, Riga, Latvia, 6-8 September 1998, Eds. Merkuryev, Y., Bruzzone, A. & Novitsky, L., Society for Computer Simulation International, Ghent, Belgium

Peterkofsky, R. I. & Daganzo, C. F. (1990). A branch and bound solution method for the crane scheduling problem. *Transportation Research Part B: Methodological*, Vol. 24B, No. 3, pp. 159-172, ISSN 0191-2615

Schattenberg, B. & Uhrmacher, A. M. (2001). Planning Agents in James. *Proceedings of the IEEE*, Vol. 89, No. 2, pp. 158-173, ISSN 0018-9219

Shen, W. & Norrie, D. H. (1999). Agent-Based Systems for Intelligent Manufacturing: A State-of-the-Art Survey. *Knowledge and Information Systems, An International Journal*, Vol. 1, No. 2, pp. 129-156, ISSN 0219-1377

Steenken, D.; Voss, S. & Stahlbock, R. (2004). Container terminal operation and operations research - a classification and literature review. *OR Spectrum*, Vol. 26, No. 1, pp. 3-49, ISSN 0171-6468

Vis, I. F. A. & de Koster, R. (2003). Transshipment of containers at a container terminal: An overview. *Europ. Jour. Operational Research*, Vol. 147, No. 1, pp. 1-16, ISSN 0377-2217

Yun, W. Y. & Choi, Y. S. (1999). A simulation model for container-terminal operation analysis using an object-oriented approach. *International Journal of Production Economics*, Vol. 59, No. 1-3, pp. 221-230, ISSN 0925-5273

Zeigler, B. P.; Praehofer, H. & Kim, T. G. (2000). *Theory of Modelling and Simulation*, Academic Press, ISBN 0-12-778455-1, New York, 2nd edition

# 4

# Inclusion of Expert Rules into Normalized Management Models for Description of MIB Structure

Antonio Martin and Carlos Leon
*Dpto. Tecnología Electrónica. University of Sevilla*
*Spain*

## 1. Introduction

With the technology of telecommunication network developing and the size of communication network enlarging constantly, network management needs additional requirements. To satisfy these needs of quality and performance, it is necessary to achieve an attended management network by advance software. The Artificial Intelligent is incorporated to the network management to provide the task of control and administration.
OSI and Internet are dominant network management models, which have been used for administration and control of the most of existing networks. Network management activities, such as fault, security, performance, etc. have performed for network management models. These activities are becoming supplementary functionalities in the network management, direct expert system involvement. Traditional intelligent management technologies can play an import role in the problems solving and reasoning techniques employed in network management. However, these systems are not flexible enough for today's evolving network needs.
It is necessary to develop new models, which offer more possibilities. We propose a hybrid solution that employs both managed model and AI reasoning technique for the management of heterogeneous distributed information networks. We present a new concept called Integrated Management Expert System that employs both managed model and AI reasoning techniques for the intelligent management of heterogeneous networks. This new paradigm contains aspects which clearly make the difference from the former management techniques which uses separately expert systems and management platforms.
We propose the normalization of the knowledge management necessary to administrate the resources that exist in the networks independently from the builder of the management resources. The goal is to get a syntactically uniformed definition of all the management knowledge supplied by the expert, independently from the maker. The novelty comes from the fact that the employed knowledge for the networks management (conditions and management operations to achieve on the different resources) is included and normalised itself in the definitions of the network elements and is treated as if it were a propriety.
So, we realise syntactically uniformed normalization of intelligence applied to the management. This technique integrates the Expert System within the Management Information Base. The advantage is that a large problem can be broken down into smaller

and manageable sub-problems/modules. For this goal, an extension of OSI and SNMP management framework specifications language has been investigated. A new property named RULE has been added into the Management Information Base (MIB), which gathers important aspects of the facts and the knowledge base of the embedded expert system.

## 2. Related work

The work that we present has two parts, figure 1. In the first part we analyse the current management models, its evolution and the applications of the expert systems in the network management. It offers a general vision of the traditional management expert system, analyzing its deficiencies and discovering the needs that push us toward new management paradigms.

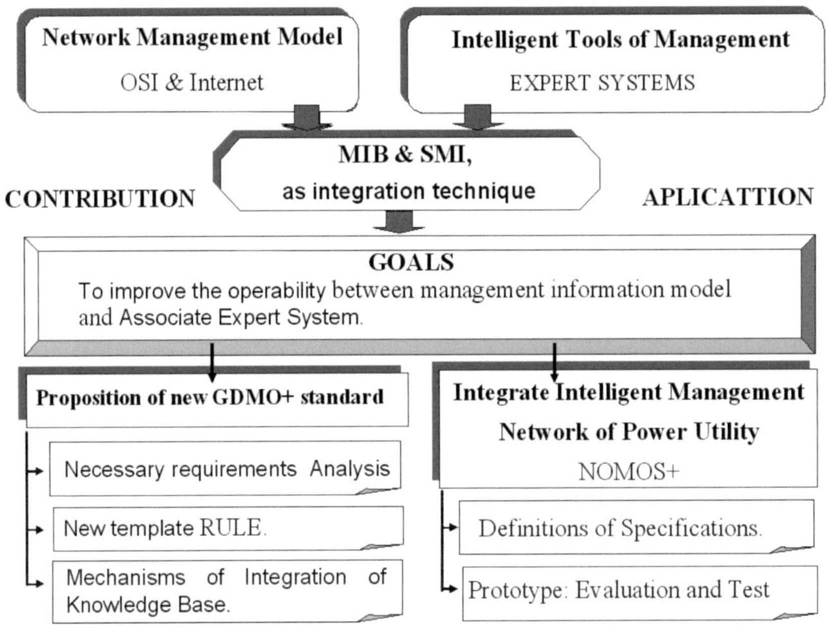


Fig. 1 Synopsis of the research.

The second step in the modelling sequence includes an introduction to standard Structure of Management Information (SMI) and intelligent management, showing the advantages and problems of integrated intelligent management. Next we present the extension of the standard GDMO (ISO/ITU-T, 1993), to accommodate the intelligent management requirements. We describe how the templates are modified using the new extension called GDMO+ and present the new template RULE, which defines the knowledge base of the management expert system. We study the characteristics of this new property RULE: template structure, behaviour, priority, inheritance of expert rules, etc.

As the last step in order to show the viability of our proposal, we perform a practical demonstration in which the information and the management knowledge is unified in a unique specification. We present a tool based on a GDMO+ standard. It shows an expert system prototype of integration rules. This prototype gives service at the moment to a power utility.

## 3. Network management

Once we have presented the scope and goals of the work, this section presents most important management models and we make a brief overview of relevant topics of the network management. There are several organizations which have developed services, protocols and architectures for network management. At the moment there are two main management models for computer communication: Open System Interconnection OSI and Internet. Those are standardized models for computer communication and starting point for understanding network management. It provides for common understanding between communicating processes in a heterogeneous environment, servers as the basis for a precise specification of network services and protocols and constitutes a vendor-independent concept (ITU-T, 1992).
ISO was the first which started, as part of its Open Systems Interconnection (OSI) program. The term OSI systems management actually refers to a collection of standards for network management that include a management service and a protocol. Common Management Information Protocol (CMIP), which provides the information exchange capability to support the Common Management Information Service (CMIS) and offers management services to management applications.
ISO issued a set of standards and draft standards for network management. A subset of these standards provides the foundation for Telecommunication Management Network (TMN) developed by the organization called Telecommunication Union (ITU). TMN conceptually a separate network that interfaces a telecommunications network at several different points (ITU-T, 1996).
Internet Model by the Internet Engineering Task Force (IEFT). Internet model is a structured and standardized approach to Internet management. This uses the Simple Network Management Protocol (SNMP) (Black, 1995), figure 2.

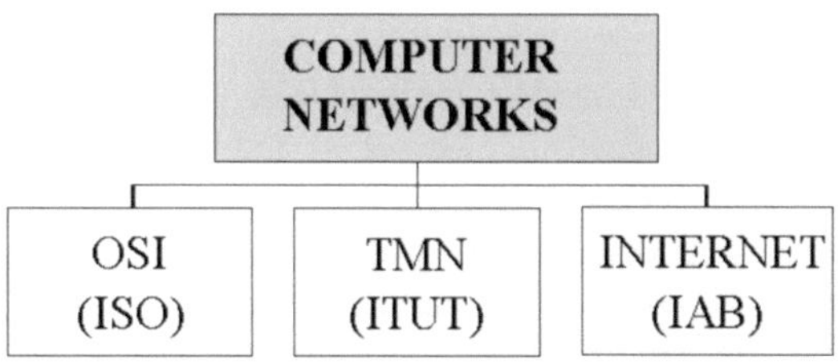


Fig. 2. Management Models

### 3.1 Network management elements

These network management systems operate using client/server architecture. Four fundamental concepts of these models are (ISO/ITU-T, 1998):

- Manager or Manager Role: In the network management model a manager is an unit that provides information to users, issues requests to devices in a network, receives responses to the requests and receives notifications.
- Agent or Agent Role: It is an unit that is part of a device in the network that monitors and maintains status about that device. It can act and respond to requests from a manager.
- Network Management Protocols: Managers and agents require some form of communication to issue their requests and responses. The Common Management

Information Protocol (CMIP) is the protocol used in management model ISO and TMN. SNMP is the protocol used to issue requests and receive responses in a management model Internet. The combination of Internet and OSI management requires the use of different network management protocols working with different levels of modelling complexity. CMIP requires the use of full OSI stack, for implementation while the SNMP protocol operates with the lower layers of stack. CMIP and SNMP have been designed to scale as network grows, i.e. the ability to perform "manager" or "agent".

- Management Information Base (MIB): In this information Base, we add the knowledge management. In addition to being able to pass information back and forth, the manager and the agent need to agree on and understand what information the manager and agent receive in any exchange. This information varies for each type of agent. The collection of this information is referred to as the management information base. A manager normally contains management information describing each type of agent the manager is capable of managing. This information would typically include ISO and Internet MIB definitions for managed objects and agents (Morris, 2003).

### 3.2 Management information

Information modelling plays a large part in modern network management systems. It is a way to represent the capabilities of a Managed object interfacing directly to the equipment's native command. This abstraction has given rise to management information models. This is the information associated with a managed object that is operated on by the OSI and Internet Management protocols to control and monitor that object.

The description of management information has two aspects. First, a Structure of Management Information (SMI) defines the logical constitution of management information and how it is identified and described. Second, the MIB, which is specified using the SMI, defines the actual objects to be managed. The MIB can be seen as kind of database. It is a conceptual repository of management information.

The elements that make up a management information model are referred to as managed objects. The content of this database is not set of managed objects themselves, but the information that is associated with the managed objects. These resources are viewed as 'managed objects'. A managed object is the abstract view of a logical or physical system resource to be managed. Thus, a managed object is the abstraction of such a resource that represents its properties as been by, and for purpose of, management. These special elements provide the necessary operations for the administration, monitoring and control of the telecommunications network. A managed object is a management view of a resource that is subject to management, such as a layer entity, a connection or an item of physical communications equipment. Managed objects are used for management functions, performance monitoring and analysis, and the setup and teardown of virtual circuits and virtual paths.

The managed object concept is refined in a number or additional standards, which are called the Structure of Management Information (SMI) standards. The managed objects are defined according to the SMI, which defines how network objects and their behaviour are to be specified, including the syntax and semantics (Clemm, 2006).

In the Open System Interconnection (OSI) and TMN systems management the information architecture is based on an object-oriented approach and the agent/manager concepts that are of paramount importance. In OSI, SMI provides the Guidelines for Definition of Managed Objects (GDMO), for definition objects contained in the MIB.

Internet management model doesn't use the Object Oriented Programming such as it is used by the OSI model. This is one of the reasons for the Internet model simplicity. The definitions contain objects, specified with ASN.1 macros. We are studying the way to integrate the expert knowledge in the management Internet model. The resources specifications can only be groups of scalar variables and cells tables in spite of not being an Object Oriented Programming model, we can use the tables as classes where the attributes are the table columns and every file contains an instance of the class. The same as in OSI every object has an OID associated identifier.

## 4. Overview on expert systems and telecommunications management

After this analysis of management elements in common to OSI an Internet models we present an overview of the state art in traditional expert management system. ISO classifies the systems management activities into five functional areas: fault management, accounting management, configuration management, performance management and security management. This specific functions performed by OSI are also applicable in the Internet model. We can categorize the expert systems within these five groups. In the next table 1 we can observe a synopsis of some expert systems applied of network management. This table indicates the applications area and technique use.

| | MANAGEMENT DOMAIN | | | | | |
|---|---|---|---|---|---|---|
| **TECHNIQUE** | **Fault** | **Accounting** | **Configuration** | **Performance** | **Security** | **Expert Interface** |
| **Expert Rules** | Max & Opti<br>ANSWER | ESS-ES<br>ECXpert | ACE<br>XCON<br>NMCS | TASA | NIDES<br>P-BEST<br>NIDX | NetHELP |
| **Bayesian Network** | Trouble Locator | APRI | | | | |
| **Case-Based Reasoning (CBR)** | CRITTER | | EXSim | NETTRA | | |
| **Blackboard** | | | | Scout | | |

Table 1. Expert system applications developed for network management.

We can observe that most of expert systems are built for fault management and use models based in expert rules. These rules obtain an experience accumulated by human experts. The last years new techniques are applied: Bayesian networks, blackboard, the systems based in case based reasoning, which make an approach to most similar case of the all cases existing in the knowledge base, etc. (Negnevitsky, 2002).

In traditional intelligent management these expert systems can be treated as a black box which receives the events coming from the elements that compose the managed system. There must exist between them a certain level of compatibility and interoperability that makes sure that the stream of data and information control flow in both directions. An unique interface must be developed, according to the possibilities and tools that the system possesses for the external communication with these applications: sockets, CORBA, etc.

The traditional expert management presents inconvenients, which show the limitations of current systems management. Some disadvantages of traditional Expert Management are the next:

- There are disadvantages caused by the existence of two independent elements, the management platform and the expert system, with a permanent and continuous dialogue between them.
- Restrictions may appear when choosing a determined managed platform.
- This presets incompatibility problems between management platform and expert system.
- Because of the need to use heterogeneous networks, it is difficult to normalize the knowledge for an adequate management of resources.

This disadvantage can be solved by using the integration management that we propose. In the integrate intelligent management, it is not necessary to interconnect the expert system and the management platform. Both components are defined using the same syntax, being finally both completely integrated in some unique specifications.
In next sections we will approach our research in the integration of knowledge management into MIB of OSI and Internet management models.

## 5. Knowledge integration in OSI and internet models

This paper focuses on a framework and a language for formalizing knowledge management descriptions and combining them with existing managed object definitions. To find a solution, we must develop an interface that makes it possible to introduce expert rules of management within the definition of the MIB objects. This is a solution that will be generally more flexible than the one found by traditional methods (formal language). The solution is the inclusion of formal knowledge descriptions in regular RULE-templates.
An essential part of the definition of an intelligent managed object is the relationship between their properties and the management knowledge of the resource. This relationship is not modelled in a general way. Unfortunately, the knowledge of managed objects is defined in another way using programming language. This results in resources specifications which are not often information about the knowledge base of expert system, and increasing the possibility of different intelligent implementations not being interoperable. To achieve consistent, clear, concise, and unambiguous specifications, a formal methodology has to be used.
OSI and Internet Management models are faced with several impediments to improve the integrated intelligent management. Management models explain how individual management operations should be performed. The current management standards do not specify however the sequence in which intelligent operations should be performed to solve specific management problems. OSI and Internet management is rather complicated. The model has introduced several new concepts, which are sometimes difficult to comprehend. This makes if difficult the construction of intelligent platforms that work with the model. The basic questions raised are:

1. Formulate, describe, distribute and assign the knowledge between the different intelligent agents defined.
2. Communicate the knowledge between different objects and what communications language or protocol must they use.
3. Ensure that the objects act coherently when they make a decision or conduct.

4. It enables the objects to be reason about the actions, plans and knowledge of other objects.

To make this possible, it is essential to explore the capacities that have the management information models. In particular, the OSI GDMO standard specifies the proprieties and characteristics that make easy the inclusion of the expert management rules as a part of the defined managed resources. In this way the intelligent network managers can interpretate the rules. Finished by the expert system and achieve an intelligent treatment of the management information. The integrated framework that we propose has numerous advantages. Among them:

- Abstraction of the users from the management systems and good compatibility between different management platforms.
- The ability of different network management system to communicate the knowledge management of the expert systems without overloading the management applications.
- The flexibility to define new managed objects than contain knowledge management without modifying all the system that needs to interact with them.
- Possibility of standardization of the new expert rules, by an organism such as the ISO and a common set of network management knowledge based in standards.
- Reusability of the knowledge management. GDMO has an object oriented syntax. Its classes inheritance mechanisms allows management expert rules reuse.
- Easily to build new expert management systems. A general top-down view of the integrated multi-vendor network and structured list of network management objects that contains some set of expert rules for the administration.

This structure also provides a richer level of abstraction facilitating the coexistence knowledge management, allowing different levels of modelling complexity, and organizing the knowledge management of the managed objects, figure 3.

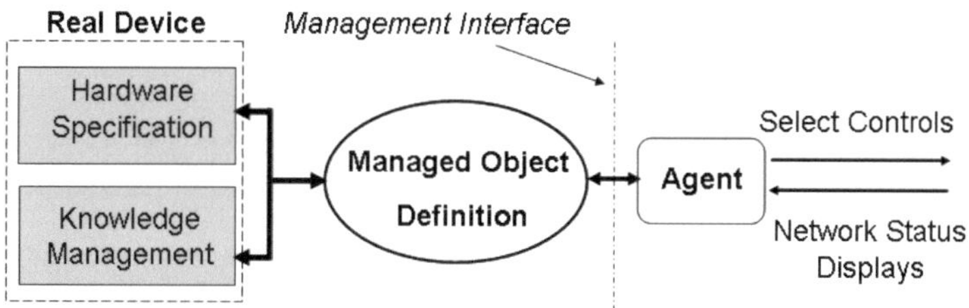


Fig. 3. Integration of Knowledge management and Resources properties in one single specification.

## 6. Including formal knowledge management in OSI

To allow deployment of equipment from different vendors, the managed models, OSI framework define the language GDMO (Guidelines for the Definition of Managed Objects). GDMO has been standardized by ITU (International Telecommunication Union) in ITU-T X.722 and is now widely used to specify interfaces between different components of the TMN architecture. This section introduces a framework for the inclusion of formal Knowledge management descriptions into GDMO specifications and focuses on the syntax and semantics of the language GDMO and the extension GDMO+.

According to OSI's Management Information Model, the management view of a managed object is viable at the managed object boundary. Managed object can be viewed as mediator between the network management interface and the hardware in the networks. Managed Object is modelled by attributes, actions and notifications, figure 4.

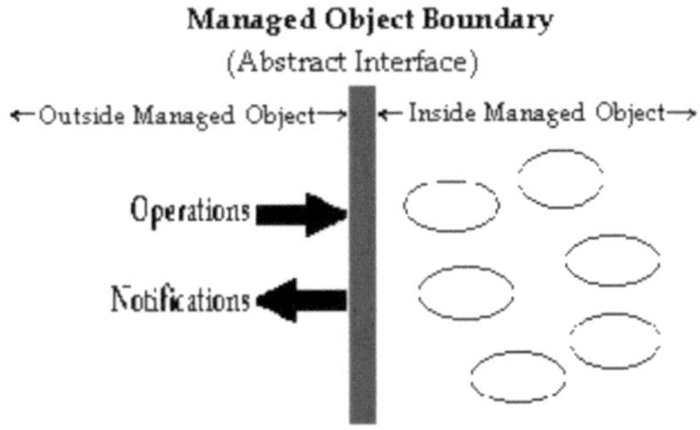


Fig. 4. Managed Object Boundary

They can usually represent a certain part of the internal state of an element. Actions invoke certain functions which a device can perform. Notifications are spontaneous messages emitted if certain events occur. Managed objects which are alike are grouped together to form Managed Object Classes. Classes can inherit their appearance from other classes and add new features (Goleniewski & Jarrett, 2006).

The appearance of managed objects can be formally described using the language GDMO. This language defines a number of so-called templates, which are standard formats used in the definition of a particular aspect of a real device in the networks. A complete object definition is a combination of interrelated templates. The nine templates that conform the actual GDMO standard are next:

- Managed Object Class Template references all other templates either directly or indirectly to make up the managed object class definition: Package, Parameter, Attribute, Attribute Group, Behaviour, Action and Notification.
- Package Template, defines a combination of behaviour definitions, attributes, attribute groups, operations, actions, notifications and parameters for later inclusion in a managed object class template. A package can be referenced by more than one managed object class definition.
- Attribute Template defines individual attribute types used by managed object classes. A single attribute can be referenced by more than one managed object class definition. If desired these attribute types can be combined in a group by using the Attribute Group template.
- Action Template defines actions that can be performed by a managed object class. These actions can be executed by using the Common management information service (CMIS). In particular the M-ACTION service that requests an action to be performed on a managed object. An action can be referenced by more than one managed object class. The Action template defines the behaviour and syntax of an action. The syntax definitions specify the contents of the action information and action reply fields in CMIS action requests and responses.
- The Notification template defines the behaviour and syntax of a notification that can be emitted by a managed object class. A notification can be referenced by more than one managed object class. The syntax definitions specify the contents of the event information and event reply fields in CMIP, event report requests and responses.

- The Parameter template specifies and registers the parameter syntaxes and associated behaviour that may be associated with particular attributes, operations, and notifications within Package, Attribute, Action, and Notification templates. A parameter can be referenced by one or more of each of these templates. The type specified in a Parameter template is used to fill in the ANY DEFINED BY oid construct.
- Attribute Group Template defines one or more attributes that can be referenced as a group. A managed object class definition can include all attributes of a group by referencing the group, rather than referencing each attribute individually. More than one managed object class definition can reference an attribute group. Attribute groups make it easier to collectively perform operations on a large number of individual attribute.
- Behaviour Template describes the expected behaviour of another element of the standard: Managed object classes, Name bindings, Parameters, Attributes, Actions and Notifications. The behaviour may be defined by readable text, high level languages, formal description techniques, references to standard constructs, or by any combination of the preceding definition methods.
- Name Binding Template, specifies instantiation and legal parameters for managed objects. Containment, creation, and deletion constraints are initiated from this template.

The elements that at the moment form the GDMO standard do not make a reference to the knowledge base of an expert system. To answer these questions, it will be necessary to make changes on the template of the GDMO standard. We present an extension of the standard GDMO, to accommodate the intelligent management requirements, figure 5.

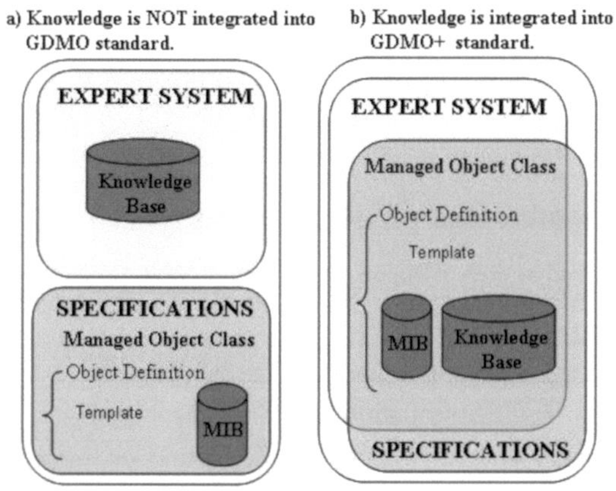


Fig. 5. Extension of GDMO+

Before considering a concrete language for specifying managed object's knowledge, a methodology for combining knowledge management with GDMO+ definitions is presented. To improve the quality of the descriptions and the resulting implementations, a formal method for specifying knowledge is desirable. We describe how to achieve this goal using a new extension called GDMO+. An object-oriented logic programming language is presented, which can be used in conjunction with the framework to specify the management knowledge of a managed object. The methodology is independent of the language used and can be combined with other approaches for formalizing knowledge It is based on the notion of events and supports the object-oriented features of GDMO (Hebrawi, 1995).

Management knowledge is introduced in GDMO+ which defines a number of new templates that contain certain aspects of the expert rules. This extension presents a new element RULE, which defines the knowledge base of the management expert system. This template groups the knowledge base supplied by an expert in a specific management dominion. It allows the storage of the management knowledge in the definition of the resources that form the system to be managed, figure 6.

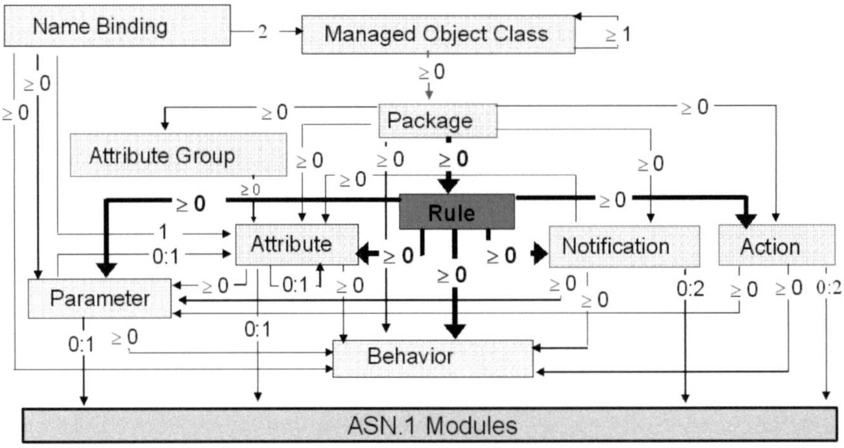


Fig. 6. Relations between proposed standard templates

The standard we propose contains the singular template RULE and its relations to other templates. Two relationships are essential for the inclusion of knowledge in the component definition of the network: Managed Object Class and Package Template. In the standard we propose, both templates have the new property RULES. Let us study both relationships.

## 7. Template for management of object classes

This template is used to define the different kinds of objects that exist in the system.

```
<class-label> MANAGED OBJECT CLASS
    [DERIVED FROM <class-label> [,<class-label>]*;]
    [CHARACTERIZED BY <package-label> [,<package-label]*;]
    [CONDITIONAL PACKAGES <package-label> PRESENT IF condition;
                                  ,<package-label>] PRESENT IF condition]*;]
 REGISTERED AS object-identifier;
```

DERIVED FROM plays a very important role, when determining the relations of inheritance which makes it possible to reutilize specific characteristics in other classes of managed objects. In addition, a great advantage is the reusability of the object classes and therefore of the expert rules which are defined.

This template also can contain packages and conditional packages, including the clauses CHARACTERIZED BY and CONDITIONAL PACKAGES. A package is used to define a package that contains a combination of many characteristics of a managed object class: behaviours, attributes, groups of attributes, operations, parameters, actions and notifications. The structure of package template is next:

```
<package-label> PACKAGE
    [BEHAVIOUR <behaviour-label> [,<behaviour-label>]*;]
    [ATTRIBUTES <attribute-label>propertylist[,<parameter-label>]*
                [,<attribute-label>propertylist[,<parameter-label>]*]*;]
    [ATTRIBUTE GROUPS <group-label> [<attribute-label>]*
                      [<group-label> [<attribute-label>]*]* ;]
    [ACTIONS <action-label> [<parameter-label>]*
             [<action-label> [<parameter-label>]*]* ;
    [NOTIFICATIONS <notification-label> [<parameter-label>]*
                   [<notification-label> [<parameter-label>]*]* ;]
    [RULES    <rule-label>  [,<rule-label>]*;]
 REGISTERED AS object-identifier;
```

In addition to the properties indicated above, we suggest the incorporation of a new property called RULES, which contains all the specifications of the knowledge base for the expert system. Next definition shows the elements of a package template, in which it is possible to observe the new property RULES.

## 8. Expert rule template

The RULE template permits the normalised definition of the specifications of the expert rule to which it is related. Figure 7 shows a UML representation of Rule template.

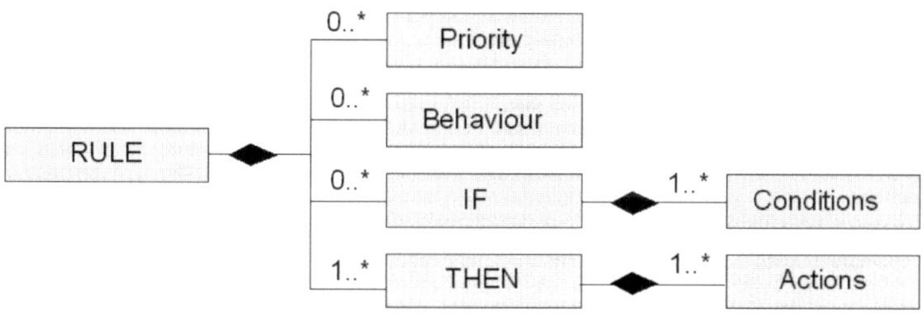


Fig. 7. RULE template in GDMO+ standard.

This template allows a particular managed object class to have properties that provide a normalised knowledge of a management dominion. A rule is an expression such as: "If the antecedent is true for facts in a list of facts, then it can carry out the actions specified in consequent" (Brachman, 2004). Each of the template type consists of a specification label, a template type, a list of keywords and an unique ASN.1 object identifier. The structure of the RULE template is shown here:

```
<rule-label> RULE
    [PRIORITY      <priority> ;]
    [BEHAVIOUR <behaviour-label> [,<behaviour-label>]*;]
    [IF occurred-event-pattern [,occurred-event-pattern]*]
    [THEN     sentence [, sentence]* ;]
 REGISTERED AS object-identifier;
```

The first element in a template definition is headed. It consists of two sections: <rule-label> this is the name of the management expert rule and RULE, a key word indicates the type of template. After the head, the following elements compose a normalised definition of an expert rule. After the head, the following elements compose the RULE template.

- BEHAVIOUR: This construct is used to extend the semantics of previously defined templates. It describes the behaviour of the rule. This element is common to the others templates of the GDMO standard.
- PRIORITY: This represents the priority of the rule, that is, the order in which competing rules will be executed.
- IF: We can add a logical condition that will be applied on the events occurred or their parameters. It contains all the events that must be true to activate a rule. Those events must be defined in the Notification template. The occurrence of these events is necessary for the activation of the rule and the execution of their associated actions. For the establishment of the conditions of fire, we must following structure:
  [IF <patter> [, <patter>]* <condition> [,condition]*;]
- THEN: This gives details of the operations performed when the rule is executed. Those operations must be previously defined in the Action template. These are actions and diagnoses that the management platform makes as an answer to network events occurred.

Our formal structure is next: [THEN <sentence> [,<sentence>]*;]

- REGISTERED AS is an object-identifier: A clause identifies the location of the expert rule on the ISO Registration Tree. The identifier is compulsory.

To finish this section next paragraph shows an expert rule integration; it defines the managed object of a resource port that belongs to a switch.

```
port MANAGED OBJECT CLASS
    DERIVED FROM "element_switch":top;
    CHARACTERIZED BY portPackage PACKAGE;
REGISTERED AS {nm-MobjectClass 1};

portPackage PACKAGE
    ATTRIBUTES portNum GET, portStatus GET, ...
    OPERATIONS diagnose, disconnect, connect, ...
    NOTIFICATIONS portFailure, portInitializated...
    RULES transmissionError, powerError, ...,
REGISTERED AS {nm-package 1};

transmissionError RULE
    PRIORITY 4;
    BEHAVIOUR transmissionErrorBehaviour;
    IF (?date ?time1 ?station portFailure ?portNum ALARM)
       (?date ?time2 ?station portFailure ?portNum ALARM
       & : (=(portstatus(? ?time1 ?time2)) OFF)
    THEN ("Severity:" PRIORITY) ,
     ("Diagnostic: "It damages in the modulate transmission of port", ?portNum),
     ("Recommendation "Revision port"),
     ("Actions "put off/on port");
  REGISTERED AS {nm-rule 2);
```

In the previous example a Class of Managed Object is port, which defines the properties corresponding to the port of a switch. This class includes the compulsory portPackage which contains all the specifications corresponding to the device. We can indicate which re the expert rules that have been associated with the defined class by means of the RULES clause. The rules are defined by using the RULE template. These rules detect anomalies or defects of operations produced in the port and suggest the necessary measures for solving the problem. The first rule transmisionError is in charge of detecting failures in the data transmission module and gives recommendations on how to solve this failure.

## 9. New expert management rules identification

The integration of the management knowledge in the proposal GDMO+, implies that the expert rules are defined together with the specifications of the managed objects. As the rest of the proprieties of the GDMO+ standard that are included in the definition of a managed object class: classes, attributes, actions, etc., the expert management rules will be attached to revision convention and analogue register structures. The CCIT Rec. X.208 | ISO/IEC 8824, provides a structure for the object identifies and the proprieties that composed it, as soon as the new property called RULES.

To make this possible, the systems management functions, function(2) and the management information model smi(3), must allow an amplification, so that these recommendations can pick up functions and mechanisms of definition that allow to contain the new property RULES. For this purpose we propose new versions of these recommendations. In the next figure 8 we show the amplification that shows the sections function(2) and smi(3) and the new elements RULES(11) is presented and gathers all necessary elements to define an associated knowledge.

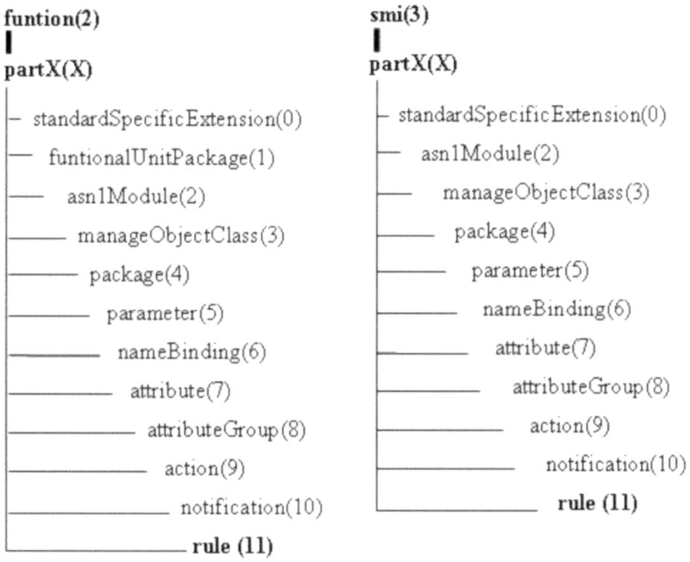


Fig. 8. Amplification of the Model of Information and Management Functions

In this way, the definition of a new rule "New_Rule" would be included in the place indicated to the recommendation X.720 of CCITT | ISO 10165-1 [ISO92B]. In this case it is to add a new management rule to the Standard GDMO+ proposal. Extended, it would be achieved in the following level: {join-iso-ccitt ms(9) smi(9) part4(4) rule(11) New_Rule(45)}

## 10. Integrate environment for internet network management.

This section discusses and analyzes the intelligent management approach that is standardized by the Internet Engineering Task Force. Internet management can be compared to OSI management model. In fact, Internet management uses many of the concepts that existed in OSI at the time SNMP started. As a result, the remarks that were made above are to some extent also applicable to Internet management. As opposed OSI management, however, Internet management uses only a small part of management functions for exchange of management information.

Note that objects in the Internet and those OSI are different. Internet objects are similar to attributes in a OSI managed object, and an Internet object group can best be described as analogous to a OSI management object class. In the Internet, an object is more like a variable found in programming languages; it has a syntax and semantics. Each object can have one or more object instances, each of which, in turn, has one or more values.

An interesting difference between Internet and OSI is that the Internet management model is more simple that OSI model. The principal characteristics of their architecture are next:

- The cost of adding network management to existing systems is minimal.
- All systems connected to the network should be manageable with SNMP. It should be noted that SNMP protocol only defines how management information should be exchanged, they do not define which management information exits. Such information is defined by the various MIB standards.
- It should be relatively easy to extend the management capabilities of existing systems, by extending the existing MIB's or adding a new MIB.

Due to these aspects, it is questionable whether OSI management will reach the dominant market position that has originally been anticipated. Simple measures that solve all the above mentioned problems are difficult to find.

### 10.1 Internet extension to integrate the knowledge management

For two systems to communicate, each must understand the data sent from one to the other. This can be achieved by using a language that has the same syntax and semantics. In the application layer, we use abstract syntax, which states only how data are arranged and what meaning they have. One of the possible abstract syntaxes is abstract syntax notation one (ASN.1). Between the application layer and the presentation layer, a local set of rules can be used to transform data; however the syntax of the data transferred between presentation entities must be understood by each end. This is known as transfer syntax. Abstract syntax and transfer syntax are negotiated at the beginning, during association time.

One transfer syntax is Basic Encoding Rules (BER). BER state how data must be transferred to the other presentation entity. The local syntax can be purely dependent on the local protocols used, in this case SNMP. Figure 9 illustrates the concept of abstract syntax and transfer syntax. ITU-T Recommendations X.208 and X.680 describe standardized ways and steps to define data types and data values (Douglas & Kevin, 2005).

The textual MIB representation is called module and consist on a plane text file. This file is used as a subset language of ASN.1 and based in the SMI specifications. The MIB-II is the most important and probably best known MIB. It contains all the variables to control the major Internet protocols: IP, ICMP, UDP, TCP, EGP and SNMP. The structure of this MIB is simple; all management variables that belong to the same protocol are grouped together. Within a protocol group there is hardly any additional structure that helps understanding the various variables within that group.

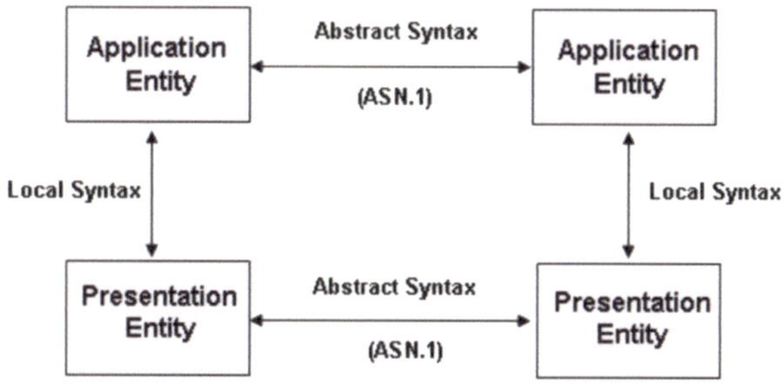


Fig. 9. Abstract Syntax Notation and Basic Encoding Rules

Soon after definition of the MIB-II other MIBs, some of these standardized MIB`s are FDDI, ATM, X.25, X.500 Directory Monitoring, etc. Next to the standardized MIBs there are also a large number of enterprise specifics MIBs. The objects to be managed in Internet must follow a certain set of rules as mentioned in the Structure of Management Information (SMI) such as an object defined by the X group is compatible with the definition of the object by the Y group.
We propose to add a new type named RULES and the incorporation of an extension of the MIB II named MIB II+. We broadly classify the ASN.1 built-in types as follows: simple types, structured types, Tagged types and subtypes. We introduced in ASN.1 a new concept denominated Expert Rule to definition the new group RULE existing in MIB II+. This group is introduced as a textual convention in this MIB II+ document. Group RULES contains all the aspect related with the expert management, figure 10.

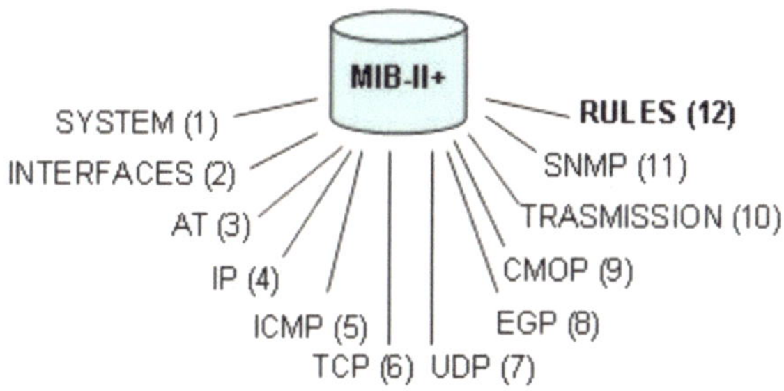


Fig. 10. MIB-II+ Objects group

To define an expert rule we use the modules definitions. Modules definitions are primarily used for grouping ASN.1 definitions. They also help in using type definitions characterized in the other places by making use IMPORT and EXPORT mechanisms. Modules are analogous to functions in C language or subroutines in PASCAL. There are module definitions in the definitions of managed object classes in standards and the others documents. The macro used for MIBs definition in SNMP was defined in RFC 1155 draft (Structure of Management Information) and later extended in the RFC 1212 (Concise MIB Definitions). RFC 1155 version is used to define objects in MIB-I. RFC 1212 version including more information and is used to define objects in MIB-II. Next definition shows OBJEC-TYPE macro in RFC 1212.
These can be enhanced by including formal descriptions. In this case, the specifications formal parts and the knowledge must be distinguishable. An easy solution is the separation by keywords. Next section shows an example of definition of an integrate expert rule in Internet model:

```
transminissionError OBJECT-TYPE
    SYNTAX SEQUENCE
    {conditionRule COMPONENTS OF transminissionErrorCondition,
    actionRule COMPONENTS OF transminissionErrorAction}
    ACCESS read-write
    STATUS mandatory
    DESCRIPTION
"Rule devoted to the detection of errors in the data transmission module"
::= { tcp 13 }

trastimissionErrorCondition OBJECT-TYPE
    SYNTAX Condition
    ACCESS read-write
    STATUS mandatory
    DESCRIPTION "Information about a particular condition"
    INDEX { date, time, local, signal, remote, alarm, operator}
::= { tcpConnTable 1 }

TcpConnEntry ::=
    SEQUENCE { date INTEGER,
    time TIMETICKS,
    local OCTETSTRING,
    signal OCTETSTRING,
    remote OCTETSTRING,
    alarm OCTETSTRING,
    operator OCTETSTRING}

transminissionErrorAction OBJECT-TYPE
    SYNTAX OCTETSTRING
    ACCESS read-only
    STATUS mandatory
    DESCRIPTION "Action execute when fire the expert rule"
::= { tcp 13 }
```

This example defines a portion of the Management Information Base (MIB) for use with network management knowledge in TCP/IP-based internets. In particular, it defines objects for managing remote network monitoring devices. This standard extends that specification by documenting the knowledge management in SMIv2 format. These groups are defined to provide a means of assigning managed object, and to provide a method for implementers of managed agents to know which objects can be administrator.

## 11. Case of study and practical experiments

At present, some reports of experience with the OSI and expert system architecture are available, although the first OSI-based platforms are already on the market. We provide a rule-based expert system applied to the fault diagnosis in telecommunication system of a power utility (Maggiora et al., 2000). The communications systems employed to implement the integrated intelligent management prototype belongs to the SEVILLANA-ENDESA a major Spanish power utility. This Telecommunications network is made of several equipments and systems, which differ from each other in terms of age, technology, network domain, etc. Each kind of equipment communicates to a particular supervisory system, which is responsible the equipment's operation and maintenance information to one or more operations centers. The supervisory are also responsible for sending commands from the operations' staff to the supervisory equipment. A supervisory system as composed from two kinds of equipment: Remote Terminal Units (UTR) and central units (UC). UTRs function is to gather, code and transmit information to the managed equipment and distribute command signals to those equipment (Boyer, 1999), figure 11.

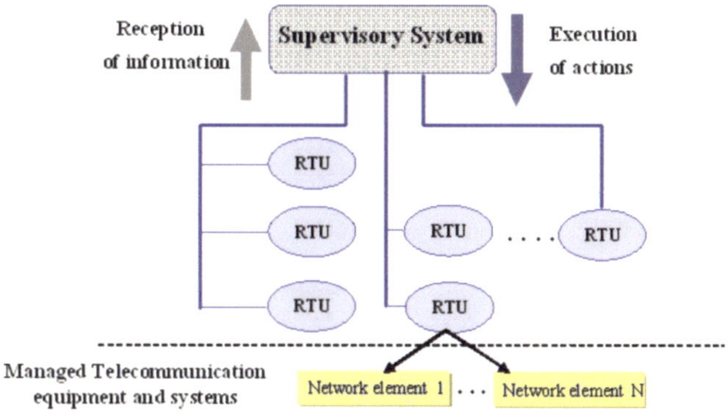


Fig. 11. Network framework

Management and control of that network as based on an Expert System called NOMOS+ developed by the Electronic Technology Department in the University of Seville. This tool understands transceivers and multiplex equipment. This expert system that integrates the knowledge base and the resources definition into an unique specifications. The knowledge base of this system is integrated in the specifications of the resources using for that purpose our GDMO+ proposal. It has been employed Workstation to program the expert system. The resultant expert system has about 200 rules.

NOMOS+ is implemented in Brightware's ART*Enterprise, an expert system shell. ART*Enterprise is a set of programming paradigms and tools that are focused on the development of efficient, flexible, and commercially deployable knowledge-based systems. Expert system shells simplify developer interactions by eliminating the developer's concern with operating system requirements. Its use can therefore reduce the design and implementation time of a program considerably.

### 11.3 Expert system architecture

The integrate expert system we propose is composed of three major components: a knowledge base, an inference engine and a user interface, figure 12.

The knowledge base is the core of the system, which is a collection of facts and if-then production rules that represent stored knowledge about the problem domain. The inference engine is the processing unit that solves any given problems by making logical inferences on the given facts and rules stored in the knowledge base. In our tool we used the ART*Enterprise. The user interface controls the inference engine and manages system input and output. The user interface of our tool contains a preprocessor for parsing GDMO+ specification files, a set of input and output handling routines for managing the system. Also, the user interface components allow administrators to inspect the definitions of management object classes interactively (Giarratano & Riley, 2005). When new knowledge is uncovered, it will need to be incorporated into the system to keep it updated. The user interface allows to modify or include new expert management rules in the managed objects definition. The System Management window allows a system administrator to perform the following functions:

- Adding new Management Object Class and change the definitions of the existing management object class.
- Configure the Nomos+ workstation's for the appropriate function and actions.
- Set up alarm actions to be executed automatically when specific alarms occur.
- Examine and modify the GDMO+ specifications and log files, configuration files, and other text files using an online visual editor.

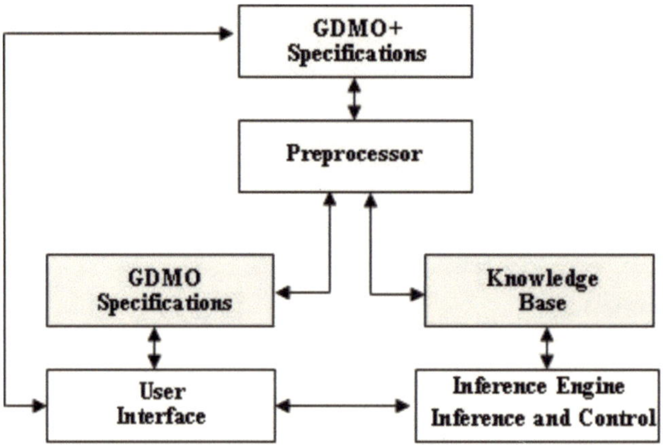


Fig. 12. Elements of the prototype NOMOS+

Since the window is used to modify system files, only one administrator can use it. Figure 12a shows an example of a web based management interface where the system Object Class of a network is defined in NOMOS+. There is an Object Class called muxtiplexCTR190. If the "Modify" button is pressed then the definition of an Object Class dialog box, Figure 12b, will show. That dialog box allows describing different characteristics of an object class, such as attributes, actions and expert rules.

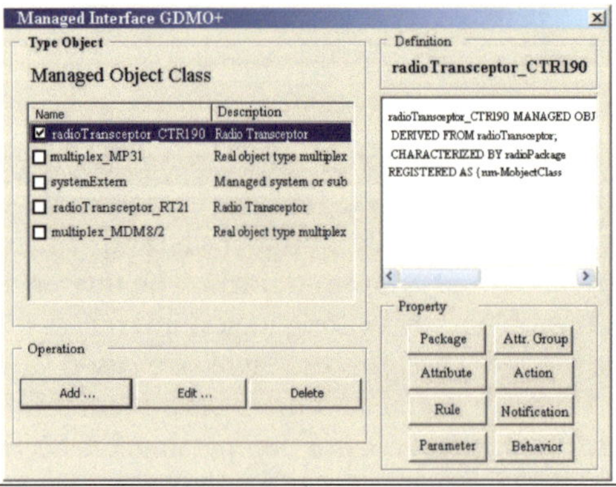


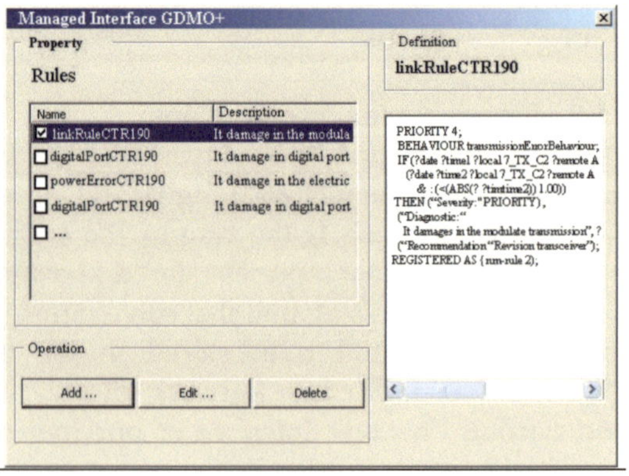


Fig. 13a System Object Classes Dialog Box Fig. 13b System Object Classes Update

## 12. Conclusion

In this paper we showed possibilities to apply and integrated the artificial intelligence techniques in network management and supervision, using ISO and Internet network management models. In fact we believe that these kinds of applications underline the power of CMIS as both simple and powerful knowledge modeling languages, offering possibilities that simpler protocols such as SNMP and CMIP do not offer.

We have seen that the current management systems are not able to solve questions shown in the initial parts of this work. Until now the managed objects did not contain the knowledge coming from base of the expert systems. The managed objects are not able to use the knowledge management, which collects the management operations and control of a management domain. The point is to solve the current problem to undertake an intelligent integrated management. We offer an original contribution to include expert rules in the specifications of the network features, a new model named Integrated Intelligent Management an extension of standard GDMO+ and MIB II+.

This paper presented a language for formalizing the knowledge base of the expert systems descriptions in OSI and Internet telecommunications management network framework. A number of questions which arise when designing a language have been discussed and a general framework for the inclusion of formal knowledge management in MIB specifications has been presented. The proposed model was used to formally specify the expert knowledge. An expert system has been implemented and used to manage the specification to the language used by the simulation environment. This demonstrates that expert systems is capable of specifying the knowledge of a reasonably sized information model. A large amount of the knowledge could be described in a surprisingly short and easy to understand manner. The specification of the NOMOS+ information model showed that a large part of the knowledge management was specified in a rather imperative fashion.

Our research has demonstrated an useful and interesting modular approach in the development of a knowledge based integrated expert system which can be quite powerful in tackling the huge and enormously wide subject on diagnosis of common problems in management network. It is suggested that future work should aim to:

- Further development of this prototype system by adding more modules based on the framework provided by NOMOS+ so that more in-depth knowledge and specialized subjects may be captured; in particular the following are of great interest: Development of a design module, possibly a large system, for identifying specific areas as accounting management, configuration management, performance management and security management.
- Enhancement work in combining and integrating the various modules will be required as the number of modules increases with the growth of the knowledge base.
- Use of external programs and graphics interface to enhance the functions of the system will be desirable. A graphics interface has not been adopted in NOMOS+ but is an option that can be added in future enhancements to the system.

## 13. Acknowlegment

The work described in this paper has been supported by the Spanish Ministry of Education and Science (MEC: Ministerio de Educación y Ciencia) through project reference number DPI2006-15467-C02-02.

## 14. References

Boyer, Stuart A. (1999). *Supervisory control and data acquisition.* Research Triangle Park, NC: Instrument Society of America, cop.

Black, U.D. (1995). *Network Management Standards.* McGraw Hill

Brachman, Ronald J. & Levesque, Hector J. (2004) *Knowledge representation and reasoning.* San Francisco, CA: Elsevier/Morgan Kaufmann.

Clemm, Alexander. (2006). *Network Management Fundamentals.* Cisco Press

Douglas, Mauro and Kevin, Schmidt. (2005). *Essential SNMP*, 2nd Edition. O'Reilly.

Giarratano, Joseph C. & Riley, Gary D. (2005). *Expert Systems: Principles and Programming.* Book, Brooks/Cole Publishing Co.

Goleniewski, L. and Jarrett, K. W. (2006). *Telecommunications Essentials, Second Edition: The Complete Global Source.* Addison Wesley Professional.

Hebrawi, B. (1995). *GDMO, Object modelling and definition for network management.* Technology appraisals

ISO/IEC DIS 10165-4 / ITU-T. (1993). *Recommendation X.722, Information Technology - Part 4: Guidelines for the Definition of Managed Objects (GDMO)*, International Organization for Standardization and International Electrotechnical Committee.

ISO/IEC and ITU-T. (1998). *Information Processing Systems - Open Systems Interconnection - Systems Management Overview.* Standard 10040-2, Recommendation X.701.

ITU-T. (1992). *Recommendation X.700, Management Framework for Open Systems Interconnection (OSI).* CCITT Applications.

ITU-T (1996). *Rec. M.3010, Principles for a Telecommunications Management Network (TMN).* Study Group IV.

Maggiora, Paul L. Della., Elliott, Christopher E., Pavone, Robert L., Phelps, Kent J., Thompson, James M. (2000). *Performance and Fault Management.* Cisco Press

Morris, Stephen B. (2003). *Network Management, MIBs and MPLS: Principles, Design and Implementation by Publisher*: Addison Wesley.

Negnevitsky, Michael. (2002). *Artificial intelligence: a guide to intelligent systems.* New York: Addison Wesley.

5

# Robust and Active Trajectory Tracking for an Autonomous Helicopter under Wind Gust

Adnan Martini, François Léonard and Gabriel Abba
*Industrial Engineering and Mechanical Production Lab, Ecole Nationale d'Ingénieurs de Metz*
*France*

## 1. Introduction

High levels of agility, maneuverability and the capability of operating in degraded visual environments and adverse weather conditions are the new trends of helicopter design nowadays. Helicopter flight control system should make these performance requirements achievable by improving tracking performance and disturbance rejection capability. Robustness is one of the critical issues which must be considered in the control system design for such highperformance autonomous helicopter, since any mathematical helicopter model, especially those covering large flight envelope, will unavoidably have uncertainty due to the empirical representation of aerodynamic forces and moments.

The purpose of this chapter is to present the stabilization (tracking) with motion planning of a reduced-order helicopter model having 3DOF (Degrees Of Freedom) (see Fig.1). This last one represents a scale model helicopter mounted on an experimental platform. It deals with the problem of disturbance reconstruction acting on the autonomous helicopter, the disturbance consists in vertical wind gusts. The objective is to compensate these disturbances and to improve the performances of the control. Consequently, a nonlinear simple model with 3DOF of a helicopter with unknown disturbances is used. Three approaches of robust control are then compared via simulations: a robust nonlinear feedback control, an active disturbance rejection control based on a nonlinear extended state observer and a backstepping control.

Design of control of autonomous flying systems has now become a very challenging area of research, as shown by a large literature (Beji & Abichou, 2005) (Frazzoli et al., 2000) (Koo & Sastry, 1998). Many previous works focus on (linear and nonlinear, robust, ...) control, including a particular attention on the analysis of the stability (Mahony & Hamel, 2004), but very few works have been made on the influence of wind gusts acting on the flying system, whereas it is a crucial problem for out-door applications, especially in urban environment: as a matter of fact, if the autonomous flying system (especially when this system is relatively slight) crosses a crossroads, it can be disturbed by wind gusts and leave its trajectory, which could be critical in a highly dense urban context.

In (Martini et al., 2005) and (Martini et al., 2007a), three controllers (nonlinear, $H_\infty$ and robust nonlinear feedback) are designed for a nonlinear reduced-order model of a 3 DOF helicopter. In (Pflimlin et al., 2004), a control strategy stabilizes the position of the flying vehicle in wind gusts environment, in spite of unknown aerodynamic efforts and is based on robust backstepping approach and estimation of the unknown aerodynamic efforts.

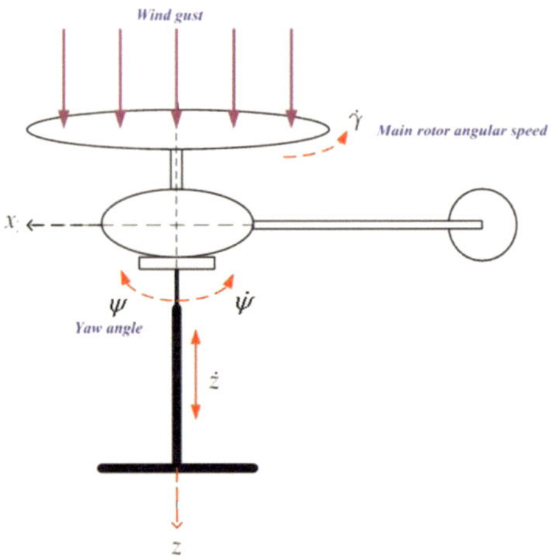


Fig.1. Helicopter-platform with wind gust

In recent papers, feedback linearization techniques have been applied to helicopter models. The main difficulty in the application of such an approach is the fact that, for any meaningful selection of outputs, the helicopter dynamics are non-minimum phase, and hence are not directly input-output linearizable. However, it is possible to find good approximations to the helicopter dynamics (Koo & Sastry, 1998) such that the approximate system is inputoutput linearizable, and bounded tracking can be achieved.

Nonlinear control designs previously attempted include neural network based controllers (McLean & Matsuda, 1998), fuzzy control (Sanders et al., 1998), backstepping designs (Mahony & Hamel, 2004), and adaptive control (Dzul et al., 2004). These methods either assume feedback linearizability, which in turn restricts the motion to be around hover, or do not include parametric uncertainties, or realistic aerodynamics. Specific issues such as unknown trim conditions that degrade the performance of the helicopter have not been addressed. While adaptive control schemes have been proposed in the aircraft and spacecraft control context, there is a lack of similar work on helicopter control. The non-minimum phase nature of the helicopter dynamics adds to the challenge of finding a stable adaptive controller. (Wei, 2001) showed the control of nonlinear systems with unknown disturbances, using a disturbance observer based control (DOBC). In (Ifassiouen et al., 2007) a robust sliding mode control structure is designed using the exact feedback linearization procedure of the dynamic of a small-size autonomous helicopter in hover.

This chapter is organized as follows. In section 2, a 3DOF Lagrangian model of the disturbed helicopter mounted on an experimental platform is presented. This model can be seen as made of two subsystems (translation and rotation). In section 3 two approaches of robust control design for the reduced order model are proposed. The application of three approaches of robust control on our disturbed helicopter is analyzed in section 4. Section 6 is devoted to simulation results and the study of model stability is carried out in section 5. Finally some conclusions are presented in Section 7.

## 2. Model of the disturbed helicopter

Helicopters operate in an environment where task performance can easily be affected by atmospheric turbulence. This chapter discusses the airborne flight test of the *VARIO Benzin*

*Trainer* helicopter in turbulent conditions to determine disturbance rejection criteria and to develop a low speed turbulence model for an autonomous helicopter simulation. A simple approach to modeling the aircraft response to turbulence is described by using an identified model of the *VARIO Benzin Trainer* to extract representative control inputs that replicate the aircraft response to disturbances. This parametric turbulence model is designed to be scaled for varying levels of turbulence and utilized in ground or in-flight simulation. Hereafter the nonlinear model of the disturbed helicopter (Martini et al., 2005) starting from a non disturbed model (Vilchis, 2001) is presented. The *Vario* helicopter is mounted on an experimental platform and submitted to a vertical wind gust (see Fig.1). It can be noted that the helicopter is in an Out Ground Effect (OGE) condition. The effects of the compressed air in take-off and landing are then neglected. The Lagrange equation, which describes the system of the helicopter-platform with the disturbance, is given by:

$$M(q)\ddot{q} + C(q,\dot{q}) + G(q) = Q(q,\dot{q},u,v_{raf}) \tag{1}$$

where the input vector of the control $u = [u_1\ u_2]^T$ and $q = [z\ \ \psi\ \ \gamma]^T$ is the vector of generalized coordinates. The first control $u_1$ is the collective pitch angle (swashplate displacement) of the main rotor. The second control input $u_2$ is the collective pitch angle (swashplate displacement) of the tail rotor. The induced gust velocity is noted $v_{raf}$. The helicopter altitude is noted $z$, $\psi$ is the yaw angle and $\gamma$ is the main rotor azimuth angle. $M \in R^{3\times3}$ is the inertia matrix, $C \in R^{3\times3}$ is the Coriolis and centrifugal forces matrix, $G \in R^3$ represents the vector of conservative forces, $Q(q,\ \dot{q},\ u,\ v_{raf}) = [f_z\ \tau_z\ \tau_\gamma]^T$ is the vector of generalized forces. The variables $f_z$, $\tau_z$ and $\tau_\gamma$ represent respectively, the total vertical force, the yaw torque and the main rotor torque in presence of wind gust.

Finally, the representation of the reduced system of the helicopter, which is subjected to a wind gust, can be expressed as (Martini et al., 2005) :

$$\begin{array}{rcl}
\dot{x}_1 & = & x_2 = \dot{z}, \\
\dot{x}_2 & = & \frac{1}{c_0}[c_8\dot{\gamma}^2 u_1 + c_9\dot{\gamma} + c_{10} - c_7] + \frac{1}{c_0}c_{16}\dot{\gamma}v_{raf}, \\
\dot{x}_3 & = & x_4 = \dot{\psi}, \\
\dot{x}_4 & = & \frac{1}{c_1c_5 - c_4^2}[c_5c_{11}\dot{\gamma}^2 u_2 - c_4(c_{14}\dot{\gamma}^2 + c_{15})] \\
 & & -\frac{c_4}{c_1c_5 - c_4^2}[(c_{12}\dot{\gamma} + c_{13} + c_8\dot{\gamma}v_{raf})u_1 + (\frac{5}{2}c_9 + c_{17}v_{raf})v_{raf}], \\
\dot{x}_5 & = & x_6 = \dot{\gamma}, \\
\dot{x}_6 & = & \frac{c_4}{c_1c_5 - c_4^2}[c_{11}\dot{\gamma}^2 u_2 + c_1(c_{14}\dot{\gamma}^2 + c_{15})] \\
 & & +\frac{1}{c_1c_5 - c_4^2}[c_1c_4(c_{12}\dot{\gamma} + c_{13} + c_8\dot{\gamma}v_{raf})u_1 + (\frac{5}{2}c_9 + c_{17}v_{raf})v_{raf}].
\end{array} \tag{2}$$

where $c_i$ (i =0,...,17) are numerical aerodynamical constants of the model given in table 1 (Vilchis, 2001). For example $c_0$ represents the helicopter weight, $c_{15} = 2ka_{1s}b_{1s}$ where $a_{1s}$ and $b_{1s}$ are the longitudinal and lateral flapping angles of the main rotor blades, $k$ is the blades stiffness of main rotor.

Table 2 shows the variations of the main rotor thrust and of the main rotor drag torque (variations of the helicopter parameters) operating on the helicopter due to the presence of wind gust. These variations are calculated from a nominal position defined as the equilibrium of helicopter when $v_{raf} = 0$: $\dot{\gamma} = -124.63 rad/s$, $u_1 = -4.588 \times 10^{-5}$, $u_2 = 5 \times 10^{-7}$, $T_{Mo} = -77.3N$ and $C_{Mo} = 4.6N.m$.

| | | | |
|---|---|---|---|
| $c_0$ | $7.5\,kg$ | $c_9$ | $0.6004\,kg\,m/s$ |
| $c_1$ | $0.4305\,kg\,m^2$ | $c_{10}$ | $3.679\,N$ |
| $c_2$ | $3\times10^{-4}\,kg\,m^2$ | $c_{11}$ | $-0.1525\,kg\,m$ |
| $c_3$ | -4.143 | $c_{12}$ | $12.01\,kg\,m/s$ |
| $c_4$ | $0.108\,kgm^2$ | $c_{13}$ | $1\times10^5\,N$ |
| $c_5$ | $0.4993\,kgm^2$ | $c_{14}$ | $1.206\times10^{-4}\,kg\,m^2$ |
| $c_6$ | $-6.214\times10^{-4}\,kg\,m^2$ | $c_{15}$ | $2.642\,N$ |
| $c_7$ | $-73.58\,N$ | $c_{16}$ | 0.1706 kg |
| $c_8$ | 3.411 kg | $c_{17}$ | $0.2132\,N\,s^2/m$ |

Table 1. 3DOF model parameters

| $v_{raf}(m/s)$ | $\delta T_M(N)$ | $\frac{\delta T_M}{T_{M_o}}\%$ | $\delta C_M(N.m)$ | $\frac{\delta C_M}{C_{M_o}}\%$ |
|---|---|---|---|---|
| 0.68 | -14.5 | 19 % | 1.1 | 25 % |
| 3 | -64 | 83 % | 6.5 | 141% |

Table 2. Variation of forces and torques for different wind gusts

Three robust nonlinear controls adapted to wind gust rejection are now introduces in section 4.1, 4.2 and 4.3 devoted to control design of disturbed helicopter.

## 3. Control design

### 3.1 Robust feedback control

Fig.2 shows the configuration of this control (Spong & Vidyasagar, 1989) based on the inverse dynamics of the following mechanical system:

$$M(q)\ddot{q} + h(q,\dot{q}) = u. \tag{3}$$

Since the inertia matrix $M$ is invertible, the control $u$ is chosen as follows:

$$u = M(q)v + h(q,\dot{q}). \tag{4}$$

The term $v$ represents a new input to the system. Then the combined system (3-4) reduces to:

$$\ddot{q} = v. \tag{5}$$

Equation (5) is known as the double integrator system. The nonlinear control law (4) is called the inverse dynamics control and achieves a rather remarkable result, namely that the new system (5) is linear, and decoupled.

$$u = \widehat{M}(q)v + \widehat{h}(q,\dot{q}), \tag{6}$$

where $\widehat{M},\widehat{h}$ represent nominal values of $M$, $h$ respectively. The uncertainty or modeling error, is represented by: $\Delta M = \widehat{M}(q) - M(q)$; $\Delta h = \widehat{h}(q,\dot{q}) - h(q,\dot{q})$., with system equation (3) and nonlinear law (6), the system becomes:

$$\widehat{M}v + \widehat{h} = M\ddot{q} + h. \tag{7}$$

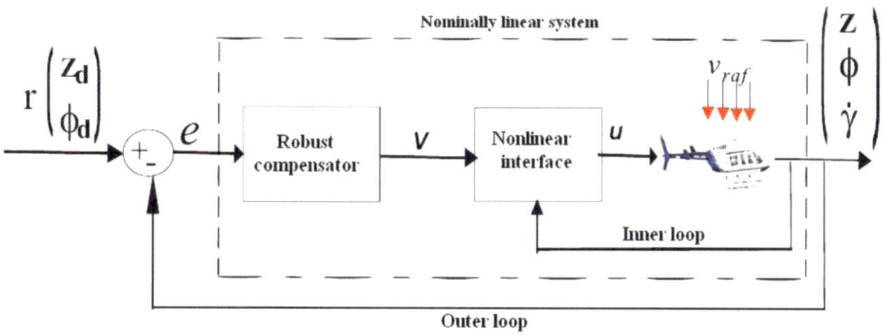


Fig.2. Architecture of robust feedback control

Thus $\ddot{q}$ can be expressed as

$$\ddot{q} = M^{-1}\widehat{M}v + M^{-1}\Delta h = v + (M^{-1}\widehat{M} - I)v + M^{-1}\Delta h. \tag{8}$$

Defining $E = M^{-1}\widehat{M} - I$ and $\eta = Ev + M^{-1}\Delta h$, then in state space the system (8) becomes:

$$\dot{\zeta} = A\zeta + B(v + \eta), \tag{9}$$

where: $A = \begin{bmatrix} 0 & I \\ 0 & 0 \end{bmatrix}$, $B = \begin{bmatrix} 0 \\ I \end{bmatrix}$, $\zeta = \begin{bmatrix} q \\ \dot{q} \end{bmatrix}$.

Using the error vectors $e = [e_1 \; e_2]^T$, $e_1 = q - q_d$ and $e_2 = \dot{q} - \dot{q}_d$, leads to:

$$\dot{e} = Ae + B\{v + \eta - \ddot{q}_d\}. \tag{10}$$

Therefore the problem of tracking the desired trajectory $q_d(t)$ becomes one of stabilizing the (time-varying, nonlinear) system (10). The control design to follow is based on the premise that although the uncertainty $\eta$ is unknown, it may be possible to estimate "worst case" bounds and its effects on the tracking performance of the system. In order to estimate a worst case bound on the function $\eta$, the following assumptions can be used (Spong & Vidyasagar, 1989) :

- Assumption 1: $\sup_{t\geq 0}\|\ddot{q}^d\| < Q_1 < \infty$.
- Assumption 2: $\|E\| = \|M^{-1}\widehat{M} - I\| \leq \alpha < 1$for some $\alpha$, and for all $q \in R^n$.
- Assumption 3: $\|\Delta h\| \leq \psi(e,t)$ for a known function $\psi$, bounded in $t$.

Assumption 2 is the most restrictive and shows how accurately the inertia of the system must be estimated in order to use this approach. It turns out, however, that there is always a simple choice for $\widehat{M}$ satisfying Assumption 2. Since the inertia matrix $M(q)$ is uniformly positive definite for all $q$ there exist positive constants $\underline{M}$ and $\overline{M}$ such that:

$$\underline{M} \leq \|M^{-1}(q)\| \leq \overline{M} \quad \forall q \in R^n. \tag{11}$$

If we therefore choose: $\widehat{M} = \frac{1}{c}I$ where $c = \frac{\overline{M}+\underline{M}}{2}$., it can be shown that: $\|M^{-1}\widehat{M} - I\| \leq \frac{\overline{M}-\underline{M}}{\overline{M}+\underline{M}} = \alpha < 1$. Finally, the following algorithm may now be used to generate a stabilizing control $v$:

Step 1 : Since the matrix $A$ in (9) is unstable, we first set:

$$v = \ddot{q}_d - K_1 e_1 - K_2 e_2 + \Delta v, \tag{12}$$

where $K = [K_1\, K_2]$:, and : $K_1 = \text{diag}\{\omega_1^2, \ldots, \omega_n^2\}$, $K_2 = \text{diag}\{2\zeta_1\omega_1, \ldots, 2\zeta_n\omega_n\}$. The desired trajectory $q_d(t)$ and the additional term $\Delta v$ will be used to attenuate the effects of the uncertainty and the disturbance. Then we have:

$$\dot{e} = \overline{A}e + B\{\Delta v + \overline{\eta}\}, \tag{13}$$

where $\overline{A} = A - BK$ is Hurwitz and $\overline{\eta} = E\Delta v + E(\ddot{q}_d - Ke) + M^{-1}\Delta h$.
Step 2: Given the system (13), suppose we can find a continuous function $\rho(e, t)$, which is bounded in $t$, satisfying the inequalities:

$$\|\Delta v\| < \rho(e,t), \qquad \|\overline{\eta}\| < \rho(e,t). \tag{14}$$

The function $\rho$ can be defined implicitly as follows. Using Assumptions 1-3 and (14), we have the estimate:

$$\begin{aligned} \|\overline{\eta}\| &\leq \|E\Delta v + E\ddot{q}_d - EKe + M^{-1}\Delta h\| \\ &\leq \alpha\rho(e,t) + \alpha Q_1 + \|K\|.\|e\| + \overline{M}\psi(e,t) = \rho(e,t). \end{aligned} \tag{15}$$

This definition of $\rho$ makes sense since $0 < \alpha < 1$ and we may solve for $\rho$ as:

$$\rho(e,t) = \frac{1}{1-\alpha}\{\alpha Q_1 + \|K\|.\|e\| + \overline{M}\psi(e,t)\}. \tag{16}$$

Note that whatever $\Delta v$ is now chosen must satisfy (14).
Step 3: Since $A$ is Hurwitz, choose a $n \times n$ symmetric, positive definite matrix $Q$ and let $P$ be the unique positive definite symmetric solution to the Lyapunov equation:

$$\overline{A}^T P + P\overline{A} + Q = 0. \tag{17}$$

Step 4: Choose the outer loop control $\Delta v$ according to:

$$\Delta v = \begin{cases} -\rho(e,t)\frac{B^T Pe}{\|B^T Pe\|} & \text{if} \quad \|B^T Pe\| \neq 0, \\ 0 & \text{if} \quad \|B^T Pe\| = 0, \end{cases} \tag{18}$$

that satisfy (14). Such a control will enable us to remove the principal influence of the wind gust.

### 3.2 Active disturbance rejection control

The primary reason to use the control in closed loop is that it can treat the variations and uncertainties of model dynamics and the outside unknown forces which exert influences on the behavior of the model. In this work, a methodology of generic design is proposed to treat the combination of two quantities, denoted as disturbance. A second order system described by the following equation is considered (Gao et al., 2001) (Hou et al., 2001):

$$\ddot{y} = f(y, \dot{y}, p) + bu, \tag{19}$$

where $f(.)$ represents the dynamics of the model and the disturbance, $p$ is the input of unknown disturbance, $u$ is the input of control, and $y$ is the measured output. It is assumed that the value of the parameter $b$ is given. Here $f(.)$ is a nonlinear function. An alternative method is presented by (Han, 1999) as follows. The system in (19) is initially increased:

$$\dot{x}_1 = x_2; \quad \dot{x}_2 = x_3 + bu; \quad \dot{x}_3 = \dot{f} \tag{20}$$

where $x_1 = y$, $x_2 = \dot{y}$, $x_3 = f(y, \dot{y}, p)$. $f(.)$ is treated as an increased state. Here $f$ and $\dot{f}$ are unknown. By considering $f(y, \dot{y}, p)$ as a state, it can be estimated with a state estimator. Han in Han (1999) proposed a nonlinear observer for (20):

$$\begin{aligned} \dot{\hat{x}} &= A\hat{x} + Bu + Lg(e, \alpha, \delta), \\ \hat{y} &= C\hat{x}, \end{aligned} \tag{21}$$

where:

$$A = \begin{bmatrix} 0 & 1 & 0 \\ 0 & 0 & 1 \\ 0 & 0 & 0 \end{bmatrix}, B = \begin{bmatrix} 0 \\ b \\ 0 \end{bmatrix}, C = \begin{bmatrix} 1 & 0 & 0 \end{bmatrix}, L = \begin{bmatrix} L_1 \\ L_2 \\ L_3 \end{bmatrix}. \tag{22}$$

The observer error is $e = y - \hat{x}_1$ and:

$$g_i(e, \alpha_i, \delta)_{|i=1,2,3} = \begin{cases} |e|^{\alpha_i} sign(e) & |e| > \delta \\ \frac{e}{\delta^{1-\alpha_i}} & |e| \leq \delta \end{cases} \qquad \delta > 0. \tag{23}$$

The observer is reduced to the following set of state equations, and is called extended state observer (*ESO*):

$$\begin{cases} \dot{\hat{x}}_1 = \hat{x}_2 + L_1 g_1(e, \alpha, \delta), \\ \dot{\hat{x}}_2 = \hat{x}_3 + L_2 g_2(e, \alpha, \delta) + bu, \\ \dot{\hat{x}}_3 = L_3 g_3(e, \alpha, \delta). \end{cases} \tag{24}$$

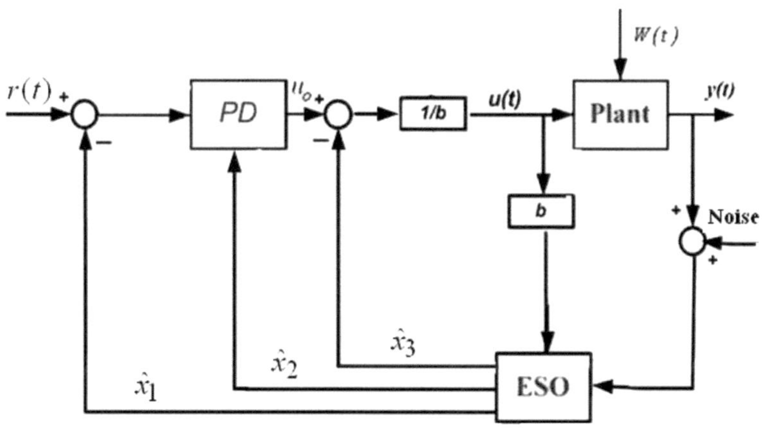


Fig.3. ADRC structure

The active disturbance rejection control (*ADRC*) is then defined as a method of control where the value of $f(y, \dot{y}, p)$ is estimated in real time and is compensated by the control signal $u$. Since $\hat{x}_3 \to f$, it is used to cancel actively $f$ by the application of: $u = (u_0 - \hat{x}_3)/b$. This expression reduces the system to: $\ddot{y} = (f - \hat{x}_3) + u_0 \approx u_0$. The process is now a double integrator with a unity gain, which can be controlled with a PD controller. $u_0 = k_p(r - \hat{x}_1) - k_d\hat{x}_2$, where $r$ is the reference input. The observer gains $L_i$ and the controller gains $k_p$ and $k_d$ can be calculated by a pole placement. The configuration of *ADRC* is presented in fig.3.

## 4. Control of disturbed helicopter

### 4.1 Robust feedback control

#### 4.1.1 Control of altitude z

We apply this robust method to control the altitude dynamics $z$ of our helicopter. Let us remain the equation which describes the altitude under the effect of a wind gust:

$$\ddot{z} = \frac{1}{c_0}[c_8\dot{\gamma}^2 u_1 + c_9\dot{\gamma} + c_{10} - c_7] + \frac{1}{c_0}c_{16}\dot{\gamma}v_{raf}, \tag{25}$$

with $u_1 = M_1(q)\ddot{z} + h_1(q, \dot{q}, v_{raf})$ where $M_1(q) = \frac{c_0}{c_8\dot{\gamma}^2}$ and $h_1(q, \dot{q}, v_{raf}) = -\frac{1}{c_8\dot{\gamma}^2}[c_9\dot{\gamma} + c_{10} - c_7 + c_{16}\dot{\gamma}v_{raf}]$.

Moreover: $\underline{\dot{\gamma}} \leq \dot{\gamma} \leq \overline{\dot{\gamma}}$ and $\underline{v_{raf}} \leq v_{raf} \leq \overline{v_{raf}}$. It is noted that $\dot{\gamma}_{trim} = -124,63rad/s$ and:

$$\underline{\dot{\gamma}} = -209.4rad/s \quad \overline{\dot{\gamma}} = -99.5rad/s; \; \underline{v_{raf}} = -0.68m/s \; ; \; \overline{v_{raf}} = 0.68m/s. \tag{26}$$

The value of $|v_{raf}| = 0.68m/s$ corresponds to an average wind gust. In that case, we have the following bounds: $5 \times 10^{-5} \leq M_1 \leq 22.2 \times 10^{-5}$; $-2,2 \times 10^{-3} \leq h_1 \leq 1,2 \times 10^{-3}$.

Note: We will add an integrator to the control law to reduce the static error of the system and to attenuate the effects of the wind gust which is located in low frequency *(raf ≤7rad/s.* We then obtain (Martini et al., 2007b):

$$v_1 = \ddot{z}_d - K_1e_1 - K_2e_2 - K_3\int_0^t e_1 dt + \Delta v_1, \tag{27}$$

and the value of $\Delta v$ becomes: $\Delta v_1 = -\rho_1(e, t)\, sign\, (287e_1 + 220e_2 + 62e_3)$. Moreover $\rho_1 = 1.7\|v_1\| + 184$.

#### 4.1.2 Control of yaw angle $\psi$:

The control law for the yaw angle is:

$$\begin{aligned} u_2 = \; & \frac{1}{c_5c_{11}\dot{\gamma}^2}[(c_1c_5 - c_4^2)\ddot{\psi} + c_4((c_{12}\dot{\gamma} + c_{13} + c_8\dot{\gamma}v_{raf})u_1 \\ & + c_{14}\dot{\gamma}^2 + c_{15} + \tfrac{5}{2}c_9v_{raf} + c_{17}v_{raf}^2)] \end{aligned} \tag{28}$$

We have:

$$\begin{array}{rcl} u_2 & = & M_2(q)\ddot{\psi} + h_2(q, \dot{q}, v_{raf}), \\ M_2(q) & = & \frac{c_1c_5 - c_4^2}{c_5c_{11}\dot{\gamma}^2}, \\ h_2(q, \dot{q}, v_{raf}, u_1) & = & \frac{c_4}{c_5c_{11}\dot{\gamma}^2}[((c_{12}\dot{\gamma} + c_{13} + c_8\dot{\gamma}v_{raf})u_1 \\ & & +c_{14}\dot{\gamma}^2 + c_{15} + \frac{5}{2}c_9v_{raf} + c_{17}v_{raf}^2)]. \end{array} \quad (29)$$

Using (26) and with $\underline{u}_1 \le u_1 \le \overline{u}_1, \overline{u}_1 = 0$, $\underline{u}_1 = -0.0112m$, we find the following values : $-2.7 \times 10^{-4} \le M_2 \le -6.1 \times 10^{-5}$; $-1.3 \times 10^{-3} \le h_2 \le 0.16$.
We also add an integrator to the control law of the yaw angle (Martini et al., 2007b) :

$$v_2 = \ddot{\psi}_d - K_4e_1 - K_5e_2 - K_6\int_0^t e_1dt + \Delta v_2, \quad (30)$$

where $e_1 = \psi - \psi_d$ and $e_2 = \dot{e}_1$, $\dot{e}_3 = e_1$. We obtain : $\rho_2 = 1.7\|v_2\| + 1614.6$, the value of $\Delta v$ becomes: $\Delta v_2 = -\rho_2(e, t)sign(217e_1 + 87e_2 + 4e_3)$.
On the other hand, the variation of inertia matrices $M_1(q)$ and $M_2(q)$ from their equilibrium value (corresponding to $\dot{\gamma} = -124.63 rad/s$) are shown in table 3. It appears, in this table, that when $\dot{\gamma}$ varies from $-99.5$ to $-209$, $4 rad/s$ an important variation of the coefficients of matrices $M_1(q)$ and $M_2(q)$ of about 65% is obtained.

| $M_1(s^2/rad^2)$ | $\lvert\frac{\Delta M_1}{M_1}\rvert(\%)$ | $M_2(m.s^2/rad^2)$ | $\lvert\frac{\Delta M_2}{M_2}\rvert(\%)$ |
|---|---|---|---|
| $1.42 \times 10^{-4}$ | 65% | $-1.72 \times 10^{-4}$ | 65% |

Table 3. Variations of the inertia matrices $M_1$ and $M_2$

### 4.2 Active disturbance rejection control

Two approaches are proposed here (Martini et al., 2007a) . The first uses a feedback and supposes the knowledge of a precise model of the helicopter. For the second approach, only two parameters of the helicopter are necessary, the remainder of the model being regarded as a disturbance, as well as the wind gust.

- Approach 1 (ADRC) : Firstly, the nonlinear terms of the non disturbed model ($v_{raf} = 0$) are compensated by introducing two new controls $v_1$ and $v_2$ such as:

$$\begin{array}{rcl} u_1 & = & \frac{1}{c_8\dot{\gamma}^2}[c_0v_1 - c_9\dot{\gamma} - c_10 + c_7], \\ u_2 & = & \frac{c_1c_5 - c_4^2}{c_5c_{11}\dot{\gamma}^2}[v_2 + \frac{c_4}{c_1c_5 - c_4^2}((c_{12}\dot{\gamma} + c_{13})u_1 + c_{14}\dot{\gamma}^2 + c_{15})]. \end{array} \quad (31)$$

Since $v_{raf} \neq 0$, a nonlinear system of equations is obtained:

$$\begin{array}{rcl} \ddot{z} & = & v_1 + \frac{1}{c_0}c_{16}\dot{\gamma}\, v_{raf}, \\ \ddot{\psi} & = & v_2 - \frac{c_4c_0v_{raf}}{(c_1c_5 - c_4^2)\dot{\gamma}}v_1 - \frac{c_4v_{raf}}{c_1c_5 - c_4^2}\left[\frac{c_7 - c_{10}}{\dot{\gamma}} + \frac{5}{2}c_9 + c_{17}\, v_{raf}\right]. \end{array} \quad (32)$$

Approach 2 (ADRCM): By introducing the two new controls $\acute{u}_1$ and $\acute{u}_2$ such as:

$$\begin{array}{rcl} \acute{u}_1 & = & \frac{c_8}{c_0}\dot{\gamma}^2u_1, \\ \acute{u}_2 & = & \frac{c_5c_{11}}{c_1c_5 - c_4^2}\dot{\gamma}^2u_2, \end{array}$$

a different nonlinear system of equations is got:

$$\begin{array}{rcl}\ddot{z} & = & \acute{u}_1 + \frac{1}{c_0}\left[\, c_9\dot{\gamma} + c_{10} - c_7 + c_{16}\dot{\gamma}v_{raf}\right], \\ \ddot{\psi} & = & \acute{u}_2 - \frac{c_4}{(c_1c_5 - c_4^2)}[(c_{12}\dot{\gamma} + c_{13} + c_8\dot{\gamma}v_{raf})\,\frac{c_0\acute{u}_1}{c_8\dot{\gamma}^2} + c_{15} \\ & & +c_{14}\dot{\gamma}^2 + \frac{5}{2}c_9v_{raf} + c_{17}v_{raf}^2].\end{array} \tag{33}$$

The systems (32) and (33) can be written as the following form:

$$\ddot{y} = f(y,\dot{y},p) + bu, \tag{34}$$

with $b = 1$, $u = v_1$ or $v_2$ for the approach 1, whether:

$$\begin{array}{ll}\text{For} \quad y = z \;\Rightarrow f_z(y,\dot{y},p) = & \frac{1}{c_0}c_{16}\dot{\gamma}v_{raf}, \\ \text{For} \quad y = \psi \;\Rightarrow f_\psi(y,\dot{y},p) = & -\frac{c_4c_0v_{raf}}{(c_1c_5 - c_4^2)\dot{\gamma}}v_1 \\ & -\frac{c_4v_{raf}}{c_1c_5 - c_4^2}\left[\frac{c_7 - c_{10}}{\dot{\gamma}} + \frac{5}{2}c_9 + c_{17}v_{raf}\right],\end{array} \tag{35}$$

and $b = 1$, $u = \acute{u}_1$ or (ADRC) $\acute{u}_2$ for the approach 2, whether:

$$\begin{array}{ll}\text{For} \quad y = z \;\Rightarrow \acute{f}_z(y,\dot{y},p) = & \frac{1}{c_0}\left[c_9\dot{\gamma} + c_{10} - c_7 + c_{16}\dot{\gamma}v_{raf}\right], \\ \text{For} \quad y = \psi \;\Rightarrow \acute{f}_\psi(y,\dot{y},p) = & -\frac{c_4}{(c_1c_5 - c_4^2)}\left[(c_{12}\dot{\gamma} + c_{13} + c_8\dot{\gamma}v_{raf})\,\frac{c_0\acute{u}_1}{c_8\dot{\gamma}^2}\right. \\ & \left. +c_{15} + c_{14}\dot{\gamma}^2 + \frac{5}{2}c_9v_{raf} + c_{17}v_{raf}^2\right].\end{array} \tag{36}$$

Concerning the first approach, an observer is built:

- for altitude $z$:

$$\left\{\begin{array}{l}\dot{\hat{x}}_1 = \hat{x}_2 + L_1g_1(e_z,\alpha_1,\delta_1), \\ \dot{\hat{x}}_2 = \hat{x}_3 + L_2g_2(e_z,\alpha_2,\delta_2) + b\,v_1, \\ \dot{\hat{x}}_3 = L_3g_3(e_z,\alpha_3,\delta_3),\end{array}\right. \tag{37}$$

where $e_z = z - \hat{z}_1$ is the observer error, $g_i(e_i, \alpha_i, \delta_i)$ is defined as exponential function of modified gain:

$$g_i(e_z,\alpha_{iz},\delta_i)_{|i=1,2,3} = \left\{\begin{array}{ll}|e_z|^{\alpha_{iz}}sign(e_z), & |e_z| > \delta_i, \\ \frac{e_z}{\delta_i^{1-\alpha_{iz}}}, & |e_z| \leq \delta_i,\end{array}\right. \tag{38}$$

with $0 < \alpha_i < 1$ and $0 < \delta_i \leq 1$, a *PID* controller is used in stead of *PD* in order to attenuate the effects of disturbance:

$$v_1 = \ddot{z}_d - \lambda_1(\hat{x}_1 - z_d) - \lambda_2(\hat{x}_2 - \dot{z}_d) - \lambda_3\int_0^t(x_1 - z_d)dt - \hat{x}_3. \tag{39}$$

The control signal $v_1$ takes into account of the terms which depend on the observer $(\hat{x}_1, \hat{x}_2)$. The fourth part, which also comes from the observer, is added to eliminate the effect of disturbance in this system.

- for the yaw angle $\psi$:

$$\begin{cases} \dot{\hat{x}}_4 = \hat{x}_5 + L_4 g_4(e_\psi, \alpha_4, \delta_4), \\ \dot{\hat{x}}_5 = \hat{x}_6 + L_5 g_5(e_\psi, \alpha_5, \delta_5) + b\, v_2, \\ \dot{\hat{x}}_6 = L_6 g_6(e_\psi, \alpha_6, \delta_6), \end{cases} \tag{40}$$

where $e_\psi = \psi - \hat{x}_4$ is the observer error, with $g_i(e_\psi, \alpha_{i\psi}, \delta_i)$ is defined as exponential function of modified gain:

$$g_i(e_\psi, \alpha_{i\psi}, \delta_i)_{|i=4,5,6} = \begin{cases} |e_\psi|^{\alpha_{i\psi}} sign(e_\psi), & |e_\psi| > \delta_i, \\ \dfrac{e_\psi}{\delta_i^{1-\alpha_{i\psi}}}, & |e_\psi| \leq \delta_i, \end{cases} \tag{41}$$

and

$$v_2 = \ddot{\psi}_d - \lambda_4(\hat{x}_4 - \psi_d) - \lambda_5(\hat{x}_5 - \dot{\psi}_d) - \lambda_6 \int_0^t (x_4 - \psi_d)dt - \hat{x}_6, \tag{42}$$

$z_d$ and $\psi_d$ are the desired trajectories. *PID* parameters are designed to obtain two dominant poles in closed-loop: for $z \begin{cases} \omega_{c1} = 2rad/s \\ \xi_1 = 1 \end{cases}$ and for $\psi \begin{cases} \omega_{c2} = 5rad/s \\ \xi_2 = 1 \end{cases}$. The approach 2 uses the same observer with the same gain, simply $(-\hat{x}_3)$ and $(-\hat{x}_6)$ compensate respectively $\acute{f}_z(y, \dot{y}, p)$ and $\acute{f}_\psi(y, \dot{y}, p)$.

### 4.3 Backstepping control

To control the altitude dynamics $z$ and the yaw angle $\psi$, the steps are as follows:

1. Compensation of the nonlinear terms of the nondisturbed model ($v_{raf} = 0$) by introducing two new controls $V_z$ and $V_\psi$ such as:

$$\begin{aligned} u_1 &= \tfrac{1}{c_8 \dot{\gamma}^2}[c_0 V_z - c_9 \dot{\gamma} - c_{10} + c_7], \\ u_2 &= \tfrac{1}{c_5 c_{11} \dot{\gamma}^2}[(c_1 c_5 - c_4^2)V_\psi + c_4((c_{12}\dot{\gamma} + c_{13})u_1 + c_{14}\dot{\gamma}^2 + c_{15})], \end{aligned} \tag{43}$$

with these two new controls, the following system of equations is obtained:

$$\ddot{z} = V_z + d_1(\dot{\gamma}, v_{raf}), \tag{44}$$

$$\ddot{\psi} = V_\psi + d_2(V_z, \dot{\gamma}, v_{raf}). \tag{45}$$

2. Stabilization is done by backstepping control, we start by controlling the altitude z then the yaw angle $\psi$.

#### 4.3.1 Control of altitude z

We already saw that $\ddot{z} = V_z + d_1(\dot{\gamma}, v_{raf})$. The controller, generated by backstepping, is generally a *PD* (Proportional Derived). Such *PD* controller is not able to cancel external disturbances with non zero average unless they are at the output of an integrating process. In order to attenuate the errors due to static disturbances, a solution consists in equipping the regulators obtained with an integral action (Benaskeur et al., 2000). The main idea is to introduce, in a virtual way, an integrator in the transfer function of the process and t carry out the development of the control law in a conventional way using the method of backstepping. The state equations of $z$ dynamics which are increased by an integrator, are given by:

$$\begin{aligned} x_1 &= z, \\ \dot{x}_1 &= x_2 = \dot{z}, \\ \dot{x}_2 &= x_3 + d_1(\dot{\gamma}, v_{raf}), \\ \dot{x}_3 &= w_1, \end{aligned} \tag{46}$$

where $V_z = \int w_1 = x_3$. The introduction of an integrator into the process only increases the state of the process. Hereafter the control by backstepping is developed:

**Step 1**: Firstly, we ask the output to track a desired trajectory $x_{1d}$, one introduces the trajectory error: $\xi_1 = x_{1d} - x_1$, and its derivative:

$$\dot{\xi}_1 = \dot{x}_{1d} - \dot{x}_1 = \dot{x}_{1d} - x_2, \tag{47}$$

which are both associated to the following Lyapunov candidate function:

$$V_1 = \frac{1}{2}\xi_1^2. \tag{48}$$

The derivative of Lyapunov function is evaluated: $\dot{V}_1 = \xi_1\dot{\xi}_1 = \xi_1(\dot{x}_{1d} - \dot{x}_1) = \xi_1(\dot{x}_{1d} - x_2)$. The state $x_2$ is then used as intermediate control in order to guarantee the stability of (47). We define for that a virtual control: $\alpha_1 = a_1\xi_1 + \dot{x}_{1d}$, where: $\dot{\alpha}_1 = a_1\dot{\xi}_1 + \ddot{x}_{1d}$.

**Step 2**: It appears a new error: $\xi_2 = \alpha_1 - x_2 = a_1\xi_1 + \dot{\xi}_1$. Its derivative is written as follows:

$$\dot{\xi}_2 = \dot{\alpha}_1 - \dot{x}_2 = a_1\dot{\xi}_1 + \ddot{x}_{1d} - x_3 - d_1. \tag{49}$$

In order to attenuate this error, the precedent candidate function (48) is increased by another term, which will deal with the new error introduced previously:

$$V_2 = \frac{1}{2}\xi_1^2 + \frac{1}{2}\xi_2^2, \tag{50}$$

its derivative: $\dot{V}_2 = \xi_1\dot{\xi}_1 + \xi_2\dot{\xi}_2 = -a_1\xi_1^2 + \xi_2(\dot{\alpha}_1 - x_3 + \xi_1) - \xi_2 d_1$. The state $x_3$ can be used as an intermediate control in (49). This state is given in such a way that it must return the expression between bracket equal to $-a_2\xi_2$ for $d_1(\dot{\gamma}, v_{raf}) = 0$. The virtual control obtained is: $\alpha_2 = a_2\xi_2 + \dot{\alpha}_1 + \xi_1 = (1 + a_1a_2)\xi_1 + (a_2 + a_1)\dot{\xi}_1 + \ddot{x}_{1d}$, its derivative: $\dot{\alpha}_2 = (1 + a_1a_2)\dot{\xi}_1 + (a_2 + a_1)\ddot{\xi}_1 + x_{1d}^{(3)}$.

**Step 3:** Still here, another term of error is introduced:

$$\begin{aligned} \xi_3 &= \alpha_2 - x_3 \\ &= (1 + a_1a_2)\xi_1 + (a_2 + a_1)\dot{\xi}_1 + \ddot{\xi}_1 + d_1, \end{aligned} \tag{51}$$

and the Lyapunov function (50) is augmented another time, to take the following form:

$$V_3 = \frac{1}{2}\xi_1^2 + \frac{1}{2}\xi_2^2 + \frac{1}{2}\xi_3^2, \tag{52}$$

its derivative:

$$\begin{aligned}\dot{V}_3 = &\ \xi_1\dot{\xi}_1 + \xi_2\dot{\xi}_2 + \xi_3\dot{\xi}_3\\ = &\ -a_1\xi_1^2 + \xi_2(\dot{\alpha}_1 - x_3 + \xi_1) - \xi_2 d_1(\dot{\gamma}, v_{raf}) + \xi_3(\dot{\alpha}_2 - \dot{x}_3)\\ = &\ -a_1\xi_1^2 + \xi_2(\dot{\alpha}_1 + \xi_3 - \alpha_2 + \xi_1) + \xi_3(\dot{\alpha}_2 - \dot{x}_3) - \xi_2 d_1(\dot{\gamma}, v_{raf})\\ = &\ -a_1\xi_1^2 - a_2\xi_2^2 + \xi_3(\xi_2 + \dot{\alpha}_2 - \dot{x}_3) - \xi_2 d_1(\dot{\gamma}, v_{raf})\\ = &\ -a_1\xi_1^2 - a_2\xi_2^2 + \xi_3(\xi_2 + \dot{\alpha}_2 - w_1) - \xi_2 d_1(\dot{\gamma}, v_{raf}).\end{aligned} \tag{53}$$

The control $V_z$ should be selected in order to return the expression between the precedent bracket equal to $-a_3\xi_3$ for $d_1 = 0$:

$$\begin{aligned}w_1 &= a_3\xi_3 + \xi_2 + \dot{\alpha}_2\\ &= x_{1d}^{(3)} + (a_3 + a_1a_2a_3 + a_1)\xi_1\\ &\quad +(2 + a_1a_2 + a_1a_3 + a_2a_3)\dot{\xi}_1 + (a_1 + a_2 + a_3)\ddot{\xi}_1,\end{aligned} \tag{54}$$

with the relation (47), we obtain: $\dot{\xi}_1 = \dot{x}_{1d} - \dot{x}_1 \Rightarrow \ddot{\xi}_1 = \ddot{x}_{1d} - \ddot{x}_1$. These values, replaced in the control law, gives for $d_1(\dot{\gamma}, v_{raf}) = 0$:

$$\begin{aligned}\dot{V}_z &= w_1\\ &= x_{1d}^{(3)} + (a_3 + a_1a_2a_3 + a_1)\xi_1\\ &\quad +(2 + a_1a_2 + a_1a_3 + a_2a_3)\dot{\xi}_1 + (a_1 + a_2 + a_3)\ddot{\xi}_1.\end{aligned} \tag{55}$$

If we replace (54) in (53), we obtain finally:

$$\dot{V}_3 = -a_1\xi_1^2 - a_2\xi_2^2 - a_3\xi_3^2 - \xi_2 d_1. \tag{56}$$

**Step 4:** It is here that the design of the control law by the method of backstepping stops. The integrator, which was introduced into the process, is transferred to the control law, which gives the final following control law:

$$V_z = \ddot{x}_{1d} + T_d(\dot{x}_{1d} - \dot{x}_1) + K_c(x_{1d} - x_1) + T_i \int (x_{1d} - x_1)dt \tag{57}$$

where $s$ is the Laplace variable and: $T_d = a_1 + a_2 + a_3$; $K_c = 2 + a_1a_2 + a_1a_3 + a_2a_3$; $1/T_i = a_1 + a_3 + a_1a_2a_3$.

### 4.3.2 Control of yaw angle $\psi$:

The calculation of the yaw angle control is also based on backstepping control (Zhao & Kanellakopoulos, 1998) dealing with the problem of the attenuation of the disturbance which acts on lateral dynamics. The representation of yaw state dynamics with the angular velocity of the main rotor is:

$$\begin{aligned}\dot{x}_4 = &\ x_5 = \dot{\psi},\\ \dot{x}_5 = &\ \ddot{\psi} = V_\psi + d_2(V_z, \dot{\gamma}, v_{raf}) = w_2 + \dot{\gamma} + d_2(V_z, \dot{\gamma}, v_{raf}),\\ \dot{x}_6 = &\ \ddot{\gamma} = f_1(\dot{\gamma})V_z + f_2(\dot{\gamma})w_2 + d_3(V_z, \dot{\gamma}, v_{raf}).\end{aligned} \tag{58}$$

The backstepping design then proceeds as follows:

**Step 1:** We start with the error variable: $\xi_4 = x_4 - x_{4d}$, whose derivative can be expressed as: $\dot{\xi}_4 = x_5 - \dot{x}_{4d}$, here $x_5$ is viewed as the virtual control, that introduces the following error variable:

$$\xi_5 = x_5 - \alpha_4 \tag{59}$$

where $\alpha_4$ is the first stabilizing function to be determined. Then we can represent $\dot{\xi}_4$ as:

$$\dot{\xi}_4 = \alpha_4 + \xi_5 - \dot{x}_{4d} \tag{60}$$

In order to design $\alpha_4$, we choose the partial Lyapunov function $V_4 = \frac{1}{2}\xi_4^2$ and we evaluate its time derivative along the solutions of (60): $\dot{V}_4 = \xi_4\dot{\xi}_4 = \xi_4(\alpha_4 + \xi_5 - \dot{x}_{4d})$. The choice of: $\alpha_4 = -a_4\xi_4 + \dot{x}_{4d}$, yields: $\dot{V}_4 = -a_4\xi_4^2 + \xi_4\xi_5$ and $\dot{\xi}_4 = -a_4\xi_4 + \xi_5$.

**Step 2:** According to the computation of step 1, driving $\xi_5$ to zero will ensure that $\dot{V}_4$ is negative definite in $\xi_4$. We need to modify the Lyapunov function to include the error variable $\xi_5$:

$$V_5 = \frac{1}{2}(\xi_4^2 + \xi_5^2). \tag{61}$$

We rewrite $\dot{\xi}_5$:

$$\begin{aligned} \dot{\xi}_5 &= \dot{x}_5 - \dot{\alpha}_4 \\ &= \ddot{\psi} - \dot{\alpha}_4 \\ &= w_2 + \dot{\gamma} + d_2 - a_4^2\xi_4 + a_4\xi_5 - \ddot{\psi}_d. \end{aligned} \tag{62}$$

In this equation, $\dot{\gamma}$ is viewed as the virtual control. This is a departure from the usual backstepping design which only employs state variables as virtual controls. In this case, however, this simple modification is not only dictated by the structure of the system, but it also yields significant improvements in closed-loop system response. The new error variable is $\xi_6 = \dot{\gamma} - \alpha_5$, and $\alpha_5$ is yet to be computed. Then (62) becomes: $\dot{\xi}_5 = w_2 + d_2 - a_4^2\xi_4 + a_4\xi_5 - \ddot{\psi}_d + \xi_6 + \alpha_5$,

$$\begin{aligned} \dot{V}_5 &= \xi_4\dot{\xi}_4 + \xi_5\dot{\xi}_5 \\ &= -a_4\xi_4^2 + \xi_5\left[w_2 + \xi_4 - a_4^2\xi_4 + a_4\xi_5 - \ddot{\psi}_d + \alpha_5 + d_2 + \xi_6\right]. \end{aligned} \tag{63}$$

From (63), the choice of: $\alpha_5 = (a_4^2 - 1)\xi_4 - (a_4 + a_5)\xi_5 - w_2 + \ddot{\psi}_d$, provides:

$$\dot{\xi}_5 = -\xi_4 - a_5\xi_5 + \xi_6 + d_2 \Rightarrow \dot{V}_5 = -a_4\xi_4^2 - a_5\xi_5^2 + \xi_5\xi_6 + \xi_5 d_2. \tag{64}$$

**Step 3:** Similarly to the previous steps, we will design the stabilizing function $w_2$ in this step. To achieve that, firstly, we define the error variable $\xi_6$: $\xi_6 = \dot{\gamma} - \alpha_5$, its time derivative:

$$\begin{aligned} \dot{\xi}_6 &= \ddot{\gamma} - \dot{\alpha}_5 \\ &= f_1(\dot{\gamma})V_z + f_2(\dot{\gamma})w_2 + \tfrac{c_4}{c_5}\dot{\gamma} - \psi_d^{(3)} - (a_5a_4 + a_5^2 + a_4^2 - 1)\xi_5 \\ &\quad -(2a_4 + a_5 - a_4^3)\xi_4 + (a_4 + a_5)\xi_6 + \dot{w}_2 + (a_4 + a_5)d_2 + d_3. \end{aligned} \tag{65}$$

Therefore, along the solutions of $\dot{\xi}_4$, $\dot{\xi}_5$ and $\dot{\xi}_6$, we can express the time derivative of the partial Lyapunov function $V_6 = \frac{1}{2}(\xi_4^2 + \xi_5^2 + \xi_6^2)$ as:

$$\begin{aligned}
\dot{V}_6 &= \xi_4\dot{\xi}_4 + \xi_5\dot{\xi}_5 + \xi_6\dot{\xi}_6 \\
&= -a_4\xi_4^2 - a_5\xi_5^2 + \xi_5\xi_6 + \xi_5 d_2 + \xi_6[f_1(\dot{\gamma})V_z + f_2(\dot{\gamma})w_2 \\
&\quad + \tfrac{c_4}{c_5}\dot{\gamma} - (2a_4 + a_5 - a_4^3)\xi_4 - (a_5a_4 + a_5^2 + a_4^2 - 1)\xi_5 \\
&\quad + (a_4 + a_5)\xi_6 + \dot{w}_2 + (a_4 + a_5)d_2 + d_3 - \psi_d^{(3)}] \\
&= -a_4\xi_4^2 - a_5\xi_5^2 - a_6\xi_6^2 + (\xi_5 + \xi_6(a_4 + a_5))d_2 + \xi_6 d_3 \\
&\quad + \xi_6[(a_4 + a_5 + a_6)\xi_6 + f_1(\dot{\gamma})V_z + f_2(\dot{\gamma})w_2 + \tfrac{c_4}{c_5}\dot{\gamma} + \dot{w}_2 \\
&\quad - (2a_4 + a_5 - a_4^3)\xi_4 - (a_5a_4 + a_5^2 + a_4^2 - 2)\xi_5 - \psi_d^{(3)}].
\end{aligned} \tag{66}$$

In the above expression (66), our choice of $\dot{\omega}_2$ is:

$$\begin{aligned}
\dot{w}_2 &= -f_1(\dot{\gamma})V_z - f_2(\dot{\gamma})w_2 - \tfrac{c_4}{c_5}\dot{\gamma} - (a_4 + a_5 + a_6)\xi_6 \\
&\quad + (2a_4 + a_5 - a_4^3)\xi_4 + (a_5a_4 + a_5^2 + a_4^2 - 2)\xi_5 + \psi_d^{(3)}.
\end{aligned} \tag{67}$$

Then one replaces (67) in (65), to obtain: $\dot{\xi}_6 = -\xi_5 - a_6\xi_6 + (a_4 + a_5)d_2 + d_3$, the derivative of $V_6$ becomes:

$$\dot{V}_6 = -a_4\xi_4^2 - a_5\xi_5^2 - a_6\xi_6^2 + (\xi_5 + \xi_6(a_4 + a_5))d_2 + \xi_6 d_3 \tag{68}$$

The integral of (67) provides $w_2$ and $V_\psi = w_2 + \dot{\gamma}$. In this way, the yaw angle control is calculated.

## 5. Stability analysis of ADRC control

In this section, the stability of the perturbed helicopter controlled using observer based control law (ADRC) is considered. To simplify this study, the demonstration is done with one input and one output as in (Hauser et al. (1992)) and the result is applicable for other outputs. Let us first define the altitude error using equations (32) , (37) and the control (39): $\begin{bmatrix} e_1 & e_2 \end{bmatrix}^T = \begin{bmatrix} z & \dot{z} \end{bmatrix}^T - \begin{bmatrix} z_d & \dot{z}_d \end{bmatrix}^T$, and $e_3 = \int e_1$, we can write:

$$\begin{aligned}
\begin{bmatrix} \dot{e}_1 \\ \dot{e}_2 \\ \dot{e}_3 \end{bmatrix} &= \underbrace{\begin{bmatrix} 0 & 1 & 0 \\ -\lambda_1 & -\lambda_2 & -\lambda_3 \\ 1 & 0 & 0 \end{bmatrix}}_{A} \times \begin{bmatrix} e_1 \\ e_2 \\ e_3 \end{bmatrix} + \underbrace{\begin{bmatrix} 0 & 0 & 0 \\ +\lambda_1 & +\lambda_2 & 1 \\ 0 & 0 & 0 \end{bmatrix}}_{B} \times \begin{bmatrix} \hat{e}_1 \\ \hat{e}_2 \\ \hat{e}_3 \end{bmatrix} \\
\dot{\eta} &= \beta(z, \eta, v_{raf})
\end{aligned} \tag{69}$$

Where $A$ is a stable matrix determined by pole placement, and $\dot{\eta}$ represents the zero dynamics of our system, $\eta = \dot{\gamma} - \dot{\gamma}_{eq}$, where $\dot{\gamma}_{eq} = -124.63 rad/s$ is the equilibruim of the main rotor angular speed :

$$\begin{bmatrix} \hat{e}_1 \\ \hat{e}_2 \\ \hat{e}_3 \end{bmatrix} = \begin{bmatrix} x_1 \\ x_2 \\ x_3 \end{bmatrix} - \begin{bmatrix} \hat{x}_1 \\ \hat{x}_2 \\ \hat{x}_3 \end{bmatrix}$$

is the observer error. Hereafter, we consider the case of a linear observer, so that:

$$\begin{bmatrix} \dot{\hat{e}}_1 \\ \dot{\hat{e}}_2 \\ \dot{\hat{e}}_3 \end{bmatrix} = \begin{bmatrix} -L_1 & 1 & 0 \\ -L_2 & 0 & 1 \\ -L_3 & 0 & 0 \end{bmatrix} \begin{bmatrix} \hat{e}_1 \\ \hat{e}_2 \\ \hat{e}_3 \end{bmatrix} + \begin{bmatrix} 0 \\ 0 \\ \dot{f}(z, \eta, v_{raf}) \end{bmatrix} \quad (70)$$

which can be written as: $\dot{\hat{e}} = \hat{A}\hat{e} + \dot{f}(z, \eta, v_{raf})$. Where $\hat{A}$ is a stable matrix determined by pole placement.

***Theorem:*** Suppose that:

- The zero dynamics of the system $\dot{\eta} = \beta(z, \eta, v_{raf})$ (where is represented by the $\dot{\gamma}$ dynamics) are locally exponentially stable and
- The amplitude of $v_{raf}$ is sufficiently small and the function $\dot{f}(z, \eta, v_{raf})$ is bounded and small enough (i.e $\hat{l}_u < 1/5$, see equation (72) for definition of bound $\hat{l}_u$).

Then for desired trajectories with sufficiently small values and derivatives ($z_d$, $\dot{z}_d$, $\ddot{z}_d$), the states of the system (32) and of the observer (37) will be bounded.

***Proof:*** Since the zero dynamics of model are assumed to be exponentially stable, a conserve Lyapunov theorem implies the existence of a Lyapunov function $V_1(\eta)$ for the system: $\dot{\eta} = \beta(0, \eta, 0)$ satisfying $k_1\|\eta\|^2 \le V_1(\eta) \le k_2\|\eta\|^2$, $\frac{\partial V_1}{\partial \eta}\eta(0, \eta, 0) \le -k_3\|\eta\|^2$, $\left\|\frac{\partial V_1}{\partial \eta}\right\| \le k_4\|\eta\|$ for some positive constants $k_1$, $k_2$, $k_3$ and $k_4$. We first show that $e$, $\hat{e}$, $\eta$ are bounded. To this end, consider as a Lyapunov function for the error system ((69) and (70)):

$$V(e, \hat{e}, \tilde{\eta}) = e^T P e + \delta \hat{e}^T \hat{P} \hat{e} + \mu V_1(\eta) \quad (71)$$

where $P$, $\hat{P} > 0$ are chosen so that: $A^T P + PA = -I$ and $\hat{A}^T \hat{P} + \hat{P}\hat{A} = -I$ (possible since $A$ and $\hat{A}$ are Hurwitz), $\mu$ and $\delta$ are a positives constants to be determined later. Note that, by assumption, $z_d$ and its first derivatives are bounded: $\|z\| \le \|e\| + b_d$, $\|\hat{x}\| \le \|\hat{e}\| + \|e\| + b_d$. The functions, $\beta(z, \eta, v_{raf})$ and $\dot{f}(z, \eta, v_{raf})$ are locally Lipschitz (since $\dot{f}$ is bounded) with $\dot{f}(0, 0, 0) = 0$, we have:

$$\begin{aligned} \|\beta(z^1, \eta^1, v^1_{raf}) - \beta(z^2, \eta^2, v^2_{raf})\| &\le l_q(\|z^1 - z^2\| + \|\eta^1 - \eta^2\| + \|v^1_{raf} - v^2_{raf}\|) \\ \|2\hat{P}\dot{f}(z, \eta, v_{raf})\| &\le \hat{l}_u \end{aligned} \quad (72)$$

with $l_q$ and $\hat{l}_u$ 2 positive reals. Using these bounds and the properties of $V_1(.)$, we have:

$$\begin{aligned} \frac{\partial V_1}{\partial \eta}\beta(z, \eta, v_{raf}) &= \frac{\partial V_1}{\partial \eta}\beta(0, \eta, 0) + \frac{\partial V_1}{\partial \eta}(\beta(z, \eta, v_{raf}) - \beta(0, \eta, 0)) \\ &\le -k_3\|\eta\|^2 + k_4 l_q\|\eta\|(\|e\| + b_d + \|v_{raf}\|) \end{aligned} \quad (73)$$

Taking the derivative of $V(., ., .)$ along the trajectory, we find:

$$
\begin{aligned}
\dot{V} =& \dot{e}^T P e + e^T P \dot{e} + \delta \dot{\hat{e}}^T \hat{P} \hat{e} + \delta \hat{e}^T \hat{P} \dot{\hat{e}} + \mu \frac{\partial V_1}{\partial \dot{\gamma}} \beta(z, \eta, v_{raf}) \\
\leq& - \left[ \frac{3}{8} - \frac{\delta}{2} \hat{l}_u \right] \|e\|^2 + \left[ \frac{1}{2} - \delta + \delta \frac{5}{2} \hat{l}_u + 2(\|B\| \|P\|)^2 \right] \|\hat{e}\|^2 - \\
& - \left[ \frac{3}{4} \mu k_3 - 4 (\mu k_4 l_q)^2 + \frac{1}{2} \mu k_4 l_q \right] \|\eta\|^2 + \mu \frac{(k_4 l_q b_d)^2}{k_3} + \frac{1}{2} \left( \delta b_d \hat{l}_u \right)^2 + \\
& + \left( \delta \hat{l}_u + \frac{1}{2} \mu k_4 l_q \right) \|v_{raf}\|^2
\end{aligned}
$$

Define: $\mu_0 = \frac{k_3}{16(k_4 l_q)^2}$, $\delta_1 = \frac{2}{8\hat{l}_u}$, $\delta_2 = \frac{1+4(\|B\|\|P\|)^2}{1-5\hat{l}_u}$ and $\hat{l}_u < 1/5$. Then, for all $\mu \leq \mu_0$ and for $\delta_2 \leq \delta \leq \delta_1$, we have:

$$
\begin{aligned}
\dot{V} \leq& - \frac{\|e\|^2}{4} - \frac{\delta}{2} \|\hat{e}\|^2 - \frac{\mu k_3 \|\eta\|^2}{2} + \left[ \mu \frac{(k_4 l_q)^2}{k_3} + \right. \\
& \left. + \frac{1}{2} (\delta \hat{l}_u)^2 \right] b_d^2 + \left( \delta \hat{l}_u + \frac{1}{2} \mu k_4 l_q \right) \|v_{raf}\|^2
\end{aligned}
$$

Thus, $\dot{V} < 0$ whenever $\|e\|$, $\|\hat{e}\|$ and $\|\eta\|$ is large which implies that $\|\hat{e}\|$, $\|e\|$ and $\|\eta\|$ and, hence, $\|z\|$, $\|\hat{x}\|$ and $\|\eta\|$ are bounded. The above analysis is valid in a neighborhood of the origin. By choosing $b_d$ and $v_{raf}$ sufficiently small and with appropriate initial conditions, we can guarantee the state will remain in a small neighborhood, and which implies that the effect of the disturbance on the closed-loop can be attenuated. Moreover, if $v_{raf} \rightarrow 0$ then $\hat{l}_u \rightarrow 0$ and $\delta_1 \rightarrow \infty$; $\delta_2 \rightarrow 1 + 4(\|B\|\|P\|)^2$, so that the constraint $\hat{l}_u < 1/5$ is naturally satisfied for small $v_{raf}$.

## 6. Results in simulation

Robust nonlinear feedback control (*RNFC*), active disturbance rejection control based on a nonlinear extended state observer (*ADRC*) and backstepping control (*BACK*) are now compared via simulations.

1. *RNFC*: The various numerical values for the (*RNFC*) are the following:
   - For state variable z: $\{K_1 = 84, K_2 = 24, K_3 = 80\}$ for $\omega_1 = 2$rad/s which is the bandwidth of the closed loop in $z$ (the numerical values are calculating by pole placement).
   - For state variable $\psi$: We have $\{K_4 = 525, K_5 = 60, K_6 = 1250\}$ for $\omega_2 = 5$rad/s which is the bandwidth of the closed loop in $\psi$.
2. *ADRC*: The various numerical values for the (*ADRC*) are the following:
   a. For state variable z: $k_1 = 24$, $k_2 = 84$ and $k_3 = 80$ (the numerical values are calculating by pole placement ). Choosing a triple pole located in $\omega_{0z}$ such as $\omega_{0z} = (3 \sim 5)\,\omega_{c1}$, one can choose $\omega_{0z} = 10$ *rad/s*, $\alpha_1 = 0.5$, $\delta_1 = 0.1$, and using pole placement method the gains of the observer for the case $|e| \leq \delta$ (i.e linear observer) can be evaluated:

$$
\begin{aligned}
\frac{L_1}{\delta_1^{1-\alpha_1}} &= 3\,\omega_{0z}, \\
\frac{L_2}{\delta_1^{1-\alpha_1}} &= 3\,\omega_{0z}^2, \\
\frac{L_3}{\delta_1^{1-\alpha_1}} &= \omega_{0z}^3,
\end{aligned}
\tag{74}
$$

which leads to: $L_i = \{9.5, 94.87, 316.23\}$, $i \in [1, 2, 3]$.

b. For state variable $\psi$: $k_4 = 60$, $k_5 = 525$, $k_6 = 1250$, $\omega_{0\psi} = 25$ *rad/s*, '$_2 = 0.5$ and $\alpha_2 = 0.025$. And by the same method in (74) one can find the observer gains: $L_i = \{11.86, 296.46, 2.47 \times 10^3\}$, $i \in [4, 5, 6]$.

3. *BACK:* The regulation parameters ($a_1$; $a_2$; $a_3$; $a_4$; $a_5$; $a_6$) for the (*BACK*) controller was calculated to obtain two dominating poles in closed-loop such as $\omega_1 = 2$ *rad/s*, which defines the bandwidth of the closed-loop in $z$, and $\omega_2 = 5$ *rad/s* for $\psi$.

a. The closed-loop dynamics of the z-dynamics with $d_1(\dot{\gamma}, v_{raf}) = 0$ is given by (Benaskeur et al., 2000):

$$\begin{bmatrix} \dot{\xi}_1 \\ \dot{\xi}_2 \\ \dot{\xi}_3 \end{bmatrix} = \begin{bmatrix} -a_1 & 1 & 0 \\ -1 & -a_2 & 1 \\ 0 & -1 & -a_3 \end{bmatrix} \begin{bmatrix} \xi_1 \\ \xi_2 \\ \xi_3 \end{bmatrix} = A_0 \begin{bmatrix} \xi_1 \\ \xi_2 \\ \xi_3 \end{bmatrix}. \tag{75}$$

Eigenvalues of $A_0$ can be calculated solving:

$$\begin{aligned} |Is - A_0| &= s^3 + (a_1 + a_2 + a_3)s^2 + (a_1a_2 + a_1a_3 \\ &\quad + a_2a_3 + 2)s + a_1 + a_3 + a_1a_2a_3 = 0. \end{aligned} \tag{76}$$

If one gives as a desired dynamics specification, one dominant pole in $-\kappa$ and the two other poles in $-10\kappa$, one must solve:

$$\beta_d(s) = (s + \kappa)(s + 10\kappa)^2 = s^3 + 21\kappa s^2 + 120\kappa^2 s + 100\kappa^3, \tag{77}$$

which leads to:

$$\begin{aligned} 21\kappa &= a_1 + a_2 + a_3, \\ 120\kappa^2 &= 2 + a_1a_2 + a_1a_3 + a_2a_3, \\ 100\kappa^3 &= a_1 + a_3 + a_1a_2a_3. \end{aligned}$$

For . $= \omega_1 = 2$ *rad/s*, and resolving the above equations, we find 4 positive solutions for every parameter (see Table 1). The solution: $a_1 = 21$, $a_2 = 19$, $a_3 = 1.95$ has been used for simulation.

| $a_2$ | $a_3$ | $a_1$ | $a_5$ | $a_6$ | $a_4$ |
|---|---|---|---|---|---|
| 21 | 19 | 1.95 | 51 | 49 | 5 |
| | 1.95 | 19 | | 4.97 | 49 |
| 19 | 1.95 | 21 | 49 | 51 | 4.97 |
| | 21 | 1.95 | | 4.97 | 51 |
| 1.88 | 20+0.4i | 20-0.4i | 4.95 | 50+0.54i | 6-0.54i |
| | 20-0.4i | 20+0.4i | | 50-0.54i | 6+0.54i |

Table 1. Regulation parameters of z and $\psi$-dynamics

b. The closed-loop dynamics of the $\psi$-dynamics with $d_3(V_z, \dot{\gamma}, v_{vraf}) = 0$ is given by:

$$\begin{bmatrix} \dot{\xi}_4 \\ \dot{\xi}_5 \\ \dot{\xi}_6 \end{bmatrix} = \begin{bmatrix} -a_4 & 1 & 0 \\ -1 & -a_5 & 1 \\ 0 & -1 & -a_6 \end{bmatrix} \begin{bmatrix} \xi_4 \\ \xi_5 \\ \xi_6 \end{bmatrix} = B_0 \begin{bmatrix} \xi_4 \\ \xi_5 \\ \xi_6 \end{bmatrix}. \tag{78}$$

Eigenvalues of $B_0$ can be calculated by solving:

$$\begin{aligned} |Is - B_0| &= s^3 + (a_4 + a_5 + a_6)s^2 + (a_4a_5 + a_4a_6 \\ &\quad + a_5a_6 + 2)s + a_4 + a_6 + a_4a_5a_6 = 0. \end{aligned} \tag{79}$$

By using the same development as for z-dynamics, one can write:

$$\begin{aligned} a_4 + a_5 + a_6 &= 21\kappa, \\ 2 + a_4a_5 + a_4a_6 + a_5a_6 &= 120\kappa^2, \\ a_4 + a_6 + a_4a_5a_6 &= 100\kappa^3. \end{aligned}$$

For $\kappa = \omega_2 = 5$ *rad/s* and resolving the above equations, we find again 4 positive solutions for every parameter (see Table 1). As justified in annex ?? the solution: $a_4 = 4.97$, $a_5 = 49$, $a_6 = 51$ has been used for simulation.

The induced gust velocity operating on the principal rotor is chosen as (G.D.Padfield, 1996):

$$\begin{aligned} v_{raf} &= v_{gm}\sin(0.042t_{d1}) & 70 \le t \le 24\pi + 130 \\ v_{raf} &= -v_{gm}\sin(0.042t_{d2}) & 220 \le t \le 297 \end{aligned} \tag{80}$$

where $t_{d1} = t - 70$ and $t_{d2} = t - 220$, the value of 0.042 represent $\frac{2\pi V}{L_u}$ where $V$ in *m/s* is the height rise speed of the helicopter and $v_{gm} = 0.68m/s$ is the gust density. This density corresponds to an average wind gust, and $L_u = 1.5m$ is its length (see Fig.5). The take-off time at $t = t_{off} = 50$ *s* is imposed and the following desired trajectory is used (Vilchis et al., 2003):

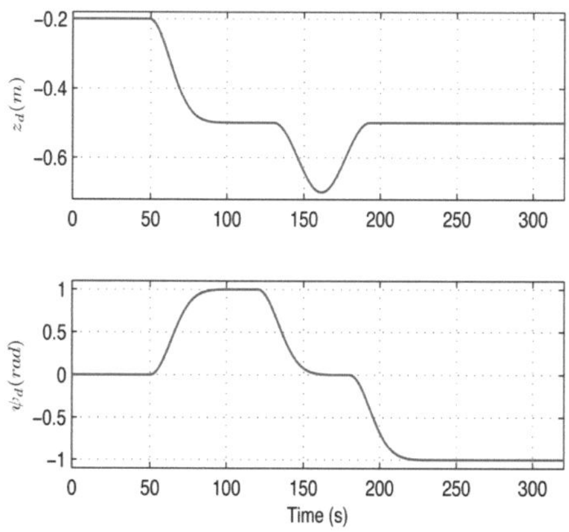


Fig. 4. Trajectories in $z$ and $\psi$

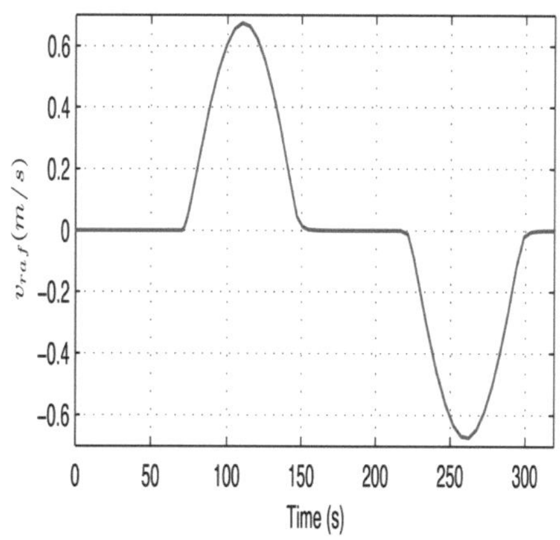


Fig. 5. Induced gust velocity $v_{raf}$

$$\begin{aligned} z_d &= -0.2, & 0 \le t \le t_{off}, \\ z_d &= -0.2 + 0.3\left[\exp^{-(t-t_{off})^2/350} - 1\right], & t_{off} < t \le t_a, \\ z_d &= -0.6 + 0.1\cos\left[(t-130)/10\right], & t_a < t < t_b, \\ z_d &= -0.5, & t \ge t_b, \end{aligned} \tag{81}$$

where $t_a = 130s$ and $t_b = 20\pi + 130s$,

$$\begin{aligned} \psi_d &= 0, & t &< t_{off}, \\ \psi_d &= 1 - \exp^{-(t-t_{off})^2/350}, & t_{off} &\le t < t_c, \\ \psi_d &= \exp^{-(t-t_{off})^2/350}, & t_c &\le t < t_d, \\ \psi_d &= -1 + \exp^{-(t-180)^2/350}, & t &\ge t_d, \end{aligned} \tag{82}$$

and $t_c = 120$ s and $t_d = 180$ s. The following initial conditions are applied: $z(0) = -0.2m$, $\dot{z}(0) = 0$, $\psi(0) = 0$, $\dot{\psi}(0) = 0$ and $\dot{\gamma}(0) = -99.5$ *rad/s*. A band limited white noise of variance $3mm$ for $z$ and $1^{\circ}$ for $\psi$, has been added respectively to the measurements of $z$ and $\psi$ for the three controls. The compensation of this noise is done using a Butterworth second-order low-pass filter. Its crossover frequency for $z$ is $\omega_{cz} = 12$ *rad/s* and for $\psi$ is $\omega_{c\psi} = 20$ *rad/s*. Fig.4 shows the desired trajectories in $z$ and $\psi$.

One can observe that $\dot{\gamma} \rightarrow -124.6$ *rad/s* remains bounded away from zero during the flight. For the chosen trajectories and gains $\dot{\gamma}$ converges rapidly to a constant value (see Fig.7). This is an interesting point to note since it shows that the dynamics and feedback control yield flight conditions close to the ones of real helicopters which fly with a constant $\dot{\gamma}$ thanks to a local regulation feedback of the main rotor speed (which does not exist on the VARIO scale model helicopter). One can also notice that the main rotor angular speed is similar for the three controls as illustrated in Fig.7. The difference between the three controls appears in Fig.6 where the tracking errors in $z$ are less significant by using the (*BACK*) and (*ADRC*) control than (*RNFC*) control. For $\psi$ it is the different. This is explained by the use of a *PID* controller for the (*RNFC*) and (*ADRC*) but a *PD* controller for the (*BACK*) controller of $\psi$ (Fig.6). Here, the (*ADRC*) and (*BACK*) controls show a robust behavior in presence of noise.

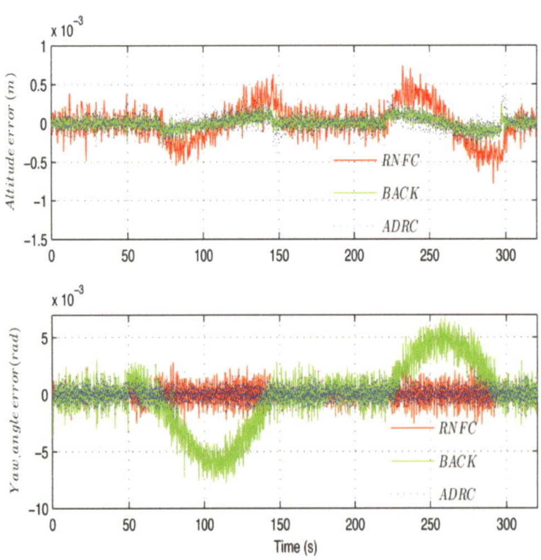


Fig. 6. Tracking error in $z$ and in $\psi$.

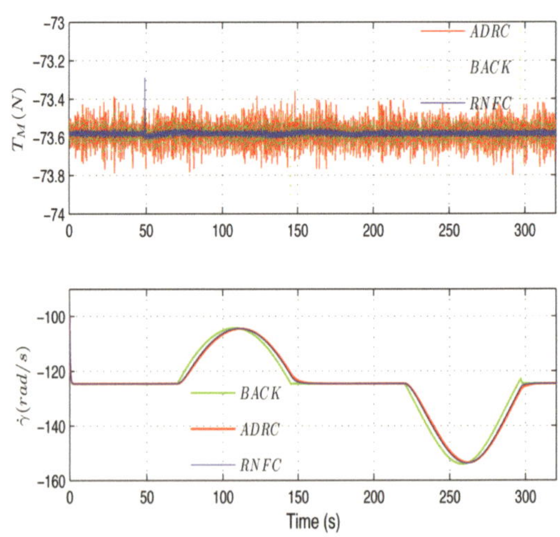


Fig. 7. Variations of the main rotor thrust $T_M$ and the main rotor angular speed $\dot{\gamma}$.

One can see in Fig.7 that the main rotor thrust converges to values that compensate the helicopter weight, the drag force and the effect of the disturbance on the helicopter. The (*RNFC*) control allows the main rotor thrust $T_M$ to be less away from its balance position than the other controls, where the *RNFC* control is less sensitive to noise. Fig.9 represent the effectiveness of the observer: $\hat{x}_3$ and $f_z(y, \dot{y}, w)$ are very close and also $\hat{x}_6$ and $f_\psi(y, \dot{y}, w)$. Observer errors are presented in the Fig.8.

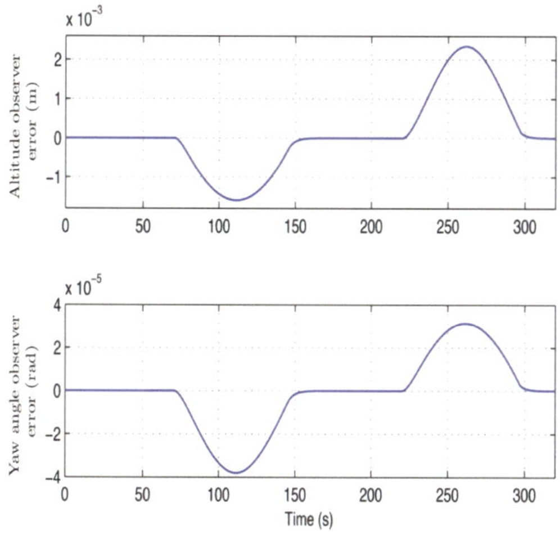


Fig. 8. Observer error in $z$ and in $\psi$

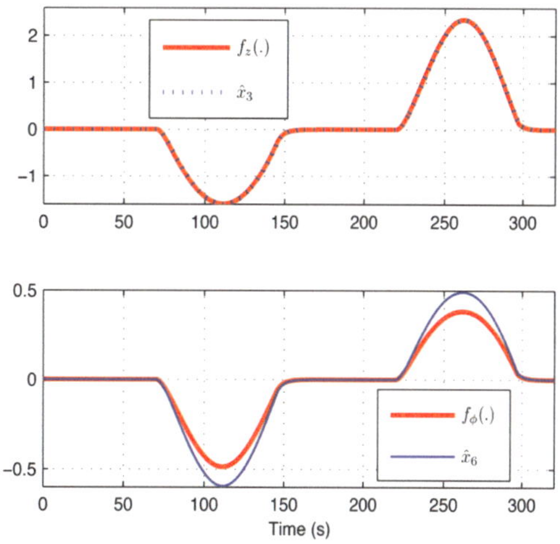


Fig. 9. Estimation of $f_z$ and of $f_\$$

If one keeps the same parameters of adjustment for the three controls and using a larger wind gust ($v_{raf}$ = $3m/s$), we find that the control (*BACK*) give better results than the two controls (*ADRC*) and (*RNFC*) (see Fig.10).

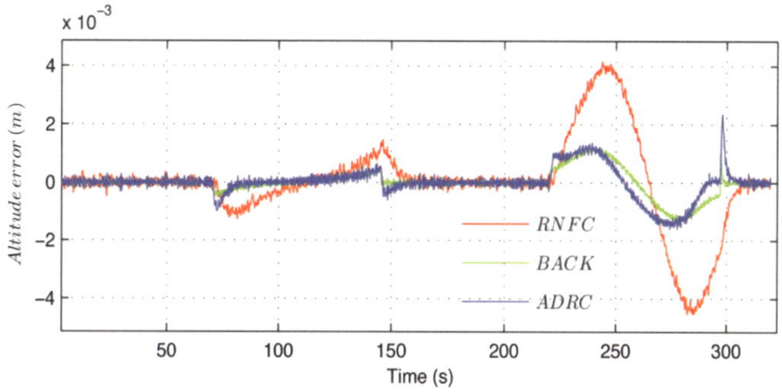


Fig. 10. Large disturbance $v_{raf} = 3m/s$

Fig.11 shows the tracking error in z and $\psi$ for two different *ADRC* controls. These errors are quite simular for approach 1 (*ADRC*) and approach 2 (*ADRCM*). Nevertheless *ADRCM* induces larger error at the take off, which can be explained by the fact that the control depends directly on the angular velocity of the main rotor: this last one need a few time to reach its equilibrium position as seen in Fig.6. The same argument can be invoked to explain the saturation of *ADRCM* control $u_1$ and $u_2$ as illustrated in Fig.12.

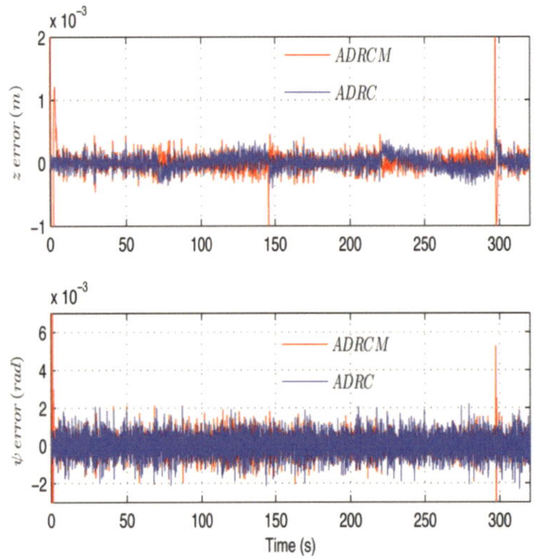


Fig. 11. Tracking error in $z$ and $\psi$ for both approachs 1 and 2 of *ADRC* control.

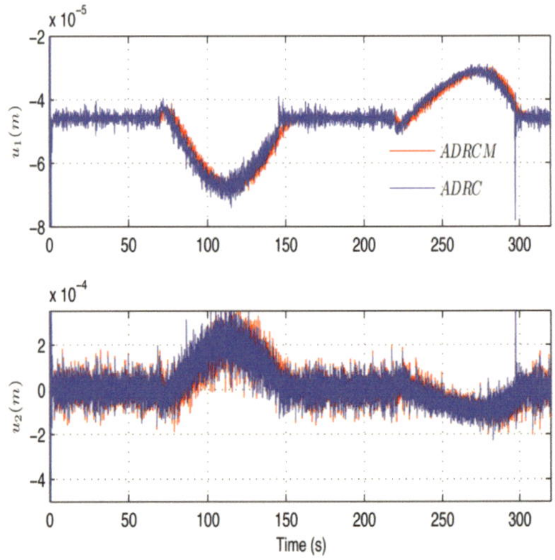


Fig. 12. Inputs $u_1$ and $u_2$ for both approachs 1 and 2 of *ADRC* control.

## 7. Conclusion

In this chapter, a robust nonlinear feedback control (*RNFC*), an active disturbance rejection control based on a nonlinear extended state observer (*ADRC*) and backstepping control (*BACK*) have been applied for the drone helicopter control disturbed by a wind gust. The technique of a robust nonlinear feedback control use the second method of Lyapunov and an additional feedback provides an extra term $\Delta v$ to overcome the effects of the uncertainty and disturbances. The basis of *ADRC* is the extended state observer. The state estimation and compensation of the change of helicopter parameters and disturbance variations are implemented by *ESO* and *NESO*. By using *ESO*, the complete decoupling of the helicopter is obtained. The major advantage of the proposed method is that the closed loop characteristics of the helicopter system do not depend on the exact mathematical model of the system.

The backstepping technique should not viewed as a rigid design procedure, but rather as a design philosophy which can be bent and twisted to accommodate the specific needs of the system at hand. In the particular example of an autonomous helicopter, we were able to exploit the flexibility of backstepping with respect to the selection of virtual controls, initial stabilizing functions and Lyapunov functions. Comparisons were made in detail between the three methods of control.

It is concluded that the three proposed controls algorithms produces satisfactory dynamic performances. Even for large disturbance, the proposed backstepping (*BACK*) and (*ADRC*) control systems are robust against the modeling uncertainties and external disturbance in various operating conditions. It is also indicated that (*BACK*) and (*ADRC*) achieve a better tracking and stabilization with prescribed performance requirements.

For practical reasons, the second *ADRC* approach is the best one because it only requires to know some aerodynamic parameters of the helicopter (dimensions of the blades of the main and tail rotor and the helicopter weight), whereas the other approaches (first *ADRC* approach, *RNCF* and *BACK*) depend on all the aerodynamic parameters which generate the forces and the couples that act on the helicopter. For first *ADRC* control, a stability analysis has been carried out where boundness of states of helicopter and observer are proved in spite of the presence of wind gust.

As illustrated in tables 2 and 3, wind gust induces large variation of helicopter parameters, and the controls quoted in this work can efficiently treat these parameter deviations.

As perspective, this work is carried on a model of a 7DOF VARIO helicopter, where *ADRC* and linearizing control will be tested in simulation. The first results using *ADRC* control on this 7DOF helicopter have been recently obtained (see (Martini et al., 2008) ). Moreover, our control methodologies will be also implemented on a new platform to be built using a Tiny CP3 helicopter.

## 8. References

Beji, L. and A. Abichou (2005). Trajectory generation and tracking of a minirotorcraft. *Proceedings of the 2005 IEEE International Conference on Robotics and Automation*, Spain, 2618–2623.

Benaskeur, A., L. Paquin, and A. Desbiens (2000). Toward industrial control applications of the backstepping. *Process Control and Instrumentation*, 62–67.

Dzul, A., R. Lozano, and P. Castillo (2004). Adaptive control for a radio-controlled helicopter in a vertical flying stand. *International journal of adaptive control and signal processing 18*, 473–485.

Frazzoli, E., M. Dahleh, and E. Feron (2000). Trajectory tracking control design for autonomous helicopters using a backstepping algorithm. *Proceedings of the American Control Conference Chicago, Illinois*, 4102–4107.

Gao, Z., S. Hu, and F. Jiang (2001). A novel motion control design approach based on active disturbance rejection. pp. 4877–4882. Orlando, Florida USA: *Proceedings of the 40th IEEE Conference on Decision and Control.*

G.D.Padfield (1996). Helicopter Flight Dynamics: The Theory and Application of Flying Qualities and Simulation Modeling. Blackwell Science LTD.

Han, J. (1999). Nonlinear design methods for control systems. Beijing, China: *The Proc of the 14th IFAC World Congress.*

Hauser, J., S. Sastry, and G. Meyer (1992). Nonlinear control design for slightly non-minimum phase systems: Applications to v/stol aircraft. *Automatica 28* (4), 665–679.

Hou, Y., F. J. Z. Gao, and B. Boulter (2001). Active disturbance rejection control for web tension regulation. *Proceedings of the 40th IEEE Conference on Decision and Control, Orlando, Florida USA*, 4974–4979.

Ifassiouen, H., M. Guisser, and H. Medromi (2007). Robust nonlinear control of a miniature autonomous helicopter using sliding mode control structure. *International Journal Of Applied Mathematics and Computer Sciences 4* (1), 31–36.

Koo, T. and S. Sastry (1998). Output tracking control design of a helicopter model based on approximate linearization. *The 37th Conference on Decision and Control (Florida, USA) 4*, 3636–3640.

Mahony, R. and T. Hamel (2004). Robust trajectory tracking for a scale model autonomous helicopter. *Int. J. Robust Nonlinear Control 14*, 1035–1059.

Martini, A., F. Léonard, and G. Abba (2005). Suivi de trajectoire d'un hélicoptère drone sous rafale de vent[in french]. *CFM 17ème Congrès Français de Mécanique. Troyes, France, CD ROM.N⁰.467.*

Martini, A., F. Léonard, and G. Abba (2007a). Robust and active trajectory tracking for an autonomous helicopter under wind gust. *ICINCO International Conference on Informatics in Control, Automation and Robotics, Angers, France 2*, 333–340.

Martini, A., F. Léonard, and G. Abba (2007b). Suivi robuste de trajectoires d'un hélicoptère drone sous rafale de vent. *Revue SEE e-STA 4*, 50–55.

Martini, A., F. Léonard, and G. Abba (2008, 22 -26 Septembre). Robust nonlinear control and stability analysis of a 7dof model-scale helicopter under wind gust. In *IEEE/RSJ, IROS, International Conference of Intelligent Robots and Systems, to appear*. NICE, France.

McLean, D. and H. Matsuda (1998). Helicopter station-keeping: comparing LQR, fuzzy-logic and neural-net controllers. *Engineering Application of Artificial Intelligence 11*, 411–418.

Pflimlin, J., P. Soures, and T. Hamel (2004). Hovering flight stabilization in wind gusts for ducted fan uav. *Proc. 43 rd IEEE Conference on Decision and Control CDC, Atlantis, Paradise Island, The Bahamas 4*, 3491- 3496.

Sanders, C., P. DeBitetto, E. Feron, H. Vuong, and N. Leveson (1998). Hierarchical control of small autonomous helicopters. *37th IEEE Conference on Decision and Control 4*, 3629 - 3634.

Spong, M. and M. Vidyasagar (1989). *Robot Dynamics and Control*. John Willey and Sons.

Vilchis, A. (2001). *Modélisation et Commande d'Hélicoptère*. Ph. D. thesis, Institut National Polytechnique de Grenoble.

Vilchis, A., B. Brogliato, L. Dzul, and R. Lozano (2003). Nonlinear modeling and control of helicopters. *Automatica 39*, 1583 -1596.

Wei, W. (2001). Approximate output regulation of a class of nonlinear systems. *Journal of Process Control 11*, 69–80.

Zhao, J. and I. Kanellakopoulos (1998). Flexible backstepping design for tracking and disturbance attenuation. *International journal of robust and nonlinear control 8*, 331–348.

6

# An Artificial Neural Network Based Learning Method for Mobile Robot Localization

Matthew Conforth and Yan Meng
*Department of Electrical and Computer Engineering*
*Stevens Institute of Technology*
*Hoboken, NJ 07030, USA*

## 1. Introduction

One of the most used artificial neural networks (ANNs) models is the well-known Multi-Layer Perceptron (MLP) [Haykin, 1998]. The training process of MLPs for pattern classification problems consists of two tasks, the first one is the selection of an appropriate architecture for the problem, and the second is the adjustment of the connection weights of the network.

Extensive research work has been conducted to attack this issue. Global search techniques, with the ability to broaden the search space in the attempt to avoid local minima, has been used for connection weights adjustment or architecture optimization of MLPs, such as evolutionary algorithms (EA) [Eiben & Smith, 2003], simulated annealing (SA) [Jurjoatrucj et al., 1983], tabu search (TS) [Glover, 1986], ant colony optimization (ACO) [Dorigo et al., 1996] and particle swarm optimization (PSO) [Kennedy & Eberhart, 1995]. The NeuroEvolution of Augmenting Topologies (NEAT) [Stanley & Miikkulainen, 2002] method turns the neural networks topology and connect weight simultaneioulsy using an evolutionary computation method. It evolves efficient ANN solutions quickly by complexifying and optimizing simultaneously; it achieves performance that is superior to comparable fixed-topology methods. In [Patan & Parisini, 2002] the stochastic methods Adaptive Random Search (ARS) and Simultaneous Perturbation Stochastic Approximation (SPSA) outperformed extended dynamic backpropagation at training a dynamic neural network to control a sugar factory actuator. Recently, artifical neural networks based methods are applied to robotic systems. In [Racz & Dubrawski, 1994], an ANN was trained to estimate a robot's position relative to a particular local object. Robot localization was achieved by using entropy nets to implement a regression tree as an ANN in [Sethi & Yu, 1990]. An ANN was trained in [Choi & Oh, 2007] to correct the pose estimates from odometry using ultrasonic sensors.

In this paper, we propose an aritifucal neural networks learning method for mobile robot localization, which combines the two popular swarm inspired methods in computational intelligence areas: Ant Colony Optimization (ACO) and Particle Swarm Optimization (PSO) to train the ANN models. ACO was inspired by the behaviors of ants and has many successful applications in discrete optimization problems. The particle swarm concept originated as a simulation of a simplified social system. It was found that the particle swarm model could be used as an optimizer. These algorithms have been applied already to solving

problems of clustering, data mining, dynamic task allocation, and optimization [Lhotska et al., 2006].
The basic idea of the proposed SWarm Intelligence-based Reinforcement Learning (SWIRL) method is that ACO is used to optimize the topology structure of the ANN models, while the PSO is used to adjust the connection weights of the ANN models based on the optimized topology structure. This is designed to split the problem such that ACO and PSO can both operate in the environment they are most suited for. ACO is ideally applied to finding paths through graphs. One can treat the ANN's neurons as vertices and its connections as directed edges, thereby transforming the topology design into a graph problem. PSO is best used to find the global maximum or minimum in a real-valued search space. Considering each connection plus one associated fitness score as orthogonal dimensions in a hyperspace, each possible weight configuration is merely a point in that hyperspace. Finding the optimal weights is thus reduced to finding the global maximum of the fitness function in that hyperspace.
The chapter is organized as follows. Section 2 introduces two swarm intelligence based methods: ant colony optimiztion and particle swarm optimization. The proposed SWIRL method is described in Section 3. Simulation results and discussions using SWIRL method for mobile robot localization are presented in Section 4. Conclusions are given in Section 5.

## 2. Swarm intelligence

### 2.1 Ant colony optimization

Dorigo et al. [11] proposed an Ant Colony Optimization (ACO). ACO is essentially a system that simulates the natural behavior of ants, including mechanisms of cooperation and adaptation. The involved agents are steered toward local and global optimization through a mechanism of feedback of simulated pheromones and pheromone intensity processing. It is based on the following ideas. First, each path followed by an ant is associated with a candidate solution for a given problem. Second, when an ant follows a path, the amount of pheromone deposit on that path is proportional to the quality of the corresponding candidate solution for the target problem. Third, when an ant has to choose between two or more paths, the path(s) with a larger amount of pheromone are more attractive to the ant. After some iterations, eventually, the ants will converge to a short path, which is expected to be the optimum or a near-optimum solution for the target problem.

### 2.2 Particle swarm optimization

The PSO algorithm is a population-based optimization method, where a set of potential solutions evolves to approach a convenient solution (or set of solutions) for a problem. The social metaphor that led to this algorithm can be summarized as follows: the individuals that are part of a society hold an opinion that is part of a "belief space" (the search space) shared by every possible individual. Individuals may modify this "opinion state" based on three factors: (1) The knowledge of the environment (explorative factor); (2) The individual's previous history of states (cognitive factor); (3) The previous history of states of the individual's neighborhood (social factor).
Therefore, the basic idea is to propel towards a probabilistic median, where explorative factor, cognitive factor (local robot respective views), and social factor (global swarm wide views) are considered simultaneously and try to merge these three factors into consistent behaviors for each robot. The exploration factor can be easily emulated by random movement.

## 3. The SWIRL approach

In the SWIRL approach, the ACO algorithm is utilized to select the topology of the neural network, while the PSO algorithm is utilized to optimize the weights of the neural network. The SWIRL approach is modeled on a school, with the ACO, PSO, and neural networks taking on the roles of administrator, teacher, and student respectively. Students learn, teachers train students, and administrators allocate resources to teachers. In the same fashion, the ACO algorithm allocates training iterations to the PSO algorithms. The PSO algorithms then run for their allotted iterations to train their neural networks. The global best score for all the neural networks trained by a particular PSO instance is then used by the ACO algorithm to reallocate the training iterations.

### 3.1 ACO-based topology optimization

ACO is used to allocate training iterations among a set of candidate network topologies. The desirability in ACO is defined as:

$$d(i) = \frac{1}{h+1} \tag{1}$$

where $h$ is the number of hidden nodes in neural network $i$. $\tau$ (pheromone concentration) is initialized to 0.1, so that the ants' initial actions are based primarily on desirability. $\tau$ is then updated according to:

$$\tau(i,t+1) = \rho \cdot \tau(i,t) + n_a(i) \cdot \frac{g(i)}{g_{sum}} \tag{2}$$

where $\rho$ is the rate of evaporation, $n_a$ is the number of ants returning from neural network $i$, $g(i)$ is the global best for $i$, and $g_{sum}$ is the sum of all the current global bests.
Each ant represents one training iteration for the PSO teacher. During each major iteration (i.e. ACO step), the ants go out into the topology space. The probability ant $k$ goes to neural network $i$ is given by:

$$p(i) = \frac{[\tau(i,t)]^{\alpha}[d(i)]^{\beta}}{\sum_{j=1}^{i}\{\ [\tau(j,t)]^{\alpha} \cdot [d(j)]^{\beta}\}} \tag{3}$$

where $\alpha$ and $\beta$ are constant factors that control the relative influence of pheromones and desirability, respectively.

### 3.2 PSO-based weight adjustment

The ANNs are then trained via the PSO algorithm for a number of iterations determined by the number of ants at that node. Each PSO teacher starts with a group of networks whose connection weights are randomly initialized. The student ANNs are tested on the chosen problem and each receives a score. The PSO teacher keeps track of the global best configuration and each student's individual best configuration. A configuration for an ANN with $n$ connections can be considered as a point in an $n$+1 dimensional space, where the extra dimension is for the reinforcement score. After every round of testing, the teacher updates the connection weights of the student ANNs according to the following equations:

$$\mathbf{v}_{t+1} = c_{inr}\mathbf{r}_1 \bullet \mathbf{v}_t + c_{cgn}\mathbf{r}_2 \bullet \left(\mathbf{x}_{pb} - \mathbf{x}_t\right) + c_{scl}\mathbf{r}_3 \bullet \left(\mathbf{x}_{gb} - \mathbf{x}_t\right) \quad (4)$$

$$\mathbf{x}_{t+1} = \mathbf{x}_t + \mathbf{v}_{t+1} \quad (5)$$

where position and velocity vectors are denoted by **x** and **v**, respectively. The big dot symbol is for Hadamard multiplication. The $\mathbf{r}_i$ represents vectors where each element is a new sample from the unit-interval uniform random variable. Personal best, $\mathbf{x}_{pb}$, is the point in the solution-space where that particular student received its highest score so far. Global best, $\mathbf{x}_{gb}$, is the point with the highest score achieved by any student of this PSO teacher. The three constants, $c_{inr}$, $c_{cgn}$, and $c_{scl}$, allow the adjustment of the relative weighting for the inertial, cognitive, and social components of the velocity, respectively.
After exhausting its allotted training iterations, the PSO teacher reports the global best to the ACO administrator. If/when it is allocated additional training iterations, the PSO teacher resumes training from exactly where it left off.

### 3.3 The SWIRL method summary

Pseudo-code for the SWIRL algorithm follows:

```
procedure SWIRL
        initialize ACO_Administrator
        for(candidate topology i=1...N)
                create PSO_Teacher(i)
                for(ANN_Student j=1...M)
                        initialize ANN_Student
                end for
        end for
        while(solution not found)
                compute ant movement CDF
                ants allocate training iterations
                for(PSO_Teacher i=1...N)
                        while(iterations < allocation)
                                for(ANN_Student j=1...M)
                                        test ANN_Student(j)
                                end for
                                for(ANN_Student j=1...M)
                                        update weights ANN_Student(j)
                                end for
                        end while
                        return global best
                end for
                update pheromone concentrations
        end while
end
```

## 4. Simulation

The SWIRL system is implemented in Java for simulation testing. There is a 5:1 ratio of ants to candidate network topologies. The candidates are 1 through 5 hidden nodes. The pheromone influence factor, desirability influence factor, and rate of pheromone evaporation are set to 2, 1, and 0.5 respectively. The initial pheromone level is 0.1 for all topologies. Each PSO teacher has 100 students. The PSO particle velocity is capped at 5. The velocity factors are 0.8 for the inertial constant, 2 for the cognitive constant, and 2 for the social constant. The neural networks are fully connected, with initial connection weights uniformly random in the range (-5,5). Hyperbolic tangent is used as the transfer function.

The SWIRL system was retrofitted into an existing Markov localization simulator, which required reimplementing SWIRL in C++. Markov localization systems are typically divided into an odometry component and a sensory component which alternate updating the belief values for each location. The SWIRL system was challenged with generating an ANN that could replace the sensory component in the Markov localization system.

Fig. 1 shows a series of snapshots of a robot localization simulation using the SWIRL system. The red pentagon is the robot. The pale blue wedges represent the robot's sonar sensors and the green rays are the robot's laser rangefinders. The large green square is the goal to which the robot must navigate. The large orange square represents the location where the robot most strongly believes itself to be. The map is divided into a grid of squares, which are each divided into a black and red triangle. The black triangle and red triangle indicate the belief of the Markov algorithm and current best ANN, respectively, that the robot may be located in that particular square. Note that the red triangles do not show up underneath the pale blue wedges; this is merely a color layering issue in the simulator's screen painter. The map is 1086 by 443 pixels in size. The Markov grid squares are 7 pixels on a side, and rotations are multiples of 10°. Note that this quantization applies only to the pose estimation in the localization system. The robot is simulated with a full range of motion using floating point $x$, $y$, and $\theta$ values. The robot has 4 sonar sensors and 19 laser rangefinders. The sonar sensor range is 75 pixels in the map, whereas the range of the laser rangefinders is 100 pixels.

Fig. 1(a) shows the localization simulator at the beginning of a simulation run. Fig. 1(b) shows an early state of the simulation. The blue pentagons are stamped periodically on the map to indicate the path taken by the robot. At this point, the robot has narrowed down its location to a few regions on the map. In Fig. 1(c), near the end of the simulation, the robot is certain of its general location. Finally, Fig. 1(d) shows the robot when it has successfully reached the goal.

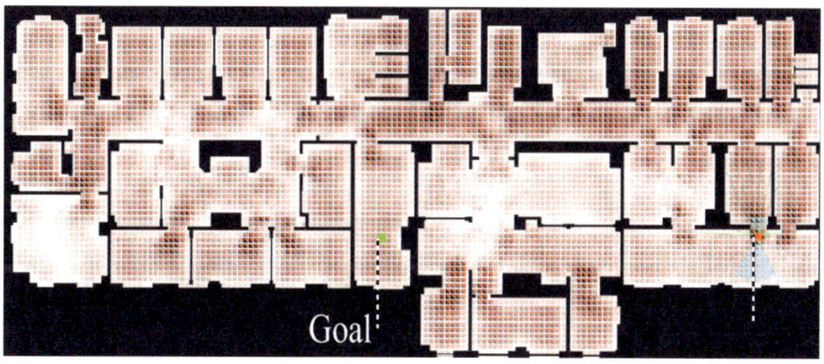


Fig. 1(a). The initialization state of a simulation.

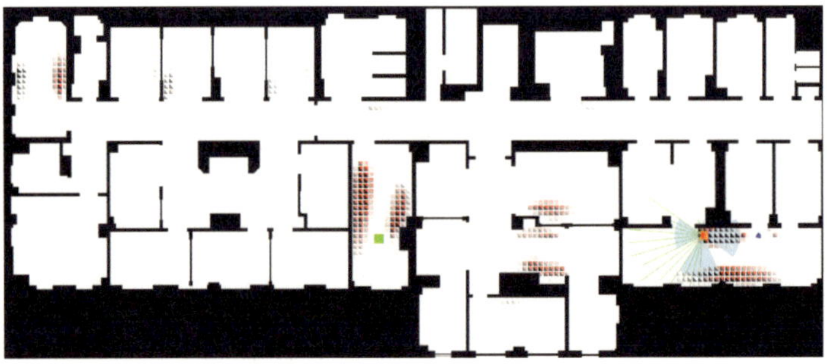

Fig. 1(b). An early state of a simulation.

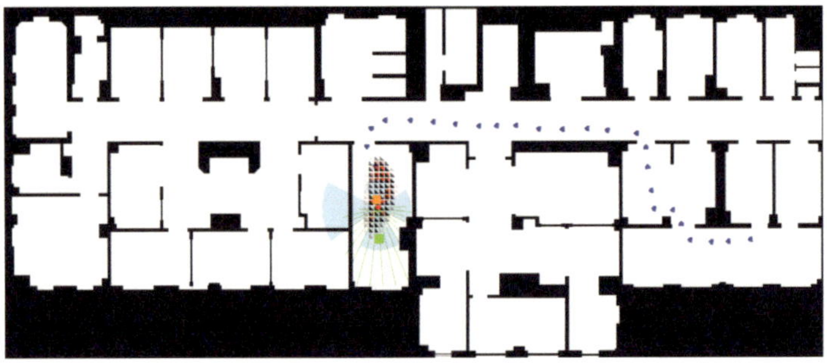

Fig. 1(c). A late state of a simulation.

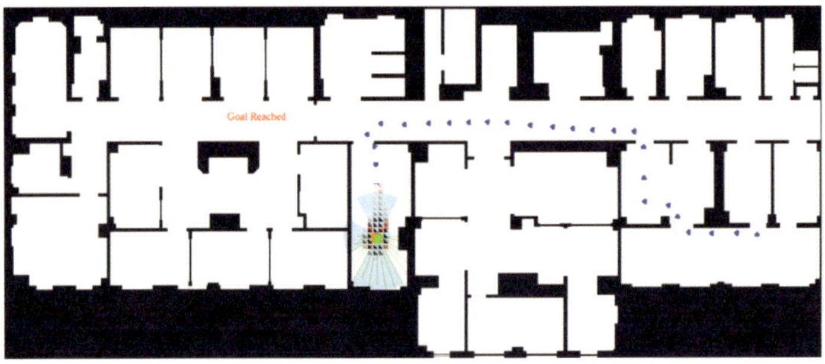


Fig. 1(d). The end of a simulation.

Fig. 1. Start of a simulation. Black = Markov value; Red = ANN value. The robot, shown as a red pentagon, must navigate from its initial position to the goal, the green square. The triangles indicate the likelihood that the robot is in that location according to the beliefs of the localization systems. The green lines show the robot's laser rangefinders and the pale blue wedges show its sonar sensors. The orange square is the robot's most likely position according to whichever localization system is currently in use for navigation. Blue pentagons are periodically stamped on the map to show the path taken by the robot.

The sensory component of the Markov localization system computes the likelihood the robot is at a particular location by comparing the current sensor readings to the predicted sensor readings for that location which are generated from the map. Sensor noise is the main source of difficulty. The following pseudo-code describes this function:

```
function probability(location x)
        prob_match=1.0
        for(i=1...Number_Sensors)
                j=normalize(reading(i)-predict(x,i))
                prob_match=prob(j)*prob_match
        end for
        return prob_match
end
```

These match probabilities are then used to update the robot's location belief matrix. In place of this, the ANN generated by SWIRL should take a vector of j's as the input, and give prob_match as the output.

The obvious source of concern in this scenario is that ANNs inherently deal in weighted sums, whereas joint probability calls for a product. This is solved by taking advantage of the fact that j > 0 and $e^{\ln(x)+\ln(y)} = xy$ for $x,y > 0$. Thus, a vector of ln(j)'s is used as the input, and $e^{\text{output}}$ is used as prob_match.

In the simulation, a robot must navigate from an unknown starting position to a goal position that is specified on the map. Sensor noise is the main source of difficulty. The sensor noise has 3 components: bias, skew, and incidental. Each sensor has its own bias and skew values that are randomly initialized at the beginning of the simulation, but remain fixed thereafter. The incidental noise is a new Gaussian value generated each time the sensor is read.

The SWIRL system is first trained over one or more training runs. The best ANN produced is then used to replace the sensory component of the Markov localization system for the testing run. For the purpose of training, another Markov localizer with access to noise-free sensors is used to produce the "solution set" for the fitness function. This enables the ANN (ideally) to weigh the sensors according to their relative accuracy. The simulation results are shown in Fig. 2, Fig. 3, and Fig. 4. The results are produced for 1 and 3 training runs.

Unfortunately, the wavefront navigation system used by the Markov localization simulator introduces substantial variability into the robot's performance, as demonstrated by the large standard deviations. Consequently, it would require an exorbitantly large number of tests to establish with a high degree of certainty whether the average SWIRL solution is slightly better or slightly worse than the Markov method. In any case, improving Markov localization is not the goal here. It is clear from the results that SWIRL can indeed produce ANNs that function comparably to the sensory component of the Markov localization system.

SWIRL demonstrates its ability to generate effective solutions in the face of real-world complications such as noise and poor calibration. After only a single test run, the SWIRL solution is already performing comparably to the Markov method. The reason more fine-grained progress is not presented is that the nature of Markov localization is such that early mistakes are compounded each cycle. No matter how good the SWIRL solution gets, its mistakes from earlier in the run will prevent it from making accurate predictions. To make the graphical display useful, every so often during training (**only** during training) SWIRL's old belief matrix is overwritten with "correct" values so that one can see an estimate of how accurate the best ANN is at that point. However, this obviously gives a rough estimate only, and is not appropriate for a numerical comparison with the Markov predictions.

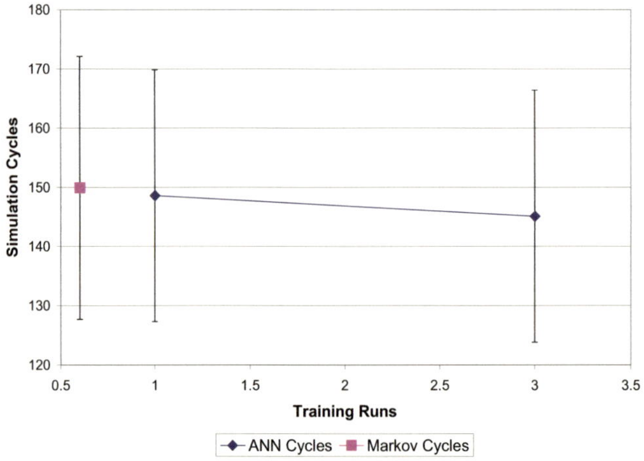


Fig. 2. Simulation cycles versus training runs.

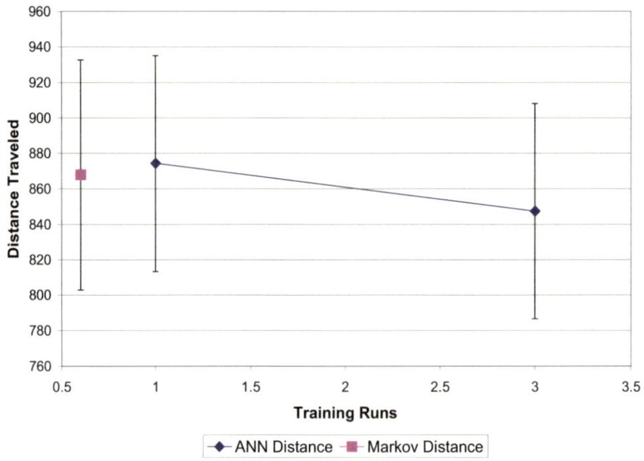


Fig. 3. Distance traveled versus training runs

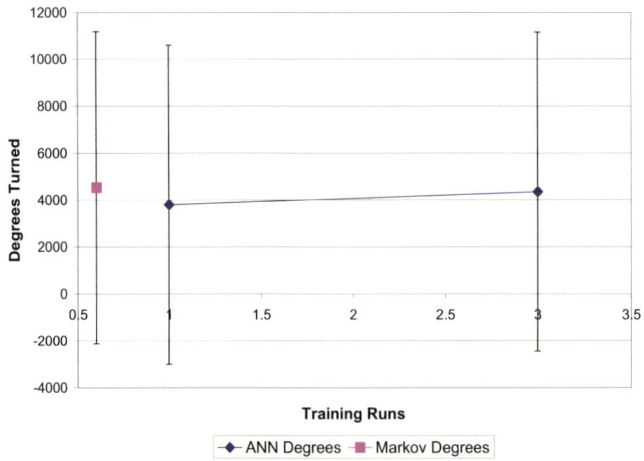


Fig. 4. Degrees turned versus training runs

## 7. Conclusion

In this paper, a SWIRL algorithm is proposed to generate ANN solutions to tasks/problems amenable to reinforcement learning. Basically, the ACO algorithm is applied to select the neural network topology, while the PSO algorithm is utilized to adjust the connection weights of the selected topology. The robot localization using the SWIRL algorithm have been efficiently conducted in a noisy, imprecise environment with large-scale number of input signals, as might be encountered in the real world. As a result of its generality, the SWIRL method is scalable, robust, and can be applied to almost any real world task. This implementation is merely a simple proof-of-concept where the ACO algorithm chooses the number of hidden nodes. By limiting the scope to fully-connected feedforward neural networks, this single number fully defines the topology (since the input and output nodes are determined by the test problem). These are limits of this test implementation only; not limits on the SWIRL method.

## 8. References

Haykin, S. (1998). *Neural Networks: A comprehensive Foundation*. 2nd Edition, Prentice Hall.

Eiben, E. & and Smith, J.E. (2003). *Introduction to Evolutionary Computing*. Natural Computing Series. MIT Press. Springer. Berlin.

Kirkpatrick, S.; Gellat Jr., C.D., & Vecchi, M. P., (1983). Optimization by simulated annealing, *Science*, 220: 671-680.

Glover, F. (1986). Future paths for integer programming and links to artificial intelligence, *Computers and Operation Research*, Vol. 13, pp. 533-549.

Dorigo, M.; Maniezzo, V., & Colorni. A. (1996). Ant System: optimization by a colony of cooperating agents, *IEEE Transactions on Systems, Man and Cybernetics - Part B*, vol. 26, no. 1, pp. 29-41.

Kennedy, J. & Eberhart, R. (1995). Particle Swarm Optimization, in: *Proc. IEEE Intl. Conf. on Neural Networks (Perth, Australia)*, IEEE Service Center, Piscataway, NJ, IV:1942-1948.

Stanley, K. O. & Miikkulainen. R. (2002). Evolving Neural Networks through Augmenting Topologies, *Evolutionary Computation*, 10(2): 99-127.

Patan, K. & Parisini, T. (2002). Stochastic learning methods for dynamic neural networks: simulated and real-data comparisons, *Proceedings of American Control Conference*.

Racz. J. & Dubrawski, A. (1994). Mobile Robot Localization with an Artificial Neural Network, International Workshop on Intelligent Robotic Systems IRS '94, Grenoble, France.

Sethi, I. K. & Yu, G. (1990). A Neural Network Approach to Robot Localization Using Ultrasonic Sensors, *Proceedings of 5th IEEE International Symposium on Intelligent Control*, 1990. pp. 513-517 vol. 1, 5-7.

Choi, W. S. & Oh, S. Y. (2007). Range Sensor-based Robot Localization Using Neural Network, *International Conference on Control, Automation and Systems*, pp. 230-234, 17-20.

Lhotská, L.; Macaš, M., & Burša, M. (2006). *PSO and ACO in Optimization Problems*, E. Corchado et al. (Eds.): IDEAL 2006, LNCS 4224, pp. 1390 - 1398.

7

# The Identification of Models of External Loads

Yuri Menshikov
*Dnepropetrovsk University*
*Ukraine*

## 1. Introduction

One of the important problems of mathematical modelling of dynamic systems is the coincidence of modelling results with experimental measurements. Such a coincidence is being attained by construction of "correct" mathematical model (MM) of the dynamical system and the choice of a "good" model of external load (MEL). MM of object the motion of which coincides with experimental measurements with acceptable accuracy under action of MEL (or external impact) is understood by us as "correct" model. Thus the degree of "correctness" of MM depends directly on the chosen model of EL and required accuracy of the coincidence with experiment. It is formally possible to write as an inequality for models with the concentrated parameters the following:

$$F\,(A_p z,\, u_\delta) \le \varepsilon, \tag{1}$$

where $A_p$ is an operator of the certain structure which carries out the connection of EL ($z$) and the response of MM ($u$, $A_p z = u$) and which depends on vectors - parameters $p$; $\varepsilon = Const > 0$ is the required accuracy of the coincidence of experiment with results of mathematical modelling; $z$ is the function of model of EL, $z \in Z$; $u$, $u_\delta$ are the vector-functions of the response of researched object on external load, $u \in U$, $u_\delta \in U$. One of possible variants of an inequality (1) can be the following inequality

$$\| A_p z - u_\delta \|_U \le \varepsilon, \tag{2}$$

where $\| \cdot \|_U$ there is a norm in functional space $U$.
Characteristic feature for problems of the considered type is that the operator $A_p$ is compact operator (Tikhonov & Arsenin ,1979).
The value $\varepsilon$ is set a priori and characterizes desirable quality of mathematical modeling. The vector-function $u_\delta$ is obtained from experiment with the know error $\delta$:

$$\| u_T - u_\delta \|_U \le \delta_0, \tag{3}$$

where $u_T$ is an exact response of object on real EL.
It is obvious that in the case of performance of an inequality (2) operators $A_p$ and function $z$ are connected. It is easy to show that at the fixed operator $A_p$ in (2) exists infinite set of various among themselves functions z_which satisfies to an inequality (2) (Tikhonov & Arsenin, 1979). And, on the contrary, at the fixed function $z$ there are infinite many various

operators for which an inequality (2) is valid. Thus, there are no opportunities of a choice of good model of system (of process) separately from a choice of correct model of external load.

As a rule the check of inequality (2) is not executed in the practice of mathematical modelling, but its performance is meant. The error of the measuring equipment $\delta_0$ is contained in value $\varepsilon$ as obligatory component and therefore the inequality $\delta_0 \leq \varepsilon$ is always takes place. It occurs for the reason, that the accuracy of experimental measurements is higher as required accuracy of modelling as a rule. Frequently only qualitative coincidence of results of mathematical modelling with experiment satisfies.

At research of real dynamic systems the structure of the mathematical description, as a rule, is fixed. For example, at research of dynamics rolling mills (Menshikov, 1976, 1985), at the solution of a problem of unbalance diagnostics (Menshikov,2004) it is possible to use the models with the concentrated parameters. Proceeding from design features of real systems or devices, it is possible to determine parameters of the mathematical description (parameters of operators) precisely enough. However these parameters are believed to be given approximately. The error of definition of parameters depends on a way of reduction of dynamic systems to more simple systems, from a various sort of the conditions and assumptions, from the account of those or other factors (Menshikov, 1994). This error can be appreciated from above and it does not surpass 10 % as a rule.

Two approaches exist to problem of construction of couple MM and model of EL:

1. MM is given a priori with inexact parameters and then the model of EL is being determined for which the inequality (2) is valid;
2. Some model of EL is given a priori and then MM is being chosen for which the inequality (2) is satisfied.

For example for the operators which are given with an error it is necessary to construct models of external load under use of which the results of mathematical modeling will coincide with the certain accuracy with results of experiment. Such algorithms of construction of pair (mathematical description + model of external load) are not unique.

## 2. Statement of synthesis of external loads by identification method

Let's consider now opportunities of the first algorithm on an example of dynamic system $\Sigma$ with the concentrated parameters, the motion of which is described by ordinary differential equations of n-order. It is suggested that the records of all external loads $f_2(t), f_3(t),..., f_m(t)$ (except only one $f_1(t)$) and one state variable, for example $x_1(t)$, are obtained in experimental way during motion of system for some interval of time $t \in [0,T]$.

It is necessary to find the model $z(t)$ of external load $f_1(t)$ after the action of which the mathematical model of system $\Sigma$ (MM$\Sigma$) moves in a such way that the state variable $x_1(t)$ coincides with experimental record $\tilde{x}_1(t)$ of $x_1(t)$. The rest of external loads coincide with external loads $f_2(t), f_3(t),..., f_m(t)$ known from experiment. The problems of such a type were named the problems of external loads identification (Gelfandbein & Kolosov, 1972), (Ikeda et al., 1976).

The model $z(t)$ which was obtained by such a method depends on chosen MM$\Sigma$ and on goals of the use at mathematical modeling in future.

If the initial dynamic system does not satisfy the condition as have been specified above then this system can be reduced to system $\Sigma$ with the help of additional measurements (Menshikov, 2004).
Let us assume that the MM$\Sigma$ is linear and that the connection between unknown function $z(t)$ and functions $f_2(t), f_3(t),..., f_m(t)$, $x_1(t)$ has the form:

$$A_p z = B_p x, \tag{4}$$

where $A_p$ is linear integral operator ($A_p : Z \to U$) which depends continuously on vector parameters $p$ of mathematical model of system (MM$\Sigma$), $p = (p_1, p_2,..., p_N)^T, (\cdot)^T$ is the sign of transposition, $p \in R^N, R^N$ is the Euclidean vector space with norm $\| p \|^2 = (p, p)$; $B_p$ is linear bounded operator ($B_p : X \to U$) which depends continuously on vector parameters $p$; $x = (x_1(t), f_2(t),..., f_m(t))^T$; $z \in Z, x \in X$; $Z, X, U$ are Gilbert spaces. The functions $x_1(t), f_2(t),..., f_m(t)$ are given with known inaccuracy $\tilde{x} = (\tilde{x}_1(t), \tilde{f}_2(t),... .\tilde{f}_m(t))^T$ as these functions had been obtained from experimental measurements:

$$\| x(t) - \tilde{x}(t) \|_X \le \delta, \tag{5}$$

where $x(t)$ is the exact vector function of initial data, $\delta$ - given value.
Besides it is supposed that the vector of parameters $p$ given inexactly. So vector $p$ can have values in some closed domain $D$: $p \in D \subset R^N$. Two operators $A_p, B_p$ correspond to each vector from $D$. The set of possible operators $A_p$ has been denoted as class of operators $K_A$, the set of possible operators $B_p$ has been denoted as class of operators $K_B$. So we have $A_p \in K_A, B_p \in K_B$. The maximal deviations of operators $A_p$ from class $K_A$ and operators $B_p$ from class $K_B$ are equal:

$$\left\| A_{p_\alpha} - A_{p_\beta} \right\|_{Z \to U} \le h, \ \left\| B_{p_\eta} - B_{p_\gamma} \right\|_{X \to U} \le d.$$

Denote by $Q_{\delta,p}$ the set of the possible solutions of equation (4) with account of experimental measurements inaccuracy only:

$$Q_{\delta,p} = \{ z : z \in Z, A_p z \in U_{\delta,p}, p \in D \},$$

where $U_{\delta,p} = \{ u = B_p x : u \in U, \ x \in X_\delta, p \in D \}$, $X_\delta = \{ x : x \in X, \| x - \tilde{x} \|_X \le \delta \}$.
Any function $z$ from set $Q_{\delta,p}$ simulates of the motion of dynamic system MM$\Sigma$ with the inaccuracy of experimental measurements only.
The operator $A_p$ in equation (4) is a completely continuous operator for overwhelming majority of cases and so the set $Q_{\delta,p}$ is unbounded set in space $Z$ as a rule (ill-posed problem) (Tikhonov et al., 1990).

The regularization method for equations (4) was used for obtaining of stable solutions of denoted above problems (Tikhonov & Arsenin ,1979).

Let us consider the stabilizing functional $\Omega[z]$ which has been defined on set $Z_1$, where $Z_1$ is everywhere dense in $Z$ (Tikhonov & Arsenin ,1979). Consider now the extreme problem I:

$$\Omega[z_p] = \inf_{z \in Q_{\delta,p} \cap Z_1} \Omega[z], \; p \in D. \tag{6}$$

It was shown that under certain conditions the solution of the extreme problem I exists as unique and stable to small change of initial data $\tilde{x}_1(t), \delta$ (Tikhonov & Arsenin ,1979).

The function $z_p$ is named *the stable model of EL* after taking into account experimental measurements inaccuracy only. Such a model can be used for modeling of initial system motion with operators $A_p, B_p$ only.

## 3. Synthesis of external loads for class of mathematical descriptions

But according to the first approach it is necessary to take into account the inaccuracy of operators $A_p$, $B_p$. Let us now consider the problem of EL identification in this case.

The set of possible solutions of equation (4) $Q_{\delta,p}$ has to expand to the set $Q_{\delta,D}$ if we additionally take into account the inaccuracy of operators $A_p$, $B_p$:

$$Q_{\delta,h,d} = \{z : z \in Z, A_p \in K_A, B_p \in K_B, \left\| A_p z - B_p x_\delta \right\| \le \delta b_0 + d \left\| x_\delta \right\|_X + h \left\| z \right\|_Z \},$$

where $b_0 = \sup_{p \in D} \left\| B_p \right\|$.

Any function $z$ from set $Q_{\delta,h,d}$ simulates the motion of initial system with the inaccuracy of experimental measurements and inaccuracy of operators $A_p$, $B_p$.

The set $Q_{\delta,h,d}$ is unbounded for any $\delta > 0, h > 0, d > 0,\; p \in D \subset R^N$ (Tikhonov & Arsenin, 1979).

The regularization method for equations with inexact given operators was used for an obtaining stable solutions of denoted above problems (Tikhonov et al., 1990).

Consider now the extreme problem II:

$$\Omega[\tilde{z}] = \inf_{z \in Q_{\delta,h,d} \cap Z_1} \Omega[z]. \tag{7}$$

It was shown that under certain conditions the solution of the extreme problem II exists as unique and stable to small change of initial data $\tilde{x}_1(t), \delta, d, h, A_p, B_p$ (Tikhonov et al., 1990).

A problem of finding $\tilde{z} \in Q_{\delta,h,d}$ was named as *problem of synthesis of external load for a class of models* (Menshikov, 2002, 2004).

Let's consider the union of sets of the possible solutions $Q_{\delta,p}$ with fixed operators $A_p, B_p$:

$$Q_\delta^* = \bigcup_{p\in D} Q_{\delta,p} \quad (\cup \text{ is the sign of union}). \tag{8}$$

In some cases as *the solution* $z_{min}$ *of a problem of synthesis of external load for a class of models* we shall accept the stable element of set $Q_\delta^*$ instead the set $Q_{\delta,h,d}$ (extreme problem III):

$$\Omega[z_{\min}] = \inf_{z\in Q_\delta^*\cap Z_1} \Omega[z]. \tag{9}$$

This problem can being reduced to more simple extreme problem:

$$\Omega[z_{\min}] = \inf_{p\in D} \inf_{z\in Q_{\delta,p}} \Omega[z]. \tag{10}$$

The model of EL $z_{min}$ will give the results of mathematical modeling which coincide with given function $B_p\tilde{x}$ with inaccuracy $\delta b_0$.

T The statement of following problem of model EL construction with the help of identification is possible (extreme problem IV):

$$\Omega[z_{\max}] = \sup_{p\in D} \inf_{z\in Q_{\delta,p}} \Omega[z]. \tag{11}$$

The model of EL $z_{max}$ will give results of mathematical modeling with inaccuracy $\delta b_0$.

The function $z_{\max}$ gives the evaluation from above of all possible solutions of identification problem for all operators $A_p$,$B_p$ from classes $K_A$,$K_B$.

Then the stable model $z_{bel}$ which gives the evaluation from below of the selected response $B_p\tilde{x}$ of dynamic system for all possible operators $A_p$, $B_p$ can been defined as result of the solution of the following extreme problem V:

$$\left\| A_{b_{bel}} z_{bel} \right\|_U^2 = \inf_{A_b\in K_A, B_b\in K_B} \inf_{z_p} \left\| A_b z_p \right\|_U^2, \ b, p \in D, \tag{12}$$

where $z_p$ is the solution of extreme problem (6) on set $Q_{\delta,p}$.

The stable model $z_{ab}$ which gives the evaluation from above of the selected response $B_p\tilde{x}$ of dynamic system for all possible operators $A_p$, $B_p$ can been defined as result of the solution of the following extreme problem VI:

$$\left\| A_{b_{ab}} z_{ab} \right\|_U^2 = \sup_{A_b\in K_A, B_b\in K_B} \sup_{z_p} \left\| A_b z_p \right\|_U^2, \ b, p \in D. \tag{13}$$

In some cases it is necessary to synthesize model of external load by a method of identification which gives the best results of mathematical modeling for all possible mathematical descriptions of dynamic system motion. Actually such a problem is the solution of a problem of a choice of the second component (model of external load) for adequate mathematical modeling within of the first approach (Menshikov, 2008). Such kind of identification problems can find applications in different practical areas where the methods of mathematical modeling are used (Menshikov, 2004).

The stable model $z_{un}$ of external load which gives the best result of motion of dynamic system with guarantee as the solution of the following extreme problem VI is:

$$\left\| A_{p_{un}} z_{un} - \tilde{x} \right\|_U^2 = \inf_{p \in D} \sup_{c \in D} \left\| A_c z_p - B_c \tilde{x} \right\|_U^2, \quad p_{un} \in D, \tag{14}$$

where $z_p$ is the solution of extreme problem (6) on set $Q_{\delta,p}$ (Menshikov & Nakonechny, 2005).
Function $z_{un} \in Q^*_\delta$ exists and is stable to small change of initial data (function $\tilde{x}$ ), if the functional $\Omega[z]$ is stabilizing functional and the function $z_o$ is defined uniquely from (14).
The solution of extreme problem VII was named as *unitary mathematical model of external load.*
If the classes $K_A$ , $K_B$ consist of a limited number of operators $K_A = \{A_1, A_2, \dots A_N\} = \{A_i\}$, $K_B = \{B_1, B_2, \dots B_N\} = \{B_i\}$, $i = \overline{1, N}$ , then the algorithm of finding the best unitary model of external load $z_{un}$ has the form

$$\inf_{z \in Q_{D,\delta}} \sup_{p \in D} \left\| A_p z - B_p \tilde{x} \right\|_U = \left\| A_{p_{un}} z_{un} - B_{p_{un}} \tilde{x} \right\|_U =$$

$$= \min_j \max_i \left\| A_i z_j - B_i x_\delta \right\|_U , \tag{15}$$

where

$$Q_{D,\delta} = \{ z_j : \left\| A_i z_j - B_i \tilde{x} \right\|_U = \delta; j, i = 1,2,\dots,N \} .$$

The offered formulations of a problem of identification of external load can not be classified by one name as identification of external load. The additional explanations and additional assumptions are required in the solution of each particular problem. Most likely this set of problems can be united only by principle of synthesis of models - principle of use of experimental data. In other words it is possible to name all these problems as problems in which the principle of identification (comparison of results of calculations with experimental data) is used. The similar situation is present also in a rather conservative area such as identification of parameters (Menshikov, 2006).

## 4. Practical problems of identification of external load models

### 4.1 Synthesis of moment of technological resistance on rolling mill

One of the important characteristics of rolling process is the moment of technological resistance (MTR) arising at the result of plastic deformation of metal in the center of deformation. Size and character of change of this moment define loadings on the main mechanical line of the rolling mill. However the complexity of processes in the center of deformation does not allow construct the authentic mathematical model of MTR by usual methods. In most cases at research of dynamics of the main mechanical lines of rolling mills MTR is being created on basis of hypothesis and it is imitated as piecewise smooth linear function of time or corner of turn of the working barrels (Menshikov, 1985, 1994). The results of mathematical modeling of dynamics of the main mechanical lines of rolling mills with such model MTR are different among themselves (Menshikov, 1976).

In work the problem of construction of models of technological resistance on the rolling mill is considered on the basis of experimental measurements of the responses of the main mechanical system of the rolling mill under real EI (Menshikov, 1985, 1994). Such an approach allows to carry out in a future mathematical modeling of dynamics of the main mechanical lines of rolling mills with a high degree of reliability and on this basis to develop optimum technological modes.
The kinematics scheme of the main mechanical line of rolling mill was presented on Fig.1. (a).

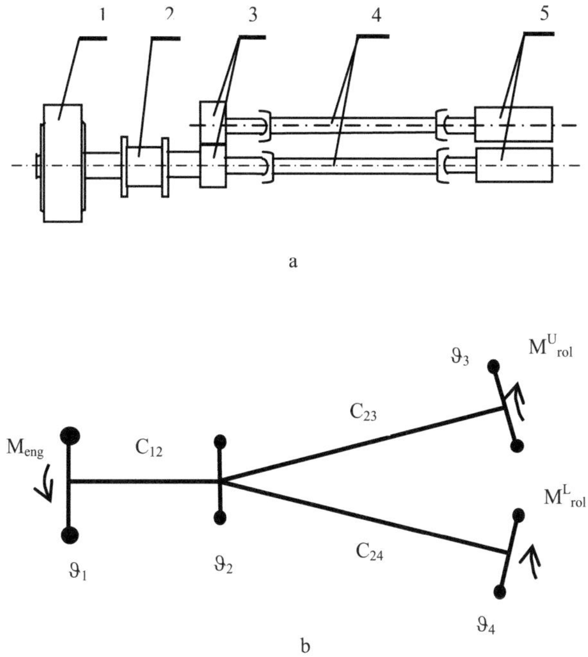


Fig.1. Kinematics scheme of the main mechanical line of rolling mill.

The four-mass model with weightless elastic connections is chosen as MM of dynamic system of the main mechanical line of the rolling mill (Menshikov, 1985, 1994):

$$\ddot{M}_{12} + \omega_{12}^2 M_{12} - \frac{c_{12}}{\vartheta_2} M_{23} - \frac{c_{12}}{\vartheta_2} M_{24} = \frac{c_{12}}{\vartheta_1} M_{eng}(t);$$

$$\ddot{M}_{23} + \omega_{23}^2 M_{23} - \frac{c_{23}}{\vartheta_2} M_{12} + \frac{c_{23}}{\vartheta_2} M_{24} = \frac{c_{23}}{\vartheta_3} M^U{}_{rol}(t); \quad (16)$$

$$\ddot{M}_{24} + \omega_{24}^2 M_{24} - \frac{c_{24}}{\vartheta_2} M_{12} + \frac{c_{24}}{\vartheta_4} M_{23} = \frac{c_{24}}{\vartheta_4} M^L{}_{rol}(t);$$

where $\omega^2{}_{ik} = \dfrac{c_{ik}(\vartheta_i + \vartheta_k)}{\vartheta_i \vartheta_k}$, $\vartheta_k$ are the moments of inertia of the concentrated weights, $c_{ik}$ are the rigidity of the appropriate elastic connection, $M^U{}_{rol}$, $M^L{}_{rol}$ are the moments of technological resistance applied to the upper and lower worker barrel, respectively, $M_{eng}(t)$ is the moment of the engine.

The problem of synthesis of model of EL can be formulated so: it is necessary to define such models of technological resistance on the part of metal which would cause in elastic connections of model fluctuations identical experimental (in points of measurements) taking into account of an error of measurements for chosen MM of the main mechanical line of rolling mill.

The information on the real motion of the main mechanical line of rolling mill is received by an experimental way (Menshikov, 1976, 1976a). Such information is understood as presence of functions $M_{12}(t)$, $M_{23}(t)$, $M_{24}(t)$.

The most typical case of rolling on a smooth working barrels was chosen for processing when the frustration of fluctuations are not observed and when a skid is absent (Menshikov, 1976, 1976a). The records of functions $M_{12}(t)$, $M_{23}(t)$, $M_{24}(t)$ by rolling process are shown on Fig.2.

Let's consider a problem of construction of models of EL to the upper working barrel. On the lower working barrel all calculations will be carried out similarly. From system (16) the equation concerning required model $M^{U}_{rol}$ can be received

$$\int_0^t \sin\omega_{23}(t-\tau)M^{U}_{rol}(\tau)d\tau = u_{\delta}(t) \text{ or } A_p z = u_{\delta}, \tag{17}$$

where $z = M^{U}_{rol}(\tau)$, $A_p$ is a linear integral operator.

The maximal deviation of the operators $A_p \in K_A$ from one another is defined by an error of parameters of mathematical model of the rolling mill. The error of definition of values of discrete weights is accepted as 8 %, the error of stiffness values - 5 %, error of values of damping factors - 30 %. The size of the maximal deviation of the operators $A_p \in K_A$ was defined by numerical methods and it is $h$ = 0.121, the size of the maximal deviation of the operators $B_p \in K_B$ was defined by numerical methods and it is $d$ = 0.11. An error initial data for a case $Z = U = C[0,T]$ is $\delta$ = 0.0665 MHм.

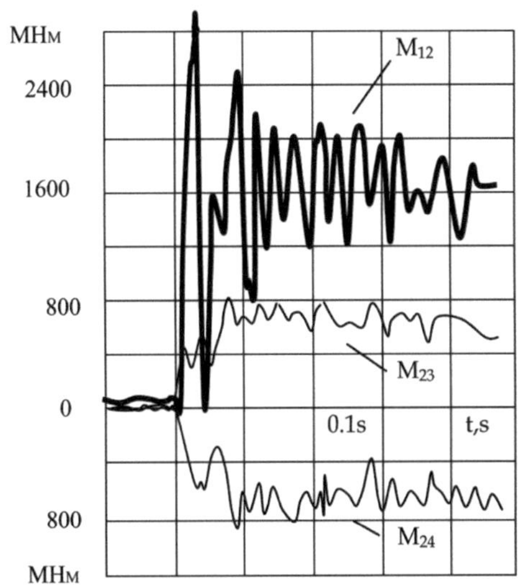


Fig. 2. The records of functions $M_{12}(t)$, $M_{23}(t)$, $M_{24}(t)$.

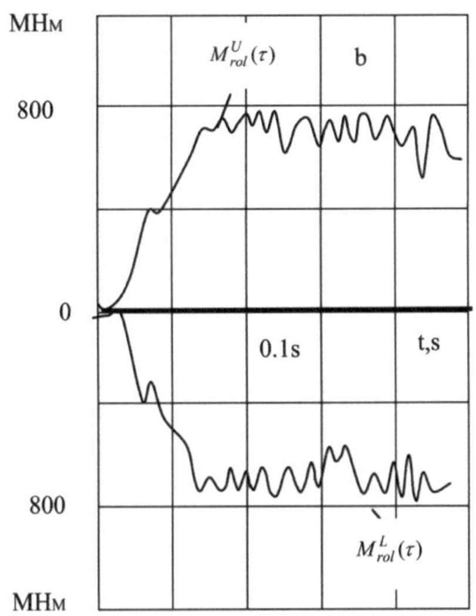


Fig. 3. The diagrams of extreme problem I solutions $M^{U}_{rol}(\tau)$, $M^{L}_{rol}(\tau)$.

We shall choose functional

$$\Omega[z]=\int_{0}^{T}(\dot{z}^{2}+z^{2})dt\,, \tag{18}$$

as the stabilizing functional (Tikhonov & Arsenin, 1979).
At first the problem of identification of the stable models of external load to working barrels was calculated with account of inaccuracy of experimental data only. The results of calculation are presented on Fig.3.

### 4.2 Unitary model of external load for rolling mill

It is evident that the results of identification will change under other parameters of mathematical descriptions. So the problem of external load identification was solved as extreme problem VI (unitary model of external load).
In Fig.4 the diagram of function $z_{un}$ for a typical case of rolling on an upper worker barrel is submitted.

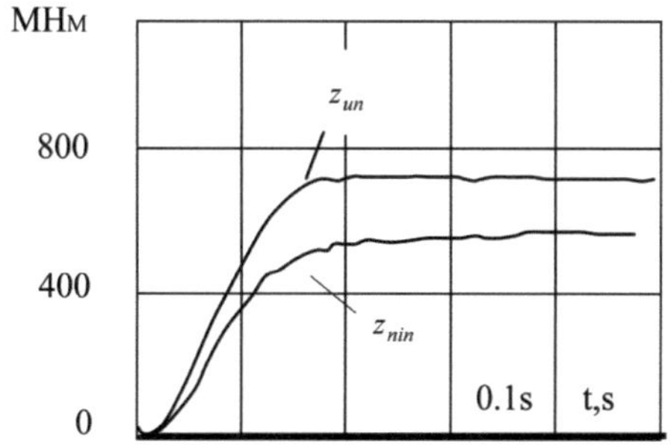


Fig. 4. The diagrams of change of models of the moment of technological resistance on the upper worker barrel of rolling mill.

The solution of extreme problem III is presented on Fig. 4. with the same conditions of initial data and inaccuracy of operators $A_p$, $B_p$.
The results of calculations show that the evaluation from above of accuracy of mathematical modeling with model $z_{un}$ for all $A_p \in K_A$ and for all $B_p \in K_B$ does not exceed 11% in the uniform metrics with error of MM parameters of the main mechanical line of rolling mill in average 10 % and errors of experimental measurements 7 % in the uniform metrics.
The calculations of model of EI $\tilde{z}$ for classes $K_A$, $K_B$ on set of the possible solutions $Q_{\delta h,d}$ was executed for comparison. The function which is the solution of a problem of synthesis in this case has the maximal deviation from zero equal 0.01 MHM. Such a model does not present interest for the purposes of mathematical modeling as it practically coincides with trivial a model.
In work (Menshikov, 1994), the comparative analysis of mathematical modeling with various known models of external load was executed. The model of external load $z_{un}$ corresponds to experimental observations in the greater degree.

## 5. Conclusion

In paper some problems of construction of external load models for dynamic systems with this case has the maximal deviation from zero equal 0.01 MHM. Such a model does not present interest for the purposes of mathematical modeling as it practically coincides with trivial a model.

In work (Menshikov, 1994), the comparative analysis of mathematical modeling with various formulations of such a problem is offered: the stable model for obtaining the best results of of mathematical modeling with guarantee, the stable model for obtaining evaluation of response from above, stable model for obtaining evaluation of response from below, stable model for mathematical modeling of the selected motion with the fixed model of dynamic system, the stable model for mathematical modeling of the selected motion of system for whole class of mathematical descriptions of system.
The offered approach to synthesis of models of external loads on dynamical system can find application in cases when the information about external impacts is absent or scarce and also for check of hypotheses on the basis of which were constructed the known models of external loads.

## 6. References

Gelfandbein, Ju. & Kolosov, L. (1972). *Retrospective identification of perturbations and interferences*, Science, Moscow.

Ikeda, S.,; Migamoto, S. & Sawaragi,Y. (1976). Regularization method for identification of distributed systems. *Proc. of IY a Symposium IFAC: Identification and evaluation of parameters of systems*, pp. 153-162, Tbilisi, USSR, v.3, 1976, Preprint. Moscow.

Menshikov, Yu. (1976). Identification of moment of technological resistance on rolling. *J. of Differential equations and their applications in Physics.* Dnepropetrovsk University, Dnepropetrovsk, Ukraine, n.1,1976, pp.22-28.

Menshikov, Yu. (1976a). About influence of external loading to dynamics of main line of rolling mill. *J. of Differential equations and their applications in Physics.* Dnepropetrovsk University, Dnepropetrovsk, Ukraine, n.1, 1976, pp. 29-33.

Menshikov, Yu. (1985). The synthesis of external impact for the class of models of mechanical objects. *J. of Differential equations and their applications in Physics.* Dnepropetrovsk University, Dnepropetrovsk, Ukraine, pp. 86-91.

Menshikov, Yu. (1994). The Models of External Actions for Mathematical Simulation. *System Analysis and Mathematical Simulation (SAMS)*, New-York, v.14, n.2, 1994, pp.139-147.

Menshikov, Yu. (2002). Identification of external impacts models. *Bulletin of Kherson State Techn. Univ.*, Cherson, Ukraine, 2(15), 2002, pp.326-329.

Menshikov, Yu. (2004). Identification of external impacts under minimum of a priori information: statement, classification and interpretation. *Bulletin of KNU, Mathematics*, Kiev, Ukraine, n. 2, pp.310-315.

Menshikov, Yu.; Polyakov N. (2004). The new statement of problem of unbalance identification, *Proceedings of ICTAM*, 4p., August, 2004, Warsaw, Poland.

Menshikov, Yu.; Nakonechny, A. (2005). Constraction of the Model of an External Action on Controlled Objects. *Journal of Automation and Information Sciences* , v.37, is. 7, 2005, pp. 20-29.

Menshikov, Yu. (2006). Parameters Identification in Minimax Statement. *Journal of Automation and Information Sciences*, v.38, is.11, 2006, pp.14-21.

Menshikov, Yu. (2008). About Adequate of Results of Mathematical Modeling, *Proceedings of Conf. Simulation-2008*, v.1, Kiev, May 2008, Kiev,Ukraine, pp.119-124.

Tikhonov, A. & Arsenin, V. (1979). *Methods of solution of incorrectly problems*, Science, Moscow.

Tikhonov, A.; Goncharsky, A.; .Stepanov, A. & Yagola, A. (1990). *Numerical methods for the solution of ill-posed problems*, Science, Moscow.

# 8

# Environment Modelling with an Autonomous Mobile Robot for Cultural Heritage Preservation and Remote Access

Grazia Cicirelli and Annalisa Milella
*Institute of Intelligent Systems for Automation (ISSIA)*
*National Research Council (CNR)*
*Italy*

## 1. Introduction

As awareness of cultural heritage raised, much effort was devoted, in the last decade, to improve accessibility and preservation of cultural assets. At present, several methods are available that generally make use of laser scanners and cameras to construct 3D photo-realistic models of different-sized items, ranging from small objects, like statues, up to large buildings and archaeological sites. These methods provide effective technological solutions for cultural heritage preservation, while guaranteeing, at the same time, their accessibility to as much people as possible. Nevertheless, the generation of models may turn into a time-consuming procedure if data acquisition and processing is done by hand, since, in order for models to be sufficiently accurate, an extremely painstaking work is required.

Contributions to implement automated model building techniques and remote access systems for cultural heritage applications may derive from experience in the mobile robotics field. Recently, several research projects have attempted to develop mobile robotic agents in museums (Burgard et al., 1999; Thrun et al., 2000; Trahanias et al., 2005), with different tasks, such as to supply remote access to distant users, accommodate and guide people in the museum, survey those areas where access is not permitted. Equipped with sensors, like cameras and laser rangefinders, mobile platforms provide a variety of viewpoints and may supply the user with dedicated tours of the exhibition and personalized tele-presence, which result in greater interaction capabilities than fixed or even remotely controllable cameras (Trahanias et al., 2005). Generally, in order for a mobile robot to perform its tasks, the knowledge of a map of the environment is needed. Hence, a number of methods for efficient environment modeling, based on information from onboard robot sensors, have been also developed. Yet, relatively little work has been done to extend these techniques to cultural heritage applications, such as modeling of historical and archaeological sites.

In this chapter, we describe our research concerning the development of methods for environment exploration and modelling by a multisensor mobile platform, in the context of cultural heritage access and preservation. Our goal is to have a system able to navigate in the environment and acquire sensorial data, in order to either construct global or local models of the site, or send information to a remote console.

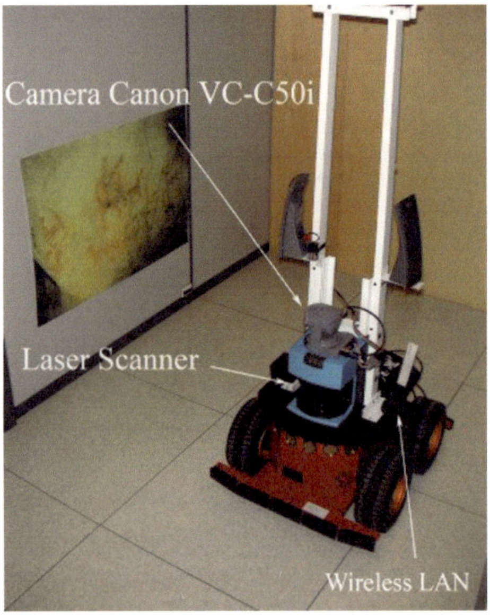


Fig. 1. The robotic mobile platform in the configuration used in the polygonal environment.

Usually, when exploring an unknown environment for the first time, a mobile robot is remotely controlled by a joystick or other tele-operation devices. Recently, a novel scenario is receiving considerable attention, which relates to the possibility of teaching the robot its environment by human interaction. This concept, introduced in (Topp & Christensen, 2005), is known as Human Augmented Mapping (HAM). Pursuing this trend, we propose a novel laser-based leg detection and tracking algorithm (Milella et al., 2007) that enables a mobile platform to follow a human user in a tour of the environment, in order to explore the surroundings, acquire sensorial data for map building, and learn particular regions or locations specified by the user.

We present two case studies. The first one deals with the problem of modelling a polygonal environment, such as a museum. The second case study is concerned with the use of a mobile robot for exploration and mapping of a pre-historical underground cave in Southern Italy, named "Grotta dei Cervi", rich in ancient wall paintings of historical and artistic relevance.

The mobile robot employed to carry out this research is shown in Fig.1. It consists of a Pioneer-P3AT by ActivMedia Robotics, equipped with a SICK LMS 200 laser range finder, sixteen forward and rear sonar sensors, encoders, gyroscope, and a monocular pan-tilt-zoom camera. The laser range finder is mounted at a height of about 30*cm* from the floor and is able to sense objects at a distance of up to 80*m* with a resolution of 0.5°. Sonar sensors can detect obstacles up to 7*m* away. Its four tractor wheels can scale a 45° gradient and sills of 9*cm*. The robot case contains four motors, a local processor, and the batteries. For experimentation in the cave, a 1*m* high aluminium support was added to the platform to carry the illumination system and the camera (see Fig. 2). Such a configuration allowed us to acquire the wall paintings from an appropriate perspective, since they are mainly located approximately between 1*m* and 2*m* above the ground. ARIA C++ libraries by ActivMedia Robotics were used for communication between sensors and the robot controller.

Fig. 2. The mobile robotic platform in the configuration used in the cave.

The results of the experimental sessions realized in both case studies show that the proposed methods are feasible and accurate. They allow us to produce detailed 2D and 3D representations that can be usefully employed to support the study of relevant historical treasures, guaranteeing, at the same time, their safety.
The remainder of this chapter is organized as follows. After a review of the literature related to this work in Section 2, the people following method for Human Augmented Mapping (HAM) is described in Section 3. Section 4 illustrates our approach to polygonal environment modelling. Section 5 describes the study conducted in the pre-historical cave. Finally, Section 6 draws some conclusions.

## 2. Related works

Several works in literature have shown mobile robots to be useful in environment modelling tasks, since, equipped with sensors, such as cameras and laser rangefinders, they can acquire and process sensorial data while navigating in the environment, with minimum human intervention. In (Biber et al., 2004), a 3D modeling method using data from a laser range finder and an omnidirectional camera mounted on a mobile robot is proposed. The method consists of manual, semi-automatic, and automatic parts. Data collection and sensor calibration is carried out manually by teleoperating the robot; wall extraction is done semi-automatically with a user interface; the rest of the processing is fully automatic. In (Nevado et al., 2004), 3D models of the inside of buildings are obtained by using points measured by a laser scanner on-board a mobile robot. The most likely orientation of the surface normal is first calculated at every point by considering also the neighbouring regions to avoid measurement noise. Similar planes are merged and each point is projected onto the planes they belong to. Finally, the contours of the points in the corresponding planes are obtained by a triangulation procedure producing a simple representation of the environment. In this case, the lack of video information limits the final 3D model to a topological representation of the environment without texture. In (Leiva et al., 2001), a 3D model of the environment is built combining sonar and video information. First, sonar sensors and odometers are used to estimate the distance of the robot from the objects in the environment. Odometric errors are corrected when image data is suitable. Then, a 2D probabilistic map of the environment is divided into segments which become planes in 3D. Finally, the texture from snapshots is assigned to each plane. A rough 3D representation of the environment is obtained, as the aim of the authors is to construct a basic 3D model by an easy and fast method using cheap sensors. Simultaneous Localization and Mapping (SLAM) approaches for concurrent robot localization and modelling of unknown environments can also be found in (Gutmann & Konolige, 1999; Se et al., 2002; Grisetti & Iocchi, 2004; Thrun et al., 2004; Ip & Rad, 2004; Stachniss et al., 2005; Wolf & Sukhatme, 2005).

Despite this high number of model building methods using mobile robots, a few authors have explicitly suggested the use of mobile robotics techniques for applications in the domain of cultural heritage, and realized field tests. Examples can be found in (Allen et al., 2001; Allen et al., 2003; Hirzinger et al., 2005). Specifically, in (Hirzinger et al., 2005), methods for 3D modelling in robotic environments are applied to digitization of cultural heritage from small to large scale objects. In (Allen et al., 2001; Allen et al., 2003), instead, the use of a mobile robot is proposed to build models of urban environments and historic sites.

In our work, we employ a multisensor mobile platform for data acquisition and processing in the context of cultural heritage, and present the results of two practical implementations of the approach, one in a typical indoor polygonal environment and the other in a pre-historical cave. In both cases, the constructed environment model is employed not only for in site navigation of the robot, but also for remote access to cultural assets. For the polygonal environment, our work is related to (Biber et al., 2004); however, we employ a monocular camera and processing is completely automatic. Furthermore, we suggest the use of a Human Augmented Mapping (HAM) (Topp & Christensen, 2005) approach to perform laser and video data acquisition, in place of usual robot teleoperation. First, an accurate 2D map of the environment is generated, based on a SLAM algorithm using laser data (Gutmann & Schlegel, 1996). Starting from this map, the wireframe model of the environment is constructed. Then, images are processed to extract the texture to be added to the wireframe in order to obtain the complete 3D model. For exploration and mapping of the cave, instead, we present an integration of two different algorithms. Specifically, the robot constructs the 2D map of the environment using laser data, and then builds the 3D model of some zones of particular interest using computer vision techniques. The combination of these two approaches makes it possible to obtain a complete knowledge of the environment in an automatic way. The proposed solution allows the access to the cave without damaging it, thus providing an effective system to monitor and preserve its relevant treasures.

## 3. People-following for Human Augmented Mapping (HAM)

There are several works in literature that concentrate on either people-tracking or following for interaction. They rely on the use of laser range finders or vision or both. (Kleinehagenbrock et al., 2002) integrate vision and laser range data to track a human user, based on a multi-modal anchoring technique. The legs of the user are extracted from laser range data, while his face is from camera images. In (Fod et al., 2002) a laser-based method for real-time tracking of multiple objects is presented, as a step towards the wider objective of identifying people and their activities. Range measurements are grouped into entities for an abstract representation of objects. A Kalman Filter is, then, associated to each object to address occlusion and sensor noise problems. (Feyrer & Zell, 2000) present a system that detects and pursues people by using both vision and laser data. Color information is used to extract faces in images, whereas convex intervals are extracted in laser scans to detect human legs. In (Pineau et al., 2003) efficient particle filter techniques are used to detect and track people.

In our work, only laser data are employed for detecting people, based on typical human leg shape and motion characteristics. Due to safety reasons, laser range sensors have to be attached near the bottom of the mobile robot; hence, laser information is merely available in a horizontal plane at leg height. In this case, legs constitute, therefore, the only part of the human body that can be used for laser-based people-tracking.

The objective of the proposed approach is to develop a Human Augmented Mapping (HAM) system, that is, to implement a human-robot interaction approach to mapping. The people detection and following method consists of two main modules:

- the **Leg Detection and Tracking** (LDT) module: this module allows the robot to detect and track people using range data based on typical shape and motion characteristics of human legs;
- the **People-Following** (PF) module: this module enables the mobile platform to navigate safely in a real indoor environment while following a human user. During the tour the robot can acquire data for environment mapping tasks.

Details about both modules are provided in the remainder of this section, along with the results of some tests performed in a real context.

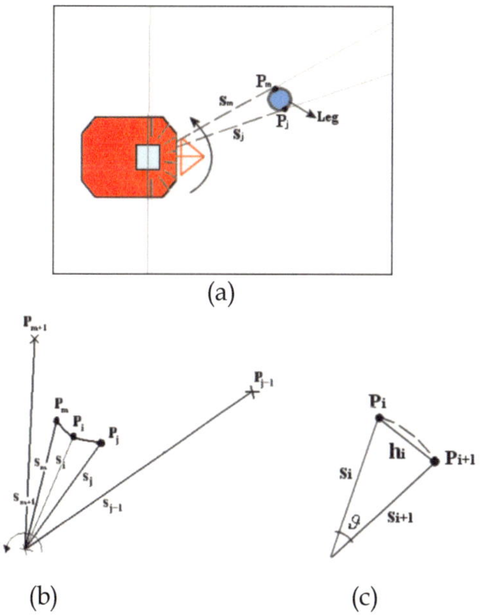


Fig. 3. (a) The robot detecting a leg-shaped object; (b) geometrical representation of a scan interval; (c) geometrical representation of two subsequent scan points.

### 3.1 Leg-detection and tracking

The Leg Detection and Tracking (LDT) method allows to detect and track legs, based on typical human leg shape and motion characteristics. The algorithm starts by acquiring a raw scan covering a 180° field of view. Laser data are analyzed to look for scan intervals with significant differences in depth at their edges (Feyrer & Zell, 2000). Specifically, let us denote

$$S = [s_1, s_2, ..., s_k, ..., s_n] \tag{1}$$

a raw scan, being $s_k$, for $k = 1, 2, \ldots n$, the laser readings ordered according to the rotation sense of the laser beam (counterclockwise sense in Fig. 3(a)). A scan interval, delimited by scan points $P_j$ and $P_m$ (see Fig. 3(b)), i.e.

$$S_{jm} = [s_j, s_i, s_{i+1}, s_{i+2}, ..., s_m] \tag{2}$$

with $S_{jm} \in S$, is selected, if it satisfies the conditions

$$s_{j-1} - s_j > \tau_j \tag{3}$$

$$|s_i - s_{i+1}| \leq \tau_i \quad \text{for } j < i < m \tag{4}$$

$$s_{m+1} - s_m > \tau_m \tag{5}$$

The thresholds $\tau_i$, for $j \leq i \leq m$, are dynamically computed at each step based on the following considerations. Let us denote $P_i$ and $P_{i+1}$ two subsequent point scan readings (see Fig. 3(c)). Since the rotational resolution $\vartheta$ of the laser beam is small (approximately 0.5°), it yields

$$h_i \cong s_i \vartheta \tag{6}$$

and

$$\vartheta \cong \sin \vartheta \tag{7}$$

so that we have

$$h_i \cong s_i \sin \vartheta \tag{8}$$

If $P_i$ and $P_{i+1}$ belong to a continuous region, then a small difference between the corresponding readings will be observed, which is

$$|s_i - s_{i+1}| << h_i \tag{9}$$

Conversely, a discontinuity between $P_i$ and $P_{i+1}$ will generate the condition

$$|s_i - s_{i+1}| >> h_i \tag{10}$$

The value of $h_i$ can be, therefore, usefully employed to define a threshold $\tau_i$, establishing whether $P_i$ and $P_{i+1}$ lie on a continuous surface or belong to different objects.
Specifically, the threshold $\tau_i$ can be expressed as

$$\tau_i = \beta \cdot h_i \tag{11}$$

where $\beta > 1$ is an empirically determined coefficient that takes into account laser noise effects.

Once a set of scan intervals has been selected, a criterion to differentiate between human legs and other similar objects, such as legs of chairs and tables and protruding door frames, must be defined. To achieve this aim, first, the width of each pattern is calculated as the Euclidean distance between its end-points and is compared with the typical diameter of a human leg (0.1*m* to 0.25*m*). Then, a Region of Interest (ROI) is fixed in the vicinity of each candidate pattern. A leg-shaped region detected within each ROI at the next scan reading is classified as a human leg if the displacement of the pattern relative to its previous position has occurred with a velocity compatible to a typical human leg velocity (0.2*m/s* to 1*m/s*). Note that if the robot is moving and thus so is the scanner, the effect of ego-motion must first be accounted for. This can be done employing the information provided by the on-board odometers or by the laser scanner.

### 3.2 People-following

The people following algorithm consists of the following steps:

1. detect human legs using the LDT module;
2. choose the closest moving person within a certain distance and angular position relative to the robot;
3. keep track and follow the target person until he/she stops or disappears from the scene.

A control loop is employed, which sets speed and turn rate of the robot, based on the distance from the person and from other objects present in the environment. An obstacle avoidance routine that uses sonar information is also implemented.

As will be shown below, the People Following (PF) module can be effectively employed in the context of Human Augmented Mapping (HAM).

### 3.3 Some tests

In order to test both the performance of the people detection and tracking algorithm and the effectiveness of the people-following method, some tests were performed in our institute. Specifically, three different test scenarios were analysed: 1) robot still, one person present; 2) robot still, two people present; 3) robot following one person for Human Augmented Mapping (HAM). Each experimental setup is discussed in the rest of this subsection.

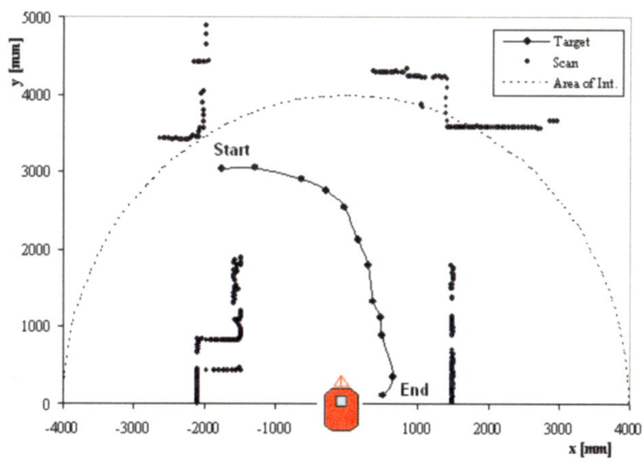


Fig. 4. The trajectory of a person moving in the area surveyed by the robot.

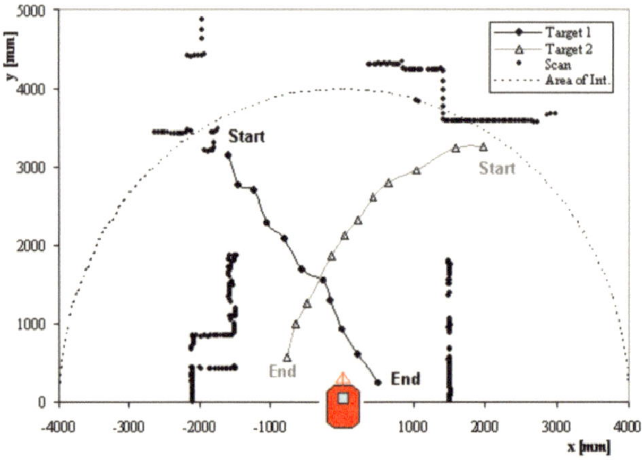


Fig. 5. The trajectories of two people crossing the area surveyed by the robot.

1. **Robot still, one person** - In this case, the robot is not moving. Only one person is present and crosses the field of interest at varying speed. Fig. 4 shows the trajectory of the target within the inspected area (i.e. the semi circumference marked by the dashed line) obtained by the tracking system during one experiment. Black dots represent laser scan readings. The target was classified at all times as a moving person.
2. **Robot still, two people** - Two people cross the scene surveyed by the robot which is not moving. Assuming that the motion direction of each person does not vary significantly, the system is able to keep track of the two trajectories separately, as is shown in Fig. 5.

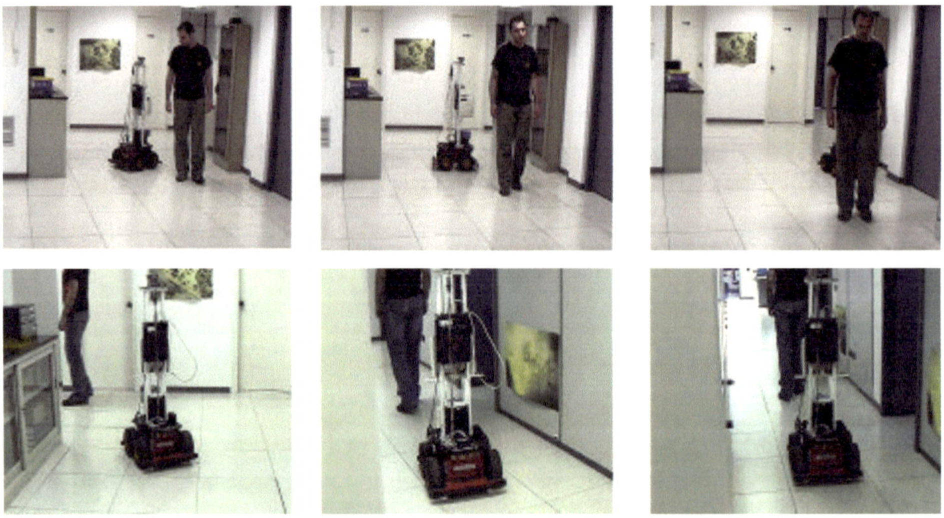

Fig. 6. Two image sequences taken during the pursuit of a person.

3. **Robot following one person for Human Augmented Mapping (HAM)** – In this case, the robot follows the user in a tour of the environment, maintaining a safety distance from him. The user is identified as the closest person in a range of $1m$. Fig. 6 shows two short image sequences taken during the experiments. While following the user, the

robot is able to acquire laser data that allows to concurrently reconstruct its trajectory and build a 2D map of the environment. The simultaneous localization and mapping process is shown in Fig. 7 for two different tests. In this figure, the continuous lines indicate the robot path, while small arcs represent the user's legs and black points the reconstructed map.

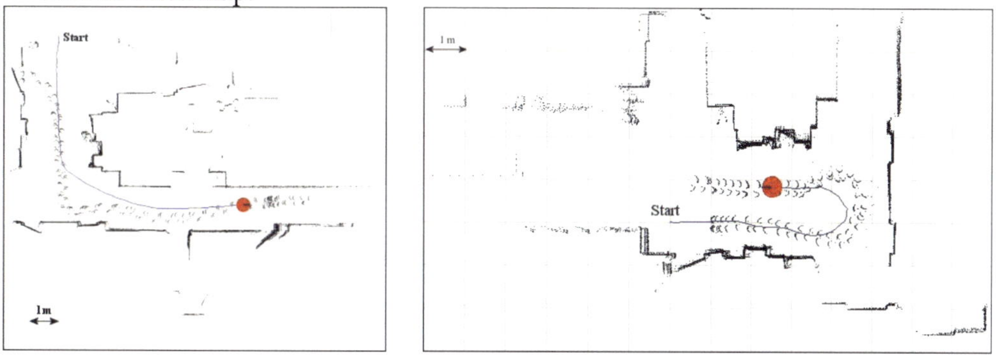


Fig. 7. Two people following experiments for Human Augmented Mapping (HAM).

In the next section, details about the 2D mapping approach will be provided. It will be also shown how, assuming the environment to be polygonal and adding visual information, the 2D map can be used to recover a full 3D model of the environment.

## 4. Modelling a polygonal environment

In this section, we illustrate our approach to build the 3D model of a polygonal environment, such as a museum, based on laser and video data acquired by a mobile platform. Experiments were carried out in a corridor of our institute. This does not lose generality, since an office environment is very similar to a museum one. First, data acquisition was performed taking the robot on a tour of the environment, and a 2D map was constructed using a laser-based Simultaneous Localization and Mapping (SLAM) algorithm. Then, a 3D wireframe model was generated, adding vertical planes to the contour defined by the planar map. Finally, the 3D model was completed, applying the texture recovered from the acquired images to the wireframe model. Note that the camera was mounted on board the robot above the laser range finder, and its optical axis was oriented 90° to the right with respect to the forward direction of the robot in order to acquire images of the lateral walls.

### 4.1 2D Map of a polygonal environment

The construction of a map of the environment is a basic step both for robot navigation and accurate knowledge of the environment. The robot can use the map to localize itself as well as to recognize places already explored. This turns the problem of building a map into a problem of on-line Simultaneous Localization and Mapping (SLAM).

In this work, the Combined Scan Matcher (CSM) method proposed in (Gutmann & Schlegel, 1996) is applied to construct a laser-based 2D map and simultaneously estimate the robot trajectory. This approach integrates the IDC algorithm by Lu and Milios (Lu & Milios, 1997) with the method proposed by Cox (Cox, 1991). Fig. 8 shows the point map, with overlaid the robot trajectory, obtained applying the CSM algorithm followed by a statistical filtering for noise removal. This map is composed of 112 frames scanned by the laser and covering an area of approximately $20 \times 2m^2$. The next step is to fit the points with line segments. The result of the line fitting procedure is a map to which a gap removing algorithm has been applied. The final map is shown in Fig. 9. Each couple of consecutive line segments are intersected to obtain the corners in the map. These points will be useful for building the 3D model of the environment.

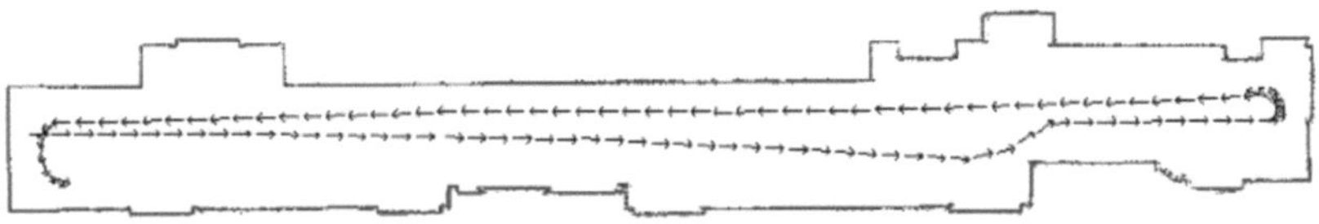

Fig. 8. Planar point map. The path of the vehicle is also drawn.

Fig. 9. Final planar map characterized by line segments.

### 4.2 Building the 3D Model of a polygonal environment

Once the planar map of the polygonal environment has been obtained, a 3D model can be constructed. The first step is to build the wireframe of the 3D model. The wireframe is generated by adding one vertical plane to each line segment in the 2D map. This operation is possible since the explored environment is polygonal. Each segment in the 2D map refers to the contour of walls, cupboards, doors, and other objects with a polygonal shape. Fig. 10 shows the resulting wireframe.

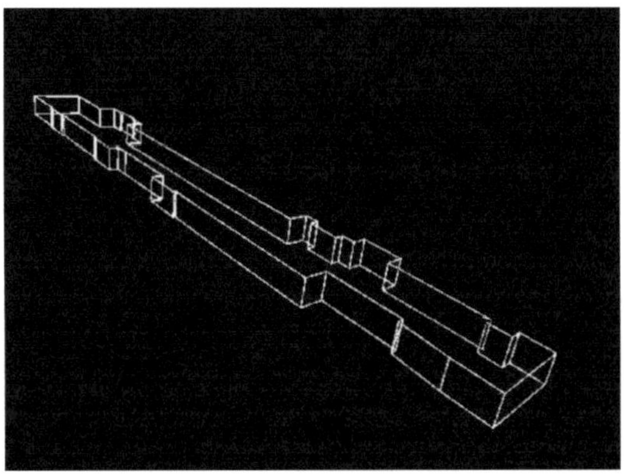

Fig. 10. Wireframe model.

The next step to obtain the complete 3D model is the addition of texture to improve the visual appearance of the model. The texture is obtained by using the images acquired by the camera, on-board the vehicle, during the acquisition phase. Note that images were taken at a fixed horizontal translation distance from each other and the positions of the vehicle, from which the images were acquired, are known. The process for adding the texture can be divided into the following fundamental steps: 1) generation of the entire image covering the whole contour of the environment, i.e. image registration or mosaicking; 2) correct positioning of the registered image to the wireframe.

1. **Image registration**. The fundamental step in image mosaicking is to find correspondences between different views, and estimate the homographies between the reference image and all other images. Each pair of consecutive images is processed as follows: feature extraction; feature matching; selection of correct matches; final image generation. Specifically, the Harris corner detector is, first, applied to extract corners. Fig. 11 shows the extracted corners on two adjacent images. A simple cross-correlation algorithm is applied to find matches. False matches are then removed using RANSAC (Hartley & Zisserman, 2003). Fig. 12 shows the correct correspondences after applying the RANSAC algorithm.

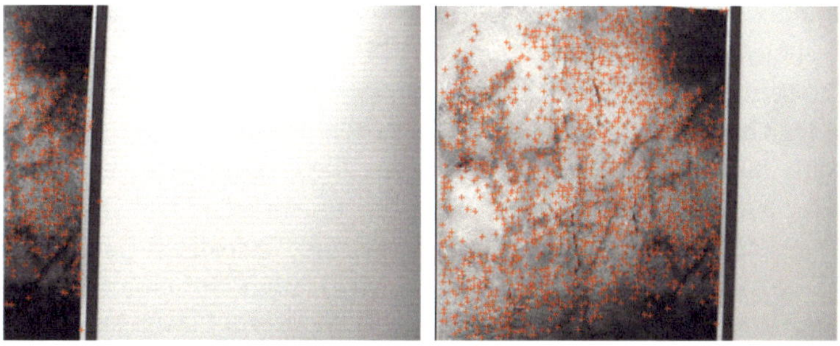

Fig. 11. Corner points extracted by the Harris operator on two adjacent images.

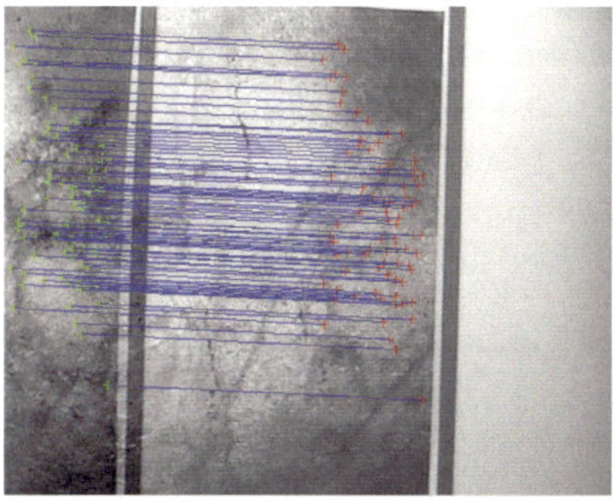

Fig. 12. Correct correspondences of the extracted corners after the application of RANSAC algorithm.

Successively, homographies are estimated by solving overdetermined systems (number of points > 4) using the linear-least squares method. By using the homographies, it is possible to estimate the horizontal translation between two consecutive images and then the images can be aligned correctly. The resulting image, of $38981 \times 480 pixel^2$, is obtained aligning 123 images of $640 \times 480 pixel^2$. Fig. 13 shows a portion of the registered image.

The described procedure works properly when images present sufficient texture. In our experiment, regions with pictures on the walls have enough texture. Problems arise, instead, in homogenous areas corresponding to walls, doors, etc. In these cases, feature extraction is difficult and, therefore, homographies cannot be estimated automatically. To solve this problem, the horizontal translations estimated between textured images are used. They are correlated to the known translation distances, performed by the robot during the acquisition phase and estimated by using the odometers. The median of the translations, estimated by using RANSAC and homographies, among textured images is then used to align those images without texture.

Fig. 13. Portion of the mosaic image.

2. **Adding texture to the wireframe.** The final step in building the 3D model is to add the texture, defined by the mosaic image, to the wireframe model. To achieve this aim, the idea is to match correctly the corners of the wireframe model to those in the mosaic image. The corners in the wireframe model are easily detected since they are known on the 2D map. The detection of the corners in the mosaic image needs some additional elaborations. For the sake of simplicity, processing is done on the single images forming the mosaic. First, the Sobel operator is applied to extract edges. Then, the vertical lines in the images are determined by using the Hough Transform (Hough, 1962). Vertical lines can be relative to door edges, picture edges, cupboard edges and so on. Only the ones relative to the corners of the environment are useful for the correct mapping of the texture on the wireframe. The knowledge of the robot positions corrected by using the scan-matching algorithm and the knowledge of the field of view of the camera allow us to find which corner falls in which image. In this way, it is possible to connect each corner to each image acquired during the acquisition phase. The result of the described procedure is the 3D model complete with texture shown in Fig. 14. For a more detailed illustration, Fig. 15 shows some portions of the same model from different points of view. The Virtual Reality Modelling Language (VRML) was used to obtain the model.

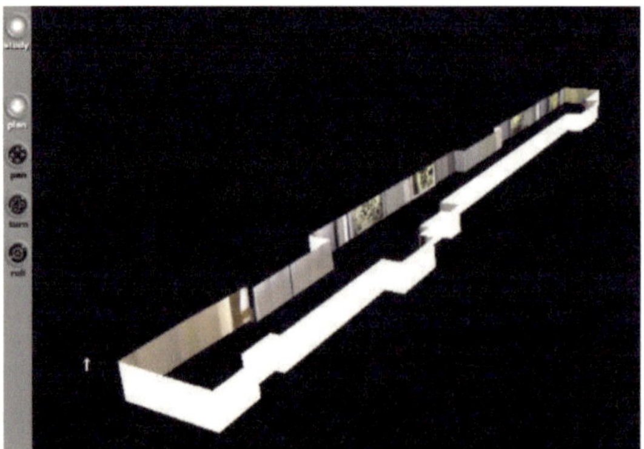

Fig. 14. Textured 3D model.

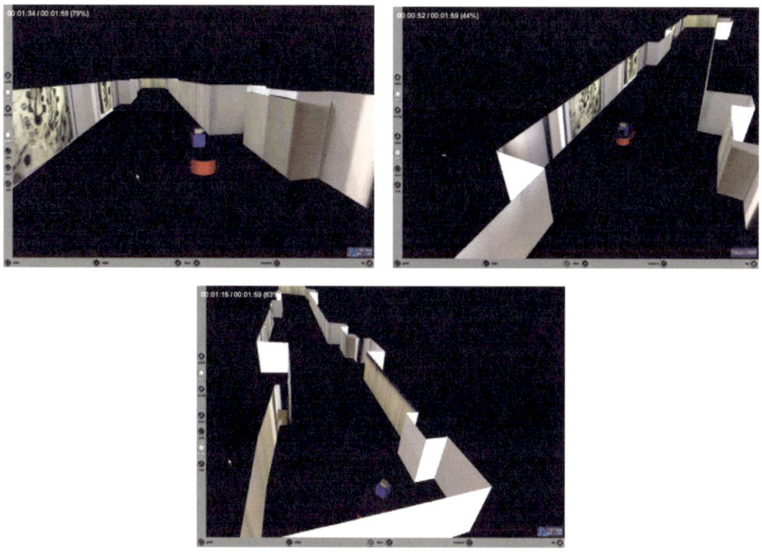

Fig. 15. Portions of the 3D model.

## 5. Reconstruction of a pre-historical cave

In the south of Italy, along the Adriatic coast, a cave, named "Grotta dei Cervi", holds a pre-historical treasure remarkable for its complexity, and artistic and historical relevance. The cave has on its walls a huge collection of paintings of hunting scenes, stags, men, and small animal groups, realized with red ochre and bat guano, dated to the Middle Neolithic period (see Fig. 16). The access to the cave is restricted to a few authorized people. Care must be taken to guarantee their safety and to prevent polluting elements from being introduced in this particular and valuable environment. The application of a technological solution seems to be the best way to allow remote access to the archaeological site, thus satisfying the need for cave preservation and safety.

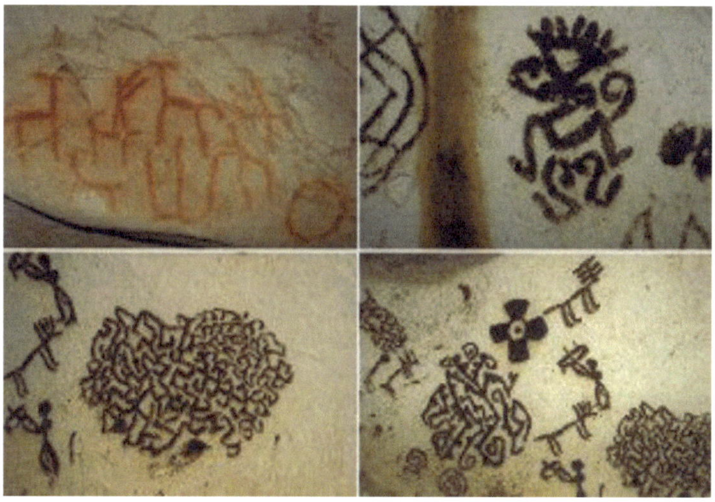

Fig. 16. Some paintings present on the walls of the cave.

In this section, we describe the study conducted in the cave, using a mobile robot able to navigate throughout the cave and acquire useful data by means of its on-board sensors. This solution reduces the risk of damaging the cave, as it does not require the installation of invasive infrastructures. The only hand-drawn map of the cave available up to now is illustrated in Fig. 17. The cave is formed by a series of narrow and twisting corridors. The area inspected by the robot is the corridor highlighted in the figure. First, a planar map of the corridor was constructed; then, the 3D model of particularly interesting areas, rich in paintings, was generated. Details for each phase are provided in the rest of this section.

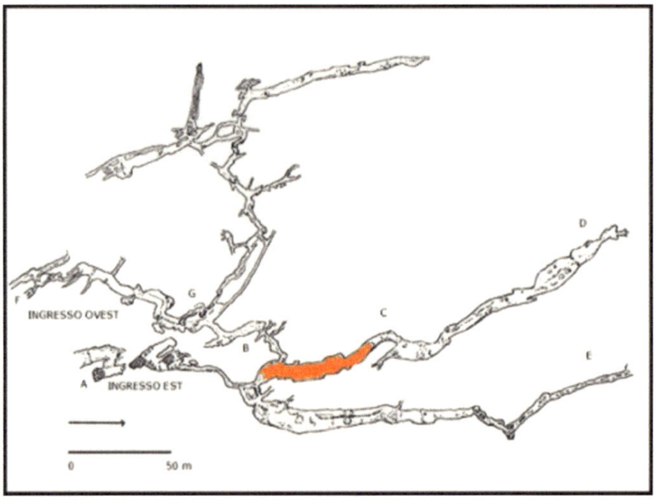


Fig. 17. The handmade map of the cave "Grotta dei Cervi". The highlighted area represents the corridor explored by the mobile robot.

### 5.1 2D map of the cave

Generally, the high accuracy of laser data allows to build accurate planar maps especially when the robot moves in a plane. The cave environment, instead, presents a rough terrain characterized by depressions and bumps (see Fig. 2). In this case, laser data, that supply planar information, must be integrated with data from an inclinometer in order to obtain accurate information about the scanned environment.

Fig. 18(a) shows the robot path and laser data during the scanning procedure. This map is composed of 189 frames scanned by the laser and covering an area of approximately 15x40$m^2$. The CSM algorithm was applied to reconstruct both the robot trajectory and the map of the corridor. Fig. 18(b) shows the point map obtained after the application of the CSM algorithm and noise removal. Four zones are also indicated. They are of particular interest for the presence of pre-historical paintings. In order to detect these zones, four artificial, non-invasive landmarks, distinguishable on the map by the laser, have been placed near those areas. Using these landmarks the robot can plan the path to these regions enabling the acquisition of images from the onboard camera of the pre-historical paintings, and the subsequent construction of a 3D model of the observed area.

The robot positions estimated on the map after the CSM application were compared to those provided by the odometers. The estimated errors on the $x$, $y$, and $\theta$ components of the robot pose are plotted in Fig. 19(a), (b), and (c) respectively. The errors on the $x$, $y$ coordinates are

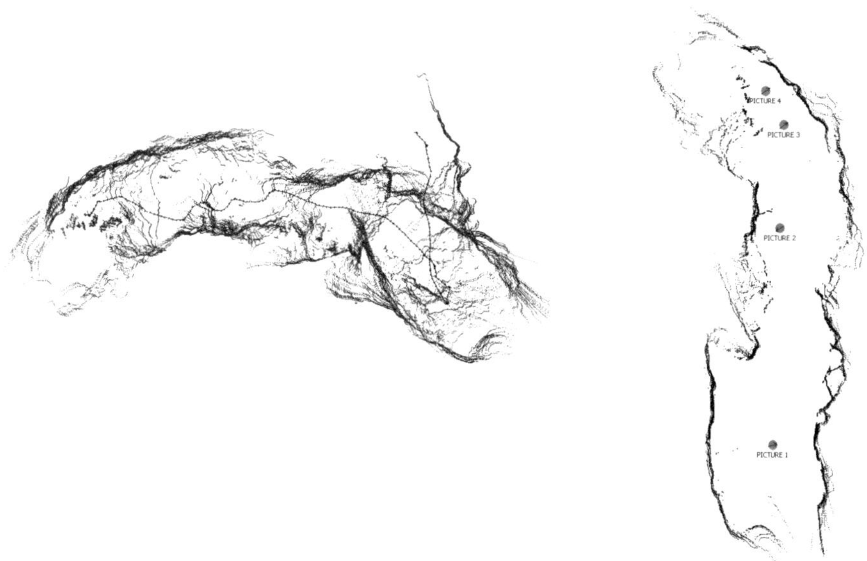


(a) (b)

Fig. 18 (a) Laser readings and odometer data after the scanning procedure. The odometer data are represented by little arrows. (b) 2D map of the explored corridor of the cave after the application of the scan-matching algorithm. Four zones rich in interesting paintings are also highlighted.

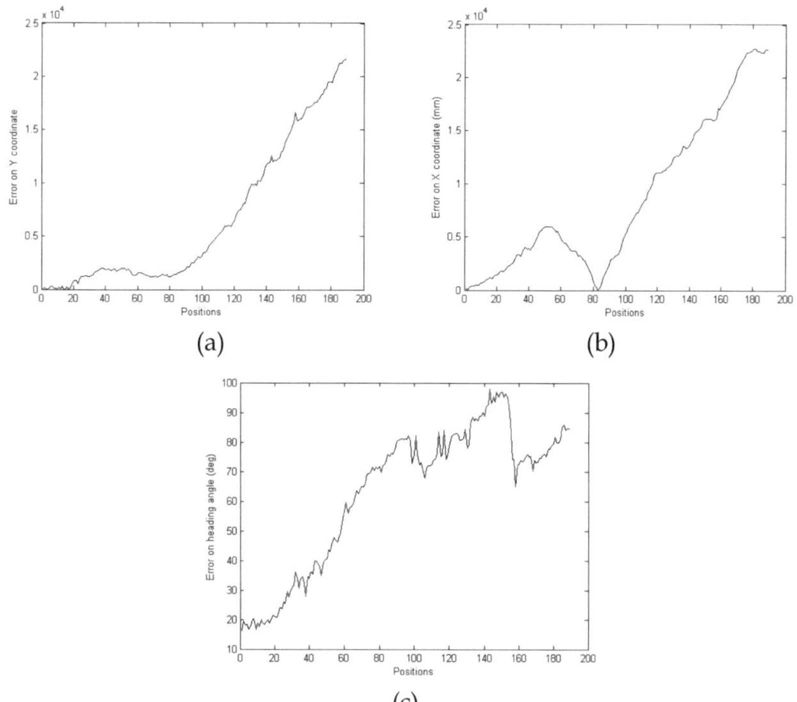


(a) (b)

(c)

Fig. 19. Estimated errors on (a) $x$, (b) $y$, and (c) $\theta$ components of the robot pose.

expressed in millimetres, whereas the error on the robot orientation $\theta$ is expressed in angle degrees. As it can be noticed, errors are considerable because of the wheel slipping on the rough terrain causing high inaccuracy in odometer position estimates. Nevertheless, the CSM algorithm is able to correct them producing an accurate map. The map obtained using the laser scanner supplies new and useful, although still approximate, information about the structure of the cave: such information was not available before our visit. It is important to note that the structure of the cave, supplied by the map, is very important for the knowledge and the study of the archaeological site, as it describes the morphology of the whole environment placing each painting inside its context and facilitating a better understanding of its role and meaning. Furthermore, the planar map is useful for robot navigation inside the cave.

### 5.2 Building the 3D model of particular areas

Reconstructing 3D models using computer vision techniques generally requires to extract the features (points, lines, target objects) and match them (Aggarwal & Vemuri, 1986; Biber et al., 2004; Gramegna et al., 2005). Moreover, it is important to determine the correspondence between features in different images, since the accuracy of the resulting model depends directly on the accuracy of the feature correspondence. The method described here uses as the only geometrical constraint the correspondence between corners in different images. A complex 3D scene is reconstructed using a set of three images acquired from three different viewpoints of the same scene. The only requirement is that images must be acquired by the same camera with a fixed focal length.

After image acquisition, feature points that correspond to high curvature points (corners), are extracted in each image using the Harris corner detector (Sequeira et al., 1999). The maximum number of corners to be extracted in each image is fixed a priori. A matching procedure is then applied to each couple of images. A classical correlation technique is first used to establish the matching candidates between two images by determining a correlation score for each couple of points. If the correlation score is higher than a given threshold, the related couple of points is considered as a candidate match.

In order to verify the candidate matches, a parameter counts the number of similar candidate matches found in the neighbourhood of each candidate matched point. The sum of these parameters for all candidate matches defines an energy function. The minimization of the energy function through a relaxation technique solves the ambiguity problem (Zhang et al., 1994). After the determination of the corner correspondence for each couple of images, the set of correct matches for all the three images is determined.

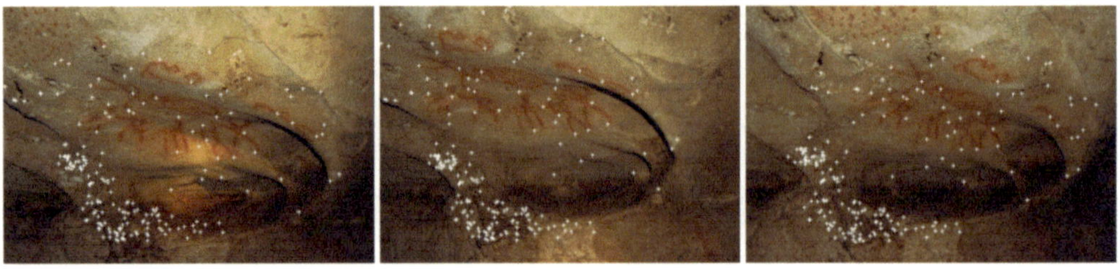

Fig. 20. Three images of an area of the wall rich in paintings. White crosses represent the correct point matches.

Fig. 20 shows three images with the matched points. Knowing the corresponding corners it is possible to determine the Fundamental Matrix and the intrinsic parameters of the camera (Hartley, 1997). At this point, all the necessary data to reconstruct the 3D scene are known. The 3D model is reconstructed through the application of the polygonal mesh technique. The 3D model of the scene was made by using the VRML. Fig. 21 shows the 3D model of the scene acquired in the cave.

Fig. 21. 3D model of an area of the cave.

## 6. Conclusions

In order to promote preservation of cultural heritage though guaranteeing accessibility to as much people as possible, novel technological solutions need to be researched. Methods from the mobile robotics field supply effective contributions to the development of environment modelling techniques that can be potentially used to either support the study of historically and artistically relevant assets or provide remote access to museums and archaeological sites, thus satisfying the need for cultural heritage conservation and accessibility.

In this chapter, we presented the results of our research in the field of cultural heritage, concerning the use of a multisensor mobile platform for data acquisition and processing. First, we described a laser-based people-following approach that enables a mobile robot to keep track of and pursue a human user in a tour of the environment. During the tour, the robot can acquire sensorial data to be used for environment modelling. This generates what is usually referred to as Human Augmented Mapping (HAM). Then, we presented the results of two case studies. The first one was related to the problem of constructing a model of a polygonal environment, such as a museum. Data acquisition was performed using a mobile robot, equipped with a 2D laser rangefinder and a CCD camera. Specifically, laser information was employed to simultaneously reconstruct the robot trajectory and build a planar map of the environment. From this map, a wireframe model was recovered. Finally, images were used to generate the texture to be added to the wireframe. Experimental results obtained for tests performed in our institute demonstrated the effectiveness of the proposed methods.

The second case study presented in the chapter focused on the application of a technological solution for remote access and mapping of a pre-historical cave in Southern Italy, named "Grotta dei Cervi". A multisensor mobile robot platform was used to explore the cave and send useful information to a remote console. Based on sensor data, the two-dimensional map of the site was reconstructed, along with the 3D model of zones of particular interest. Despite the structural complexity of the site, the proposed technological solution proved to be effective for making the archaeological cave available without damaging it. It was shown that the use of such a solution allows the growth of knowledge of this kind of sites, and improves the capability of monitoring and preserving their relevant archaeological treasures.

## 7. References


Aggarwal, J. K. & Vemuri, B. C. (1986), 3-D model construction from multiple views and intensity data, *Proceedings of IEEE Conf. Computer Vision and Pattern Recognition (CVPR)*, pp. 435-437, Miami Beach, June 1986, IEEE Computer Society, Washington

Allen, P., Stamos, I., Gueorguiev, A., Gold, E. & Blaer, P. (2001), AVENUE: Automated Site Modeling in Urban Environments, *Proceedings of IEEE International Conference on 3-D Digital Imaging and Modeling*, pp. 357-364, ISBN 0-7695-0984-3, Québec City, Canada, 28 May-1 June 2001, IEEE Computer Society, Washington

Allen, P. K., Stamos, I., Troccoli, A., Smith, B., Leordeanu, M. & Hsu, Y. C. (2003), 3D Modeling of Historic Sites using Range and Image Data, *Proceedings of IEEE International Conference on Robotics and Automation*, pp. 145–150, ISBN 0-7803-7736-2, Taipei, Taiwan, 14-19 September 2003, IEEE Computer Society, Washington

Biber, P.; Andreasson, H., Duckett, T. & Schilling, A. (2004), 3D Modeling of indoor environments by a mobile robot with a laser scanner and panoramic camera, *Proceedings of IEEE/RSJ Int. Conference on Intelligent Robots and Systems*, pp. 3430-3435, ISBN 0-7803-8464-4, Sendai, Japan, September 2004, IEEE Computer Society, Los Alamitos

Burgard, W.; Cremers, A. B., Fox, D., Hahnel, D., Lakemeyer, G., Schulz, D., Steiner, W. & Thrun, S. (1999), Experiences with an interactive museum tour-guide robot, *Artificial Intelligence*, Vol. 114, No. 1-2, (October 1999), (3-55), ISSN 0004-3702

Cox, I.J. (1991), Blanche-an experiment in guidance and navigation of an autonomous robot vehicle, *IEEE Transactions on Robotics and Automation*, Vol.7, No. 2, (April 1991), (193-204), ISSN 1042-296

Feyrer, S. & Zell, A. (2000), Robust Real-Time Pursuit of Persons with a Mobile Robot using Multisensor Fusion, *Proceedings of Int. Conference on Intelligent Autonomous Systems (IAS6)*, pp. 710-715, ISBN 1-58603-078-7, Venice, Italy, July 2000, IOS Press, Amsterdam

Fod A.; Howard, A. & Mataric, M. (2002), Laser-Based People Tracking, *Proceedings of IEEE Int. Conference on Robotics and Automation*, pp. 3024-3029, ISBN 0-7803-7273-5, Washington, DC, USA, May 2002, IEEE Computer Society, Los Alamitos

Gramegna, T., Attolico, G., Distante, A. (2005), Different approaches to improve the construction of 3D models using computer vision techniques, *Proceedings of AI*IA Workshop for Cultural Heritage*, ISBN 88-900910-0-2, Milan, Italy

Grisetti, G. & Iocchi, L. (2004), Map building in planar and non planar environments, *Proceedings of The 2nd International Workshop on Synthetic Simulation and Robotics to Mitigate Earthquake Disaster*, Lisbon, Portugal, June 2004

Gutmann, J. S. & Schlegel, C. (1996), AMOS: Comparison of scan matching approaches for self-localization in indoor environments, *Proceedings of First Euromicro Workshop on*

*Advanced Mobile Robots*, pp. 61–67, Kaiserlautern, Germany, ISBN 0-8186-7695-7, October 1996, IEEE Computer Society, Los Alamitos

Gutmann, J. S. & Konolige, K. (1999), Incremental mapping of large cyclic environments, *Proceedings of IEEE International Symposium on Computational Intelligence in Robotics and Automation,* pp. 318-325 , Monterey, CA, ISBN 0-7803-5806-6, November 1999, IEEE Computer Society, Los Alamitos

Hartley, R., (1997), Kruppa's equations derived from the fundamental matrix, *IEEE Transactions on pattern analysis and machine intelligence,* Vol. 19, No. 2, (February 1997), (133-135), ISSN 0162-8828

Hartley, R. & Zisserman, A. (2003). *Multiple View Geometry in Computer Vision*, 2nd edition, Cambridge University Press, ISBN 0521540518

Hirzinger, G., Bodenmüller, T., Hirschmüller, H., Liu, R., Sepp, W., Suppa, M., Abmayr, T. & Strackenbrock, B. (2005), Photo-realistic 3D modelling - From robotics perception towards cultural heritage. *Proceedings of International Workshop on Recording, Modeling and Visualization of Cultural Heritage,* Ascona, Switzerland, May 2005

Hough, P. V. C, (1962), Method and means for recognizing complex patterns, U.S. Patent 3069654

Ip Y. L. & Rad A. B. (2004), Incorporation of feature tracking into Simultaneous Localization and Mapping building via sonar data, *Journal of Intelligent and Robotic Systems,* Vol. 39, No. 2, (February 2004), (149-172), ISSN 0921-0296

Kleinehagenbrock, M.; Lang S., Fritsch J., Lomker F., Fink G. & Sagerer G. (2002), Person tracking with a mobile robot based on multi-modal anchoring, *Proceedings of IEEE Int. Workshop on Robot and Human Interactive Communication*, pp. 423-429, ISBN 0-7803-7545-9, Berlin, Germany, September 2002, IEEE Computer Society, Los Alamitos

Leiva, J. M.; Martinez, P., Perez, E. J., Urdiales, C. & Sandoval, F. (2001), 3D Reconstruction of static indoor environment by fusion of sonar and video data, *Proceedings of Int. Symposium on Intelligent Robotics Systems,* pp.179-188, ISBN 2-907801-01-5, Toulouse, France, July 2001, LAAS-CNRS, Toulouse

Lu, F. & Milios, E. (1997), Robot pose estimation in unknown environments by matching 2D range scans, *Journal of Intelligent and Robotic Systems,* Vol. 18, No. 3, (March 1997), (249–275), ISSN 0921-0296

Milella, A., Dimiccoli, C., Cicirelli, G. & Distante, A. (2007), A., Laser-based people-following for human-augmented mapping of indoor environments, *Proceedings of the 25th IASTED International Multi-Conference: artificial intelligence and applications,* pp. 151-155, ISBN 978-0-88986-631-7, Innsbruck, Austria, February 12-14, 2007, ACTA Press Anaheim, CA, USA

Nevado, M. M; Garcia-Bermejo, J. G., Casanova, E. Z. (2004), Obtaining 3D models of indoor environments with a mobile robot by estimating local surface directions, *Robotics and Autonomous Systems,* Vol. 48, No. 2-3, (September 2004), (131–143), ISSN 0921-8890

Pineau J.; Montemerlo, M., Pollack, M., Roy, N. & Thrun, S. (2003), Towards robotic assistants in nursing homes: challenges and results, *Robotics and Autonomous Systems,* Vol. 42, No. 3-4, (March 2003), (271-281), ISSN 0921-8890

Se, S., Lowe, D. G. & Little, J. (2002), Mobile robot localization and mapping with uncertainty using scale-invariant visual landmarks, *International Journal of Robotics Research*, Vol. 21, No. 8, (August 2002), (735-758), ISSN 0278-3649

Sequeira, V., Ng, K., Wolfart, E., Gonçalves, J. G. M., Hogg, D. C. (1999), Automated Reconstruction of 3D Models from Real Environments, *ISPRS Journal of Photogrammetry and Remote Sensing*, Vol. 54, No. 1, (February 1999), (1-22), ISSN 0924-2716

Stachniss, C; Hanhel, D., Burgard W. & Grisetti, G. (2005), On actively closing loops in grid-based fast-slam, *Advanced Robotics*, Vol. 19, No. 10, (1059-1079), ISSN 0169-1864

Thrun, S.; Beetz, M. Bennewitz, M., Burgard, W., Cremers, A. B., Dellaert, F., Fox, D., Hahnel, D., Rosenberg, C., Roy, N., Schulte, J. & Schulz, D. (2000), Probabilistic algorithms and the interactive museum tour-guide robot minerva, *The International Journal of Robotics Research*, Vol. 19, No. 11, (November 2000), (972-999), ISSN 0278-3649

Thrun, S; Liu, Y., Koller D., Ng, A. Y., Ghahramani, Z., Durrant-Whyte, H., (2004), Simultaneous Mapping and Localization with Sparse Extended Information Filters: Theory and Initial Results, *International Journal of Robotics Research*, Vol. 23, No. 7-8 (July-August 2004), (693-716), ISSN 0278-3649

Topp, E. A. & Christensen, H. I. (2005). Tracking for Following and Passing Persons, *Proceedings of IEEE/RSJ Int. Conference on Intelligent Robots and Systems (IROS)*, pp. 2321-2327, ISBN 0-7803-8913-1, Edmonton, Alberta, Canada, August 2005, IEEE Computer Society, Los Alamitos

Trahanias, P.; Burgard, W., Argyros A., Hahnel, D., Baltzakis, H., Pfaff, P. & Stachniss, C. (2005), TOURBOT and WebFAIR: Web-operated mobile robots for telepresence in populated exhibitions, IEEE *Robotics & Automation Magazine*, Vol. 12, No. 2, (June 2005), (77-89), ISSN 1070-98932

Wolf, D.F. & Sukhatme, G. S. (2005), Mobile robot simultaneous localization and mapping in dynamic environment", *Autonomous Robots*, Vol. 19, No. 1, (July 2005), (53-65), ISSN 0929-5593

Zhang, Z., Deriche, R., Faugeras, O., Luong, Q. (1994), A robust technique for matching two uncalibrated images trough the recovery of the unknown epipolar geometry, *Technical report N° 2273*, Institut national de recherche en informatique et en automatique.

9

# On-line Cutting Tool Condition Monitoring in Machining Processes using Artificial Intelligence

Antonio J. Vallejo[1], Rubén Morales-Menéndez[2] and J.R. Alique[3]
[1]*Visiting scholar at the Instituto de Automática Industrial, Madrid, Spain*
[2]*Tecnológico de Monterrey, Monterrey NL,*
[3]*Instituto de Automática Industrial, Madrid,*
[1,3]*Spain*
[2]*México*

## 1. Introduction

High Speed Machining (*HSM*) has become one of the leading methods in the improvement of machining productivity. The term *HSM* covers high spindle speeds, high feed rates, as well as high acceleration and deceleration rates. Furthermore, *HSM* does not imply only working with high speeds but also with high levels of precision and accuracy.
Additional to the *HSM*, many companies producing machine tools are interested in new technologies which provide intelligent features. Several research works (Koren et al., 1999; Erol et al., 2000; Liang et al., 2004) predict that future manufacturing systems will have intelligent functions to enhance their own processes, and the ability to perform an effective, reliable, and superior manufacturing procedures. In the areas of process monitoring and control, these new systems will also have a higher process technology level.
In any typical metal-cutting process, the key indexes which define the product quality are dimensional accuracy and surface roughness; both directly influenced by the cutting tool condition. One of the main goals in a Computer Numerically Controlled (*CNC*) machining centre is to find an *appropriate trade-off* among cutting tool condition, surface quality and productivity. A cutting tool condition monitoring system which optimizes the operating cost with the same quality of the product would be widely appreciated, (Saglam & Unuvar, 2003; Haber & Alique, 2003). For example, in (Tönshoff et al., 1988), it has been demonstrated that effective machining time of the *CNC* milling centre could be increased from 10 to 65% with a monitoring and control system. Also, (Sick, 2002) mentions that any manufacturing process can be significantly optimized using a reliable and flexible tool monitoring system.
The system must develop the following tasks:

- Collisions detection as fast as possible.
- Tool fracture identification.
- Estimation or classification of tool wear caused by abrasion or other influences.

While collision and tool fracture are sudden and mostly unexpected events that require reactions in real-time, the development of wear is a slow procedure. This section focuses on

the estimation of wear. The importance of tool wear monitoring is implied by exchanging worn tools in time, and tool costs can be reduced with a precise exploitation of the tool's lifetime.
However, cutting tool monitoring is not an easy task for several reasons. First, the machining processes are non-linear, and time-variant systems, which makes them difficult to model. Secondly, the acquired signals from sensors are dependent on other kind of factors, such as machining conditions, cutting tool geometry, workpiece material, among others. There is not a direct method for measuring the cutting tool wear, so indirect measurements are needed for its estimation. Besides, signals coming from machine tools sensors are disturbed by many other reasons such as cutting tool outbreaks, chatter, tool geometry variances, workpiece material properties, digitizers noise, sensor nonlinearity, among others. There is not a straightforward solution.

| Symbol | Description | Symbol | Description |
|---|---|---|---|
| A | State transition probability distribution | *MFCC* | Mel Frequency Cepstrum Coeff. |
| AC | Accelerometer | *MR* | Multiple Regression |
| *AE* | Acoustic Emission | M | Number of distinct obs. symbols |
| $a_e$ | Radial depth of cut (mm) | N | Spindle speed (rpm) |
| $a_{ij}$ | Elements of the transition matrix | $N_s$ | Number of states in the model |
| *ANN* | Artificial Neural Networks | $N_f$ | Number of bandpass filters |
| $a_p$ | Axial depth of cut (mm) | $n_p$ | Number of passes over workpiece |
| *BN* | Bayesian Networks | O | Observation sequence of model |
| B | Obs. symbol probability distribution | $q_t$ | State at time t |
| *CNC* | Computer Numerically Controlled | S | State sequence in the model |
| Curv | Machining geometry curvature($mm^{-1}$) | *SOFM* | Self-Organizing Feature Maps |
| DY | Dynamometer | *SP* | Spindle Power |
| *DOE* | Design Of Experiments | T | Length of observation sequence |
| $D_{tool}$ | Diameter of the cutting tool (mm) | $T_c$ | Tool life (min) |
| *FFT* | Fast Fourier Transform | $T_{mach}$ | Machining time (min) |
| *FAR* | False Alarm Rate | Tr | Training dataset |
| *FFR* | False Fault Rate | Ts | Testing dataset |
| $f_{HZ}$ | Sampling frequency (Hz) | V | Set of individual symbols |
| $f_{Mel}$ | Scale Mel frequency | VB | Flank wear (mm or μm) |
| $f_z$ | Feed per tooth (mm/rev/tooth) | VB1 | Uniform flank wear (mm o μm) |
| Fx | Cutting force in x-axis (N) | VB2 | Non-uniform wear (mm o μm) |
| Fy | Cutting force in y-axis (N) | VB3 | Localized flank wear (mm o μm) |
| Fz | Cutting force in z-axis (N) | Vol | Volume of removal metal ($mm^3$) |
| HB | Brinell Hardness Number of the workpiece (*BHN*) | x | Sample |
| *HMM* | Hidden Markov Models | z | Number of teeth of cutting tool |
| *HSM* | High Speed Machining | λ | *HMM* model specification |
| *LVQ* | Learning Vector Quantization | π | Initial state distribution for *HMM* |
| L | Machining length (mm) | μ | Mean value |
| M | Log bandpass filter output amplitude | σ | Standard deviation |

Table 1. Nomenclature.

This work proposes new ideas for the cutting tool condition monitoring and diagnosis with intelligent features (i.e. pattern recognition, learning, knowledge acquisition, and inference from incomplete information). Two techniques will be applied using Artificial Neural

Networks and Hidden Markov Models. The proposal is implemented for peripheral milling process in *HSM*. Table 1 presents all the symbols and variables used in this chapter.

## 2. State of the art

The cutting tool wear condition is an important factor in all metal cutting processes. However, direct monitoring systems are not easily implemented because their need of ingenious measuring methods. For this reason, indirect measurements are required for the estimation of cutting tool wear. Different machine tools sensors signals are used for monitoring and diagnosing the cutting tool wear condition.

There are important contributions for cutting tool monitoring systems based on Artificial Neural Networks (*ANN*), Bayesian Network (*BN*), Multiple Regression (*MR*) approaches and stochastic methods.

In (Owsley et al., 1997), the authors presented an approach for monitoring the cutting tool condition. Feature extraction from vibrations during the drilling is generated by Self-Organizing Feature Maps (*SOFM*). The signals processing implies a spectral feature extraction to obtain the time-frequency representation. These features are the inputs of a *HMM* classifier. The authors demonstrated that *SOFM* are an appropriated algorithm for vibration signals feature extraction.

A methodology based on frequency domain is presented by (Chen & Chen, 1999) for on-line detection of cutting tool failure. At low frequencies, the frequency domain presents two important peaks, which are compared to compute a ratio that could be an indicator for monitoring tool breakage.

In (Atlas et al., 2000), the authors used *HMM* for the evaluation of tool wear in milling processes. The feature extraction from vibrations signals were the root mean squared, the energy and its derivative. Two cutting tool conditions were defined: worn and no-worn condition. The reported success was around 93%.

In (Sick, 2002a), a new hybrid technique for cutting tool wear monitoring, which fuses a physical process model with an *ANN* model is proposed for turning. The physical model describes the influence of cutting conditions on measure force signals and it is used to normalize them. The *ANN* model establishes a relationship between the normalized force signals and the wear state of the cutting tool. The performance for the best model was 99.4% for the learning step, and 70.0% for the testing step.

In (Haber & Alique, 2003) is developed an intelligent supervisory system for cutting tool wear prediction using a model-based approach. The dynamic behavior of the cutting force is associated with the cutting tool and process conditions. First, an *ANN* model is trained considering the cutting force, the feed rate, and the radial depth of the cut. Secondly, the residual error obtained from the measure and predicted force is compared with an adaptive threshold in order to estimate the cutting tool condition. This condition is classified as new, half-worn, or worn cutting tool.

In (Saglam & Unuvar, 2003), the authors worked with multilayered *ANN* for the monitoring and diagnosis of the cutting tool condition and surface roughness. The obtained success rates were of 77% for tool wear and 80% for surface roughness.

In (Dey & Stori, 2004), a monitoring and diagnosis approach based on a *BN* is presented. This approach integrates multiple process metrics from sensor sources in sequential machining operations to identify the causes of process variations. It provides a probabilistic

confidence level of the diagnosis. The *BN* was trained with a set of 16 experiments, and the performance was evaluated with 18 new experiments. The *BN* diagnosed the correct state with a 60% confidence level in 16 of 18 cases.
In (Haber et al., 2004) is introduced an investigation of cutting tool wear monitoring in a *HSM* process based on the analysis of different signals signatures in time and frequency domains. The authors used sensorial information from dynamometers, accelerometers, and acoustic emission sensors to obtain the deviation of representative variables. The tests were designed for different cutting speeds and feed rates to determine the effects of a new and worn cutting tool. Data was transformed from time to frequency domain using the Fast Fourier Transform (*FFT*) algorithm. They concluded that second harmonics of tooth path excitation frequency in the vibration signal are the best indicator for cutting tool wear monitoring.
A proposal to exploit speech recognition frameworks in monitoring systems of the cutting tool wear condition is presented in (Vallejo et al., 2005). Also, (Vallejo et al., 2006) presented a new approach for online monitoring the cutting tool wear condition in face milling. The proposal is based on continuous *HMM* classifier, and the feature vectors were computed from the vibration signals between the cutting tool and the workpiece. The feature vectors consisted of the Mel Frequency Cepstrum Coefficients (*MFCC*). The success to recognize the cutting tool condition was 99.86% and 84.55%, for the training and testing dataset, respectively. Also, in (Vallejo et al., 2007) an indirect monitoring approach based on vibration measurements during the face milling process is proposed. The authors compared the performance of three different algorithms: *HMM, ANN,* and Learning Vector Quantization (*LVQ*). The *HMM* was the best algorithm with 84.24% accuracy, followed by the *LVQ* algorithm with 60.31% accuracy. Table 2 summarizes all works discussed in this section.

## 3. Experimental set-up

This research work was focused on covering a domain in mold and die industry with different aluminium alloys. In this industry, the peripheral milling process is of great importance, its geometry can be defined as a simple straight line or even as a different geometry path including concave and convex curvatures.
The experiments took place in a *HSM* centre HS-1000 Kondia, with 25 KW drive motor, three axis, maximum spindle speed 24,000 rpm, and a Siemens open Sinumerik 840D controller, as shown in Figure 1. During the experiment several HSS end mill cutting tools (25° helix angle, and 2-flute) from Sandvik Coromant were selected for the end milling process, and different workpiece materials (Aluminium with hardness from 70 to 157 HBN) were used. These materials were selected because they have important applications in the aeronautic and mold manufacturing industry. Also, several cutting tool diameters (from 8 to 20 mm) were employed.

### 3.1 Design of experiments

Currently, the most of the research experiments are related to surface roughness and flank wear (VB). In machining processes they only consider a specific combination of cutting tool and workpiece material. Therefore, several authors have pointed out the importance of building databases with information of different materials and cutting tools that allow computing models by considering a complete domain in the machining process. The *DOE* was defined to consider the most important factors affecting the surface roughness during the peripheral end milling process, see (Vallejo et al., 2007a). Therefore, its results are relevant to compute a surface roughness model as well as and a model to predict the cutting tool condition.

| Process | Monitoring States | Sensor Signals | Recognition methods | References |
|---|---|---|---|---|
| Drilling | Tool wear | AC | HMM | (Owsley et al., 1997) |
| End Milling | Tool Breakage (Normal, Broke) | AC | FFT | (Chen & Chen, 1999) |
| End Milling | Tool wear (Worn-no worn) | AC | HMM | (Atlas et al., 2000) |
| Turning | Tool wear (Wear value) | Process parameters | ANN | Sick, 2002 |
| Turning | Tool wear (New, half worn, worn) | Process parameters | ANN | (Haber & Alique, 2003) |
| Face Milling | Tool wear (Flank wear) | DY | ANN | (Saglam & Unuvar, 2003) |
| Face Milling | Tool wear (Low-high) | AE, SP | BN | (Dey & Stori, 2004) |
| Milling | Tool wear (New, worn) | AE, DY, AC | FFT | (Haber et al., 2004) |
| Face Milling | Tool wear (New, half-new, half-worn, worn) | AC | HMM | (Vallejo et al., 2006) |
| Face Milling | Tool wear (New, half-new, half-worn, worn) | AC | HMM, ANN, LVQ | (Vallejo et al., 2007) |

Table 2. Comparison of different research efforts for monitoring the cutting tool condition. The recognition method is defined by considering the machining process, sensor signals, and the classification method.

The factors and levels were defined via the application of a screening factorial design over the most important factors affecting the surface roughness. These factors and levels were the following: feed per tooth ($f_z$), cutting tool diameter ($D_{tool}$), radial depth of cut ($a_e$), hardness of the workpiece material (HB), and the machining geometry curvature (Curv). Table 3 shows the factors and levels defined for the experiments. Table 4 presents the selected aluminium alloys with the different cutting tools used in the experiments. The dimensions of the workpiece were 100x170x25 mm, and they were designed to allow the machining of four replicates. The designed geometries are depicted in Figure 2a, and the cutting tools are shown in Figure 2b.

The machining domain in *HSM* was characterized by using different aluminium alloys, cutting tools and several geometries (concave, convex and straight path) in peripheral milling process, and the *DOE* considered the following steps:

1. Run a set of experiments with the cutting tool in sharp condition. During the experimentation the process variables were recorded.
2. Wear the cutting tool with the harder aluminium alloys until reaching a specific flank wear in agreement with ISO-8688 Tool life testing in milling.
3. Run other set of experiments with a different cutting tool wear condition.
4. Repeat the steps 2 and 3 until the cutting tool reaches the tool-life criteria.

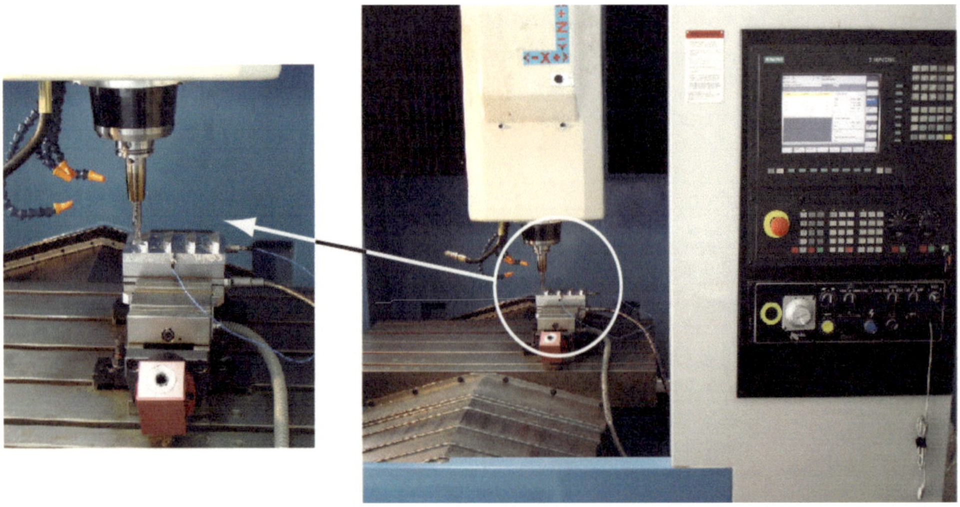

Fig. 1. Experimental Set-up. *CNC* machining centre HS-1000 Kondia (Right side), and the workpiece fixed to the table after the machined process (left side).

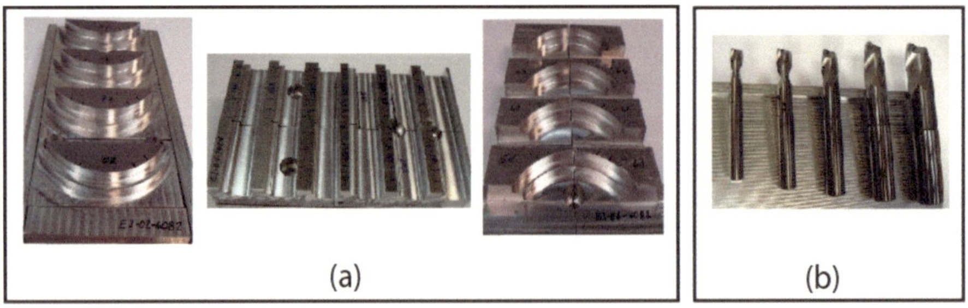


Fig. 2. a) Aluminium workpieces and geometries. b) Cutting tools for the experimentation.

| Levels | $f_z$ (mm/rev/tooth) | $D_{tool}$ (mm) | $a_e$ (mm) | HB (BHN) | Curv ($mm^{-1}$) |
|---|---|---|---|---|---|
| -2 | 0.025 | 8 | 1 | 71 | -0.05 |
| -1 | 0.05 | 10 | 2 | 93 | -0.025 |
| 0 | 0.075 | 12 | 3 | 110 | 0 |
| 1 | 0.1 | 16 | 4 | 136 | 0.025 |
| 2 | 0.13 | 20 | 5 | 157 | 0.05 |

Table 3. Factors and levels defined for the experimentation.

| Workpiece material Hardness (HB) | Cutting tools Diameter (mm) |
|---|---|
| 5083-H111 (71 HB) | R216.32-08025-AP12AH10F (8 mm) |
| 6082-T6 (93 HB) | R216.32-10025-AP14AH10F (10 mm) |
| 2024-T3 (110 HB) | R216.32-12025-AP16AH10F (12 mm) |
| 7022-T6 (136 HB) | R216.32-16025-AP20AH10F (16 mm) |
| 7075-T6 (157 HB) | R216.32-20025-AP20AH10F (20 mm) |

Table 4. Aluminium alloys and specifications of the cutting tools used in the experimentation.

### 3.2 Tool life evaluation

In practical workshop environment, the time at which a tool ceases to produce workpieces of the desired size or surface quality usually determines the end of useful tool life. It is essential to define tool life as the total cutting time to reach a specified value of tool-life criterion. Here, it is necessary to identify and classify the cutting tool deterioration phenomena, and where it occurs at the cutting edges. The main numerical values of tool deterioration used to determine tool life are the quantity of testing material required and the cost of testing. The following concepts are given to explain the deterioration phenomena in the cutting tool:

- *Tool wear*. Change in shape of the cutting edge part of a tool from its original shape, resulting from progressive loss of tool material during cutting.
- *Brittle fracture (chipping)*. Cracks occurrence in the cutting part of a tool followed by the loss of small fragments of tool material.
- *Tool deterioration measure*. Quantity used to express the magnitude of a certain aspect of tool deterioration by a numerical value.
- *Tool-life criterion*. Predetermined value of a specified tool deterioration measure indicating the occurrence of a specified phenomenon.
- *Tool life ($T_c$)*. Total cutting time of the cutting part required to reach a specified tool-life criterion.

In Figure 3, terms related to the tool deterioration phenomena on end milling cutters are shown. These terms include:

- *Flank wear* (VB): Loss of tool material from the tool flanks, resulting in the progressive development of the flank wear land.
- *Uniform flank wear* (VB1): Wear land which is normally of constant width and extends over the tool flanks of the active cutting edge.
- *Non-uniform wear* (VB2): Wear land which has an irregular width and the original flank varies at each position of measurement.
- *Localized flank wear* (VB3): Exaggerated and localized form of flank wear which develops at a specific part of the flank.

The tool-life criterion can be a predetermined numerical value for any type of tool deterioration that can be measured. If there are different forms of deterioration, they should be recorded so when any so when any of the deterioration phenomena limits has been attained, we can say the end of the tool life has been the end of the tool life has been reached.

Predetermined numerical values of specific types of tool wear are recommended:

- For a width of the flank wear land (VB) the following tool life end points are recommended:
  1. Uniform wear: 0.3 mm averaged over all teeth.
  2. Localized wear: 0.5 mm maximum on any individual tooth.
- When chipping occurs, it is to be treated as localized wear using a VB3 value equal to 0.5 mm as a tool-life end point.

Finally, flank wear measurement is carried out parallel to the surface of the wear land and in a perpendicular direction to the original cutting edge. Although the flank wear land on a significant portion of the flank wear may be of uniform size, there will be variations in its value at other portions of the flank, depending on the tool profile and edge chipping. Values of flank wear measurements are related to the area or position along the cutting edges at which the measurement is made.

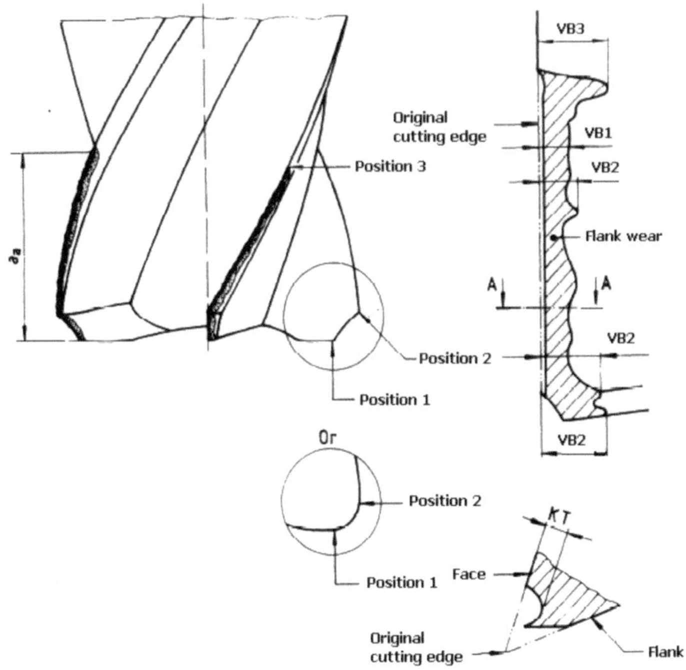


Fig. 3. Different terms in the flank wear are depicted for an end milling cutter (Taken from ISO 8088-2, 1989).

Therefore, it was necessary to define a methodology to wear the cutting tool, and to use the total tool-life during the experimentation. The assessment of the flank wear was taken as tool-life criterion. The applied methodology considers the following steps:

1. The new cutting tools are specified and the *DOE* with the four replicates is made.
2. The flank wear is assessed and registered at the end of the experimentation.
3. The cutting tools are worn by using several workpiece materials, and during the process the flank wear was observed until specific flank wear is reached.
4. The *DOE* is repeated with the new cutting tools conditions.
5. The steps 2, 3 and 4 are repeated (two more times), and the flank wear is measured and registered at the end of the process.

Figure 4 shows the evolution of the tool wear during the experimentation until the maximum tool-life criterion is reached. The experiments were interrupted at regular intervals for measurement of the flank wear (VB). The flank wear pattern along the cutting edge is showed as uniform wear over the surface (see Figure 5). In all cases, the tool wear data corresponds to localized wear.

Milling is an interrupted operation, where the cutting tool edge enters and exits the workpiece several times. The machining time of the tool in minutes was computed by Equation (1):

$$T_{mach} = \frac{L \times n_p}{f_z \times z \times N} \tag{1}$$

The volume of removed material volume was computed by Equation (2):

$$Vol = a_e a_p n_p L \tag{2}$$

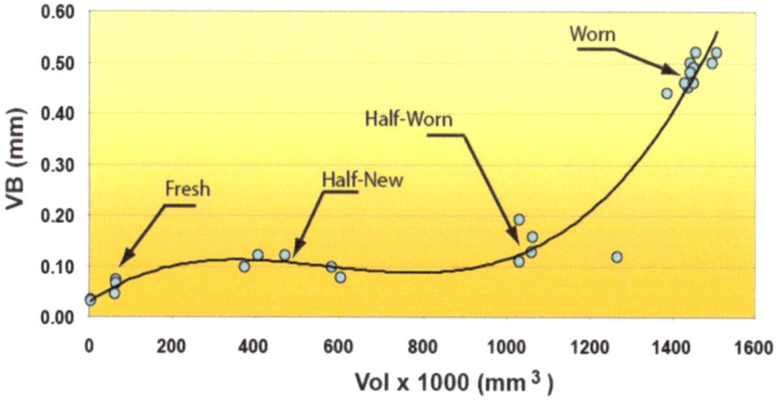


Fig. 4. Evolution of flank wear versus the volume of removal metal. The figure shows the behavior of the five cutting tools.

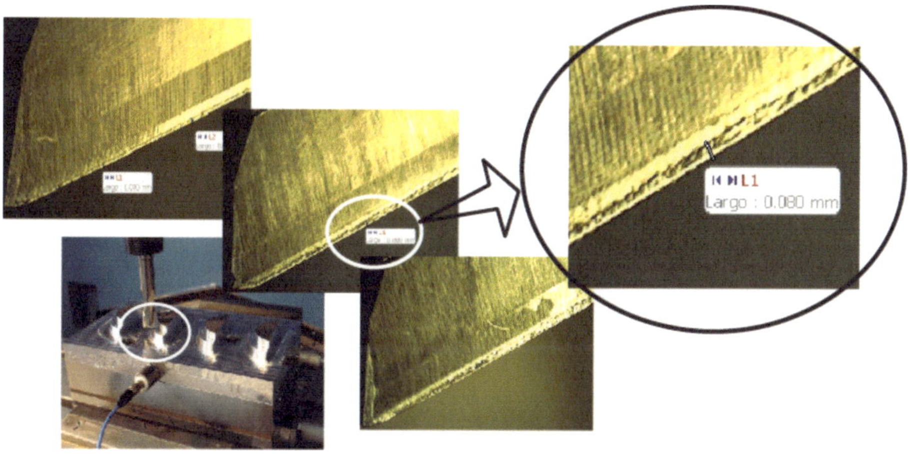


Fig. 5. Evolution of flank wear on the cutting edge. The images were taken throught a stereoscopic microscope. The cutting tool diameter is 12 mm.

The *VB* was selected as the criterion to evaluate the tool's life and its measurement was carried out according to ISO 8688-2, 1989. These two variables, *Vol* and *VB*, define the evolution of the cutting tool wear. The range of the flank wear was selected so that four cutting tool conditions were defined. They are shown in Table 5.

| Cutting tool wear condition | Flank wear (mm) |
|---|---|
| New | $0 \leq VB < 0.08$ |
| Half-new | $0.08 \leq VB < 0.1$ |
| Half-worn | $0.1 \leq VB < 0.3$ |
| Worn | $0.3 \leq VB < 0.5$ |

Table 5. Cutting tool wear conditions and the flank wear observed during the experimentation.

### 3.3 Data acquisition system

The Data Acquisition System consists of several sensors that were installed in the *CNC* machine (see Figure 6). For measuring the vibration, 2 PCB Piezotronics accelerometers model 353B04 were fixed in x and y-axis directions on the workpiece. These instruments have a sensitivity of 10 mV/g, in a frequency range from 0.35 to 20,000 Hz. Measurement range is ±500g. Other 2 Bruel and Kjaer piezoelectric accelerometers model 4370, and another model 4371, with a charge sensitivity of 98±2% pC/g, were installed on a ring fixed to the spindle. Also, these sensors allow the recording of vibration in x, y, and z-axis, during the cutting process.

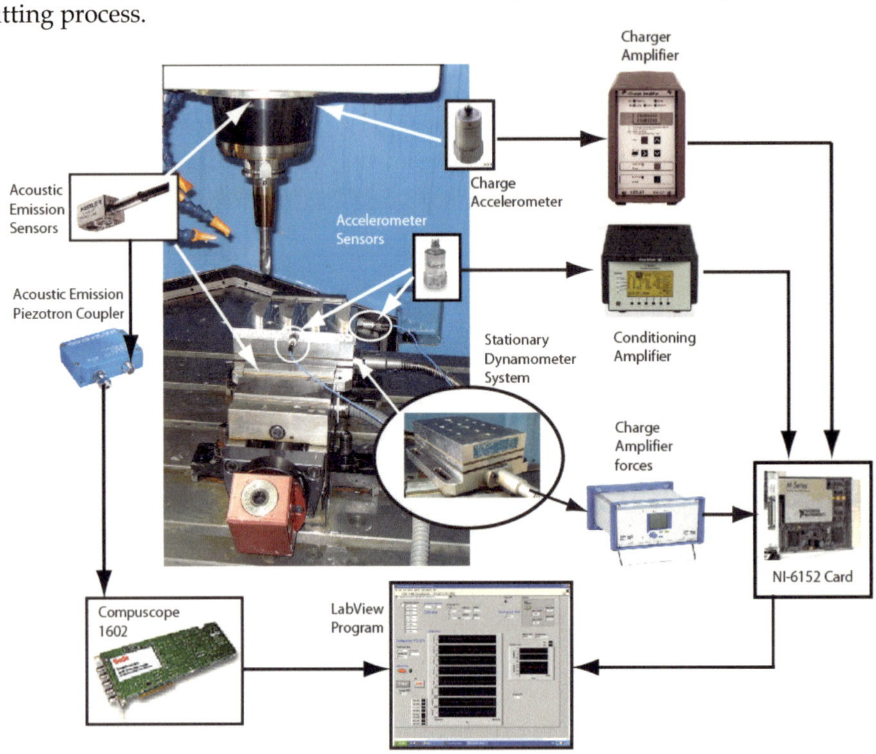


Fig. 6. Experimental Set-up. CNC machining centre and data acquisition system (sensors, amplifiers, boards and LabView interface). The vibration signals of the spindle and workpiece, and forces during machining process were acquired with the NI-6152 board. The acoustic emission signals were acquired with 1602 CompuScope board.

The dynamic cutting force components (Fx, Fy, Fz) were sensed with a 3 component force dynamometer, on which the workpiece was mounted. All the signals were acquired with a high speed multifunction DAQ NI-6152 card, which ensures 16-bit accuracy at a sampling rate of 1.25 MS/s. The system was configured to obtain the signals with a sampling rate of 40,000 samples/s.

The acoustic emissions were recorded with 2 Kistler Piezotron *AE* sensors model 8152B1, with frequency range from 50 to 400 KHz, and sensitivity of 700 V/(m/s). One was installed on a ring fixed to the spindle, and another was installed on the table of the machining centre. The AE signals were acquired with a CompuScope 1602 card for PCI bus, with 16 bit resolution. It provides a dual-channel simultaneous sampling rate of 2.5 MS/s. This board was configured to obtain signals with a sampling rate of 1,000,000 samples/s. The acquisition system was controlled with a LabView program. This program was used to control the start and end of the recorded signal and storage the information in specific files.

## 4. Processing of the process variables

Signals from the sensors must be processed to obtain the relevant features which identify the cutting tool condition. Basically, the raw signals undergo three steps in the signal processing:

1. Signal segmentation. During the machining process only one specific segment of the signal was selected and processed. This signal segment was divided into 20 small frames, which correspond to 0.15 (approximately) seconds of the machining time.
2. Features extraction. The feature vectors were computed for all the frames of each signal.
3. Average value. An average value was computed for all frames.

### 4.1 Feature extraction

The acquired signals during the machining process contain abundant information of the tool status, such as, fundamental frequencies related with the spindle speed and number of inserts, wide band frequency, amplitude of vibration signal, the sensitivity to detect the tool condition, the chatter, and so forth. The different signals are pre-processed calculating their *MFCC* representation, (Deller et al., 1993). This common transformation has shown to be more robust and reliable than other techniques, (Davis & Mermelstaein, 1980). There is a mapping between the real frequency scale ($f_{HZ}$) and the perceived frequency scale ($f_{Mel}$). The Mel scale is defined by the following equation

$$f_{Mel} = 2595 \times \log\left(1 + \frac{f_{Hz}}{700}\right) \tag{3}$$

The process to calculate the *MFCC* is shown in Figure 7. In this process, we must define the number of filters ($N_f$), sampling frequency ($f_{HZ}$), filters amplitude, and the configuration of the filter banks (triangular or rectangular shape). At the end, the MFCC are computed using the Inverse Discrete Cosine Transform:

$$MFCC_i = \sqrt{\frac{2}{N_f}} \sum_{j=1}^{N} m_j \cos\left(\frac{\pi i}{N_f}(j - 0.5)\right) \tag{4}$$

The result is a seven-dimension vector, where each dimensions correspond to one parameter. *MFCC* were computed by using the VOICEBOX: Speech Processing Toolbox for MatLab, and written by (Brookes, 2006). The routines taken from Speech Recognition module were: (a) The routine *melcepst,* which implements a mel-cepstrum front end for a recognizer; and (b) The routine *melbankm,* which generates the associated bandpass filter matrix.

### 4.2 MFCC for vibrations and force signals

Specifically for vibrations and force signals, the *MFCC* were computed by considering the following parameters: number of filters 20, sampling rate 40,000 Hz, and a bandpass filter with a triangular shape. The feature vector was of 7 dimensions (1 energy coefficient and 6 MFCC coefficients).

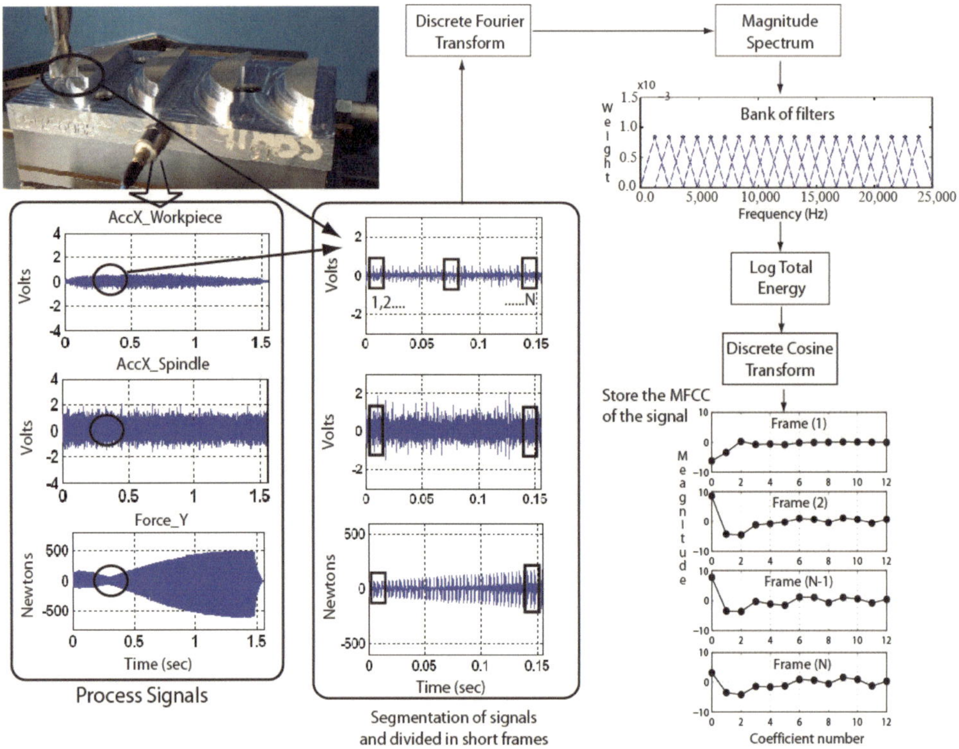


Fig. 7. Feature extraction process. The process variables (signals) are segmented and divided in short frames. A Discrete Fourier Transform and a mapping between the real frequency and the Mel frequency are computed. Then, a bandpass filters bank is applied for smoothing the scaled spectrum. Finally, the *MFCC* are computed using the discrete cosine transform.

### 4.3 MFCC for acoustic emission signals

MFCC were computed by considering the following parameters: number of filters 20, sampling rate 1,000,000 Hz, and a triangular shape bandpass filter. The feature vector was of 7 dimensions (1 energy coefficient and 6 MFCC coefficients).

## 5. Monitoring and diagnose the cutting tool wear condition with HMM

Real world processes generally produce observable outputs which can be characterized as signals. The signals can be discrete in nature (e.g., characters from a finite alphabet, quantized vectors from a codebook, etc.), or continuous in nature (e.g., speech samples, temperature measurements, vibration signals, music, etc.). They can be stationary or non-stationary, pure or corrupted from other signal sources. A problem of fundamental interest is characterizing such real-world signals in terms of signal models.

There are many reasons to consider this issue. First, a signal model can provide the basis for the theoretical description of a signal processing system that can be used to process the signal so as to provide a desired output. A second reason why signal models are important is that they are potentially capable of letting us learn a great deal about the signal source. But, the most important reason why signal models are significant is that they often work

extremely well in practice, and enable us to realize important practical systems (e.g. prediction systems, recognition systems, identification systems, among others.).
Signal models can be divided into deterministic and statistical models. Deterministic models generally exploit some known specific properties of the signal, and we only need to determine the values of the signal model parameters (e.g., amplitude, frequency, phase, etc.). On the other hand, statistical models use the statistical properties of the signal. Examples of such statistical models include Gaussian, Poison, Markov, and Hidden Markov processes. In this section, we are going to describe one type of stochastic signal model, namely *HMM*. A complete description of the *HMM* can be found in (Rabiner, 1989; Mohamed & Gader, 2000).

### 5.1 Discrete Markov Processes

Consider a system which may be described at any time as being in one of a set of $N_s$ distinct states, $S_1$, $S_2$, $S_3$, ..., $S_N$, as depicted in Figure 8 (where $N_s$=3). At regularly spaced discrete times, the system undergoes a change of state (possibly back to the same state) according to a set of probabilities associated with the state.
The time instants associated with the state changes are t = 1, 2,...., and the actual state at time t, as $q_t$. A full probabilistic description of the above system would, in general, require specification of the current state (at time t), as well as all the predecessor states. For the special case of a discrete, first order, Markov chain, this probabilistic description is reduced to just the current and the predecessor state, as shown in the following equation,

$$P[q_t = S_j | q_{t-1} = S_i, q_{t-2} = S_k, \ldots] = P[q_t = S_j | q_{t-1} = S_i] \tag{5}$$

Furthermore we only consider those processes in which the right-hand side of (5) is independent of time, thereby leading to the set of state transition probabilities $a_{i,j}$ of the form

$$a_{ij} = P[q_t = S_j | q_{t-1} = S_i], 1 \le i, j \le N \tag{6}$$

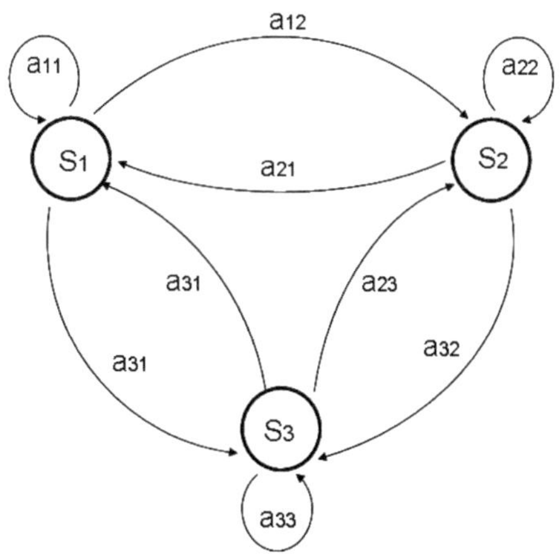


Fig. 8. Representation of a *HMM* with three states and the probabilities of the transition matrix ($a_{ij}$).

with the state transition coefficients having the properties

$$a_{ij} \geq 0$$

$$\sum_{j=1}^{N} a_{ij} = 1 \qquad (7)$$

Because, they obey standard stochastic constraints. The above stochastic process could be called an observation Markov model since the output of the process is the set of states at each instant of time, where each state corresponds to a physical event.

### 5.2 Extension to Hidden Markov Process

In this part we extend the concept of Markov models to include the case where the observation is a probabilistic function of the state, and the resulting model (which is called a *HMM*) is a doubly embedded stochastic process with an underlying stochastic process that is not observable, but can only be observed through another set of stochastic processes that produce the sequence of observations. To explain this concept, the following example is presented.

*Coin Toss Models.* Assume that somebody is in a room behind the wall, and he can not see what is happening inside. On the other side of the wall is another person who is performing a coin tossing experiment. The other person will not tell you anything about what he is exactly doing; he will only tell you the result of each coin flip. After a sequence of hidden coin tossing experiments is performed, the observation sequence consisting of a series of heads and tails, would be

$$O = O_1 O_2 O_3 \ldots\ldots O_T \qquad (8)$$

$$= H\,H\,J\,J\,J\,H\,J\,J\,H \ldots\ldots H$$

where H stands for heads and J stand for tails. Given the above scenario, the problem of interest is how do we build an *HMM* to explain the observed sequence of heads and tails. The first faced problem is deciding what states in the model correspond with what was observed. Then we should decide how many states should be in the model. One possible choice would be to assume that only a single biased coin was being tossed. In this case we could model the situation with a two-state model where each state corresponds to a side of the coin (i.e., heads or tails). This model is depicted in Figure 9a.

A second form of *HMM* for explaining the observed sequence of coin toss outcomes is given in Figure 9b. In this case there are 2 states in the model and each state corresponds to a different, biased coin being tossed. Each state is defined by a probability distribution of heads and tails. Transitions between states are characterized by a state transition matrix. The physical mechanism which accounts for how state transition is selected could be itself a set of independent coin tosses, or some other probabilistic event.

A third model of *HMM* for explaining the observed sequence of coin toss outcomes is defined very similarly to the *HMM* in Figure 8. This model corresponds to using 3 biased coins, and choosing among them a probabilistic event. Given the opportunity to choose among the three models in Figures 8 and 9 for the explanation of the observed sequence of heads and tails, a natural question would be which model matches the bets the actual observations.

It should be clear that the simple 1-coin model of Figure 9a has only 1 unknown parameter, the model of Figure 9b has four unknown parameters, and the model of Figure 8 has nine unknown parameters. Thus, with the greater degrees of freedom, the larger *HMMs* would seem to inherently be more capable of modeling a series of coin tossing experiments than it would be equivalent smaller models.

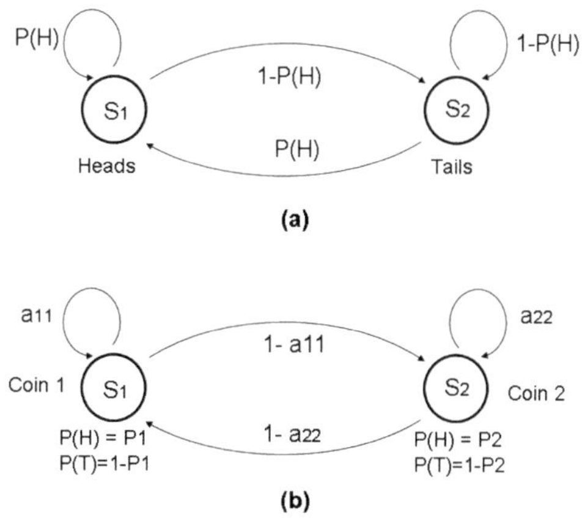


Fig. 9. (a) *HMM* with one coin and the two states. (b) *HMM* with two coins and each state with two observations.

An *HMM* is characterized by the following:

- The number of states in the model, $N_s$. Generally the states are interconnected in such as way that any state can be reached from any other state. We denote the individual states as $S=S_1,S_2,......,S_N$, and the state at time t as $q_t$.
- The number of distinct observation symbols per state, M. The individual symbols such as $V = v_1,v_2,.....,v_M$ (i.e., the symbols in the last example were H (heads) and J (tails)).
- The state transition probability distribution $A = a_{ij}$, where

$$a_{ij} = P[q_t = S_j | q_{t-1} = S_i], 1 \le i, j \le N \tag{9}$$

- The observation symbol probability distribution in state j, $B = b_j(k)$, where

$$b_j(k) = P[v_k \text{ at } t | q_t = S_j],\ 1 \le j \le N,\ 1 \le k \le M \tag{10}$$

The initial state distribution $\pi = \pi_i$ where

$$\pi_i = P[q_1 = S_i], 1 \le i \le N \tag{11}$$

Given appropriate values for $N_s$, M, A, B, and $\pi$, the HMM can be used as a generator of an observation sequence

$$O = O_1 O_2,......,O_T$$

It can be seen from the above discussion that a complete specification of an HMM requires specification of two model parameters ($N_s$, and M), observation symbols, and three probability measures A, B, and $\pi$. For convenience, the compact notation is used,

$$\lambda = (A, B, \pi) \tag{12}$$

to indicate the complete parameter set of the model.

### 5.3 Baum-Welch algorithm to train the model

The Baum-Welch algorithm, (Rabiner, 1989), is used to adjust the model parameters to maximize the probability of the observation sequence given by the model. The observation sequence used to compute the model parameters is called a training sequence. The training problem is crucial in the applications of the *HMMs*, because it allows us to optimally adapt model parameters to observed training data. The Baum-Welch algorithm is an iterative process that uses the forward and backward probabilities to solve the problem. The goal is to obtain a new model $\bar{\lambda} = (\bar{A}, \bar{B}, \bar{\pi})$ to maximize the function,

$$Q(\lambda, \bar{\lambda}) = \sum_{Q} \frac{P(O, Q|\lambda)}{P(O|\lambda)} \log\left[P(O, Q|\bar{\lambda})\right] \tag{13}$$

First, a current model is defined as $\lambda = (A, B, \pi)$, and used to estimate a new model as $\bar{\lambda} = (\bar{A}, \bar{B}, \bar{\pi})$. The new model must present a better likelihood than the first model to reproduce the observation sequence. Based on this procedure, if we iteratively use $\bar{\lambda}$ in place of $\lambda$ and repeat the calculus, then we can improve the probability of O being observed from the model until some limiting point is reached.

The result of the recalculation procedure is called a maximum likelihood estimate of the *HMM*. At the end, the new set of parameters (means, variance, and transitions) is obtained for each *HMM*.

### 5.4 Viterbi algorithm

In pattern recognition applications, it is useful to associate an *optimal* sequence of states to a sequence of observations, given the parameters of the model. In pattern recognition, the feature vector, representing the observations, is known, but the sequence of states that defines the model is unknown. A "reasonable" optimality criterion consists of choosing the state sequence (or path) that brings a maximum likelihood with respect to a given model (i.e., best "explains" the observation). This sequence can be determined recursively via the Viterbi algorithm. This algorithm identifies the single best state sequence, $Q=\{q_1 q_2 \ldots q_T\}$ for the given observation sequence $O=\{O_1 O_2 \ldots O_T\}$, and makes use of two variables:

- The highest likelihood $\delta_t(i)$ along a single path among all the paths ending in state i at time t:

$$\delta_t(i) = \max_{q_1, q_2 \ldots q_{t-1}} P[q_1 q_2 \ldots q_t = i, O_1 O_2 \ldots O_t|\lambda] \tag{14}$$

- A variable $\psi_t(i)$ which allows to keep track of the "best path" ending in state j at time t.

Using these two variables, the algorithm implies the following steps:

1. Initialization

$$\begin{aligned} &\delta_1(i) = \pi_i b_i(O_1) \quad 1 \le i \le N \\ &\psi_i = 0 \end{aligned} \tag{15}$$

2. Recursion

$$\delta_t(j) = \max_{1 \le i \le N_s}[\delta_{t-1}(i) a_{ij}] b_j(O_t), \quad 2 \le t \le T, \; 1 \le j \le N_s$$

$$\psi_t(j) = \arg \max_{1 \le i \le N_s}[\delta_{t-1}(i) a_{ij}], \; 2 \le t \le T, \; 1 \le j \le N_s \tag{16}$$

3. Termination:

$$P^* = \max_{1 \le i \le N}[\delta_T(i)]$$
$$q_T^* = \arg \max_{1 \le i \le N}[\delta_T(i)] \tag{17}$$

4. Path (state sequence) backtracking:

$$q_t^* = \psi_{t+1}(q_{t+1}^*), \quad t = T-1, T-2, \ldots, 1 \tag{18}$$

The Viterbi algorithm delivers the best states path, which corresponds to the observations sequence. This algorithm also computes likelihood along the best path.
The HMM models were computed by using the Hidden Markov Model Toolbox for MatLab. The routines were written by (Murphy, 2005).

## 6. Results

This section presents the results that were obtained by applying two different artificial intelligence techniques for monitoring and diagnosing the cutting tool condition during the peripheral end milling process in *HSM*: (1) Artificial Neural Network, and (2) Hidden Markov Models. In agreement with the experiments, a database was built with 441 experiments: 110 experiments used a new cutting tool, 112 a half-new cutting tool, 110 a half-worn cutting tool, and 109 a worn cutting tool. A MonteCarlo simulation for the training/testing steps was implemented due to the stochasticity of the approach. The results correspond to the average of 10 runs, where every time a different training data set (*Tr*) and testing data set (*Ts*) was generated (Figure 10).

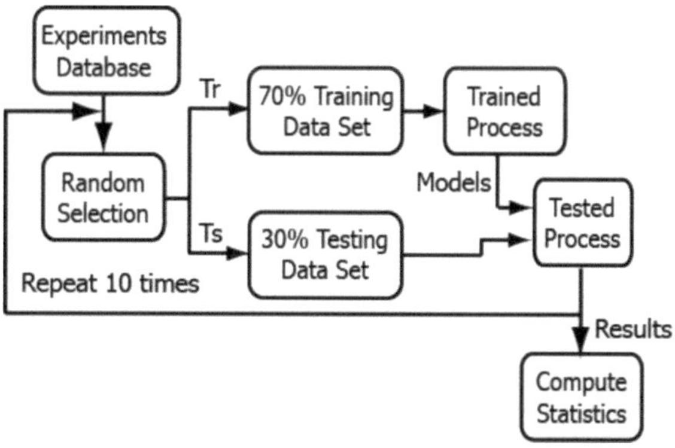


Fig. 10. Procedure for computing the approach performance. A random simulation for splitting the experimental dataset in training/testing sets was implemented due to the stochastic nature of the approaches.

### 6.1 Artificial neural network

To compare our results with classical approaches, the cutting tool wear condition was modeled with an *ANN* model. The application of *ANN* to on-line process monitoring systems has attracted great interest due to their learning capabilities, noise suppression, and parallel computability. A complete recopilation of research works in on-line and indirect tool wear monitoring with ANN are presented in (Sick, 2002). *ANN* is often defined as a computing

system made up of a number of simple elements called neurons, which possesses information by its dynamic state response to external inputs. The neurons are arranged in a series of layers. Multi-layer feed-forward networks are the most common architecture. Furthermore, there are several learning algorithms for training neural networks. Backpropagation has proven to be successful in many industrial applications and it is easily implemented.
The proposed architecture implies 12 input neurons, one hidden layer with 12 neurons, and 1 output neuron. Figure 11 shows the *ANN* model, where the input neurons represent the following information: feed per tooth, tool diameter, radial depth of cut, workpiece material hardness, curvature, and the *MFCC* vector (7 dimensions).

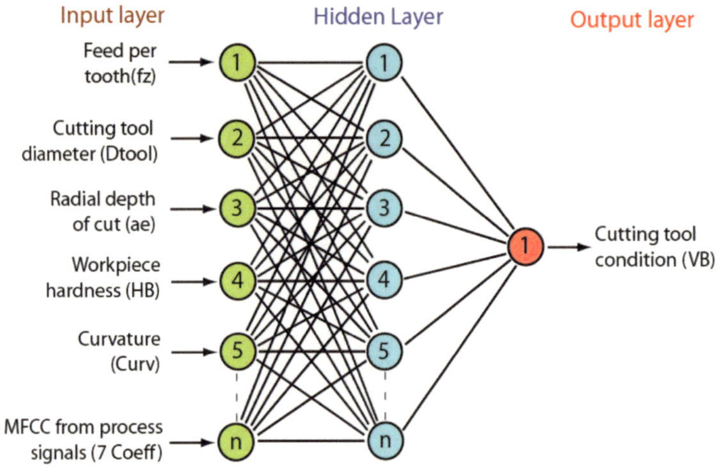


Fig. 11. ANN model implemented for monitoring and diagnosis the on-line cutting tool condition.

We used a feedforward *ANN* model and "tanh" activation function. The trained algorithm was classical backpropagation. For computing, input data ($f_z$, $D_{tool}$, $a_e$, HB, Curv, and *MFCC* vector) was normalized and output data was mapped to [-1, 1]. All the experimental dataset was normalized to avoid numerical inestability. First, the dataset was normalized by considering the mean value (μ), and standard deviation (σ) with the following equation,

$$f(x) = \frac{x - \mu}{\sigma} = \overline{x} \tag{19}$$

A second normalized method was applied: *bipolar sigmoidal*. This method was used because the minimum and maximum values are unknown in real-time. The non-linear transformation prevent most values from being compressed into essentially the same values, and it also compress the large outlier values. The *bipolar sigmoidal* was applied with the following equation:

$$f(\overline{x}) = \frac{1 - e^{(-\overline{x})}}{1 + e^{(-\overline{x})}} \tag{20}$$

With respect to the output neuron, the cutting tool condition, these values were mapped between the normalized tool-wear and tool-wear condition (see Table 6). Finally, the dataset was randomly divided into two sets, training (70%), and testing (30%) sets, in order to measure their generalization capacity.

| Normalized tool condition | Cutting tool condition |
|---|---|
| From +0.66 to +1.00 | New |
| From 0.0 to +0.66 | Half-new |
| From -0.66 to 0.0 | Half-worn |
| From -1.00 to -0.66 | Worn |

Table 6. Tool-wear from *ANN* model is mapped with the tool-wear Cutting.

The performance of the *ANN* model was computed for ten different sets of data, which were selected in random form. The training and testing processes were programmed by using MatLab software. The obtained results correspond to 8 different ANN models, all of them with the same architecture but different MFCC vector. The MFCC were computed for each of the process signals (accelerometers, forces, and acoustic emission). Table 7 shows the results computed with different process signals. The obtained performance corresponds to an avarage value from the ten data sets.

| Data sets | Workpiece Acc-X | Workpiece Acc-Y | Spindle Acc-X | Spindle Acc-Y | X Force | Y Force | AE Spindle | AE Workp. |
|---|---|---|---|---|---|---|---|---|
| Training | 90.2% | 94.5% | 97.8% | 98.7% | 94.2% | 97.6% | 99.9% | 99.2% |
| Testing | 31.3% | 33.8% | 40.4% | 47.2% | 48.5% | 48.0% | 89.9% | 69.7% |

Table 7. Performance for the training and testing data sets of the *ANN* model. The first two columns define the success of the accelerometers on the workpiece. The next two, the accelerometers installed on the spindle. The last two columns correspond with the Acoustic Emission sensors.

Table 7 shows that *ANN* model with acoustic emission signal (AE-Spindle) represents the best model for testing dataset, with a performance of 89.9% and Mean Squared Error (*MSE*) of 0.10075. Figure 12 plots the obtained results of the diagnosis system, when the *ANN* model was tested for the prediction of the cutting tool condition.

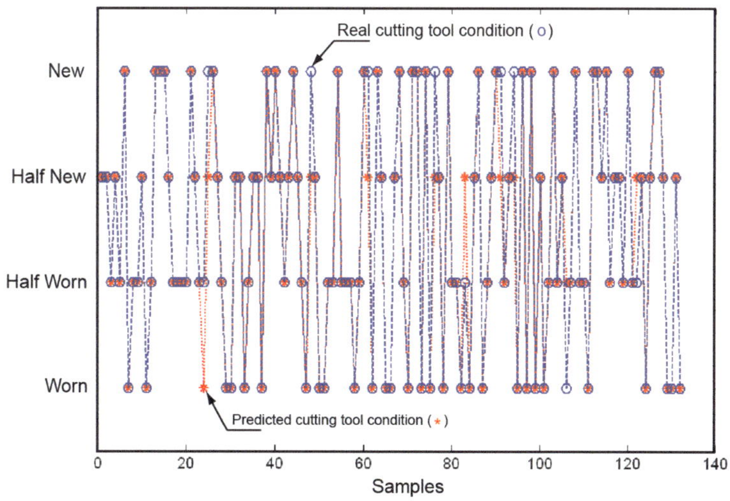


Fig. 12. Diagnosis the cutting tool condition with the ANN(12,12,1) model. The MFCC were computed for the acoustic emission signal (AE-Spindle).

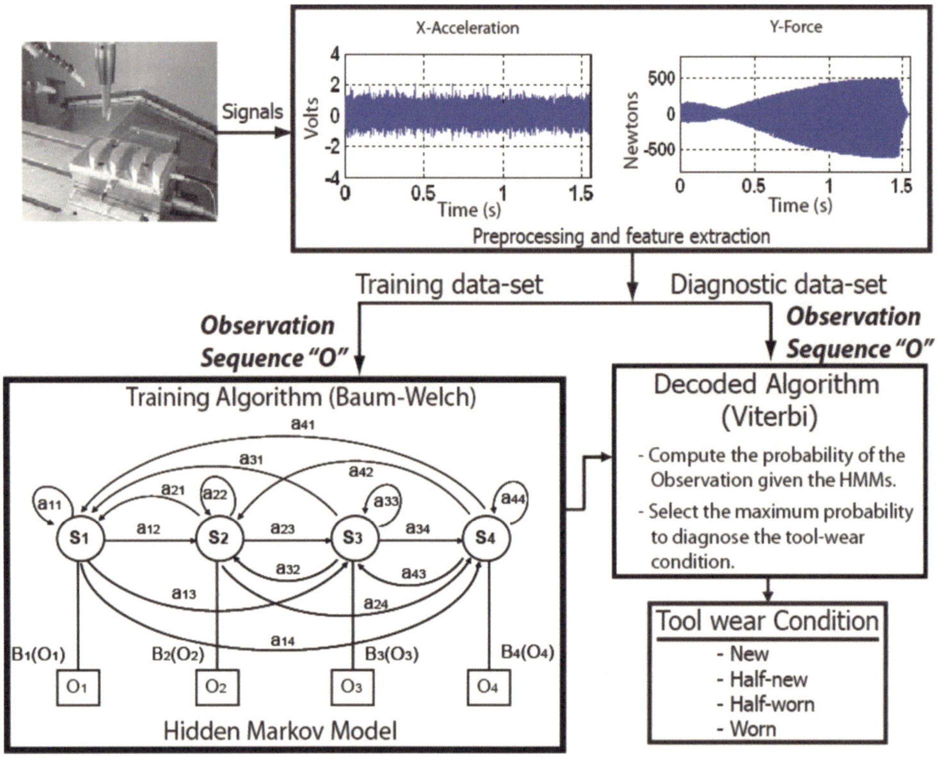


Fig. 13. Flow diagram for monitoring and diagnosis the cutting tool wear condition with continuous HMM. The features from signals are separeted into 2 branches. The training branch leads leads to *HMM*, and the diagnose branch uses the new observations and *HMMs* to recognize the cutting tool condition.

### 6.2 Hidden Markov Model

Figure 13 shows the flow diagram implemented for monitoring and diagnosing the cutting tool wear condition on-line with the *HMM* model. First, the signals are processed and splited into two: training and testing branches. Second, the training branch produces the *HMM* parameters by using the Baum-Welch algorithm. In this case, four models were computed to represent the four cutting tool conditions. Third, the testing branch uses the preprocessed signals and the HMM to compute the $P(O/\lambda)$ using the Viterbi algorithm for each model. The model with higher probability is selected as result.

The *HMM* framework was evaluated for different states and Gaussians in order to find the optimum performance results. Three different configurations were defined with seven MFCC:

1. *HMM* with 3 states and 2 Gaussians
2. *HMM* with 4 states and 2 Gaussians
3. *HMM* with 4 States and 4 Gaussians.

Figure 14 shows how the performance increases by increasing the states and Gaussians in the *HMM* approach. Based on this result, the selected configuration was with 4 states and 2 Gaussians, where the average performance was 77.51% for testing dataset. The process signal was the *AE* installed over the table.

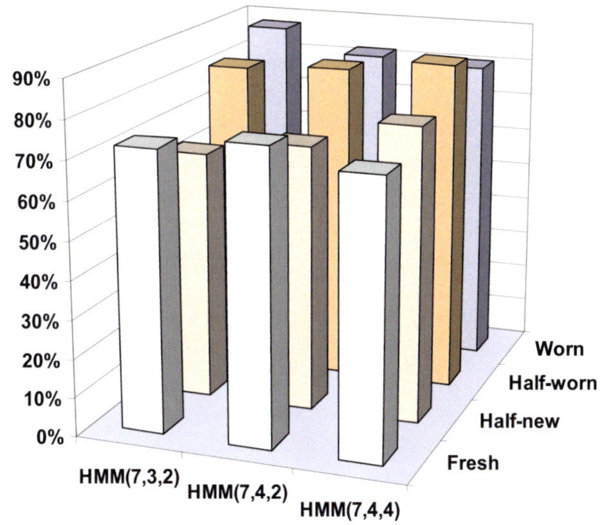


Fig. 14. Perfomance of the HMM with different configuration. The *HMM* were computed with different number of states (3, 4) and Gaussians (2,4).

Figure 15 shows the performance of the *HMMs* with the different signals. The acoustic emission signals present the best performance. For the AE-Spindle signal, the average performance was 99.4% for training dataset, and 95.1% for testing dataset.

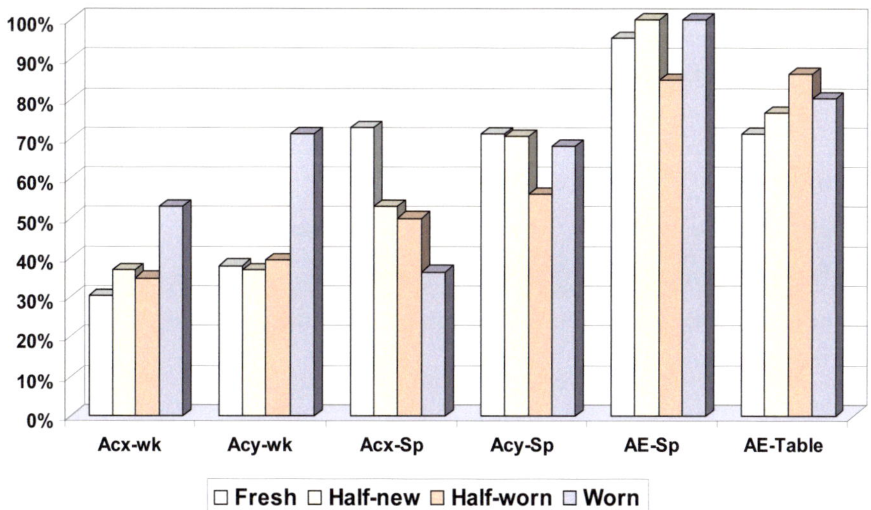


Fig. 15. Performance of the *HMM* for the different process signals. The results correspond to the obtained success for the testing dataset.

A classical test in a diagnosis system is to identify two alarms due to a false classification of cutting tool condition. These alarms are: False Alarm Rate (*FAR*), and False Fault Rate (*FFR*).

*FAR* condition represents a damage tool, but it is not true. *FFR* condition corresponds to a good state of the tool, but it is really damaged. The *FAR* condition is not a problem for diagnosis, but it reduces the productivity. However, the *FFR* condition might represent an expensive problem when the rate is high, because the tool can break before it is replaced. Figure 16 shows the misclassification percentage due to the *FFR* condition. The classifier with the lower percentage of the *FFR* was the *HMM* using the acoustic emission sensor. Once again, the AE-spindle does not produce any *FFR* condition.

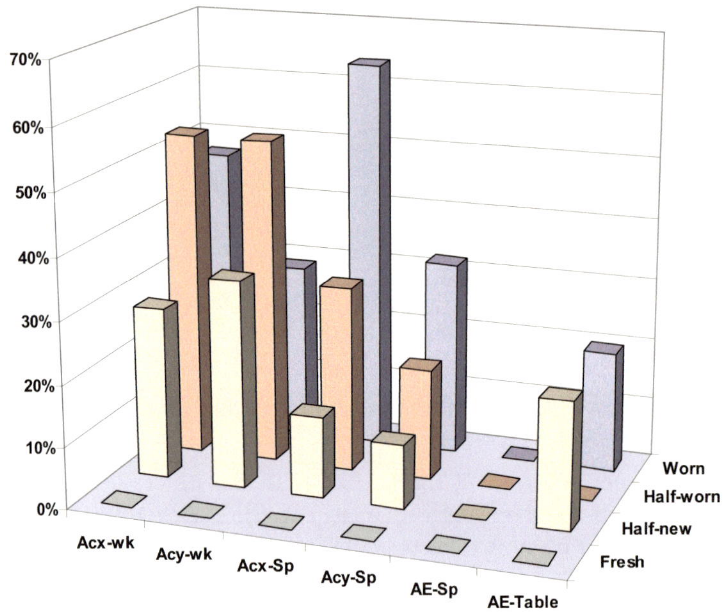


Fig. 16. Misclassification percentage in FFR alarms, for the *HMM* with different process signals.

## 7. Conclusions

This chapter presented new ideas for monitoring and diagnosis of the cutting tool condition with two different algorithms for pattern recognition: *HMM*, and *ANN*. The monitoring and diagnosis system was implemented for peripheral milling process in *HSM*, where several Aluminium alloys and cutting tools were used. The flank wear (VB) was selected as the criterion to evaluate the tool's life and four cutting tool conditions were defined to be recognized: New, half new, half worn, and worn condition.

Several sensors were used to record important process variables; accelerometer, dynamometer, and acoustic emission. Feature vectors, based on the Mel Frequency Cepstrum Coefficients, were computed to characterize the process signals during the machining processes. First, with the cutting parameters and *MFCC*, the cutting tool condition was modeled with an *ANN* model. The feedfoward *ANN* model and backpropagation algorithm were used to define the *ANN* model. The proposed architecture implies 12 input neurons and one output neuron (cutting condition). The best results were obtained by using the signals from the Acoustic Emission installed on the machine spindle. The success rate for the *ANN* model was 89.9% for the testing dataset.

Second, the *HMM* approach was configured with four states and two Gaussians, and the *HMM* models were computed with each one of the process signals. The best result was obtained with the signals coming from *AE*-Spindle. The performance was 95.08% for testing dataset, and 0.0% in the *FFR* condition. It is very important to mention, that *HMM* approach only uses one sensor to classify the cutting tool condition, while the ANN approach uses sensor fusion of five cutting parameters and one process variable to get the reported performance.

## 8. References


Atlas, L., Ostendorf, M., and Bernard, G. D., (2000). Hidden Markov Models for Machining Tool-Wear. *IEEE*, pp. 3887-3890.

Brookes, M., (2006). VOICEBOX: Speech Processing Toolbox for MatLab (http://www.ee.ic.ac.uk/hp/staff/dmb/voicebox/voicebox.html), Exhibition Road, London SW7 2BT, UK.

Chen, J.C., and Chen, W. (1999). A Tool Breakage Detection System using an Accelerometer Sensor. *J. of Intelligent Manufacturing*, 10(2), pp. 187-197.

Davis, B.S., and Mermelstaein P., (1980). Comparison of Parametric Representation for Monosyllabic Word Recognition in Continuously Spoken Sentences. *IEEE Transactions on Acoustic, Speech, and Signal Processing*, 4(28), pp. 357-366.

Deller, J.R., Hansen, J.H., and Proakis, J.G., (1993). Discrete-Time Processing of Speech Signals. *IEEE pres*, NJ 08855-1331.

Dey, S., and Stori, J.A. (2004). A Bayesian Network Approach to Root Cause Diagnosis of Process Variations. *International Journal of Machine Tools & Manufacture*, (45), pp. 75-91.

Erol, N.A., Altintas, Y., and Ito, M.R. (2000). Open System Architecture Modular Tool Kit for Motion and Machining Process Control. *IEEE/ASME Transactions on Mechatronics*, 5(3), pp. 281-291.

Haber, R.E. and Alique, A. (2003). Intelligent Process Supervision for Predicting Tool Wear in Machining Processes. *Mechatronics*, (13), pp. 825-849.

Haber, R.E., Jiménez, J.E., Peres, C.R., and Alique, J.R. (2004). An Investigation of Tool-Wear Monitoring in a High-Speed Machining Process. *Sensors and Actuators A*, (116), pp. 539-545.

ISO 8688-2, (1989). Tool Life Testing in Milling - Part 2: End Milling. *International Standard*, first edition.

Korem, Y., Heisel, U., Jovane, F., Moriwaki, T., Pritschow, G., Ulsoy, G. and Van Brussel, H. (1999). Reconfigurable Manufacturing Systems. *Annals of the CIRP* 48(2) , pp. 527-540.

Liang, S.Y., Hecker, R.L., and Landers, R.G. (2004). Machining Process Monitoring and Control: the State of the Art. *Manufacturing Science and Engineering*, 126, pp. 297-310.

Mohamed, M.A., and Gader, P., (2000). Generalized Hidden Markov Models - Part I: Theoretical Frameworks. IEEE Transactions on Fuzzy Systems, 8(1), pp. 67-81.

Murphy, K., (2005). Hidden Markov Model Toolbox for MatLab (http://www.cs.ubc.ca/~murphyk/Software/HMM/hmm.html).

Owsley, L.M., Atlas, L.E., and Bernard, G.D. (1997). Self-Organizing Feature Maps and Hidden Markov Models for Machine-Tool Monitoring. *IEEE Transactions on Signals Processing*, 45(11), pp. 2787-2798.

Rabiner, L.R., (1989). A Tutorial on Hidden Markov Models and Selected Applications in Speech Recognition. *Proceedings of the IEEE.* 77(2), pp. 257-286.

Saglam, H., and Unuvar, A. (2003). Tool Condition Monitoring in Milling based on Cutting Forces by a Neural Network. *International Journal of Production Research,* 41(7), pp. 1519-1532.

Sick, B. (2002). On-Line and Indirect Tool Wear Monitoring in Turning with Artificial Neural Networks: A review of more than a decade of research. *Mechanical Systems and Signal Processing,* 16(4), pp. 487-546.

Sick, B. (2002a). Fusion of Hard and Soft Computing Techniques in Indirect, Online Tool Wear Monitoring. *IEEE Transactions of Systems, Man, and Cybernetics,* 32(2), pp. 80-91.

Tönshoff, H.K., Wulfsberg, J.P., Kals, H.J., König, W., and Van Luttervelt, C.A. (1988). Developments and Trends in Monitoring and Control of Machining Processes. *Annals of the CIRP,* 37(2), pp. 611-622.

Vallejo, A.J., Nolazco-Flores, J.A., Morales-Menéndez, R., Sucar, L.E., and Rodríguez, C.A., (2005). Tool-wear Monitoring based on Continuous Hidden Markov Models. *LNCS 3773 Springer-Verlag, X CIARP.,* pp. 880-890.

Vallejo, A.J., Morales-Menéndez, R., Rodríguez, C.A., and Sucar, L.E., (2006). Diagnosis of a Cutting Tool in a Machining Center. *IEEE International Joint Conference on Neural Networks,* pp. 7097-7101.

Vallejo, A.J., Morales-Menéndez, R., Garza-Castañon, L.E., and Alique, J.R., (2007). Pattern Recognition Approaches for Diagnosis of Cutting Tool Wear Condition. *Transactions of the North American Manufacturing of Research Institution of SME,* 35, pp. 81-88.

Vallejo, A.J., Morales-Menéndez, and Alique, J.R., (2007a). Designing a Cost-effective Supervisory Control System for Machining Processes. *IFAC-Cost effective Automation in Networked Product Development and Manufacturing,* IFAC-PapersOnLine.

Vallejo, A.J., Morales-Menéndez, R., and Alique, J.R., (2008). Intelligent Monitoring and Decision Control System for Peripheral Milling Process. *To appear in IEEE International Conference on Systems, Man, and Cybernetics,* October, 2008.

# 10

# Controlled Use of Subgoals in Reinforcement Learning

Junichi Murata
*Kyushu University*
*Japan*

## 1. Introduction

Reinforcement learning (Kaelbling et al., 1996; Sutton & Barto, 1998) is a machine learning technique that automatically acquires a good action policy, i.e. a mapping from the current state to a good action to take, through trials and errors. A learning agent observes the state of its environment, chooses an action based on its current policy and executes the action. Responding to the action, the environment transitions to a new state, and a reword is given to the agent when applicable. The reward indicates how good or how bad the new state is, and the agent uses it to improve its policy so that it can obtain more rewards. Since reinforcement learning (abbreviated as RL hereafter) requires no other information, e.g. a model of environment, than the perceived states and rewards, it can be applied to a class of problems where the environment is complex or uncertain. The applications of RL include control of multi-legged robots (Kimura et al., 2001; Zennir at al., 2003), navigation of mobile robots (Millan, 1995), elevator hall call assignment (Crites & Barto, 1998; Kamal et al., 2005; Kamal and Murata, 2008) and board games (Tesauro, 1994). However, the learning agent must perform a large number of action trials in order to collect sufficient information about the unknown environment and organize it into a good policy. This takes a long time and therefore puts a limit on practical application of RL.

A number of techniques have been proposed to make RL faster. Most of them utilize *a priori* information on the target problem to compensate shortage of information and thus save the agent from a large number of trials. Smart and Kaelbling proposed using records of successful state-action pairs achieved by a human operator or a controller (Smart & Kaelbling, 2000). In the user-guided reinforcement learning of robots by Wang et al., a user teaches the robot a good action in real time (Wang et al., 2003). Driessens and Džeroski used a policy given by the human designer (Driessens & Džeroski, 2004). In all of the above research, the *a priori* information is in the form of good actions. Good states, however, are easier to find than good actions; you only need to find what the environment should be like but not how the agent can achieve that.

The technique described in this chapter focuses on *subgoals* as a form of good-state-related *a priori* information. A subgoal is a state or a subset of states that must be visited on the way from the initial state to the final goal state. With subgoals, the original problem can be divided into a set of small subproblems and this makes the learning faster. Use of subgoals in RL has been proposed by researchers, and it is closely related to (hierarchical) task partition and abstraction of actions. Singh used known subgoals to accelerate RL (Singh,

1992). Wiering and Schmidhuber proposed HQ-Learning which learns subgoals to convert a POMDP (partially observable markov decision process) problem to a sequence of MDP problems (Wiering & Schmidhuber, 1996). HASSLE by Bakker and Schmidhuber learns subgoals for efficient RL (Bakker & Schmidhuber, 2004). A method was proposed by McGovern and Barto for discovering subgoals to create 'options' or temporally-extended actions (McGovern & Barto, 2001), and an improved method by Kretchmar et al. (Kretchmar et al., 2003). Subgoal information is either given by human designers/operators or acquired by the agent itself through learning. The idea of automatic acquisition of subgoals by learning (Wiering & Schmidhuber, 1996; McGovern & Barto, 2001; Kretchmar et al., 2003; Bakker & Schmidhuber, 2004) is very interesting. This learning itself, however, requires a considerable amount of time, and therefore is not very fascinating from the overall learning time point of view. On the other hand, it is not unrealistic that human designers/operators give the agents their *a priori* knowledge since there is usually some sort of *a priori* information available about the target problems. This kind of *a priori* information is usually ready before starting learning and does not require additional learning time. But, instead, there is a possibility that the humans' *a priori* knowledge is not perfect; it may contain errors and/or ambiguity.

In this chapter a technique is proposed that utilizes human-supplied subgoal information to accelerate RL. The most prominent feature of the proposed technique is that it has a mechanism to control use of subgoals; it keeps watching the effect of each subgoal on the action selection, and when it detects redundancy or harmfulness of a subgoal, it gradually stops using the subgoal. The technique is an extension of the one proposed by the author previously (Murata et al., 2007). Unlike the previous one, the redundancy and the harmfulness of subgoals are treated separately and thus finer control of use of subgoals is achieved. Moreover, after a sufficient number of learning iterations, we can tell which subgoal is useful and which is harmful by looking at the values of parameters that control use of subgoals.

These features are illustrated by simulation results using grid worlds with doors as example environments where the learning agent tries to find a path from the start cell to the goal cell. It is shown that use of exact subgoal information makes RL ten times faster in these rather simple problems. More acceleration can be expected in more complex problems. When a wrong subgoal is given and used without proper control, the agent can hardly find the optimal policy. On the other hand, with the proposed control mechanism enabled, the wrong subgoal is properly controlled, and its existence does not obstruct acquiring optimal actions. In this case, the learning is delayed, but the delay is not significant. It is also confirmed that the values of parameters introduced to control use of subgoals successfully display harmfulness/usefulness of subgoals. A wrong subgoal is not necessarily harmful; in some problem settings, it helps the agent find the optimal policy, and this is also properly reflected in the derived parameter values.

The remaining part of the chapter is organized as follows. In the next section, more descriptions will be given to the subgoals in RL, and a technique will be proposed to use them with proper control. In Section 3, several examples will be shown that illustrate the validity of the proposed method, which is followed by conclusions in Section 4.

## 2. Controlled use of subgoals in reinforcement learning

### 2.1 Subgoals in reinforcement learning problems

A subgoal is a state or a subset of states that must be visited on the way from the initial state to the final goal state. Since we assume subgoal information is provided by humans it should be more adequate to define a subgoal here as 'a state or a subset of states that *the human designer/operator thinks* must be visited on the way from the initial state to the final goal state', which implies that the subgoals can be erroneous.

A subgoal divides the original problem into two subproblems with itself as the boundary: one subproblem where the agent is to find a policy that leads the agent/environment from the initial state to the subgoal state, and the other subproblem from the subgoal to the final goal state. Each subproblem is smaller than the original problem, and therefore it is easier for the agent to reach the goal of the subproblem by random actions, which results in a smaller number of trials to find a good policy. In this way, with subgoals, the original problem can be divided into a set of small subproblems, and this makes the learning faster. Imagine, for example, that you have an assistant robot in your office which learns its action policy by RL. Suppose that you order the robot to deliver a document to your colleague in an office upstairs. The robot tries to find a path from your office to the colleague's office by random wandering. But it will take a very long time to find the office upstairs by chance. Alternatively, you can give an additional instruction 'take an elevator', which means a subgoal 'being in an elevator' is given. With this instruction, the robot will first try to find an elevator and then try to find a path from the elevator to your colleague's office. This will be much easier to accomplish. The instruction 'take an elevator' can be valid in other situations like fetching a document from your colleague's office. In this sense, subgoals can be general information that is useful in different but similar problems. However, your robot may go downstairs even if it successfully arrives at the subgoal 'being in an elevator'. Or your building may have two or more elevators. Or even it may happen that you forget that your colleague has recently moved to an office on your floor. Subgoals can be a good guiding information but they may contain ambiguity and errors when given by humans.

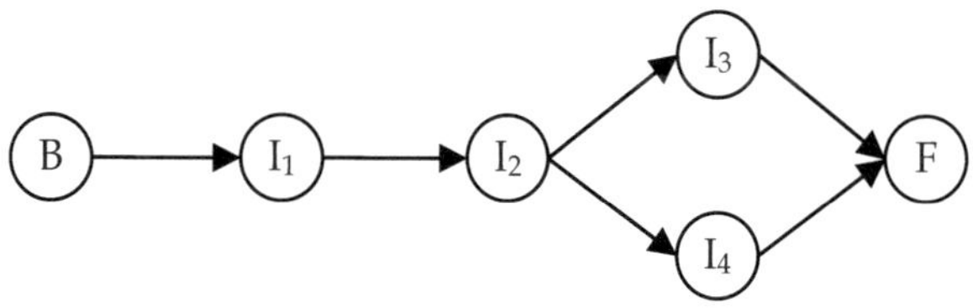


Fig. 1. Subgoals and their structure

Structure of subgoals can be represented by a directed graph as shown in Fig. 1 where B denotes the initial state and F the (final) goal state, and $I_i$, $i$=1, …, 4 stand for the subgoals. This figure indicates that, to reach the final goal F, the subgoal $I_1$ must be visited first, next the subgoal $I_2$, and then either $I_3$ or $I_4$ must be visited. Subgoals $I_3$ and $I_4$ are in parallel implying either subgoal must be visited, and $I_1$ and $I_2$ are in series indicating the subgoals must be visited in this order. In the following, terms *upstream subgoals* and *downstream subgoals* will be used. *Upstream subgoals* of a subgoal $I_i$ are those subgoals closer to the initial state B than the subgoal $I_i$ in the directed graph, and *downstream subgoals* of subgoal $I_i$ are those closer to the final goal state F. In Fig. 1, subgoals $I_1$ is an upstream subgoal of subgoal $I_2$, and $I_3$ and $I_4$ are downstream subgoals of $I_2$.

Subgoals are presumably *good* states and the learning agent must be aware of this. Here, the goodness of a subgoal is represented and conveyed to the agent as a reward. This reward is not the real reward given by the environment but a *virtual* reward given by the agent itself. The received virtual rewards must be used to guide the action selection through value functions. We have to be careful because the subgoals may contain errors and ambiguity and also because an already achieved subgoal becomes no longer useful. A wrong or useless subgoal can mislead the agent in the wrong direction or does not play any significant role in policy update. If a subgoal turns out to be harmful, we have to stop using it in learning. If a subgoal is found to be redundant, it is better to cease to use it. Therefore, we need to treat the real reward and the virtual reward associated with each subgoal separately so that we can control use of each subgoal independently. Besides, by dealing with harmfulness and redundancy separately, we can detect harmful subgoals that have been mistakenly included by human designer/operator.

### 2.2 Controlled use of subgoals

Let us assume that we have $n$ subgoals $I_1, \ldots, I_n$. The agent gives a virtual reward $r_{i,t}$ with respect to subgoal $I_i$ to itself at time $t$ which takes a positive value when state at $t$ is in subgoal $I_i$ and is equal to zero otherwise. The real reward given by the environment at time $t$ is denoted by $r_t$. To treat these real and virtual rewards separately, let us define a distinct action-value function based on each of them as follows:

$$Q(s_t, a_t) = \sum_{k=0}^{\infty} \gamma^k r_{t+1+k}, \tag{1}$$

$$Q_i(s_t, a_t) = \sum_{k=0}^{\infty} \gamma^k r_{i,t+1+k}, \quad i = 1, \cdots, n, \tag{2}$$

where $s_t$ and $a_t$ are state and action at time $t$, respectively, and $\gamma \in (0,1)$ is discount factor. Before learning, these action-value functions are initialized as zero, and then updated by the standard Q-learning:

$$Q(s_t, a_t) := (1-\alpha)Q(s_t, a_t) + \alpha\left[r_{t+1} + \gamma \max_a Q(s_{t+1}, a)\right], \tag{3}$$

$$Q_i(s_t, a_t) := (1-\alpha)Q_i(s_t, a_t) + \alpha\left[r_{i,t+1} + \gamma \max_a Q_i(s_{t+1}, a)\right], \quad i = 1, \cdots, n, \tag{4}$$

where operator ':=' means substitution and $\alpha > 0$ is learning rate.
Usually, when a subgoal is achieved, the agent can concentrate on seeking the next subgoal and then the next. For example, in Fig. 1, when subgoal $I_1$ is achieved, the agent only need to try to find subgoal $I_2$, and when this is also achieved, subgoal $I_3$ or $I_4$ is the next target. In this way, the target subgoal is switched from one to another. In other words, the original problem is hierarchically divided into distinct subproblems. However this is relevant only when the given subgoal information is correct. In the situations of our interest, the subgoal information is not perfect, and therefore the next subgoal can be a wrong subgoal. If $I_2$ is a wrong subgoal, it will be better, once $I_1$ is achieved, to seek $I_3$ or $I_4$ than $I_2$. In some cases, the order of subgoals may be wrong. In Fig. 1., for example, subgoal $I_2$ may be actually the first subgoal to be achieved before $I_1$. Therefore, we cannot employ the hierarchical problem partition. We must consider not the *next* subgoal only but *all* the subgoals in learning. In accordance with the above discussion, the actions are selected based on the following

composite action-value function that consists of the action-value function calculated from the real rewards and all the action-value functions computed from the virtual rewards:

$$Q^A(s,a) = Q(s,a) + \sum_{i=1}^{n} c_i d_i Q_i(s,a), \tag{5}$$

where suffix $t$ is dropped for simplicity. Positive coefficients $c_i$ and $d_i$, $i = 1, \ldots, n$ are introduced to control use of subgoals. Coefficient $c_i$ is used specifically to control use of subgoal when it is redundant while $d_i$ is for regulating subgoal when it is harmful. They are initialized to 1.0, i.e. at the beginning, all the virtual rewards are considered as equally strongly as the real reward in action selection. Actual action is derived by applying an appropriate exploratory variation such as ε-greedy and softmax to the action that maximizes $Q^A(s,a)$ for the current state $s$. Therefore, learning of $Q(s,a)$ by equation (3) is an *off-policy* learning, and its convergence is assured just like the ordinary Q-learning on the condition that all state-action pairs are visited infinitely often. However, our interest is not in the convergence of $Q(s,a)$ for all state-action pairs but in avoiding visiting unnecessary state-action pairs by appropriately controlled use of subgoals.

When a subgoal $I_i$ is found to be either redundant or harmful, its corresponding coefficient $c_i$ or $d_i$ is decreased to reduce its contribution to the action selection.

A subgoal $I_i$ is redundant in state $s$ when the optimal action in state $s$ towards this subgoal $I_i$ is identical to the optimal action towards the final goal F or towards another subgoal $I_j$, $j \in R_i$, where $R_i$ is the set of suffixes of subgoals that are reachable from subgoal $I_i$ in the directed graph. In other words, subgoal $I_i$ is redundant if, without help of the subgoal, the agent can find the optimal action that leads to the final goal or a downstream subgoal of subgoal $I_i$ which is closer to the final goal and thus more important. Let us define $\tilde{Q}(s,a)$ as a sum of $Q(s,a)$ and those $Q_j(s,a)$ associated with the downstream subgoals of subgoal $I_i$,

$$\tilde{Q}_i(s,a) = Q(s,a) + \sum_{j \in R_i} c_j d_j Q_j(s,a). \tag{6}$$

Then the optimal action in state $s$ towards the downstream subgoals and the final goal is given by

$$\tilde{a}_i^*(s) = \arg\max_a \tilde{Q}_i(s,a), \tag{7}$$

and the optimal action towards subgoal $I_i$ in state $s$ by

$$a_i^*(s) = \arg\max_a Q_i(s,a). \tag{8}$$

The relationship between subgoals and action-value functions is illustrated in Fig. 2. If $Q_i(s,a)$ or $\tilde{Q}_i(s,a)$ is zero or negative for any $a$, it means that sufficient positive real rewards or sufficient virtual rewards associated with $I_j$, $j \in R_i$ have not been received yet and that the *optimal actions* given by equations (7) and (8) are meaningless. So, we need the following preconditions in order to judge redundancy or harmfulness of a subgoal in state $s$:

$$\exists a, Q_i(s,a) > 0 \quad \text{and} \quad \exists a, \tilde{Q}_i(s,a) > 0. \tag{9}$$

Now, we can say that subgoal $I_i$ is redundant in state $s$ when the following holds:

$$a_i^*(s) = \tilde{a}_i^*(s). \tag{10}$$

When subgoal $I_i$ is found to be redundant in state $s$, its associated coefficient $c_i$ is reduced by a factor $\beta \in (0,1)$:

$$c_i := \beta c_i. \tag{11}$$

Coefficient $c_i$ is not set to zero at once because we have found that subgoal $I_i$ is redundant in this particular state $s$ but it may be useful in other states. Note that another coefficient $d_i$ is kept unchanged in this case. Although the composite action-value function $Q^A(s,a)$ used for the action selection includes the terms related *upstream* subgoals of subgoal $I_i$, we do not consider them in reducing $c_i$. The upstream subgoals are less important than subgoal $I_i$. Preconditions (9) mean that subgoal $I_i$ has already been achieved in the past trials. Then, if subgoal $I_i$ and any of the less important subgoals play the same role in action selection, i.e. either of them is redundant, then it is the coefficient associated with that less important upstream subgoal that must be decreased. Therefore the redundancy of subgoal $I_i$ is checked only against its *downstream* subgoals.

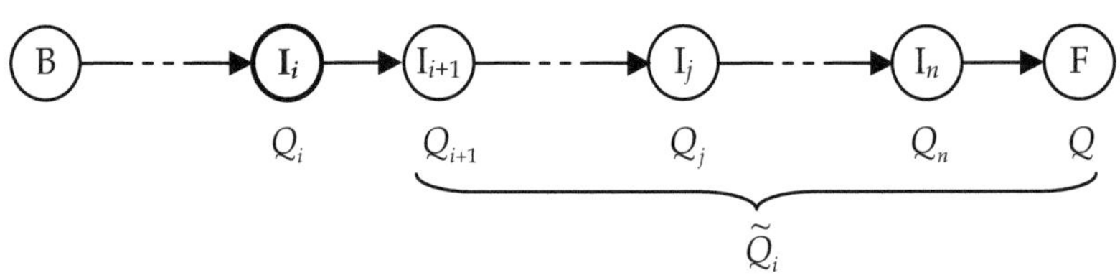


Fig. 2. Relationship between subgoals and action-value functions

A subgoal $I_i$ is harmful in state $s$ if the optimal action towards this subgoal is different from the optimal action towards the final goal or towards another subgoal $I_j$, $j \in R_i$, i.e. the action towards subgoal $I_i$ contradicts with the action towards the final goal or a downstream subgoal. This situation arises when the subgoal is wrong or the agent attempts to go back to the subgoal seeking more virtual reward given there although it has already passed the subgoal. Using $a_i^*(s)$ and $\tilde{a}_i^*(s)$ above, we can say a subgoal $I_i$ is harmful in state $s$ if

$$a_i^*(s) \neq \tilde{a}_i^*(s), \tag{12}$$

and the preconditions (9) are satisfied. When a subgoal is judged to be harmful in state $s$, its associated coefficient $d_i$ is reduced so that the subgoal does less harm in action selection. In this case coefficient $c_i$ remains unchanged. Let us derive a value of $d_i$ that does not cause the conflict (12). Such value of $d_i$, denoted by $d_i^o$, must be a value such that the action selected by maximizing $c_i d_i^o Q_i(s,a) + \tilde{Q}_i(s,a)$ does not differ from the action selected by $\tilde{Q}_i(s,a)$ only. So, the following must hold for state $s$,

$$\arg\max_a \left[ c_i d_i^o Q_i(s,a) + \tilde{Q}_i(s,a) \right] = \arg\max_a \tilde{Q}_i(s,a). \tag{13}$$

Considering equation (7), the above equation (13) holds when

$$c_i d_i^o Q_i(s,a) + \widetilde{Q}_i(s,a) \le c_i d_i^o Q_i(s,\widetilde{a}_i^*(s)) + \widetilde{Q}_i(s,\widetilde{a}_i^*(s)) \quad (14)$$

is satisfied for all $a$. Then, by straightforward calculation, the value of $d_i^o$ that assures the above inequality (14) is derived as

$$d_i^o = \min_{a\in A_i(s)} \frac{1}{c_i} \frac{\widetilde{Q}_i(s,\widetilde{a}_i^*(s)) - \widetilde{Q}_i(s,a)}{Q_i(s,a) - Q_i(s,\widetilde{a}_i^*(s))}, \quad A_i(s) = \left\{ a \middle| Q_i(s,a) > Q_i(s,\widetilde{a}_i^*(s)) \right\}. \quad (15)$$

In equation (15) we restrict actions to those belonging to set $A_i(s)$. This is because for actions which satisfy inequality $Q_i(s,a) \le Q_i(s,\widetilde{a}_i^*(s))$, inequality (14) naturally holds for any $d_i$ since $c_i d_i > 0$ and $\widetilde{Q}_i(s,a) \le \widetilde{Q}_i(s,\widetilde{a}_i^*(s))$ from the definition of $\widetilde{a}_i^*(s)$ in equation (7). Now $d_i$ is slightly reduced so that it approaches $d_i^o$ by a fraction of $\delta_i$:

$$d_i := (1-\delta_i)d_i + \delta_i d_i^o, \quad (16)$$

where $\delta_i$ is a small positive constant. There is a possibility that the original value of $d_i$ is already smaller than $d_i^o$. In that case, $d_i$ is not updated. Coefficient $d_i$ is not reduced to $d_i^o$ at once. We have observed a conflict among the subgoal $I_i$ and a downstream subgoal (or the final goal itself), and it seems that we need to reduce the coefficient $d_i$ for subgoal $I_i$ to solve the conflict. The observed conflict is genuine on the condition that the action-value functions $Q_i$, $Q_j$, $j \in R_i$ and $Q$ used to detect the conflict are sufficiently correct (in other words they are well updated). Therefore, in the early stage of learning, the observed conflict can be non-authentic. Even if the conflict is genuine, there is a situation where $d_i$ is not to be reduced. Usually a downstream subgoal of subgoal $I_i$ is more important than $I_i$, and therefore the conflict must be resolved by changing the coefficient associated with the subgoal $I_i$. However, when the downstream subgoals are wrong, reducing the coefficient associated with the subgoal $I_i$ is irrelevant. These possibilities of non-genuine conflict and/or wrong downstream subgoals demand a cautious reduction of $d_i$ as in equation (16). Moreover, to suppress possible misleading by wrong downstream subgoals, parameter $\delta_i$ is set smaller for upstream subgoals because a subgoal located closer to the initial state has a more number of downstream subgoals and therefore is likely to suffer from more undesirable effect caused by wrong subgoals.

Because update of $d_i$ depends on *downstream* coefficients $c_j$ and $d_j$, $j \in R_i$ contained in $\widetilde{Q}_i$, the update is done starting with the last subgoal namely the subgoal closest to the final goal to the first subgoal that is the closest to the initial state.

The overall procedure is described in Fig. 3. Action-values $Q$ and $Q_i$ are updated for $s_t$ and $a_t$, and then it is checked if these updates have made the subgoal $I_i$ redundant or harmful. Here the action-values for other state-action pairs remain unchanged, and thus it suffices that the preconditions (9) are checked for $s_t$ and $a_t$ only.

Each of coefficients $c_i$, $i = 1, \ldots, n$ represents non-redundancy of its associated subgoal, while $d_i$ reflects harmlessness of the subgoal. All of coefficients $c_i$ eventually tend to zero as the learning progresses since the agent does not need to rely on any subgoal once it has found an optimal policy that leads the agent and environment to the final goal. On the other hand, the value of $d_i$ depends on the property of its associated subgoal; $d_i$ remains large if its corresponding subgoal is not harmful while $d_i$ associated with a harmful subgoal decreases to zero. Therefore, by inspecting the value of each $d_i$ when the learning is complete, we can find which subgoal is harmful and which is not.

$Q(s,a) := 0$, $Q_i(s,a) := 0$, $\forall s, \forall a, i = 1, \ldots, n$.
$c_i := 1$, $d_i := 1$, $i = 1, \ldots, n$.
*Do until* the terminate condition is satisfied {
  $t := 0$, $s_t :=$ B (initial state).
  Do until $s_t =$ F (final goal state) or time is up {
    Select an action $a_t$ based on $Q^A(s_t, a)$.
    Execute the action $a_t$, observe the new state $s_{t+1}$ and receive rewards $r_{t+1}$, $r_{i,t+1}$, $i = 1, \ldots, n$.
    Update action-value functions $Q(s_t,a_t)$ and $Q_i(s_t,a_t)$, $i = 1, \ldots, n$ by

$$Q(s_t, a_t) := (1-\alpha)Q(s_t, a_t) + \alpha\left[r_{t+1} + \gamma \max_a Q(s_{t+1}, a)\right], \text{ and}$$

$$Q_i(s_t, a_t) := (1-\alpha)Q_i(s_t, a_t) + \alpha\left[r_{i,t+1} + \gamma \max_a Q_i(s_{t+1}, a)\right].$$

    *For* $I_i$ is the last subgoal *to* the first subgoal *do* {
      *If* $Q_i(s_t, a_t) > 0$ and $\tilde{Q}_i(s_t, a_t) > 0$ *then* {
        *If* $a_i^*(s) = \tilde{a}_i^*(s)$ *then*
          $c_i := \beta c_i$,
        *else*
          $d_i := (1-\delta_i)d_i + \delta_i d_i^o$.
      }
    }
    $t := t+1$.
  }
}

Fig. 3. Learning procedure

## 3. Examples

The proposed technique is tested on several example problems where an agent finds a path from the start cell to the goal cell in grid worlds. The grid worlds have several doors each of which requires a fitting key for the agent to go through it as shown in Fig. 4. The agent must pick up a key to reach the goal. Therefore having a key, or more precisely having just picked up a key, is a subgoal. The state consists of the agent's position (x-y coordinates) and which key the agent has. The agent can move to an adjacent cell in one of four directions (north, south, east and west) at each time step. When the agent arrives at a cell where a key exists, it picks up the key. Key 1 opens door 1, and key 2 is the key to door 2. The agent receives a reward 1.0 at the goal cell F and also a virtual reward 1.0 at the subgoals. When it selects a move to a wall or to the boundary, a negative reward –1.0 is given and the agent stays where it was. An episode ends when the agent reaches the goal cell or 200 time steps have passed.

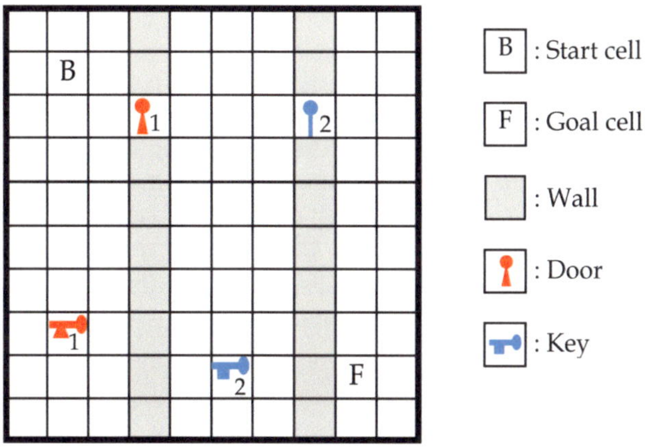


Fig. 4. Grid world 1

Fig. 5. Subgoal structure of grid world 1

### 3.1 Effect of use of correct subgoals

The subgoals in the example depicted in Fig. 4 can be represented by a directed graph shown in Fig.5. In RL, the first arrival at the goal state must be accomplished by random actions because the agent has no useful policy yet. Since the agent has to collect two keys to go through the two doors in this example, it takes a large number of episodes to arrive at the final goal by random actions only. Here we are going to see how much acceleration of RL we will have by introducing correct subgoals.

Q-learning is performed with and without taking the subgoals into consideration. The parameters used are as follows: discount factor $\gamma$=0.9, learning rate $\alpha$=0.05, $\beta$ in equation (11) is 0.99, and decreasing rates $\delta_i$ of coefficient $d_i$ is 0.005 for subgoal $I_1$ and 0.01 for $I_2$. Softmax action selection is used with 'temperature parameter' being 0.1.

The numbers of episodes required for the agent to reach the goal for the first time by greedy action based on the learnt $Q^A$ (i.e. the action that maximizes $Q^A$) and the numbers of episodes necessary to find an optimal (shortest) path to the goal are listed in Table 1. These are averages over five runs with different pseudo random number sequences. The table indicates that consideration of the correct subgoals makes the learning more than ten times faster in this small environment, which verifies the validity of introducing correct subgoals to accelerate RL. Also more acceleration can be expected for larger or more complex environments.

| | Number of episodes | |
|---|---|---|
| | First arrival at the goal | Finding an optimal path |
| Without subgoals | 11861.0 | 13295.8 |
| With subgoals | 1063.2 | 1068.0 |

Table 1. Numbers of episodes required before achieving the goal (grid world 1)

### 3.2 Effect of controlled use of subgoals

Now let us turn our attention to how the control of use of subgoals by coefficients $c_i$ and $d_i$ works. Here we consider another grid world shown in Fig. 6 where key 1 is the only correct key to the door and key 2 does not open the door. We apply the proposed method to this problem considering each of subgoal structures shown in Fig. 7. In this figure, subgoal structure (a) is the exact one, subgoal structure (b) has a wrong subgoal only, subgoal structures (c) and (d) have correct and wrong subgoals in series and subgoal structure (e) has correct and wrong subgoals in parallel. The same values are used as in the previous subsection for the parameters other than $\delta_i$. For a single subgoal in (a) and (b) $\delta_1$ is set to 0.01, for series subgoals in (c) and (d) $\delta_1$=0.005 and $\delta_2$=0.01 are used, and for the parallel subgoals in (e) 0.01 is used for both $\delta_1$ and $\delta_2$.

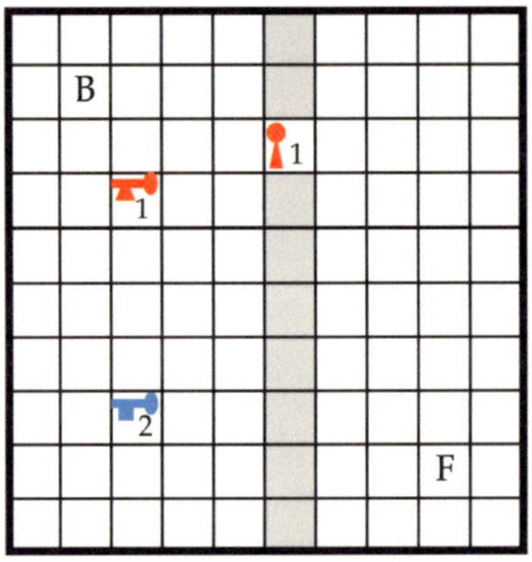


Fig. 6. Grid world 2 with a correct key and a wrong key

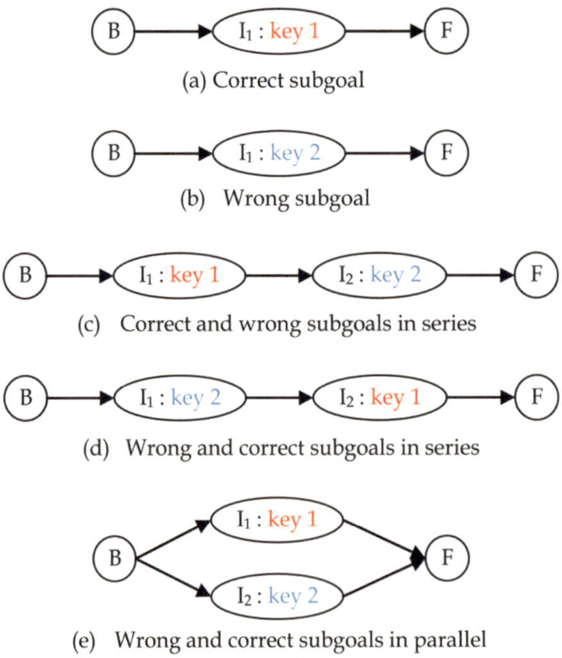


Fig. 7. Possible subgoal structures for grid world 2

The numbers of episodes before the first arrival at the goal and before finding an optimal path are shown in Table 2 together with the values of coefficients $d_i$ after learning and the ratio of $d_i$ for the correct subgoal ($d_{\mathrm{correct}}$) to $d_i$ for the wrong subgoal ($d_{\mathrm{wrong}}$) where available. All of these are averages over five runs with different pseudo random number sequences.

| Subgoals used in learning | Number of episodes | | Coefficients $d_i$ after learning | | |
|---|---|---|---|---|---|
| | First arrival at the goal | Finding an optimal path | For correct subgoal ($d_{\mathrm{correct}}$) | For wrong subgoal ($d_{\mathrm{wrong}}$) | $d_{\mathrm{correct}} / d_{\mathrm{wrong}}$ |
| None | 99.4 | 103.2 | - | - | - |
| Correct | 76.2 | 79.0 | $2.61\times10^{-1}$ | - | - |
| Wrong | 139.0 | 206.8 | - | $9.79\times10^{-5}$ | - |
| Correct & wrong in series | 116.6 | 180.0 | $3.48\times10^{-1}$ | $3.52\times10^{-5}$ | $1.30\times10^{7}$ |
| Wrong & correct in series | 87.4 | 97.8 | $4.15\times10^{-2}$ | $7.06\times10^{-3}$ | $1.37\times10^{5}$ |
| Correct & wrong in parallel | 116.8 | 163.4 | $9.85\times10^{-2}$ | $2.21\times10^{-4}$ | $1.03\times10^{8}$ |

Table 2. Numbers of episodes required before achieving the goal (grid world 2)

With the exact subgoal information given, the agent can reach the goal and find the optimal path faster than the case without considering any subgoal. When a wrong subgoal is provided in place of / in addition to the correct subgoal, the learning is delayed. However, the agent can find the optimal path anyway, which means that introducing a wrong subgoal does not cause a critical damage and that the proposed subgoal control by coefficients $c_i$ and $d_i$ works well. Finding the optimal path naturally takes more episodes than finding any path to the goal. The difference between them is large in the cases where wrong subgoal information is provided. This is because the coefficient associated with the wrong subgoal does not decay fast enough in those cases. The preconditions (9) for reducing the coefficient demand that the subgoal in question as well as at least one of its downstream subgoals have been already visited. Naturally the subgoals closer to the initial state in the state space (not in the subgoal structure graph) are more likely to be visited by random actions than those far from the initial state. In this grid world, the correct key 1 is located closer to the start cell than the wrong key 2 is, and therefore the correct subgoal decays faster and the wrong subgoal survives longer, which causes more delay in the learning.

Coefficients $d_i$ are used to reduce the effect of harmful subgoals. Therefore, by looking at their values in Table 2, we can find which subgoal has been judged to be harmful and which has not. Each of the coefficients $d_i$ for the correct subgoals takes a value around 0.1 while each of those for the wrong subgoals is around $10^{-4}$. Each ratio in the table is larger than $10^5$. Thus the coefficients $d_i$ surely reflect whether their associated subgoals are harmful or not. In Table 2, the coefficient for the wrong subgoal in the case of 'wrong and correct subgoals in series' is $7.06\times10^{-3}$ and is not very small compared with the value of $4.15\times10^{-2}$ for the correct subgoal. This has been caused by just one large coefficient value that appeared in one of the five runs. Even in this run, the learning is successfully accomplished just like in other runs. If we exclude this single value from average calculation, the average coefficient value for this subgoal is around $10^{-6}$.

To confirm the effect of subgoal control, learning is performed with the coefficient control disabled, i.e. both of $c_i$ and $d_i$ are fixed to 1.0 throughout the learning. In the case that the correct subgoal is given, the result is the same as that derived with the coefficient control. However, in other four cases where a wrong subgoal is given, the optimal path has not been found within 200000 episodes except for just one run in the five runs. Therefore, simply giving virtual rewards to subgoals does not work well when some wrong subgoals are included. When either $c_i$ or $d_i$ is fixed to 1.0 and the other is updated in the course of learning, similar results to those derived by updating both coefficients are obtained, but the learning is delayed when wrong subgoal information is provided. In composite action-value function $Q^A$ used in action selection, each action-value function $Q_i$ associated with subgoal $I_i$ is multiplied by a product of $c_i$ and $d_i$. The product decreases as the learning proceeds, but its speed is slow when either $c_i$ or $d_i$ is fixed. A large product of $c_i$ and $d_i$ makes the 'attractive force' of its corresponding subgoal strong, and the agent cannot perform a bold exploration to go beyond the subgoal and find a better policy. Then harmfulness of a subgoal cannot be detected since the agent believes that visiting that subgoal is a part of the optimal path and does not have another path to compare with in order to detect a conflict. Therefore, coefficient $c_i$ must be reduced when its associated subgoal is judged to be redundant to help agent to explore the environment and find a better policy. The above results and observation verifies that the proper control of use of subgoals is essential.

### 3.3 Effect of subgoals on problems with different properties

In the results shown in Table 2, the learning is not accelerated much even if the exact subgoal structure is given, and the results with wrong subgoal are not too bad. Those results of course depend on the problems to be solved. Table 3 shows the results for a problem where the positions of key 1 and key 2 are exchanged in grid world 2. Also the results for grid world 3 depicted in Fig. 8 are listed in Table 4. Here the correct and the wrong keys are located in the opposite directions from the start cell. The same parameter values are used in both examples as those used in the original grid world 2. The values in the tables are again averages over five runs with different pseudo random number sequences.

| Subgoals used in learning | Number of episodes | | Coefficients $d_i$ after learning | | |
|---|---|---|---|---|---|
| | First arrival at the goal | Finding an optimal path | For correct subgoal ($d_{correct}$) | For wrong subgoal ($d_{wrong}$) | $d_{correct} / d_{wrong}$ |
| None | 323.8 | 343.6 | - | - | - |
| Correct | 117.4 | 121.2 | $3.82\times10^{-3}$ | - | - |
| Wrong | 196.8 | 198.4 | - | $2.23\times10^{-2}$ | - |
| Correct & wrong in series | 188.2 | 189.6 | $6.84\times10^{-3}$ | $9.42\times10^{-3}$ | $2.53\times10^{1}$ |
| Wrong & correct in series | 117.2 | 126.8 | $4.37\times10^{-5}$ | $3.80\times10^{-1}$ | $9.47\times10^{-4}$ |
| Correct & wrong in parallel | 100.0 | 100.6 | $2.08\times10^{-2}$ | $9.57\times10^{-3}$ | 5.14 |

Table 3. Numbers of episodes required before achieving the goal (grid world 2 with keys exchanged)

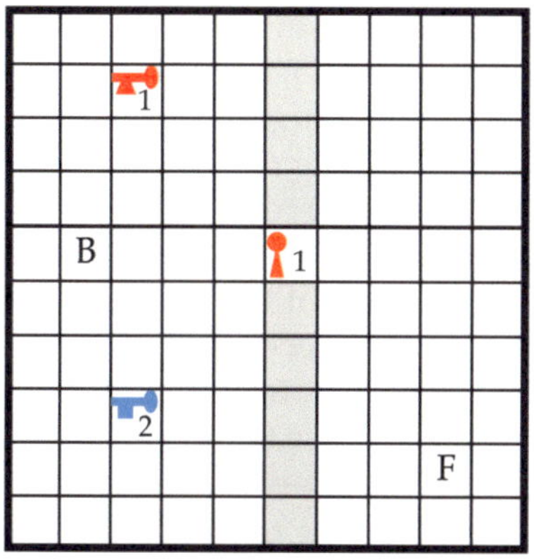


Fig. 8. Grid world 3 with two keys in opposite directions

By exchanging the two keys in grid world 2, the problem becomes more difficult than the original because the correct key is now far from the start cell. So, without subgoals, the learning takes more episodes, and introduction of subgoals is more significant than before as shown in Table 3. The wrong key is located on the way from the start cell to the correct key, and although picking up the wrong key itself has no useful meaning, the wrong subgoal guides the agent in the right direction towards the correct subgoal (correct key). Therefore the wrong subgoal information in this grid world is wrong but not harmful; it is even helpful in accelerating the learning as shown in Table 3. Also, since it is not harmful, coefficients $d_i$ corresponding to the wrong subgoals remain large after the learning.

| Subgoals used in learning | Number of episodes | | Coefficients $d_i$ after learning | | |
|---|---|---|---|---|---|
| | First arrival at the goal | Finding an optimal path | For correct subgoal ($d_{correct}$) | For wrong subgoal ($d_{wrong}$) | $d_{correct} / d_{wrong}$ |
| None | 153.8 | 155.8 | - | - | - |
| Correct | 95.8 | 99.6 | $4.80\times10^{-2}$ | - | - |
| Wrong | 150.4 | 309.8 | - | $1.08\times10^{-4}$ | - |
| Correct & wrong in series | 170.2 | 346.8 | $1.84\times10^{-3}$ | $9.83\times10^{-5}$ | $7.32\times10^{1}$ |
| Wrong & correct in series | 107.2 | 109.0 | $2.01\times10^{-3}$ | $1.22\times10^{-4}$ | $4.98\times10^{1}$ |
| Correct & wrong in parallel | 106.4 | 226.0 | $6.04\times10^{-3}$ | $4.75\times10^{-4}$ | $1.05\times10^{6}$ |

Table 4. Numbers of episodes required before achieving the goal (grid world 3)

In contrast, the wrong key in grid world 3 lies in the opposite direction from the correct key. So, this wrong subgoal has worse effect on the learning speed as shown in Table 4. Here the coefficients $d_i$ for the wrong subgoals are smaller than those for the correct subgoals.

For grid worlds 2 and 3, the actual subgoal structure is that shown in Fig. 7. (a). To investigate the performance of the proposed method on problems with parallel subgoals, key 2 in grid world 2 is changed to a key 1. So the environment now has two correct keys, and the actual subgoal structure is just like Fig. 7. (e) but both the keys are correct. Five different subgoal structures are considered here: 'near subgoal', 'far subgoal', 'near and far subgoals in series', 'far and near subgoals in series' and 'near and far subgoals in parallel' where 'near subgoal' denotes the subgoal state 'picking up key near the start cell', and 'far subgoal' refers to the subgoal 'picking up the key far from the start cell'. Note that there is no wrong subgoal in this grid world. The results shown in Table 5 are similar to those already derived. Introduction of subgoal(s) makes the goal achievement faster, but in some subgoal settings, finding the optimal path is slow. The subgoal structure 'near and far subgoals in parallel' is the exact one, but this gives the worst performance in finding the optimal path in the table. In this problem, both the keys correspond to correct subgoals, but one (near the start cell) is more preferable than the other, and the less-preferable subgoal survives longer in this setting as described in Section 3.2. This delays the learning.

| Subgoals used in learning | Number of episodes | | Coefficients $d_i$ after learning | |
|---|---|---|---|---|
| | First arrival at the goal | Finding an optimal path | For near subgoal | For far subgoal |
| None | 106.0 | 109.6 | - | - |
| Near | 76.2 | 79.0 | $2.62\times10^{-1}$ | - |
| Far | 136.2 | 203.8 | - | $2.96\times10^{-3}$ |
| Near & far in series | 126.6 | 205.2 | $4.06\times10^{-1}$ | $1.15\times10^{-5}$ |
| Far & near in series | 84.6 | 95.0 | $2.78\times10^{-2}$ | $7.06\times10^{-3}$ |
| Near & far in parallel | 116.4 | 169.6 | $7.95\times10^{-2}$ | $2.21\times10^{-4}$ |

Table 5. Numbers of episodes required before achieving the goal (grid world 2 with two correct keys)

Introduction of subgoals usually makes goal achievement (not necessarily by an optimal path) faster. But, a wrong or less-preferable subgoal sometimes makes finding the optimal path slower than the case without any subgoals considered, especially when it occupies a position far from the initial state. However, the wrong subgoals do not cause critically harmful effect such as impractically long delay and inability of finding the goal at all thanks to the proposed mechanism of subgoal control. Also we can find the harmful subgoals by inspecting the coefficient values used for subgoal control. This verifies the validity of the proposed controlled use of subgoals in reinforcement learning.

## 4. Conclusions

In order to make reinforcement learning faster, use of subgoals is proposed with appropriate control of each subgoal independently since errors and ambiguity are inevitable in subgoal information provided by humans. The method is applied to grid world examples and the results show that use of subgoals is very effective in accelerating RL and that, thanks to the proposed control mechanism, errors and ambiguity in subgoal information do not cause critical damage on the learning performance. Also it has been verified that the proposed subgoal control technique can detect harmful subgoals.

In reinforcement learning, it is very important to balance *exploitation*, i.e. making good use of information acquired by learning so far in action selection, with *exploration*, namely trying different actions seeking better actions or policy than those already derived by learning. In other words, a balance is important between what is already learnt and what is to be leant yet. In this chapter, we have introduced subgoals as a form of *a priori* information. Now we must compromise among leant information, information yet to be learnt and *a priori* information. This is accomplished, in the proposed technique, by choosing proper values for $\beta$ and $\delta_i$ that control use of *a priori* information through coefficients $c_i$ and $d_i$ as well as an appropriate choice of exploration parameter such as 'temperature parameter' used in softmax that regulates exploration versus exploitation. A good choice of parameters may need further investigations. However, this will be done using additional *a priori* information such as confidence of the human designer/operator in his/her subgoal information. Also a possible extension of the method is to combine it with a subgoal learning technique.

## 5. Acknowledgements

The author would like to acknowledge the support for part of the research by the Japan Society for the Promotion of Science, Grant-in-Aid for Scientific Research (C), 16560354, 2004-2006.

## 6. References

Bakker, B. & Schmidhuber, J. (2004). Hierarchical Reinforcement Learning with Subpolicies Specializing for Learned Subgoals, *Proc. 2nd IASTED Int. Conf. Neural Networks and Computational Intelligence*, pp. 125-130.

Crites, R.H. & Barto, A.G. (1998). Elevator Group Control Using Multiple Reinforcement Learning Agents, *Machine Learning*, Vol. 33, pp. 235-262.

Driessens, K. & Džeroski, S. (2004). Integrating Guidance into Relational Reinforcement Learning, *Machine Learning*, Vol. 57, pp. 271-304.

Kaelbling, L.P.; Litman, M.L. & Moor, A.W. (1996). Reinforcement Learning: A survey, *J. of Artificial Intelligence Research*, Vol. 4, pp. 237-285.

Kamal, M.A.S.; Murata, J. & Hirasawa, K. (2005). Elevator Group Control Using Multiagent Task-Oriented Reinforcement Learning, *IEEJ Trans. EIS*, Vol. 125, pp. 1140-1146.

Kamal, M.A.S. & Murata, J. (2008). Reinforcement learning for problems with symmetrical restricted states, *Robotics and Autonomous Systems*, to appear.

Kimura, H.; Yamashita, T. & Kobayashi, S. (2001), Reinforcement Learning of Walking Behavior for a Four-Legged Robot, *Proc. 40th IEEE Conf. Decision and Control*, pp. 411-416.

Kretchmar, R.M.; Feil, T. & Bansal, R. (2003). Improved Automatic Discovery of Subgoals for Options in Hierarchical Reinforcement Learning, *J. Computer Science & Technology*, Vol. 3, pp. 9-14.

McGovern, A. & Barto, A.G. (2001). Automatic Discovery of Subgoals in Reinforcement Learning using Diverse Density, *Proc. 18th Int. Conf. Machine Learning*, pp. 361-368.

Millan, J.D. (1995). Reinforcement learning of goal-directed obstacle-avoiding reaction strategies in an autonomous mobile robot, *Robotics and Autonomous Systems*, Vol. 15, pp. 275-299.

Murata , J.; Ota, K. & Abe, Y. (2007). Introduction and Control of Subgoals in Reinforcement Learning, *Proc. IASTED Conf. Artificial Intelligence and Applications*, pp. 329-334.

Singh, S. (1992). The Efficient Learning of Multiple Task Sequences, In: *Advances in Neural Information Processing Systems 4*, pp. 251-258, Morgan Kauffman, San Mateo, USA.

Smart, W.D. & Kaelbling, L.P. (2000). Practical Reinforcement Learning in Continuous Spaces, *Proc. 17th Int. Conf. Machine Learning*, pp. 903-910.
Sutton, R.S. & Barto, A.G. (1998). *Reinforcement Learning, An Introduction*, A Bradford Book, The MIT Press, Cambridge, USA.
Tesauro, G.J. (1994). TD-gammon, a self-teaching backgammon program, archives master-level play, *Neural Computation*, Vol. 6, pp. 215–219.
Wang, Y.; Huber, M.; Papudesi, V.N. & Cook, D.J. (2003). User-Guided Reinforcement Learning of Robot Assistive Tasks for an Intelligent Environment, *Proc. 2003 IEEE/RSJ Int. Conf. Intelligent Robots and Systems*, pp. 424-429.
Wiering, M. & Schmidhuber, J. (1996). HQ-Learning: Discovering Markovian Subgoals for Non-Markovian Reinforcement Learning, *Tech. Rep. IDSIA-95-96.*
Zennir, Y.; Couturier, P. & Temps, M.B. (2003). Distributed Reinforcement Learnig of a Six-Legged Robot to Walk, *Proc. 4th Int. Conf. Control and Automation*, pp. 896-900.

11

# Fault Detection Algorithm Based on Filters Bank Derived from Wavelet Packets

Oussama Mustapha[1], Mohamad Khalil[2,3], Ghaleb Hoblos[4], Houcine Chafouk[4] and Dimitri Lefebvre[1]
[1]*University Le Havre, GREAH, Le Havre,*
[2]*Lebanese University, Faculty of Engineering,*
[3]*Islamic University of Lebanon, Biomedical Department, Khaldé,*
[4]*ESIGELEC, IRSEEM, Saint Etienne de Rouvray,*
[1,4]*France*
[2,3]*Lebanon*

## 1. Introduction

The fault detection and isolation (FDI) are of particular importance in industry. In fact, the early fault detection in industrial systems can reduce the personal damages and economical losses. Basically, model-based and data-based methods can be distinguished.

Model-based techniques require a sufficiently accurate mathematical model of the process and compare the measured data with the estimations provided by the model in order to detect and isolate the faults that disturb the process. Parity space approach, observers design and parameters estimators are well known examples of model-based methods (Patton et al, 2000), (Zwingelstein, 1995), (Blanke et al., 2003), (Maquin & Ragot, 2000). In contrast, data-based methods require a lot of measurements and can also be divided into signal processing methods and artificial intelligence approaches. Many researchers have performed fault detection by using vibration analysis for mechanical systems, or current and voltage signature analysis for electromechanical systems (Awadallah & Morcos 2003), (Benbouzid et al., 1999). Other researchers use the artificial intelligence (AI) tools for faults diagnosis (Awadallah & Morcos 2003) and the frequency methods for faults detection and isolation (Benbouzid et al., 1999). This study continues our research in frequency domain, concerning fault detection by means of filters bank (Mustapha et al-a, 2007), (Mustapha et al-b, 2007).

The aim of this article is to propose a method for the on-line detection of changes applied after a modeling of the original signal. This modeling is based on a filters bank decomposition that is needed to explore the frequency and energy components of the signal. The filters coefficients are derived from the wavelet packets, so the wavelet packets characteristics are approximately conserved and this leads to both filtering and reconstruction of the signal.

This work is a continuity of our previous works for deriving a filters-bank from wavelet packets because the wavelet packets offer more flexibility for signal analysis and offer a lot of bases to represent the signal.

The main contributions are to derive the filters and to evaluate the error between filters bank and wavelet packets response curves. Filters bank is preferred in comparison with wavelet packets because it could be directly hardware implemented as a real time method. Then, the Dynamic Cumulative Sum detection method (Khalil, Duchêne, 1999) is applied to the filtered signals (sub-signals) in order to detect any change in the signal (figure 1). The association of filters bank decomposition and the DCS detection algorithm will be shown to be of great importance when the change is in the frequency domain.

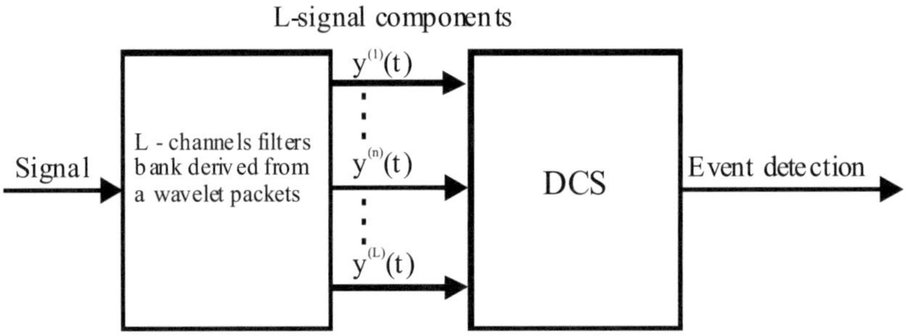


Fig. 1.Two stages algorithm for change detection (*L* is the number of filters used).

This paper is decomposed as follows. First we will explain the problem and we will present the utility of the decomposition before the detection. Then in section 3, the wavelet transform and the wavelet packets are presented and the derivation of filters bank from a wavelet packets and the problem of curve fitting are introduced, in the same section, the best tree selection based on the entropy of the signal and the filters bank channels corresponding to the suitable scale levels are discussed. In section 4, the Cumulative Sum (CUSUM), the Dynamic Cumulative Sum (DCS) algorithms and the fusion technique are detailed. Finally, the method is applied for the diagnosis of the Tennessee Eastman Challenge Process.

## 2. Filters bank for decomposition and detection

The simultaneous detection and isolation of events in a noisy non-stationary signal is a major problem in signal processing. When signal characteristics are known before and after the change, an optimal detector can be used according to the likelihood ratio test (Basseville & Nikiforov 1993). But when the signal to be detected is unknown, the Generalized Likelihood Ratio Test (GLRT) which consists of using the maximum likelihood estimate of the unknown signal will be used.

In general, the segmentation depends on the parameters that change with time. These parameters, to be estimated, depend on the choice of the signal modeling. Most authors make use of application-dependent representations, based on AR modeling or on wavelet transform, in order to detect or characterize events or to achieve edge detection in signals (Mallat, 2000).

When the change is energetic, many methods exist for detection purposes. But when the change is in frequency contents, special modeling, using a filters bank, is required before the application of the detection methods.

After this modeling, the detection algorithm (DCS) will be applied on the decomposed signals instead of applying it to the original signal (see figure 1). The motivation is that the filters bank modeling can filter the signals and transform the frequency change into energy change. Then we choose only the sub-signals which present energy changes after decomposition. Furthermore, the detectability of DCS is improved if the changes are in energy. The sub-signals can be used also to classify the detected events and this will be done after extracting the necessary parameters from isolated events and finally this aims to make diagnosis.

This work is originated from the analysis and characterization of random signals. In our case, the recorded signals can be described by a random process *x(t)* as:

$$x(t) = x_1(t) \text{ Before the point of change } t_r$$

$$x(t) = x_2(t) \text{ After the point of change } t_r$$

$t_r$ is the exact time of change. $x_1(t)$ and $x_2(t)$ can be considered as random processes where the statistical features are unknown but assumed to be identical for each segment 1 or 2. Therefore we assume that the signals $x_1(t)$ and $x_2(t)$ have Gaussian distributions. We will suppose also that the appearance times of the changes are unpredictable.

In our case, we suppose that the frequency distribution is an important factor for discriminating between the successive events. In this way, the filters bank decomposition will be useful for classification purposes.

For each component *m*, and at any discrete time *t*, the sample $y^{(m)}(t)$ is on-line computed in terms of the original signal *x(t)* using the parameters $a(i)^{(m)}$ and $b(i)^{(m)}$ of the corresponding filter according to the following difference equation:

$$y^{(m)}(t) = \sum_{i=0}^{p} b(i)^{(m)} x(t-i) - \sum_{i=1}^{q} a(i)^{(m)} y^{(m)}(t-i) \tag{1}$$

After decomposition of *x(t)* into $y^{(m)}(t)$, $m$=1,2,3,...$L$ the problem of detection can be transformed to an hypothesis test:

$$H_0 : y^{(m)}t\ ; t = \{1,...,t_r\} \text{ has a probability density function } f_0$$

$$H_1 : y^{(m)}t\ ; t = \{t_r+1,...,n\} \text{ has a probability density function } f_1$$

Figure 2 shows the original signal presenting a change in frequency content at time 1000 time units (TU). We can see that the decomposition enhances the energy changes, and it is more accurate to apply the detection on the sub-signals instead of applying it on the original signal.

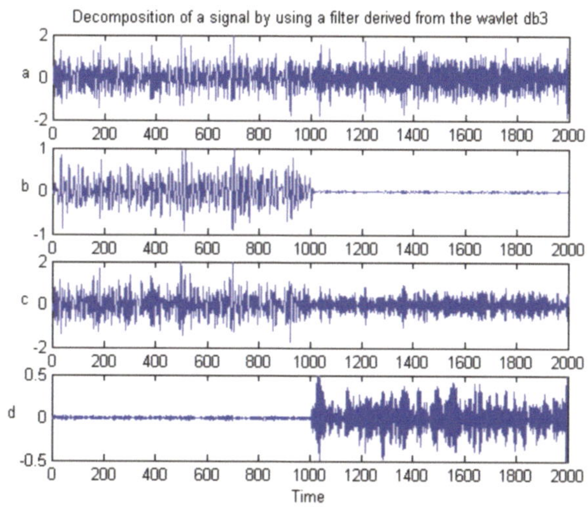


Fig. 2. a) original simulated signal presenting a frequency change at $t_r$=1000 *TU*. b,c,d) the decomposition of the signal into three sub-signals using 3 channels filters bank.

## 3. Filters bank derived from wavelet packets

### 3.1 Wavelet Transform

The Fourier analysis is the most well known mathematical tool used for transforming the signal from time domain to frequency domain. But it has an important drawback represented by the loss of time information when transforming the signal to the frequency domain. To preserve the temporal aspect of the signals when transforming them to frequency domain, one solution is to use the Wavelet Transform (Flandrin, 1993) which analyzes non-stationary signals by mapping them into time-scale and time-frequency representation.

The Wavelet Transform is similar to the Short Time Fourier Transform but provides, in addition, a multi-resolution analysis with dilated and shifted windows. The multi-resolution analysis consists of decomposing the signal *x(t)* using the wavelet $\psi(t)$ and its scale function $\phi(t)$ ,(Papalambros & Wilde, 2000).

$$\psi_{ab}(t)=\frac{1}{\sqrt{a}}\psi(\frac{t-b}{a}) \qquad (2)$$

$$T_x^{\psi}(a,b)=\int_{-\infty}^{+\infty}x(t)\psi_{ab}(t)dt \qquad (3)$$

where $a$ and $b$ are respectively the dilation and translation parameters. The filter associated to the scale function $\phi(t)$ is a low pass filter and the filter associated to the wavelet $\psi(t)$ is a band pass filter.

The relation (3) can be seen as the cross correlation function of the signal *x(t)* and $\frac{1}{\sqrt{a}}\psi(\frac{t}{a})$.

This relation can also be written as follows:

$$T(\tau,a)=x(\tau)*\frac{1}{\sqrt{a}}\psi^*(\frac{-\tau}{a}) \qquad (4)$$

But knowing that $\psi^*(-t)=\psi(t)$ , then:

$$T(\tau,a)=x(\tau)*\frac{1}{\sqrt{a}}\psi(\frac{\tau}{a}) \qquad (5)$$

The relation (5) indicates that the wavelet transform can be obtained by filtering the signal *x(t)* by using a series of band pass filters of impulse responses $\frac{1}{\sqrt{a}}\psi(\frac{t}{a})$ and of central frequencies centered at $f_0/a$ .

The band width of the band pass filters bank decreases when $a$ increases.

In order to decompose a signal into components of equal decreasing frequency intervals, we have to use a discrete time-frequency domain and the dyadic wavelet transform:

For $a=2^j$ and $b=k2^j$ (*j* and *k* are integers, the wavelets $\frac{1}{\sqrt{a}}\psi(\frac{t-\tau}{a})$ become:

$$\psi_{j,k}(t) = 2^{-j/2}\psi(2^{-j}t - k) \tag{6}$$

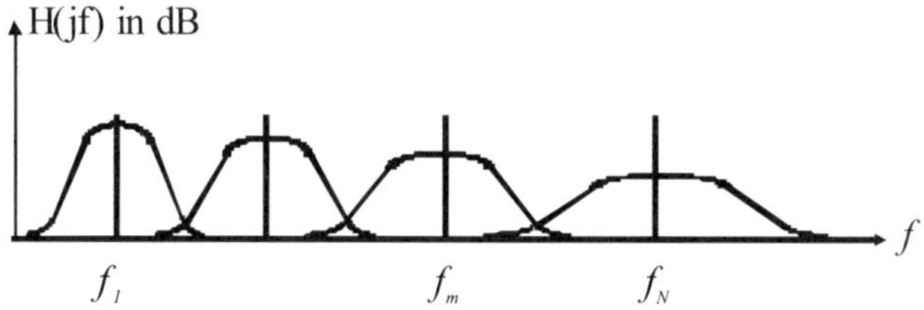


Fig. 3. Response curves of the wavelet filters.

If the signal *x(t)* is sampled, the discrete wavelet transform will be:

$$T_{j,k} = \sum_{n=-\infty}^{+\infty} x_n \psi^*_{j,k}(n) \tag{7}$$

Note that all the wavelet frequencies will be between 0 and $f_s$ (sampling frequency of the signal).
The use of discrete time-frequency domain and the dyadic wavelet transform leads to decompose the signal into components of equal decreasing frequency intervals.

### 3.1.1 Scaling function and wavelet bases

The projection of a function $f(t) \in L_2(R)$ on the orthonormal basis $\{\varphi(t-k)\}$ is a correlation between the original function *f(t)* and the scaling function $\varphi(t)$ sampled at integer intervals.

$$Approx_j(t) = (projV_j x)(t) \tag{8}$$

The approximations resulting from the projection of *f(t)* on the scaling function basis form subspaces $V_j \in L_2(R)$.
The projection of a function on the orthonormal scaling function basis results in loss of some information about the function:

$$Detail_j(t) = approx_{j-1}(t) - approx_j(t) \tag{9}$$

To obtain the detail information of the function, we use the projections on the orthonormal wavelet bases.
With Discrete Wavelet Transform (DWT), the multi-resolution analysis uses a scaling function and a wavelet to perform successive decompositions of the signal into approximations and details (Chendeb, 2002).

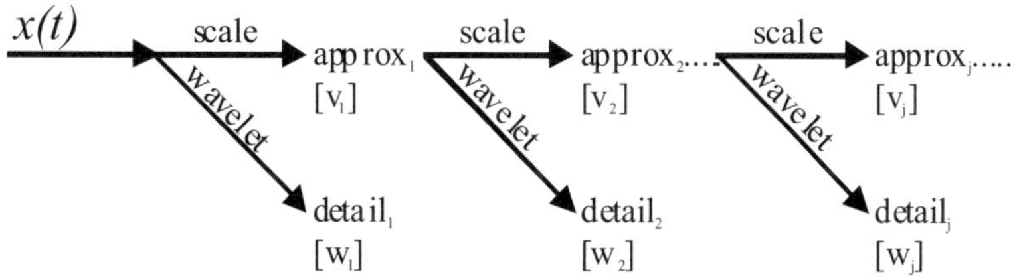


Fig. 4. Multiresolution Analysis: successive decompositions into approximations and details.

### 3.1.2 Orthogonal and bi-orthogonal analysis with Mallat algorithm

Practically, the Mallat's algorithm is used for orthogonal and bi-orthogonal analysis. This algorithm consists of low-pass filtering the signal with a filter having its impulse response *h[n]* then down-sampling the result to obtain the lower resolution approximation, and high-pass filtering the signal with a filter having its impulse response *g[n]* then down-sampling the result to obtain the lower resolution detail information.
The numerical sequence *h[n]* will be considered as the impulse response of a digital filter.
In the filters bank theory, we can find that the wavelet can be modeled by a discrete quadrature mirror filters bank *g[n]* and *h[n]* such that (Flandrin, 1993):

$$g[n] = (-1)^n h[1-n] \tag{10}$$

*g[n]* will be considered as the impulse response of a discrete low pass filter whose coefficients can represent the scale function $\varphi(t)$. And *h[n]* will be considered as the impulse response of a discrete high pass filter whose coefficients can represent the wavelet $\Psi(t)$.
Figure 5 represents the Mallat's multiresolution recursive analysis.

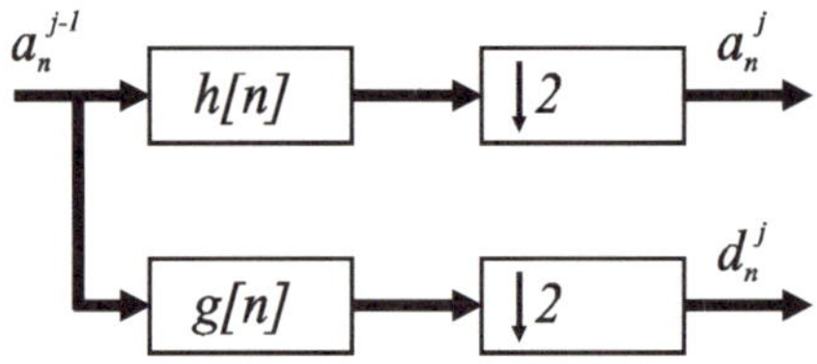


Fig. 5. Analysis Algorithm of Mallat.

The discrete wavelet transform consists of discrete filtering operation followed by down-sampling process. Figure 6 shows the response curves of the quadrature mirror filters used for three levels decomposition.

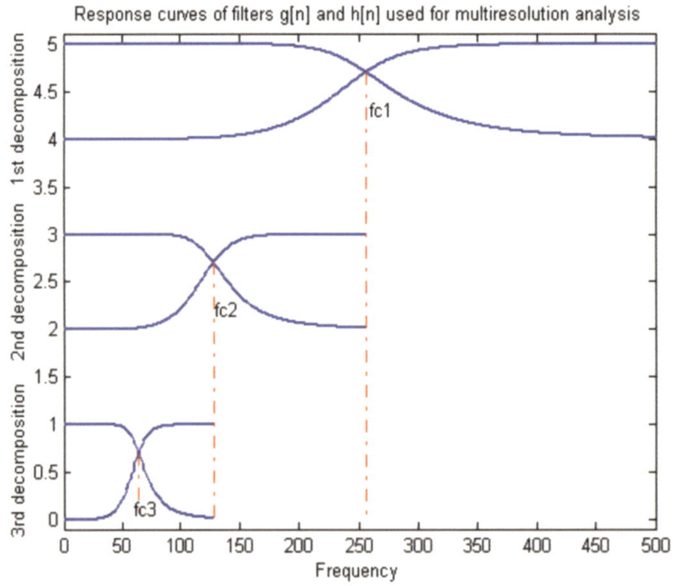


Fig.6. Response curves of the filters used in Mallat's algorithm ($f_{ci}$ are the cut off frequencies of the filters used).

### 3.1.3 Base and wavelet selection

The choice of the wavelet is a critical problem. To extract a specific signal event, the choice of the wavelet becomes important such that the wavelet must be adapted to the changes to be detected.

On the other hand, the choice of scale levels is also important. After decomposing a signal, the most suitable scale levels are chosen according to their frequency bands, and the application of the detection must be applied to the most suitable scale levels.

Most likely, the choice of the filters bank, after the derivation, depends on the original signal and its frequency band. The number of filters depends on the details that we have to extract from the signal and to the events that must be distinguished.

The result of detection depends on the number of the filters used and the cut off frequency of each channel. In practice, L-channels derived filters are used. These filters are chosen so that the cut off frequencies are between zero Hertz and the half of the sampling frequency ($f_s/2$). We will see later an automatic procedure, based on the entropy of the signal, to select the best filters bank cut off frequencies.

### 3.2 The wavelet packets

The wavelet packet decomposition is a generalization of the wavelet decomposition that offers more flexibility for signal analysis. In wavelet analysis, the approximation of the signal is decomposed into a lower-level approximation and detail. So the dyadic decomposition tree of the wavelet transform represents the signals by means of a fixed basis, and for n-level decomposition tree, we have $n+1$ possible ways to decompose the signal. In wavelet packet analysis, the details as well as the approximations can be decomposed into a lower-level approximations and details and this yields more than $2^{2^{n-1}}$ possible ways to decompose the signal. So the wavelet packets offer a lot of bases to represent the signal from which a best decomposition tree can be selected, according to a given criterion, to meet the design objectives.

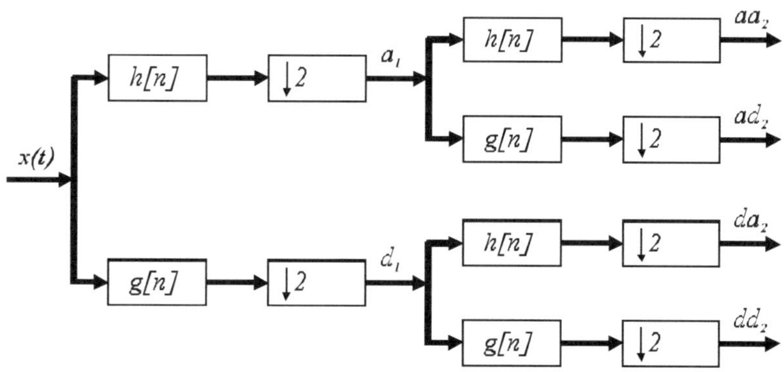


Fig. 7. Wavelet packet, decomposition of both the details and the approximations.

### 3.2.1 Wavelet packets decomposition

The wavelet packets are defined by Coifman, Meyer and Wickerhauser (Mallat, 1999) as a generalized relation between the multiresolution approximations and the wavelet. The approximation multiresolution space $V_j$ will be decomposed into a lower resolution approximation $V_{j+1}$ and details $W_{j+1}$. This is done by decomposing the orthogonal basis

$\left\{\phi_j(t-2^j n)\right\}_{n\in Z}$ of $V_j$ into two new orthogonal bases $\left\{\phi_{j+1}(t-2^{j+1} n)\right\}_{n\in Z}$ of $V_{j+1}$ and $\left\{\psi_{j+1}(t-2^{j+1} n)\right\}_{n\in Z}$ of $W_{j+1}$.

The difference between the wavelet and the wavelet packet decompositions are shown by the binary tree indicated by figure 8. And for seek of simplicity, in wavelet packet decomposition, the vector space $V$ will be replaced by $W$.

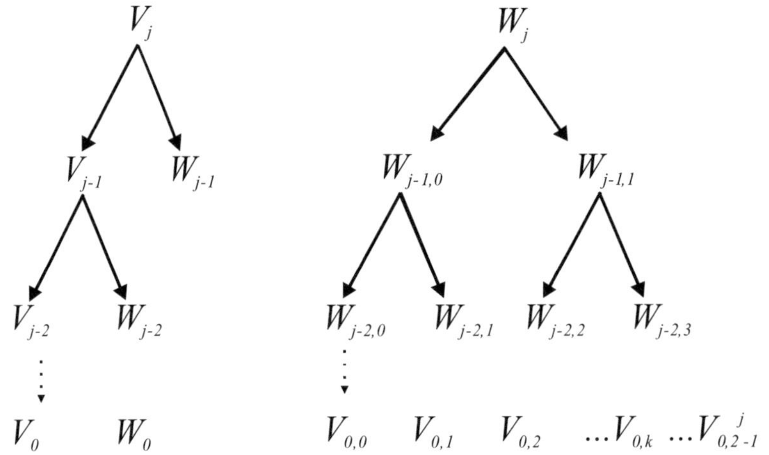


a) Wavelet decomposition b) Wavelet packet decomposition

Fig. 8. Wavelet and wavelet packet decompositions.

The wavelet packets nodes can be indexed by the scale parameter $j$, the node index $n$ at the scale level $j$ and the time parameter $k$.

For different values of $j$ and $n$, the wavelet packets decomposition can be organized as a binary tree as shown in figure (9).

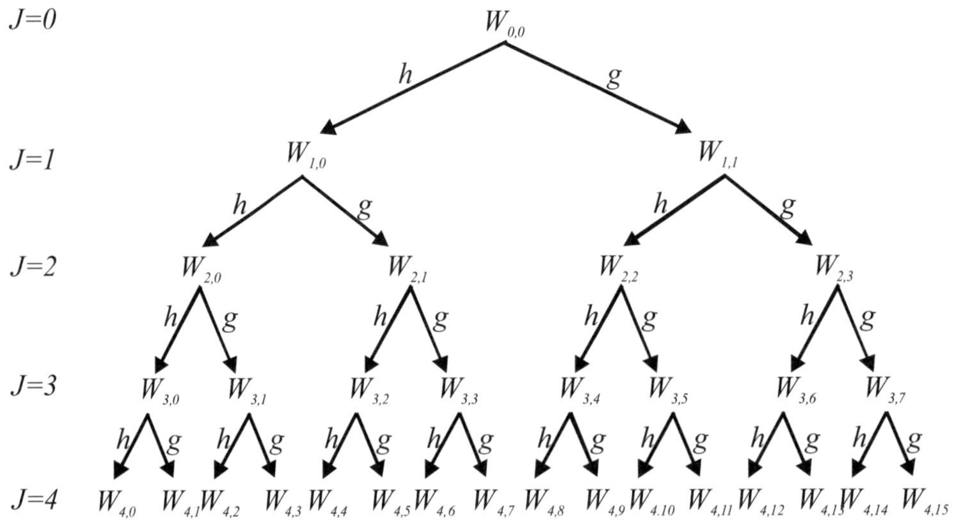


Fig. 9. Binary tree obtained after wavelet packets decomposition.

Figure 9 shows clearly that the wavelet packet decomposition offers a lot of bases to represent the signal and we have more than $2^{2^{n-1}}$ ,instead of *n+1*, possible ways to decompose the signal, from which a best decomposition tree can be selected for signal analysis.

### 3.2.2 Wavelet packets filters bank

Each node of the binary tree is indexed by *(j,n)* where *n* is the number of the node at level *j*. For each node there is an associated space $W_{j,n}$ which admits an orthogonal basis $\left\{\psi_j^n(t-2^j k)\right\}_{k\in Z}$. The relations between a node and its children nodes can be seen as follows:

$$\psi_{j+1,2n}(t) = \sum_{k=-\infty}^{+\infty} h[k]\psi_{j,n}(t-2^j k) \tag{11}$$

and

$$\psi_{j+1,2n+1}(t) = \sum_{k=-\infty}^{+\infty} g[k]\psi_{j,n}(t-2^j k) \tag{12}$$

The wavelet packets coefficients are calculated by using a filters bank algorithm. Suppose that the projection tree of a signal in the sub-space *W* is *C*, we can define a wavelet packets $C_{j,n} = \left\{C_{j,n}(k)\right\}_{k\in Z}$ as the projection of the signal in the sub-space $W_{j,n}$ where $\psi_{j,n}$ is an orthogonal basis.

For each node *(j,n)* of the tree, the coefficients of the wavelet packets of a function $f(t)$ can be calculated as follows:

$$C_{j,n}(k) = \left\langle f(t), \psi_{j,n}(t-2^j k)\right\rangle \tag{13}$$

The wavelet packets decomposition is obtained by using pyramidal recursive algorithm as follows:

$$C_{j+1,2n}(k) = \sum_l \hat{h}[2k-l]C_{j,n} \tag{14}$$

$$C_{j+1,2n+1}(k) = \sum_l \hat{g}[2k-l]C_{j,n} \tag{15}$$

The packet $C_{j+1,2n}$ (respectively $C_{j+1,2n+1}$) is obtained by low-pass (respectively high-pass) filtration of $C_{j,n}$ by using the filter *g* (resp. *h*), followed by down sampling by factor of 2 that projects the signal to sub-space $W_{j+1,2n}$ (resp. $W_{j+1,2n+1}$).

### 3.2.3 Best tree selection

The wavelet packet decomposition generates redundant representations. Each decomposition level contains all the signal information. Some combinations of packets contain non-redundant representation of the signal, from which a best tree can be selected. The selection must be based on a criterion according to the problem to be treated. In our case and for detection purposes, the Shannon entropy-based algorithm is used, where the criterion is defined to allow an optimal separation of the frequency components of the signal (in case of frequency change). The best tree selection algorithm offers optimal signal representation (Coifman & Wicherhauser, 1992).

For a series of discrete elements $X = \{x_1, x_2, ..., x_N\}$, the Shannon entropy is defined by:

$$M(\mathbf{x}) = -\sum_{i=1}^{N} (x_i)^2 \log(x_i)^2 \tag{16}$$

where $(x_i)^2 \log(x_i)^2 = 0$ if $x_i = 0$.

*M(X)*, which is the expectation of information quantity, reflects the concentration of energy in the discrete series *X*.

Based on the entropy criterion, the best tree selection is done as follows:

a. For a given node $A_{j,n}$, the energy is compared to threshold energy *(SE)* then this node is labeled by 0 if its energy is lower than *SE* and by 1 in other cases.

$$A_{j,n} = \begin{cases} 1 & if\ energy\ (C_{j,n}) > SE \\ 0 & else \end{cases} \tag{17}$$

b. Indicate the number of components present in each packet.

c. Select the single-component packets.

In (Hitti. & Eric, 1999), Hitti proposed a method to determine the threshold energy by estimating the sum of energies of packets in each level $j$ ($EnsSP_j$) and then the threshold energy can be calculated as follows:

$$S_j = \frac{EnsSP_j}{2^j} \tag{18}$$

Figure (10) shows the different steps to select the best tree.

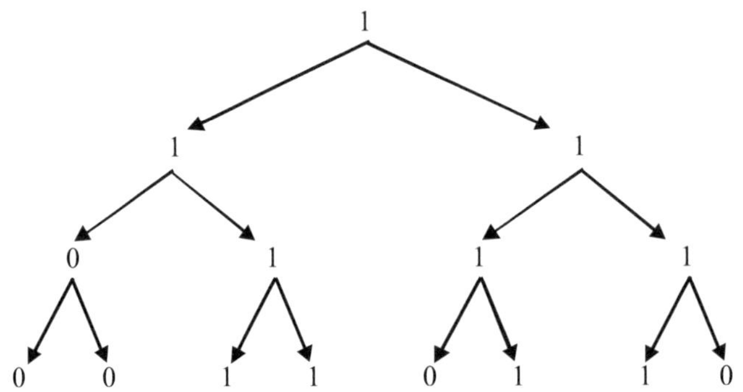


a) Tree obtained after thresholding the packets energies (the node is labeled by 0 if its energy is lower than SE and by 1 in other cases).

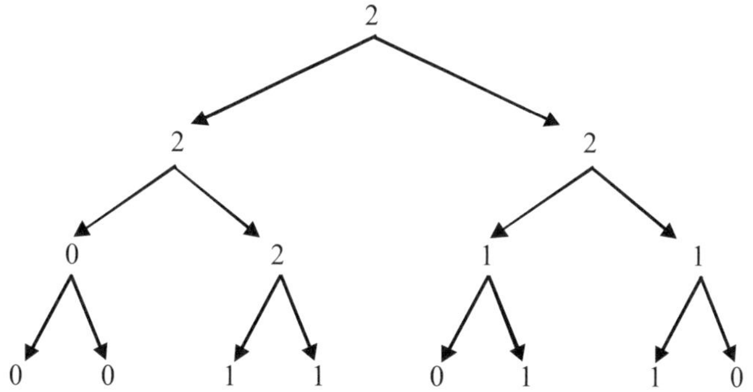


b) Number of present components per packet.

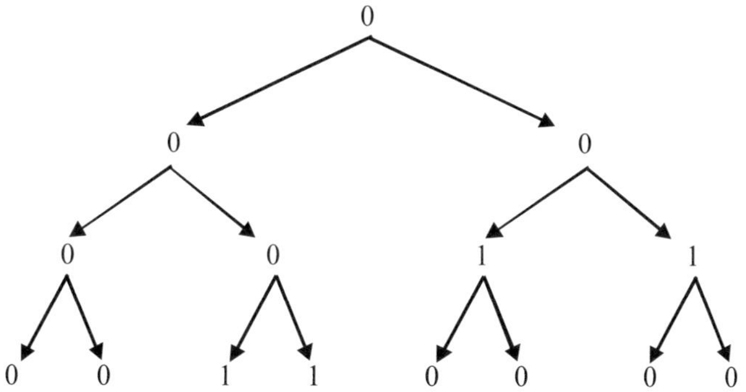


c) The best tree is determined by single-component packets (the nodes labeled 1).

Fig. 10. Steps of best tree selection according to Hitti algorithm.

### 3.3 Filters bank parameters estimation (Butterworth filters)

In order to explore the frequency and energy components of the original signal, an important pre-processing step is required before detection, feature extraction and classification. At a discrete time $t$, the signal is first decomposed by using an L-channels filters bank whose parameters are derived from wavelet packets. Each component $m \in \{1,\ldots, L\}$ is the result of filtering the original signal $x(t)$ by the derived filter for a given cut off frequency.

The filters coefficients are derived from the wavelet packets in order to use them as wavelet packets so the wavelet packets characteristics are approximately conserved and can be used for both filtering and reconstruction of the signal. Also the wavelet packets derived filters bank can be hardware implemented and can be used for online detection.

For a given wavelet, we can use its frequency domain transfer function $h_{wav}$ to derive (extract) the transfer function of a low pass Butterworth filter $h_{filt}$. After the extraction, the frequency responses of the wavelet packets-based filters bank can be determined as follows:

1. Selection of the wavelet.
2. Extractions of the transfer function of the wavelet $h_{wav}$.
3. Calculation of the cut off frequency ($f_c$) of the wavelet filter (at -3db attenuation).
4. Estimation of the order of the filter that minimizes the error between the two transfer functions $h_{wav}$ and $h_{filt}$. The estimation is done by using the least square method to calculate the optimal filter order $N$ that minimizes the error between the two transfer functions $h_{wav}$ and $h_{filt}$ (figure 11).

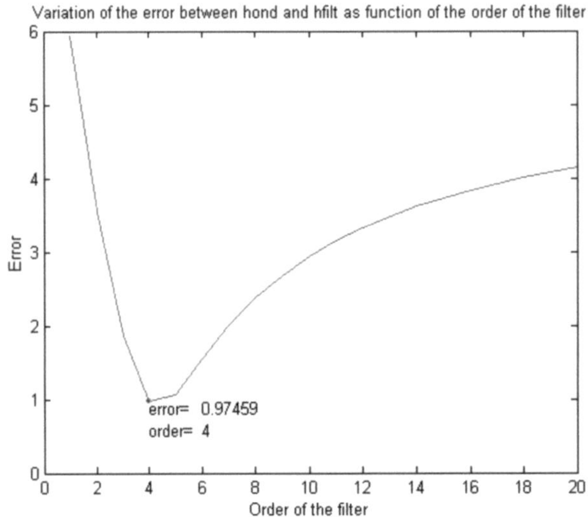


Fig.11. Curve representing the variation of the error in terms of the order of the filter.

1. Decompose the signal by using the wavelet packets.
2. Select the best tree by using an entropy-based method.
3. Determine the suitable cut off frequencies of the filters bank according to the selected best tree.
4. Design of the corresponding Butterworth filters according to the Butterworth general transfer function:

$$H(jf) = \frac{1}{\sqrt{1+(f/f_c)^{2N}}} \tag{19}$$

where *H(jf)* is the transfer function, $f$ is the frequency, $f_c$ is the cut off frequency and $N$ is the order of the filter. Practically, the coefficients *a(i)* and *b(i)* necessary to calculate the equation (19) will be determined from the Butterworth table. Note that $m$ represents the decomposition level.
To have a quadrature mirror filters, we can extract a high pass filter by using the following relation:

$$h_{filt\_H} = (-1)^k h_{filt\_L}(2N+1-k) \; for \; k = 1,2,...,2N \tag{20}$$

So we can design a filters bank that behaves as a wavelet packets. This filters bank can be used, instead of a wavelet packets, to decompose a signal into some frequency components in order to explore the signal.
The main problem is to solve nonlinear curve-fitting (data-fitting) in the least squares sense. Given input data $h_{wav}$, and the desired output data $h_{filt}$, find coefficient $N$ that "best-fit" the two curves is well known in the numerical analysis domain. This problem is named curve-fitting problem (Papalambros & Wilde, 2000), (Rustagi, 1994). This problem consists to find the parameters which permit the best fitting of the analytic model to the real data. If the model is linear, the problem is solved by using linear regression methods (Saporta, 1990), and if the model is nonlinear, the problem is solved by using nonlinear regression methods (Rardin, 1998).

The optimization consists to express the relation between the two variables $f$ and $h_{wav}$ for a function $h_{filt}$ that have non linear dependence with the vector:

$$h_{filt} = F(f, N) \tag{21}$$

In order to solve this problem we use the least squares method, we must minimize with respect to $N$ the quadratic error:

$$\min_N \frac{1}{2}\left\| h_{filt}(N, f) - h_{wav} \right\|^2 = \frac{1}{2}\sum_i h_{filt}(N, f_i) - h_{wav\,i})^2 \tag{22}$$

So the nonlinear curve-fitting problem can be solved by the classical algorithm of Gauss - Newton (Papalambros & Wilde, 2000) that is: knowing the desired output we can find the optimal coefficient $N$ that best fits the two curves.
Because $F$ is a non linear function of parameter $N$, the optimization is done by an iterative method: We start with a given value of $N$, then we continue the iterations until that the parameter $N$ can't be changed. The figure 12 shows the response curve of the wavelet 'db5' and the response curve of the filter derived from it. The order of the derived filter is 6, and the error corresponding to the minimum error and calculated by using the Euclidian distance:

$$e = \sum (h_{ond} - h_{filt})^2 \tag{23}$$

Is e=0.78869.

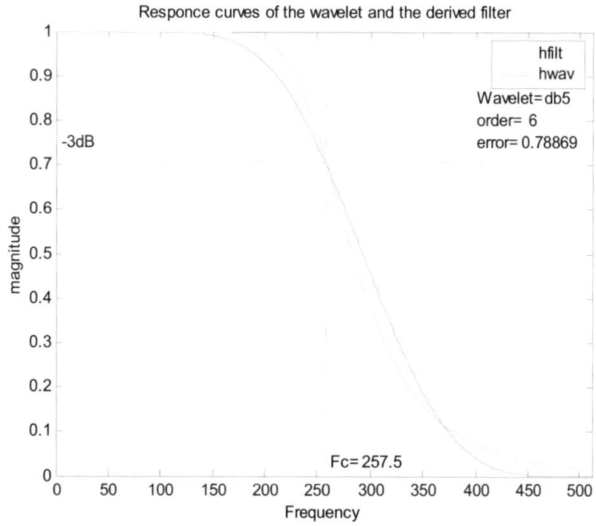


Fig.12. Response curve of a Butterworth low pass filter derived from the wavelet 'db5' (error: e = 0.78869).

## 4. Real time detection

### 4.1 Sequential algorithms

The CUSUM algorithm is based on a recursive calculation of the logarithm of the likelihood ratios. This method can be considered as a sequence of repeated tests around the point of change $t_r$ (Nikiforov, 1986), (Basseville & Nikiforov 1993). Let $x_1, x_2, x_3, \ldots, x_t$ be a sequence of observations. Let us assume that the distribution of the process $X$ depends on parameter $\theta_0$ until time $t_r$ and depends on parameter $\theta_1$ after the time $t_r$. At each time t we compute the sum of logarithms of the likelihood ratios as follows:

$$S_1^{(t,m)} = \sum_{i=1}^{t} s^{(m)}{}_i = \sum_{i=1}^{t} Ln \frac{f_{\theta_1}(x_t / x_{t-1}, \ldots, x_1)}{f_{\theta_0}(x_t / x_{t-1}, \ldots, x_1)} \tag{24}$$

where, $f_\theta$ is the probability density function. The importance of this sum comes from the fact that its sign changes after the point of change.

The CUSUM method is usually applied when we have a priori knowledge about the segments before and after the changes (energy change or probability density distribution before and after the change).

The Dynamic Cumulative Sum (DCS) technique, based on the local dynamic cumulative sum, is preferred when the parameters of the signal are unknown (Khalil & Duchêne, 1999). The DCS is a repetitive sequence around the point of change $t_r$. It is based on the local cumulative sum of the likelihood ratios between two local segments estimated at the current time $t$. These two dynamic segments $S_a^{(t)}$ (after $t$) and $S_b^{(t)}$ (before $t$) are estimated by using two windows of width $W$ (figure 13) before and after the instant t as follows:

* $S_b^{(t)} : x_i \;\; ; i = \{t-W, \ldots t-1\}$ follows a probability density function $f_{\theta_b}(x_i)$
* $S_a^{(t)} : x_i \;\; ; i = \{t+1, \ldots t+W\}$ follows a probability density function $f_{\theta_a}(x_i)$

The parameters $\hat{\theta}_b^{(t)}$ of the segment $S_b^{(t)}$, are estimated using $W$ points before the instant $t$ and the parameters $\hat{\theta}_a^{(t)}$ of the segment $S_a^{(t)}$, are estimated using $W$ points after the instant $t$.

At a time $t$, and for each level $m$, the DCS is defined as the sum of the logarithm of likelihood ratios from the beginning of the signal up to the time $t$:

$$DCS^{(m)}(S_a^{(m)}, S_b^{(m)}) = \sum_{i=1}^{t} Ln \frac{f^{(i)}_{\hat{\theta}_a}(x_i)}{f^{(i)}_{\hat{\theta}_b}(x_i)} = \sum_{i=1}^{t} \tilde{S}i \tag{25}$$

(Khalil, 1999) proves that the DCS function reaches its maximum at the point of change $t_r$. The detection function used to estimate the point of change is:

$$g^{(m)}t = \max_{1 \le i \le t}[DCS^{(m)}(S_a^{(t)}, S_b^{(t)})] - DCS^{(m)}(S_a^{(t)}, S_b^{(t)}) \tag{26}$$

The instant at which the procedure is stopped is $t_s = min\ \{t : g(m)t \ge h\}$, where $h$ is the detection threshold. The point of change is estimated as follows:

$$t_c = \max\{t > 1 : g^{(m)}t = 0\} \tag{27}$$

The application of the DCS method improves the detection when applied after filtration, specially when the signal presents no abrupt change, and the direct application of the DCS algorithm leads us to faulty results and sometimes difficult to interpret for accurate fault detection (figure 14). The use of the filters bank improves considerably the detection ability of the DCS method. And the sub-signal frequency components, constituting the frequencies in defined ranges will be transformed into energy changes.

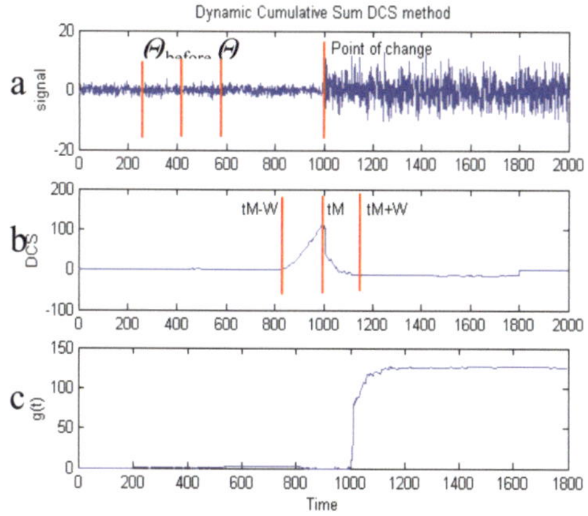


Fig. 13. Application of the DCS on a signal of abrupt change.
a) Original signal.
b) DCS function.
c) Detection function g(t).

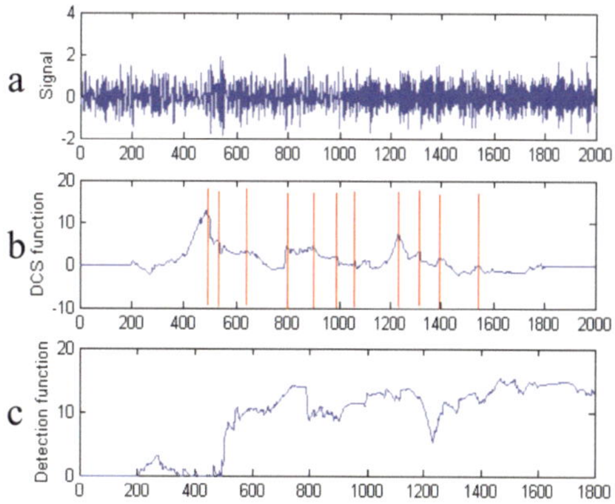


Fig. 14. Application of the DCS on a signal with frequency change.
a) Original signal.
b) DCS function.
c) Detection function g(t).

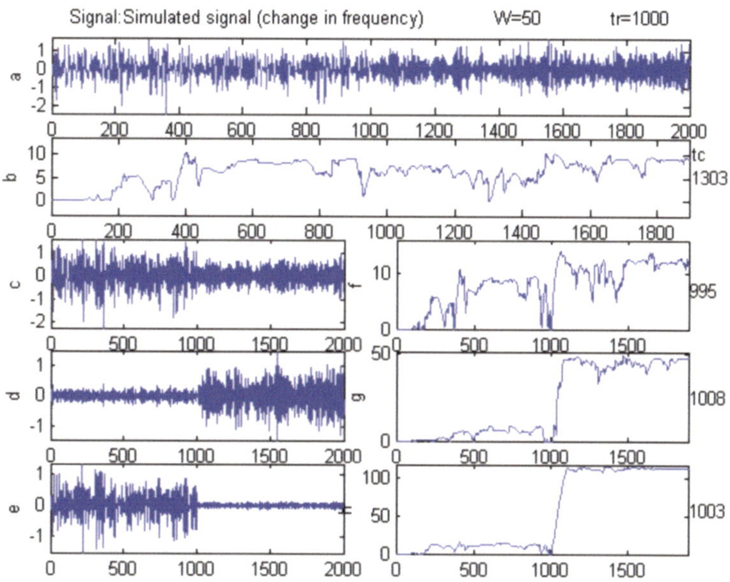


Fig. 15. Results after the decomposition. At left: decomposition of the signal into 3 components. At right: the detection functions of each component. a) Original signal presenting frequency change at 1000 *T.U.* b) DCS applied directly on the original signal.
c) d) e) Decomposition using filters bank derived from the 'Haar' wavelet packet.
f) g) h) Detection functions applied on the filtered signals (c, d, e).

Figure (15) shows clearly the improvement offered by the derived filters bank pre - processing before the application of the DCS algorithm. The original signal is a simulated signal obtained by concatenating two segments $S_1(t)_{t\in[0\ 1000s]}$ and $S_2(t)_{t\in[1001\ 2000s]}$, i.e the real point of change is $t_r = 1000$. These segments are generated by filtering a white gaussian noise by two band pass filters of central frequencies $f_1$ and $f_2$. The point of change of each component is calculated by the detection algorithm.

## 4.2 Fusion of change points

Because the detection algorithm is applied individually to each frequency component, it is important to apply a fusion technique to the resulting times of change in order to get a single value for a given fault in the system. The fusion technique is achieved as follows:

- Each point of change at a given level is considered as an interval *[tc-a, tc+a]*, where *a* is an arbitrary number of points taken before and after the point of change.
- The time intervals with common areas are considered to correspond to the same fault.
- The resulting point of change $t_f$ is calculated as the center of gravity (or mean) of the superimposed intervals.

This procedure is shown on figure 16. The signal is a simulated one with two segments (each segment has 1000 points) of different frequency. The point of change of each component is calculated by the detection algorithm, and then the fusion technique is applied to get a single point of change $t_f$. The real point of change is $t_r$ = 1000 and, after fusion, the estimated point of change is $t_f$ = 1003.

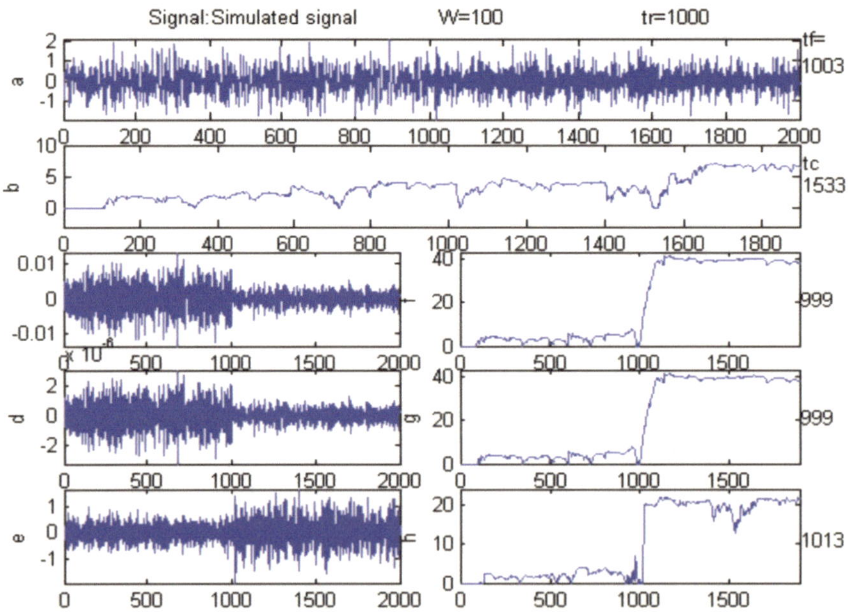


Fig. 16. Fusion of the change points. a) Original signal presenting a frequency change b) DCS applied directly on the original signal c) d) e) Decomposition using derived filters f) g) h) Detection functions applied on the filtered signals.

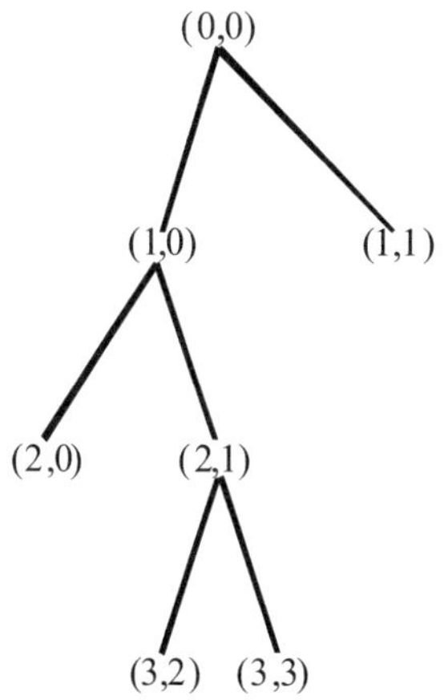


Fig.17. Obtained best tree.

Figure 17 shows the best tree obtained after decomposing the signal by using filters bank derived from the 'Haar' wavelet packet.

### 4.3 Performance evaluation

To study the performance of the detection method, the probability of detection and the probability of false alarm are calculated. Also the detection delay which is the difference between the stopping time and the real time of change is calculated. In order to evaluate the performance of our algorithm, we compare the DCS with filtration and the DCS without filtration in case of a change in variance. The comparison is done by using the Receiver Operational Curve ROC which is a curve that represents the probability of detection as function of the probability of false alarm. The two methods are tested, by using 20 signals randomly generated. The variances of the signals are 1 and 1.3.

The comparison between curves in figure 18 shows that, for the same probability of false alarm, the probability of detection, in case of DCS with filtration, is higher than that of the DCS without filtration. However, let us focus on the following remarks:

1. The use of a filters bank to decompose the signal, before the application of the DCS method, causes a delay to the detection because the current output sample is calculated in terms of current and past input samples. In the case of on-line detection, it is necessary to estimate a model on line at the same time as the input data is received, to make the decision on-line, and in order to investigate possible time variation in the signal parameters during the collection of data. Our future works include the study of detection delay in order to deal with real time applications.
2. The use of filters bank is not suitable in case of varying mean signals because the band pass filters, used to decompose the signal, will attenuate all mean variation.

In this last part, the influence of the window size is also investigated. The comparison between curves in figure 19 shows that, for the same probability of false alarm, the probability of detection increases if the width $W$ of the window increases.

These curves are plotted for $W = 25$, $W = 50$ and $W = 100$ points. It is clear that if the window becomes large, the false alarm probability decreases. But in this case small events will not be detected and the delay detection time will be increased. A trade-off must be found between the detection delay and the length of the window.

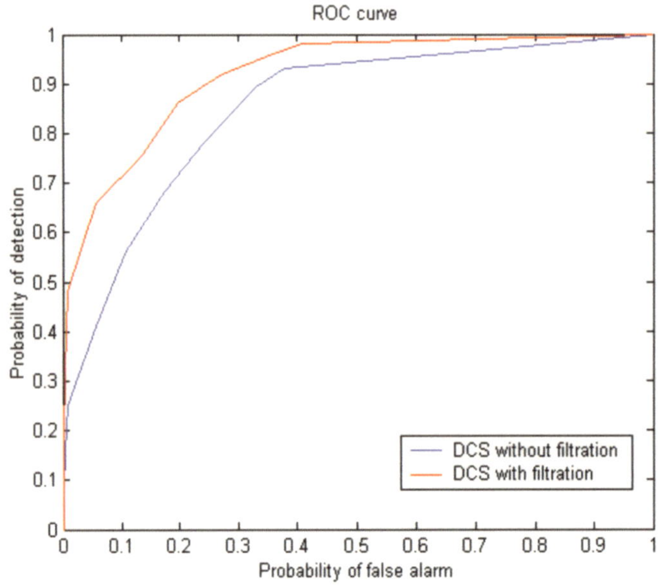


Fig. 18. Comparison between the DCS and DCS with decomposition.

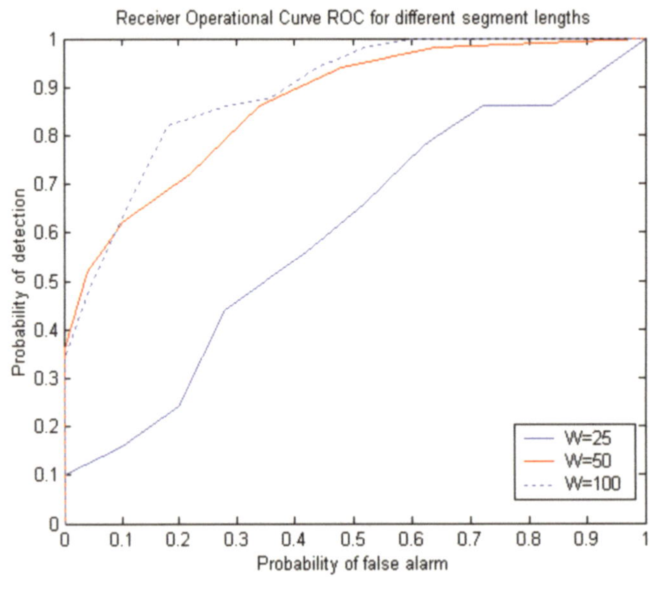


Fig. 19. Comparison between the DCS with filtration for three different window lengths.

## 5. Applications to the Tennessee Eastman Challenge Process

In this section, the method based on wavelet packet filters bank decomposition and DCS algorithm is applied to detect disturbances on the Tennessee Eastman Challenge Process (TECP, Figure 20). The TECP (Downs & Vogel, 1993) is a multivariable non-linear, high dimensionality, unstable open-loop chemical reactor that is a simulation of a real chemical plant provided by the Eastman Company. The process results in final products G and H from four reactants A, C, D and E. The plant has 7 operating modes, 41 measured variables and 12 manipulated variables. There are also 20 disturbances IDV1 through IDV20 that could be simulated (Downs & Vogel, 1993), (Singhal, 2000). The sampling period for measurements is 60 seconds. The TECP offers numerous opportunities for control and fault detection and isolation studies. In this work, we use a robust adaptive multivariable (4 inputs and 4 outputs) RTRL neural networks controller (Leclercq et al., 2005), (Zerkaoui et al, 2007) to regulate the temperature (Y1) and pressure (Y2) in reactor, and the levels in separator (Y3) and stripper (Y4). For this purpose, the controller drives the purge valve (U1), the stripper input valve (U2), the condenser CW valve (U3), and reactor CW valve (U4). The controller is presented in figure 20 (full lines represent measurements and dashed line represent actuators updating). This controller compensates all perturbations IDV1 to IDV 20 excepted IDV1, IDV6 and IDV7. Particularly, the controller is robust for perturbation IDV16 that will be used in the following.

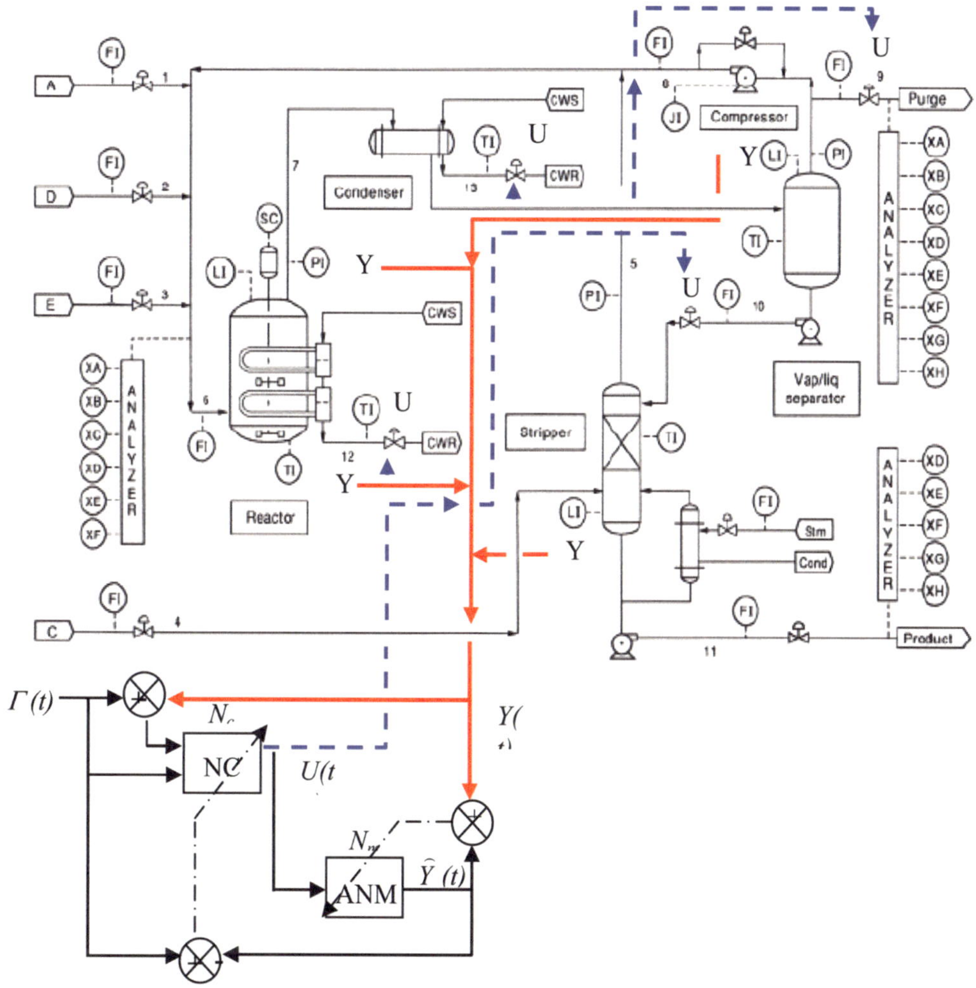


Fig. 20. Tennessee Eastman Challenge Process and robust adaptive neural networks controller (Leclercq et al., 2005), (Zerkaoui et al, 2007).

The figure 21 illustrates the advantage of our method to detect changes for real world FDI applications. Measurements of the stripper level (figure 21 a) are decomposed into 3 components by using filters bank derived from the 'Haar' wavelet packet. From time $t_r$ = 600 hours, the perturbation IDV16, that corresponds to a random variation of the A, B, C composition, modifies the dynamical behavior of the system. The detection functions applied on the 3 components (figure 21 f, g, h) can be compared with the detection function applied directly on measurement of pressure (figure 21 b). After fusion, the point of change is calculated to be $t_f$ = 659. Detection results are considerably improved by using the derived filters bank as a preprocessor.

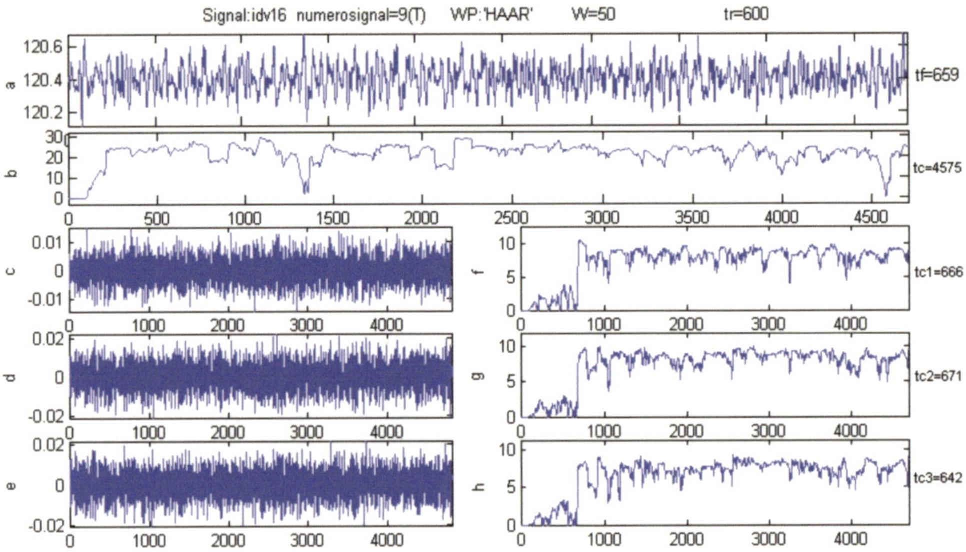


Fig. 21. Analysis of the stripper level measurements (%) for TECP with robust adaptive control and for IDV 16 perturbation from t = 600.

At left: decomposition of the signal into 3 components.
At right: the detection functions of each component.
a) Original signal b) DCS applied directly on the original signal.
c) d) e) Decomposition using filters bank derived from the 'Haar' wavelet packet.
f) g) h) Detection functions applied on the filtered signals (c, d, e).

## 6. Conclusions and perspectives

The aim of our work is to detect the point of change of statistical parameters in signals collected from complex industrial systems. This method uses a filters bank derived from a wavelet packet and combined with DCS to characterize and classify the parameters of a signal in order to detect any variation of the statistical parameters due to any change in frequency and energy. The main contribution of this paper is to derive the parameters of a filters bank that behaves as a wavelet packet. The proposed algorithm provides also good results for the detection of frequency changes in the signal. The application to the Tennessee Eastman Challenge Process illustrates the interest of the approach for on-line detection and real world applications.

In the future, our algorithm will be tested with more data issued form several systems in order to improve and validate it and to compare it to other methods. We will consider mechanical and electrical machines (Awadallah & Morcos 2003, Benbouzid et al.,1999), and as a consequence our intend is to develop FDI methods for wind turbines and renewable multi-source energy systems (Guérin et al., 2005).

## 7. References


Awadallah M., M.M. Morcos, Application of AI tools in faults diagnosis of electrical machines and drives - a review, *Trans. IEEE Energy Conversion*, vol. 18, no. 2, pp. 245-251, june 2003.

Basseville M., Nikiforov I. Detection of Abrupt Changes: *Theory and Application*. Prentice-Hall, Englewood Cliffs, NJ, 1993.

Benbouzid M., M. Vieira, C. Theys, "Induction motor's faults detection and localization using stator current advanced signal processing techniques", *IEEE Transaction on Power Electronics*, Vol. 14, N° 1, pp 14 - 22, January1999.

Blanke M., Kinnaert M., Lunze J., Staroswiecki M., *Diagnosis and fault tolerant control*, Springer Verlag, New York, 2003.

Coifman R. R., and Wicherhauser M.V. (1992): *'Entropy based algorithms for best basis selection'*, IEEE Trans. Inform. Theory, 38, pp. 713-718.

Downs, J.J., Vogel, E.F, 1993, A plant-wide industrial control problem, *Computers and Chemical Engineering*, 17, pp. 245-255.

Flandrin P. *Temps fréquence*, édition HERMES, Paris,1993.

Guérin F., Druaux F., Lefebvre D., Reliability analysis and FDI methods for wind turbines: a state of the art and some perspectives, 3ème French - German Scientific conference « *Renewable and Alternative Energies*», December 2005, Le Havre and Fécamp, France.

Hitti. Eric 3 *Sélection d'un banc optimal de filtres à partir d'une décomposition en paquets d'ondelettes. Application à la détection de sauts de fréquences dans des signaux multicomposantes* » THESE de DOCTORAT, Sciences de l'Ingénieur, Spécialité: Automatique et Informatique Appliquee, 9 novembre 1999, Ecole Centrale de Nantes.

Khalil.M, Une approche pour la détection fondée sur une somme cumulée dynamique associée à une décomposition multiéchelle. Application à l'EMG utérin. *Dix-septième Colloque GRETSI sur le traitement du signal et des images*, Vannes, France,1999.

Khalil M., Duchêne J., Dynamic Cumulative Sum approach for change detection, *EDICS* NO: SP 3.7. 1999.

Leclercq, E., Druaux, F. Lefebvre, D., Zerkaoui, S., 2005. *Autonomous learning algorithm for fully connected recurrent networks*. Neurocomputing, vol. 63, pp. 25-44.

Mallat S. (1999): *'A Wavelet Tour of Signal Processing'*, Academic Press, San Diego, CA.

Mallat S. *Une exploration des signaux en ondelettes*, les éditions de l'école
polytechnique, Paris, juillet 2000. http ://www.cmap.polytechnique.fr/~mallat/ Wavetour_fig/.

Chendeb Marwa, *Détection et caractérisation dans les signaux médicaux de longue durée par la théorie des ondelettes. Application en ergonomie*, stage du DEA modélisation et simulation informatique (AUF), octobre 2002.

Maquin D. and Ragot J., *Diagnostic des systèmes linéaires*, Hermes, Paris, 2000.

Mustapha O., Khalil M., Hoblos G, Chafouk H., Ziadeh H., Lefebvre D., About the Detectability of DCS Algorithm Combined with Filters Bank, *Qualita 2007*, Tanger, Maroc, April 2007.

Mustapha O, Khalil M., Hoblos G, Chafouk H., Lefebvre D., *Fault detection algorithm using DCS method combined with filters bank derived from the wavelet transform*, IEEE - IFAC ICINCO 2007, 09- 11 May, Angers, France, 2007.

Nikiforov I. Sequential detection of changes in stochastic systems. *Lecture notes in Control and information Sciences*, NY, USA, 1986, pp. 216-228.

Papalambros P. Y, Wilde J. D. Principles of optimal design. *Modeling and computation*. Cambridge university press, USA, 2000.

Patton R.J., Frank P.M. and Clarck R., *Issue of Fault diagnosis for dynamic systems*, Springer Verlag, 2000.

Rardin L. R. *Optimization in operation research*. Prentice-Hall, NJ, USA, 1998.

Rustagi S. J. *Optimization techniques in statistics*. Academic press, USA, 1994.

Saporta G. *Probabilités, analyse des données et statistiques*, éditions Technip, 1990.

Singhal, A., 2000. Tennessee Eastman Plant Simulation with Base Control System of McAvoy and Ye., *Research report*, Department of Chemical Engineering, University of California, Santa Barbara, USA.

Zerkaoui S., Druaux F., Leclercq E., Lefebvre D., 2007, *Multivariable adaptive control for non-linear systems : application to the Tennessee Eastman Challenge Process*, ECC 2007, Kos, Greece, July 2 - 5.

Zwingelstein G., *Diagnostic des défaillances*, Hermes, Paris, 1995.

# 12

# Pareto Optimum Design of Robust Controllers for Systems with Parametric Uncertainties

Amir Hajiloo[1], Nader Nariman-zadeh[1 2] and Ali Moeini[3],
[1]*Dept. of Mechanical Engineering, Faculty of Engineering, University of Guilan*
[2]*Intelligent-based Experimental Mechanics Center of Excellence, School of Mechanical Engineering, Faculty of Engineering, University of Tehran*
[3]*Dept. of Algorithms & Computations, Faculty of Engineering, University of Tehran*
*Iran*

## 1. Introduction

The development of high-performance controllers for various complex problems has been a major research activity among the control engineering practitioners in recent years. In this way, synthesis of control policies have been regarded as optimization problems of certain performance measures of the controlled systems. A very effective means of solving such optimum controller design problems is genetic algorithms (GAs) and other evolutionary algorithms (EAs) (Porter & Jones, 1992; Goldberg, 1989). The robustness and global characteristics of such evolutionary methods have been the main reasons for their extensive applications in off-line optimum control system design. Such applications involve the design procedure for obtaining controller parameters and/or controller structures. In addition, the combination of EAs or GAs with fuzzy or neural controllers has been reported in literature which, in turn, constitutionally formed intelligent control scheme (Porter et al., 1994; Porter & Nariman-zadeh, 1995; Porter & Nariman-zadeh, 1997). The robustness and global characteristics of such evolutionary methods have been the main reasons for their extensive applications in off-line optimum control system design. Such applications involve the design procedure for obtaining controller parameters and/or controller structures. In addition to the most applications of EAs in the design of controllers for certain systems, there are also much research efforts in robust design of controllers for uncertain systems in which both structured or unstructured uncertainties may exist (Wolovich, 1994). Most of the robust design methods such as *μ-analysis*, $H_2$ or $H_\infty$ design are based on different norm-bounded uncertainty (Crespo, 2003). As each norm has its particular features addressing different types of performance objectives, it may not be possible to achieve all the robustness issues and loop performance goals simultaneously. In fact, the difficult mixed norm-control methodology such as $H_2/H_\infty$ has been proposed to alleviate some of the issue of meeting different robustness objectives (Baeyens & Khargonekar, 1994). However, these are based on the worst case scenario considering in the most possible pessimistic value of the performance for a particular member of the set of uncertain models (Savkin et al., 2000). Consequently, the performance characteristics of such norm-bounded uncertainties robust designs often degrades for the most likely cases of uncertain models as the likelihood of the worst-case design is unknown in practice (Smith et al., 2005). Recently, there have been many efforts for designing robust control methods. In these methods for reducing the conservatism or accounting more for the most likely plants with respect to uncertainties, the probabilistic uncertainty, as a weighting factor, propagates through the uncertain parameter of plants. In fact, probabilistic uncertainty specifies set of plants as the actual dynamic system to each of which a probability density function (PDF) is assigned (Crespo & Kenny, 2005). Therefore, such additional information regarding the likelihood of each plant allows a reliability-based design in which probability is incorporated in the robust design. In this method, robustness and performance are stochastic variables (Stengel & Ryan, 1989). Stochastic behavior of the system can be simulated by Monte- Carlo Simulation (Ray & Stengel, 1993). Robustness and performance can be considered as objective functions with respect to the controller parameters in optimization problem. GAs have also been recently deployed in an augmented scalar single objective optimization to minimize the probabilities of unsatisfactory stability and performance estimated by Monte Carlo simulation (Wang & Stengel, 2001), (Wang & Stengel, 2002). Since conflictions exist between robustness and performance metrics, choosing appropriate weighting factor in a cost function consisting of weighted quadratic sum of those non-commensurable objectives is inherently difficult and could be regarded as a subjective design concept. Moreover, trade-offs existed between some objectives cannot be derived and it would be, therefore, impossible to choose an appropriate optimum design reflecting the compromise of the designer's choice concerning the absolute values of objective functions. Therefore, this problem can be formulated as a multi objective optimization problem (MOP) so that trade-offs between objectives can be derived consequently.

In this chapter, a new simple algorithm in conjunction with the original Pareto ranking of non-dominated optimal solutions is first presented for MOPs in control systems design. In this Multi-objective Uniform-diversity Genetic Algorithm (MUGA), a $\epsilon$-elimination diversity approach is used such that all the clones and/or $\epsilon$-similar individuals based on normalized Euclidean norm of two vectors are recognized and simply eliminated from the current population. Such multi-objective Pareto genetic algorithm is then used in conjunction with Monte-Carlo simulation to obtain Pareto frontiers of various non-commensurable objective functions in the design of robust controllers for uncertain systems subject to probabilistic variations of model parameters. The methodology presented in this chapter simply allows the use of different non-commensurable objective functions both in frequency and time domains. The obtained results demonstrate that compromise can be readily accomplished using graphical representations of the achieved trade-offs among the conflicting objectives.

## 2. Stochastic robust analysis

In real control engineering practice, there exist a variety of typical sources of uncertainty which have to be compensated through robust control design approach. Those uncertainties include plant parameter variations due to environmental condition, incomplete knowledge of the parameters, age, un-modelled high frequency dynamics, and etc. Two categorical types of uncertainty, namely, structured uncertainty and unstructured uncertainty are generally used in classification. The structured uncertainty concerns about the model uncertainty due to unknown values of parameters in a known structure. In conventional optimum control system design, uncertainties are not addressed and the optimization process is accomplished deterministically. In fact, it has been shown that optimization without considering uncertainty generally leads to non-optimal and potentially high risk solution (Lim et al., 2005). Therefore, it is very desirable to find robust design whose performance variation in the presence of uncertainties is not high. Generally, there exist two approaches addressing the stochastic robustness issue, namely, robust design optimization (RDO) and reliability-based design optimization (RBDO) (Papadrakakis et al., 2004). Both approaches represent non deterministic optimization formulations in which the probabilistic uncertainty is incorporated into the stochastic optimal design process. Therefore, the propagation of a priori knowledge regarding the uncertain parameters through the system provides some probabilistic metrics such as random variables (e.g., settling time, maximum overshoot, closed loop poles, ...), and random processes (e.g., step response, Bode or Nyquist diagram, ...) in a control system design (Smith et al., 2005). In RDO approach, the stochastic performance is required to be less sensitive to the random variation induced by uncertain parameters so that the performance degradation from ideal deterministic behaviour is minimized. In RBDO approach, some evaluated reliability metrics subjected to probabilistic constraints are satisfied so that the violation of design requirements is minimized. In this case, limit state functions are required to define the failure of the control system. Figure (1) depicts the concept of these two design approaches where $f$ is to be minimized. Regardless the choice of any of these two approaches, random variables and random processes should be evaluated reflecting the effect of probabilistic nature of uncertain parameters in the performance of the control system.

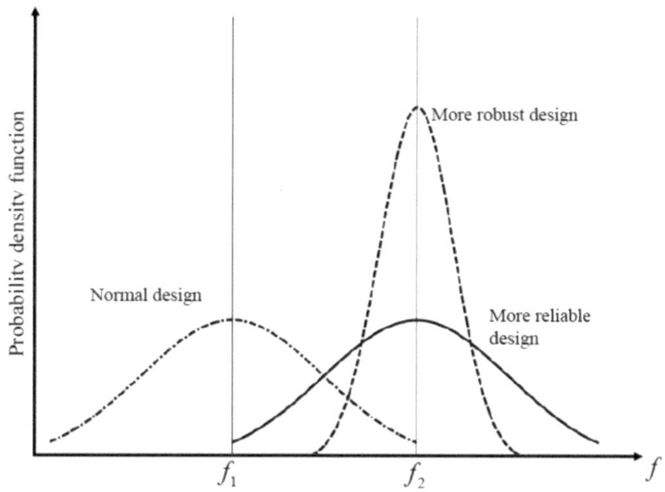


Fig. 1. Concepts of RDO and RBDO optimization

With the aid of ever increasing computational power, there have been a great amount of research activities in the field of robust analysis and design devoted to the use of Monte Carlo simulation (Crespo, 2003; Crespo & Kenny, 2005; Stengel, 1986; Stengel & Ryan, 1993; Papadrakakis et al., 2004; Kang, 2005). In fact, Monte Carlo simulation (MCS) has also been used to verify the results of other methods in RDO or RBDO problems when sufficient number of sampling is adopted (Wang & Stengel, 2001). Monte Carlo simulation (MCS) is a direct and simple numerical method but can be computationally expensive. In this method, random samples are generated assuming pre-defined probabilistic distributions for uncertain parameters. The system is then simulated with each of these randomly generated samples and the percentage of cases produced in failure region defined by a limit state function approximately reflects the probability of failure.

Let $X$ be a random variable, then the prevailing model for uncertainties in stochastic randomness is the probability density function (PDF), $f_X(x)$ or equivalently by the cumulative distribution function (CDF), $F_X(x)$, where the subscript X refers to the random variable. This can be given by

$$F_X(x) = \Pr(X \leq x) = \int_{-\infty}^{x} f_X(x)dx \tag{1}$$

where Pr(.) is the probability that an event ($X \leq x$) will occur. Some statistical moments such as the first and the second moment, generally known as mean value (also referred to as expected value) denoted by E(X) and variance denoted by $\sigma^2(X)$, respectively, are the most important ones. They can also be computed by

$$E(X) = \int_{-\infty}^{\infty} x dF_X(x) = \int_{-\infty}^{\infty} f_X(x)dx \tag{2}$$

and

$$\sigma^2(X) = \int_{-\infty}^{\infty} (x - E(X)) f_X(x) dx \tag{3}$$

In the case of discrete sampling, these equations can be readily represented as

$$E(X) \cong \frac{1}{N}\sum_{i=1}^{N} x_i \tag{4}$$

and

$$\sigma^2(X) \cong \frac{1}{N-1}\sum_{i=1}^{N} (x_i - E(X))^2 \tag{5}$$

where $x_i$ is the $i^{th}$ sample and $N$ is the total number of samples.

In the reliability-based design, it is required to define reliability-based metrics via some inequality constraints (in time or frequency domain). Therefore, in the presence of uncertain parameters of plant (**p**) whose PDF or CDF can be given by $f_{\mathbf{P}}(\mathbf{p})$ or $F_{\mathbf{P}}(\mathbf{p})$, respectively, the reliability requirements can be given as

$$P_f^i = \Pr(g_i(\mathrm{p}) \leq 0) \leq \varepsilon \quad i = 1, 2, \ldots, k \tag{6}$$

In equation (6), $P_f^i$ denotes the probability of failure (i.e., $g_i(\mathbf{p}) \leq 0$) of the $i^{th}$ reliability measure and $k$ is the number of inequality constraints (i.e., limit state functions) and $\varepsilon$ is the highest value of desired admissible probability of failure. It is clear that the desirable value of each $P_f^i$ is zero. Therefore, taking into consideration the stochastic distribution of uncertain parameters ( $\mathbf{p}$ ) as $f_p(\mathbf{p})$, equation (6) can now be evaluated for each probability function as

$$P_f^i = \Pr(g_i(\mathbf{p}) \leq 0) = \int_{g_i(\mathbf{p})\leq 0} f_{\mathbf{p}}(\mathbf{p})d\mathbf{p} \tag{7}$$

This integral is, in fact, very complicated particularly for systems with complex g($\mathbf{p}$) (Wang & Stengel, 2002) and Monte Carlo simulation is alternatively used to approximate equation (7). In this case, a binary indicator function $I_{g(\mathbf{p})}$ is defined such that it has the value of 1 in the case of failure (g($\mathbf{p}$)≤0) and the value of zero otherwise,

$$I_{g(\mathbf{p})} = \begin{cases} 0 & g(\mathbf{p}) > 0 \\ 1 & g(\mathbf{p}) \leq 0 \end{cases} \tag{8}$$

Consequently, for each limit state function, g($\mathbf{p}$), the integral of equation (7) can be rewritten as

$$P_f(\mathbf{p}) = \int_{-\infty}^{\infty} I_{g(\mathbf{p})}(G(\mathbf{p}), C(\mathbf{k})) f_{\mathbf{p}}(\mathbf{p})d\mathbf{p} \tag{9}$$

where G($\mathbf{p}$) is the uncertain plant model and C($\mathbf{k}$) is the controller to be designed in the case of control system design problems. Based on Monte Carlo simulation (Ray & Stengel, 1993; Wang & Stengel, 2001; Wang & Stengel, 2002; Kalos, 1986), the probability using sampling technique can be estimated using

$$P_f(\mathbf{p}) = \frac{1}{N}\sum_{i=1}^{N} I_{g_i(\mathbf{p})}(G(\mathbf{p}), C(\mathbf{k})) \tag{10}$$

where $G_i$ is the $i^{th}$ plant that is simulated by Monte Carlo Simulation. In other words, the probability of failure is equal to the number of samples in the failure region divided by the total number of samples. Evidently, such estimation of $P_f$ approaches to the actual value in the limit as $N \to \infty$ (Wang & Stengel, 2002). However, there have been many research activities on sampling techniques to reduce the number of samples keeping a high level of accuracy. Alternatively, the quasi-MCS has now been increasingly accepted as a better sampling technique which is also known as Hammersley Sequence Sampling (HSS) (Smith et al., 2005; Crespo & Kenny, 2005). In this paper, HSS has been used to generate samples for probability estimation of failures. In a RBDO problem, the probability of representing the reliability-based metrics given by equation (10) is minimized using an optimization method. In a multi-objective optimization of a RBDO problem presented in this paper, however, there are different conflicting reliability-based metrics that should be minimized simultaneously.

In the multi-objective RBDO of control system problems, such reliability-based metrics (objective functions) can be selected as closed-loop system stability, step response in time domain or Bode magnitude in frequency domain, etc. In the probabilistic approach, it is, therefore, desired to minimize both the probability of instability and probability of failure to a desired time or frequency response, respectively, subjected to assumed probability distribution of uncertain parameters. In a RDO approach that is used in this work, the lower bound of degree of stability that is the distance from critical point -1 to the nearest point on the open lop Nyquist diagram, is maximized. The goal of this approach is to maximize the mean of the random variable (degree of stability) and to minimize its variance. This is in accordance with the fact that in the robust design the mean should be maximized and its variability should be minimized simultaneously (Kang, 2005). Figure (2) depicts the concept of this RDO approach where $f_X(x)$ is a PDF of random variable, $X$. It is clear from figure (2) that if the lower bound of $X$ is maximized, a robust optimum design can be obtained. Recently, a weighted-sum multi-objective approach has been applied to aggregate these objectives into a scalar single-objective optimization problem (Wang & Stengel, 2002; Kang, 2005).

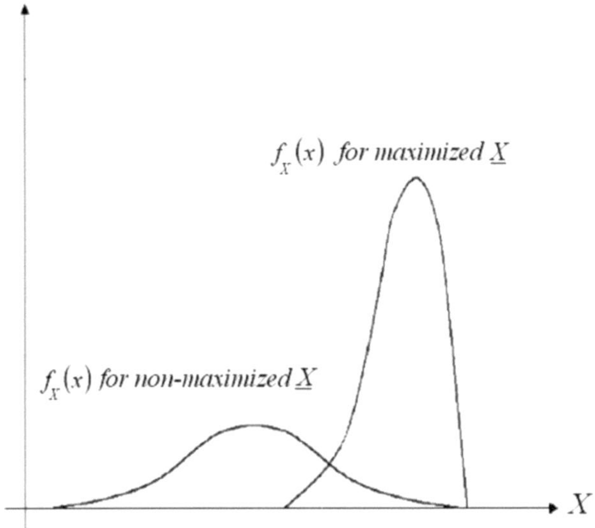


Fig. 2. Concept of RDO approach

However, the trade-offs among the objectives are not revealed unless a Pareto approach of the multi-objective optimization is applied. In the next section, a multi-objective Pareto genetic algorithm with a new diversity preserving mechanism recently reported by some of authors (Nariman-Zadeh et al., 2005; Atashkari et al., 2005) is briefly discussed for a combined robust and reliability-based design optimization of a control system.

## 3. Multi-objective Pareto optimization

Multi-objective optimization which is also called multi-criteria optimization or vector optimization has been defined as finding a vector of decision variables satisfying constraints to give optimal values to all objective functions (Atashkari et al., 2005; Coello Coello & Christiansen, 2000; Coello Coello et al., 2002; Pareto, 1896). In general, it can be mathematically defined as follows; find the vector $X^* = \left[x_1^*, x_2^*, \ldots, x_n^*\right]^T$ to optimize

$$F(X) = \left[f_1(X), f_2(X), \ldots, f_k(X)\right]^T \tag{11}$$

subject to $m$ inequality constraints

$$g_i(X) \le 0 \quad i = 1 \cdots m \tag{12}$$

and $p$ equality constraints

$$hj(X) \le 0 \quad j = 1 \cdots p \tag{13}$$

where, $X^* \in \Re^n$ is the vector of decision or design variables, and $F(X) \in \Re^k$ is the vector of objective functions. Without loss of generality, it is assumed that all objective functions are to be minimized. Such multi-objective minimization based on the Pareto approach can be conducted using some definitions.

### Pareto dominance

A vector $\mathbf{U} = [u_1, u_2, ..., u_k] \in \Re^k$ dominates to vector $\mathbf{V} = [v_1, v_2, ..., v_k] \in \Re^k$ (denoted by $\mathbf{U} \prec \mathbf{V}$) if and only if $\forall i \in \{1,2,...,k\},\ u_i \le v_i \wedge \exists j \in \{1,2,...,k\}: u_j < v_j$. It means that there is at least one $u_j$ which is smaller than $v_j$ whilst the rest $u$'s are either smaller or equal to corresponding $v$'s.

### Pareto optimality

A point $X^* \in \Omega$ ($\Omega$ is a feasible region in $\Re^n$) is said to be Pareto optimal (minimal) with respect to all $X \in \Omega$ if and only if $F(X^*) \prec F(X)$. Alternatively, it can be readily restated as $\forall i \in \{1,2,...,k\}, \forall X \in \Omega - \{X^*\},\ f_i(X^*) \le f_i(X) \wedge \exists j \in \{1,2,...,k\} \ :\ f_j(X^*) < f_j(X)$. It means that the solution $X^*$ is said to be Pareto optimal (minimal) if no other solution can be found to dominate $X^*$ using the definition of Pareto dominance.

### Pareto Set

For a given MOP, a Pareto set $\mathcal{P}^*$ is a set in the decision variable space consisting of all the Pareto optimal vectors, $\mathcal{P}^* = \{X \in \Omega |\ \nexists X' \in \Omega : F(X') \prec F(X)\}$ . In other words, there is no other $X'$ in $\Omega$ that dominates any $X \in \mathcal{P}^*$

### Pareto front

For a given MOP, the Pareto front $\mathcal{PT}^*$ is a set of vectors of objective functions which are obtained using the vectors of decision variables in the Pareto set $\mathcal{P}^*$, that is, $\mathcal{PT}^* = \{F(X) = (f_1(X), f_2(X), ...., f_k(X)) : X \in \mathcal{P}^*\}$. Therefore, the Pareto front $\mathcal{PT}^*$ is a set of the vectors of objective functions mapped from $\mathcal{P}^*$.

Evolutionary algorithms have been widely used for multi-objective optimization because of their natural properties suited for these types of problems. This is mostly because of their parallel or population-based search approach. Therefore, most difficulties and deficiencies within the classical methods in solving multi-objective optimization problems are eliminated. For example, there is no need for either several runs to find the Pareto front or quantification of the importance of each objective using numerical weights. It is very important in evolutionary algorithms that the genetic diversity within the population be preserved sufficiently (Osyezka, 1985). This main issue in MOPs has been addressed by much related research work (Nariman-zadeh et al., 2005; Atashkari et al., 2005; Coello Coello & Christiansen, 2000; Coello Coello et al., 2002; Pareto, 1896; Osyezka, 1985; Toffolo & Benini, 2002; Deb et al., 2002; Coello Coello & Becerra, 2003; Nariman-zadeh et al., 2005). Consequently, the premature convergence of MOEAs is prevented and the solutions are directed and distributed along the true Pareto front if such genetic diversity is well provided. The Pareto-based approach of NSGA-II (Osyezka, 1985) has been recently used in a wide range of engineering MOPs because of its simple yet efficient non-dominance ranking procedure in yielding different levels of Pareto frontiers. However, the crowding approach in such a state-of-the-art MOEA (Coello Coello & Becerra, 2003) works efficiently for two-objective optimization problems as a diversity-preserving operator which is not the case for problems with more than two objective functions. The reason is that the sorting procedure of individuals based on each objective in this algorithm will cause different enclosing hyper-boxes. It must be noted that, in a two-objective Pareto optimization, if the solutions of a Pareto front are sorted in a decreasing order of importance to one objective, these solutions are then automatically ordered in an increasing order of importance to the second objective. Thus, the hyper-boxes surrounding an individual solution remain unchanged in the objective-wise sorting procedure of the crowding distance of NSGA-II in the two-objective Pareto optimization problem. However, in multi-objective Pareto optimization problem with more than two objectives, such sorting procedure of individuals based on each objective in this algorithm will cause different enclosing hyper boxes. Thus, the overall crowding distance of an individual computed in this way may not exactly reflect the true measure of diversity or crowding property for the multi-objective Pareto optimization problems with more than two objectives.

In our work, a new method is presented to modify NSGA-II so that it can be safely used for any number of objective functions (particularly for more than two objectives) in MOPs. Such a modified MOEA is then used for multi-objective robust desing of linear controllers for systems with parametric uncertainties.

## 4. Multi-objective Uniform-diversity Genetic Algorithm (MUGA)

The multi-objective uniform-diversity genetic algorithm (MUGA) uses non-dominated sorting mechanism together with a ε-elimination diversity preserving algorithm to get Pareto optimal solutions of MOPs more precisely and uniformly (Jamali et.al., 2008.)

### 4.1 The non-dominated sorting method

The basic idea of sorting of non-dominated solutions originally proposed by Goldberg (Goldberg, 1989) used in different evolutionary multi-objective optimization algorithms such as in NSGA-II by Deb (Deb et al., 2002) has been adopted here. The algorithm simply compares each individual in the population with others to determine its non-dominancy. Once the first front has been found, all its non-dominated individuals are removed from the main population and the procedure is repeated for the subsequent fronts until the entire population is sorted and non-dominately divided into different fronts.

A sorting procedure to constitute a front could be simply accomplished by comparing all the individuals of the population and including the non-dominated individuals in the front. Such procedure can be simply represented as following steps:

*1-Get the population (pop)*
*2-Include the first individual {ind(1)} in the front P* as P*(1), let P*_size=1;*
*3-Compare other individuals {ind (j), j=2, Pop_size)} of the pop with { P*(K), K=1, P*_size} of the P*;*
*If ind(j)<P*(K) replace the P*(K) with ind(j)*
*If P*(K)<ind(K), j=j+1, continue comparison;*
*Else include ind(j) in P*, P*_size= P*_size+1, j=j+1, continue comparison;*
*4-End of front P*;*

It can be easily seen that the number of non-dominated solutions in $P^*$ grows until no further one is found. At this stage, all the non-dominated individuals so far found in $P^*$ are removed from the main population and the whole procedure of finding another front may be accomplished again. This procedure is repeated until the whole population is divided into different ranked fronts. It should be noted that the first rank front of the final generation constitute the final Pareto optimal solution of the multi-objective optimization problem.

### 4.2 The ε-elimination diversity preserving approach

In the ε-elimination diversity approach that is used to replaced the crowding distance assignment approach in NSGA-II (Deb et al., 2002), all the clones and ε-similar individuals are recognized and simply eliminated from the current population. Therefore, based on a value of ε as the elimination threshold, all the individuals in a front within this limit of a particular individual are eliminated. It should be noted that such ε-similarity must exist both in the space of objectives and in the space of the associated design variables. This will ensure that very different individuals in the space of design variables having ε-similarity in the space of objectives will not be eliminated from the population. The pseudo-code of the ε-elimination approach is depicted in figure (3). Evidently, the clones and ε-similar

```
ε-elim= ε-elimination(pop)                // pop includes design variables and
                                             objective function
i=1; j=1;
get K (K=1 for the first front);
While i,j <pop_size
e(i,j)= ‖ X(i,:),X(j,:) ‖ / ‖ X(i,:) ‖ ; X(i),X(j) ∈ P*_k Ụ PF*_k  //finding mean value of ε
                                                            within pop.
end
ε=mean(e);
i=1;
until i+1<pop_size;
j=i+1
        until j<pop_size
        if e(i,j)<ε
        then {pop}={pop}/ {pop(j)}         //remove the ε-similar individual
        j=j+1
        end
i=i+1
end
```

Fig. 3. The ε-elimination diversity preserving pseudo-code

individuals are replaced from the population by the same number of new randomly generated individuals. Meanwhile, this will additionally help to explore the search space of the given MOP more effectively. It is clear that such replacement does not appear when a front rather than the entire population is truncated for ε-similar individual.

### 4.3 The main algorithm of MUGA

It is now possible to present the main algorithm of MUGA which uses both non-dominated sorting procedure and ε-elimination diversity preserving approach and is given in figure (4).

```
Get N                      //population size
t=1 ;                      //set generation number
Random_N(P_t);             //generate the first population (P_1) randomly
Q_t=Recomb(P_t)            //generate population Q_t from P_t by genetic operators
R_t=P_t Ụ Q_t              //union of both parent and offspring population
R_t'=ε-elimination (R_t)   //remove ε-similar individuals in R_t
R_t''= R_t'  Ụ  Random_(R_t_size-R'_t_size) (P_t')    //add random individuals to fill R_t to 2N

Do non-dominate sorting procedure (R_t'')    //R_t''=P*_1Ụ P*_2Ụ...ỤP*_k  where k is total
                                                                    number of fronts
i=1
P_t+1=Θ
While not P_t+1_size>N               //includes fronts into new population
              P_t+1= P_t+1Ụ P*_i
              i=i+1
end
N'=N- P_t+1_size
While not (0.9 N'< P_t+1_size<1.1 N')         //remove the ε-similar individuals within
                                              the tolerance of  ±10 percent
      F'=ε-elimination (P*_i-1)
   If  F'_size< N'
      e=1.1*e
      else
      e=0.9 * e          //adjust the value of threshold  to get the right population
                         size of the last front
      end
end
t=t+1                          //Start next generation
```

Fig. 4. The pseudo-code of the main algorithm of MUGA

It first initiates a population randomly. Using genetic operators, another same size population is then created. Based on the ε-elimination algorithm, the whole population is then reduced by removing ε-similar individuals. At this stage, the population is re-filled by randomly generated individuals which helps to explore the search space more effectively. The whole population is then sorted using non-dominated sorting procedure. The obtained fronts are then used to constitute the main population. It must be noted that the front which must be truncated to match the size of the population is also evaluated by ε-elimination procedure to identify the ε-similar individuals. Such procedure is only performed to match the size of population within ±10 present deviation to prevent excessive computational effort to population size adjustment. Finally, unless the number of individuals in the first rank front is changing in certain number of generations, randomly created individuals are inserted in the main population occasionally (e.g. every 20 generations of having non-varying first rank front).

## 5. Process model and controller evaluation method

In this section, the process models and the robust PI/PID controller design methodologies are presented using some conflicting objective functions defined in both time and frequency domains.

### 5.1 The process model

Many industrial systems can be adequately presented by a first-order lag with time delay (Toscana, 2005) as

$$G(s) = \frac{ke^{-\tau s}}{1+Ts} \tag{14}$$

In the case of stochastic robust design, parameters of the plant given by equation (14) vary according to *a priori* known probabilistic distribution functions around a nominal set of parameters. In this work, beta distributions with the coefficients of 2 and 2 with the limits of ±50% of the nominal values of plant parameters, $k = \tau = T = 1$ have been selected, respectively. Stochastic step response of the 10 samples that are simulated by Monte Carlo simulation is shown in figure (5). It is clear from figure (5) that the response of the uncertain system has a large variability and the performance of the system deteriorates significantly with parameters variation. Consequently, the controller design must be accomplished robustly.

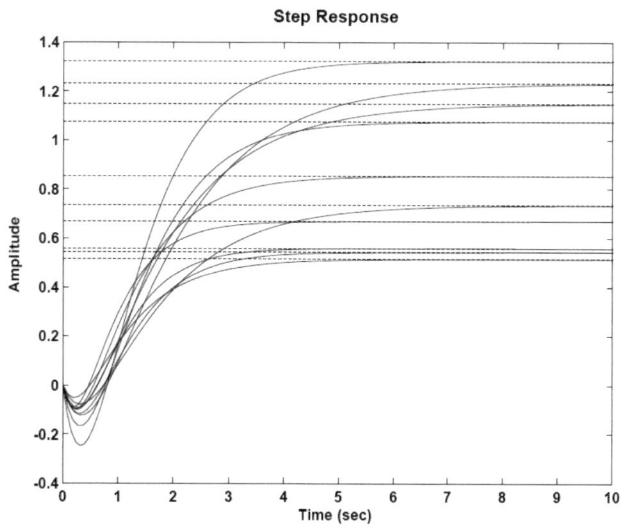


Fig. 5. Stochastic step response of the uncertain plant

### 5.2 The robust design of PI/PID controllers

Simple structure PI/PID Controllers are widely used for many industrial processes represented by the transfer function of equation (14). The transfer functions, C(s), of the standard PI/PID Controllers of the feedback control system shown in figure (6) are

$$\begin{cases} C(s) = K_p + \dfrac{K_i}{s} \\ C(s) = K_p + \dfrac{K_i}{s} + K_d s \end{cases} \tag{15}$$

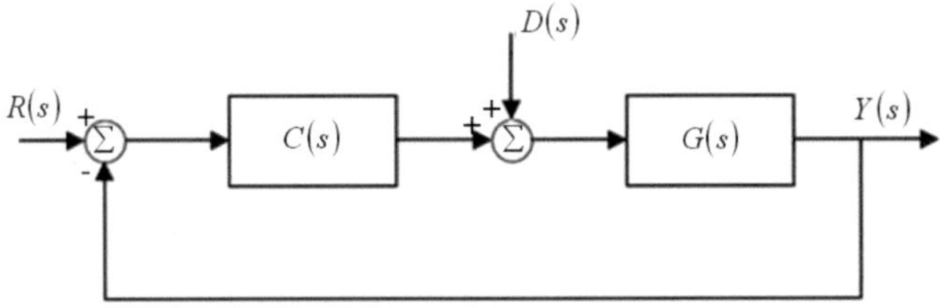


Fig. 6. Closed loop SISO system with plant G(s) and controller C(s)

The design vector of the PI and PID controllers are $\mathbf{k}_{PI} = [K_p, K_i]$ and $\mathbf{k}_{PID} = [K_p, K_i, K_d]$, respectively. They have to be optimally determined based on the mixed robust and reliability-based multi-objective Pareto approach for the uncertain first-order system using some stochastic evaluation metrics that are introduced as follows.

Two robust performance metrics have been proposed in this work, performance metrics in time domain and performance metrics in frequency domain. In this section, design vector of PI controller is obtained based on time domain performance metrics and design vector of PID controller is obtained based on frequency domain performance metrics.

The most important goal of robust controller design is the robust stability which implies that all the closed-loop poles of the system remain in the stable left half-plane ($\Re(s_i) < 0$) in the presence of any uncertainty in the nominal plant's transfer function. Thus, in the case of stochastic robust design, the limit state function to define the probability of failure of robust stability will be represented by

$$g_{ins}(\mathbf{p}) = -\max\{\Re(s_1^i), \Re(s_2^i), \ldots, \Re(s_n^i)\} \quad i = 1,2,\cdots,N \tag{16}$$

where, $g_{ins}(\mathbf{p})$ is the limit state function of the instability, $\Re(s^i)$ is the real part of the closed-loop poles of the $i^{th}$ uncertain plants, and $n$ is the order of the closed-loop plant.

The probability of failure of stochastic stability can now be computed using by equation (10)

$$\Pr_{ins} = \frac{1}{N}\sum_{i=1}^{N} I_{g_{ins}}(G(\mathbf{p}_i), C(\mathbf{k})) \tag{17}$$

in association with equation (16) employing the quasi Monte Carlo Sampling or HSS for $N$ samples. For obtaining the acceptable stability, such probability of instability should be minimized.

In addition to the minimizing the probability of instability, maximizing the stability margin in the frequency domain is another important measure of good performance of a robust controller for uncertain systems. The inclusion of the stability margin (to be maximized) in the vector of the cost functions ensures that stable PI/PID controllers having the most stability margin are obtained. Such robust stability margin, also referred to as degree of stability $S_\infty^{-1}$, can be simply computed using the sensitivity transfer function

$$S(s) = \frac{1}{1 + C(s)G(s)} = \frac{1}{1 + L(s)} \tag{18}$$

for a unity feedback control system shown in figure (6). In frequency domain, the return difference $|-1 - L(j\omega)| = |1 + L(j\omega)|$ simply represents the length of a vector drawn from critical point -1 to open-loop transfer function in the Nyquist diagram. Consequently, the inverse of $\infty$-norm of sensitivity transfer function given by equation (18)

$$S_\infty^{-1} = \|S\|_\infty^{-1} = \min_\omega |1 + L(j\omega)| \tag{19}$$

represents the minimum distance of the Nyquist diagram to point -1. In the case of stochastic robust design, the degree of stability for each stochastic system is a random variable. Therefore, in a RDO problem considered in this study the lower bound of interested random variable (degree of stability) is maximized using an optimization method. It should be noted that the degree of stability given by equation (19) also directly represents the additive disturbance rejection property as follows

$$Y(s) = \frac{G(s)}{1 + L(s)} D(s) = G(s)S(s)D(s) \tag{20}$$

where, $D(s)$ is the load disturbance transfer function. It is evident from equation (20) that maximizing the minimum value of $|1 + L(j\omega)|$ based on equation (19) will cause a better disturbance rejection according to equation (20). Therefore, systems with high degree of stability represent a good ability to reject the load disturbance (Toscana, 2005).
A good step response behavior of the system is one of the performance metrics in controller design procedure that illustrates how system acts in transient and steady state periods. Another method to obtain these properties of the step response is Bode magnitude of the close-loop or complementary transfer function. In the stochastic robust design both step response and Bode magnitude are random process.
In the reliability- based design approach, it is desired to minimize the probability of a failure of a random process as a function of $w$ ($w$ represents time or frequency) due to the uncertain probabilistic parameters. In this approach, let $h(\mathbf{p},w)$ is the random response (step response or Bode magnitude) of an uncertain plant due to uncertain parameters $\mathbf{p}$, and let define $\overline{h}(w)$ and $\underline{h}(w)$ as upper and lower failure boundary, respectively. Therefore, if the random process is held within these bounds, the uncertain system has a robust performance.
In this work, step response metrics are used to design PI controller and Bode magnitude metrics are used to design PID controllers.
The lower and upper failure boundaries to define the corresponding limit state function, $g_{resp}(\mathbf{p}) \leq 0$, in time domain is given using the Heaviside function

$$\underline{h} = -0.1\,\mathrm{H}(t) + 0.8\,\mathrm{H}(t-3) + 0.25\,\mathrm{H}(t-7) \tag{21a}$$

$$\overline{h} = 1.2\,\mathrm{H}(t) - 0.15\,\mathrm{H}(t-7) \tag{21b}$$

for a period of $t \in [0, t_f]$, $t_f = 15$. If $\underline{r}$ and $\overline{r}$ are defined as

$$\underline{r}_i = 1\,,\; h_i < \underline{h}_i \quad i = 1,2,\ldots,k_t \qquad \underline{r}_i = 0 \quad otherwise \tag{22a}$$

$$\overline{r}_i = 1\,,\; h_i > \overline{h}_i \quad i = 1,2,\ldots,k_t \qquad \overline{r}_i = 0 \quad otherwise \tag{22b}$$

where $h$ is the time response of the plant and $k_t$ is the number of sample time, the limit state function indicator can then be computed as

$$I_{g_{resp}}(\mathbf{p}) = \frac{1}{k_t} = \sum_{i=1}^{k_t} \left( \underline{r}_i + \overline{r}_i \right) \tag{23}$$

which is used in equation (10) to obtain the probability of failure to the desires time response boundaries.

The complementary transfer function $T$(s) can be used to obtain closed-loop system response which is the transfer function of the reference input $R$(s) to the output Y(s) and is given as

$$T(s) = 1 - S(s) = \frac{L(s)}{1 + L(s)} \tag{24}$$

The quantity $|T(j\omega)|$ represents the magnitude of the closed-loop frequency response. It is well known that the performance of the closed-loop system response is related to $|T(j\omega)|$. In order to select appropriate boundaries for such frequency response behavior, the relationship between peak value of the closed-loop magnitude response (Nise, 2004), $M_p = \max_\omega |T(j\omega)|$, and the damping ratio, $\zeta$ , for a second order system with nominal parameters is considered here as a reference. Such relations are given by

$$M_p = \frac{1}{2\zeta\sqrt{1-\zeta^2}} \tag{25}$$

at a frequency of $\omega_p$ given by

$$\omega_p = \omega_n \sqrt{1 - 2\zeta^2} \tag{26}$$

Since $\zeta$ and $\omega_n$ are related to maximum overshoot and settling time of step response, respectively, for a good transient and steady-state response it is required that $\zeta \geq \zeta^+$ and $\omega_n \geq \omega_n^+$. The selected values of $\zeta^+$ and $\omega_n^+$ in this work are 0.7. In order to achieve a good closed-loop performance, the complementary transfer function T(s) given by equation (24) is used in frequency domain using lower and upper failure boundaries to define the corresponding limit state function $g_{resp}(\mathbf{p}) \leq 0$. If $h \equiv |T(j\omega)|$ which is a random process having sets of CDFs varying with frequency (Crespo & Kenny, 2005, Crespo, 2003)], both the upper failure boundary defined by $\overline{h}(\omega)$ and the lower failure boundary $\underline{h}(\omega)$ are used to compute the probability of failure to a good frequency response. Based on the previous discussion, the boundaries are defined as

$$\underline{h}(\omega)=1 \qquad \omega\in\left[10^{-2},10^{-1}\right] \tag{27a}$$

$$\overline{h}(\omega)=\frac{0.7}{\omega} \qquad \omega\in\left[0.5,10^{2}\right] \tag{27b}$$

If $\underline{r}$ and $\overline{r}$ are defined as

$$\begin{aligned}\underline{r}_i &= 1\ ,\ h_i<\underline{h}_i \qquad i=1,2,\ldots,k_\omega\\ \underline{r}_i &= 0 \quad otherwise\end{aligned} \tag{28a}$$

$$\begin{aligned}\overline{r}_i &= 1\ ,\ h_i>\overline{h}_i \qquad i=1,2,\ldots,k_\omega\\ \overline{r}_i &= 0 \quad otherwise\end{aligned} \tag{28b}$$

where $h$ is the frequency response of the plant and $k_\omega$ is the number of sample frequency, the limit state function indicator can then be computed as

$$I_{g_{resp}}(\mathbf{p})=\frac{1}{k_\omega}=\sum_{i=1}^{k_\omega}\left(\underline{r}_i+\overline{r}_i\right) \tag{29}$$

which is used in equation (10) to obtain the probability of failure to the desired frequency response boundaries of the complementary transfer function.

## 6. Results

The objectives $\Pr_{ins}$, $\Pr_{resp}$ and $S_\infty^{-1}$ are now considered simultaneously in a Pareto optimization process to obtain some important trade-offs among the conflicting objectives. In a mixed robust and reliability-based design approach, the vector of objective functions to be optimized in a Pareto sense is given as follows

$$\vec{F}=[\Pr_{ins},\Pr_{resp},S_\infty^{-1}] \tag{30}$$

which are computed using equations (17), (23), (29), and (19), respectively, in the quasi-Monte Carlo simulation process. The evolutionary process of the Pareto multi-objective optimization is accomplished using MUGA (Jamali, et.al., 2008) where a population size of 45 has been chosen with crossover probability $P_c$ and mutation probability $P_m$ as 0.85 and 0.09, respectively. The optimization process of the robust PI/PID controllers given by equation (15) is accomplished by 250 Monte Carlo evaluations using HSS distribution for each candidate control law during the evolutionary process. The vector of objective functions given by equation (30) is used to obtain non-dominated optimum PI/PID controllers to represent the trade-offs among the objective functions.

### 6.1 Pareto optimum PI controllers

A total number of 80 non-dominated optimum design points have been obtained and shown in figure (7) in the plane of probability of failure to the desired time response ($\Pr_{resp}$) and the degree of stability (). The value of probability of instability ($\Pr_{ins}$) of all the non-dominated optimum points has been obtained zero which demonstrates that all optimum controllers are stable in the Monte Carlo simulation (Hajiloo et al., 2007).

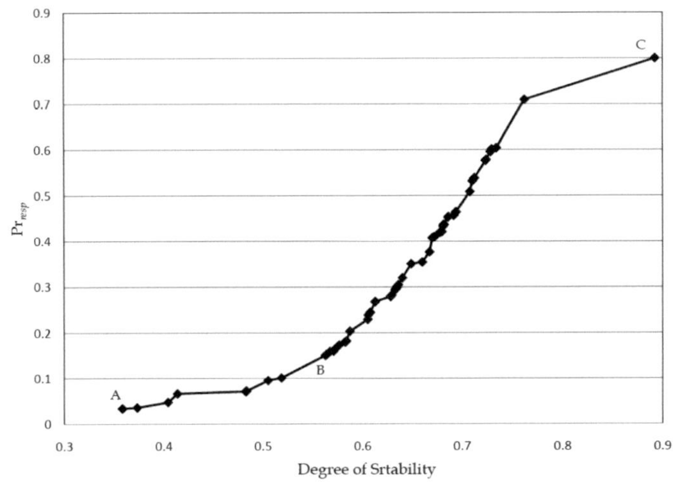


Fig. 7. Pareto fronts of $Pr_{resp}$ and degree of stability ( $S_{\infty}^{-1}$ )

Since, the value of probability of instability ($Pr_{ins}$) of all non-dominated optimum points has been found equal to zero, therefore, the result of the 3-objective optimization process corresponds to a 2-objective optimization process which is shown in figure (7). It can be observed from the Pareto front of figure (7) that improving one objective will cause another objective deteriorates accordingly.

The best point obtained for $Pr_{resp}$ is point A which corresponds to the worst value of $S_{\infty}^{-1}$. These values for time response and degree of stability are 0.0338 and 0.3577, respectively. In other words, optimum design point A represents 3.38% probability of failure to the desired time response and its minimum distance to the critical point -1+0j in the Nyquist diagram is 0.3577, representing its degree of stability for 250 Monte Carlo evaluations. Alternatively, the best value of obtained $S_{\infty}^{-1}$ is that of point C which corresponds to the worst value of $Pr_{resp}$ and are 0.8 and 0.8923, respectively. Figure (8) shows the corresponding 1, 10, 30, 50, 70, 90, 99 percentiles of time responses of both design points A and C which demonstrates the stochastic behavior of the corresponding PI controllers for 250 Monte Carlo simulations of the plant subjected to the assumed probabilistic uncertainties. An *m* percentiles curve presents a confidence limit of *m* percent probability that the time response behavior would be below that curve.

By careful investigation of figure (7) an important trade-off can be observed from the Pareto front of objectives $Pr_{resp}$ and $S_{\infty}^{-1}$. It is clear that the gradient of the Pareto from in section A-B increases noticeably in section B-C. Apparently, optimum design point B shows a significant improvement of degree of stability ($S_{\infty}^{-1}$) in comparison with that of point A whilst its probability of failure to the desired time response does not degrades significantly in section A-B as much as it does in section B-C. Thus, optimum design point B representing a PI controller with $K_p = 0.3$ and $K_i = 0.31$ can be optimally chosen from a trade-off point of view for objectives $Pr_{resp}$ and $S_{\infty}^{-1}$. Figure (9) shows percentiles of stochastic time response behavior of point B which can be compared to those of optimum design points A and C shown in figure (8).

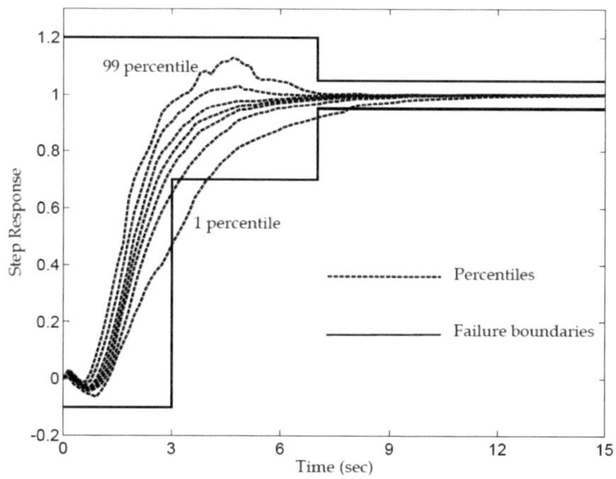


(a)

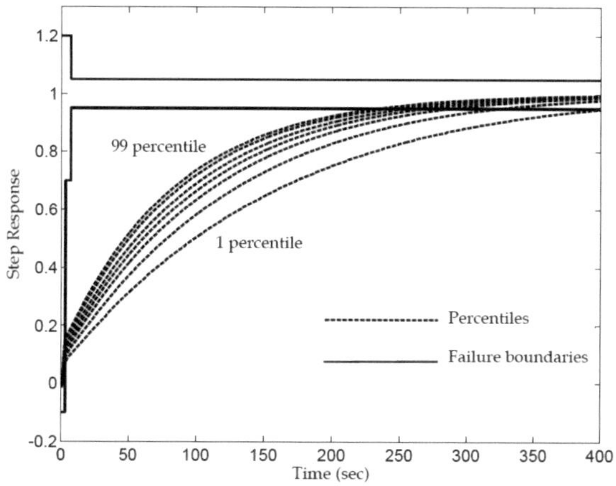


(b)

Fig. 8. Step response behaviors of optimum designs (a) point A (b) point C

Table 1 summarizes the values of those objectives together with the corresponding values of PI controller gains for three optimum design points A, B, and C shown in figure (7).

| Design points | $K_p$ | $K_i$ | $Pr_{ins}$ | $Pr_{resp}$ | $S_\infty^{-1}$ |
|---|---|---|---|---|---|
| A | 0.516 | 0.454 | 0 | 0.0338 | 0.3577 |
| B | 0.3 | 0.31 | 0 | 0.1500 | 0.5624 |
| C | 0.8 | 0.892 | 0 | 0.8 | 0.8923 |

Table 1. Optimum values of objective functions and their gains for the PI controller obtained from 250 Monte Carlo simulations

The robust stability margins of all optimum points have been shown in figure (10). In this figure, the cumulative distribution functions (CDF) have been shown for all design points. It is evident that the optimum design point C exhibits the best stability robustness, because lower bound of its degree of stability is greater than other design points and variance of the degree of stability of design point C is very small.

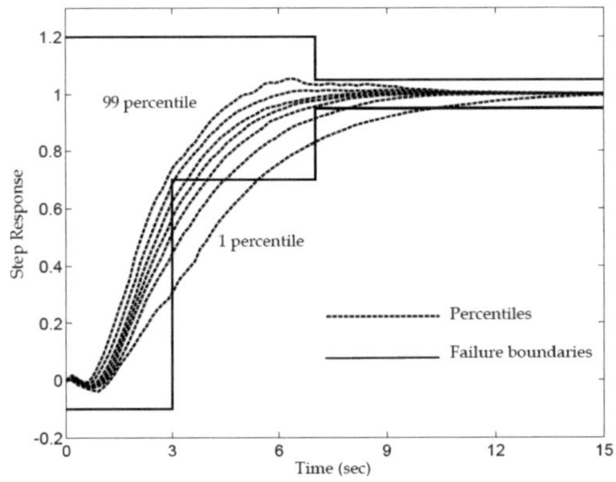


Fig. 9. Step response behaviors of optimum design B

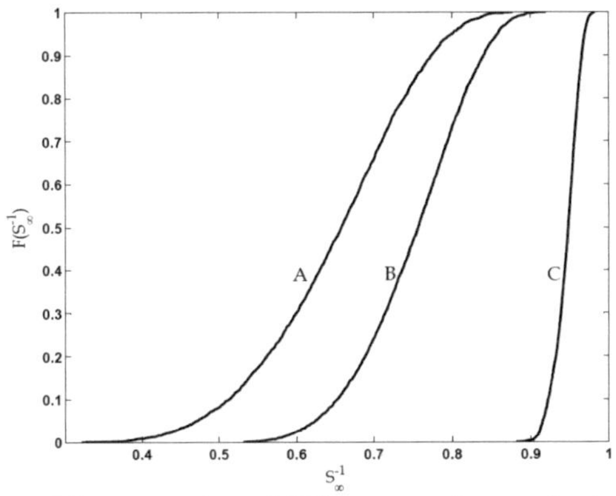


Fig. 10. CDFs for robust stability margins of different optimum designs

### 6.2 Pareto optimum PID controllers

A total number of 31 non-dominated optimum design points have been obtained and shown in figure(11) in the plane of probability of frequency response failure ($Pr_{resp}$) and the degree of stability ($S_\infty^{-1}$). The value of probability of instability ($Pr_{ins}$) of all the non-dominated optimum points has been obtained zero which demonstrates that all obtained optimum controllers are stable in the Monte Carlo simulation. Therefore, the results of the 3-objective optimization process correspond to those of a 2-objective optimization process excluding the probability of instability. It can be observed from the Pareto front of figure (11) that improving one objective will cause another objective deteriorates accordingly. The best point obtained for $Pr_{resp}$ is point A which corresponds to the worst value of $S_\infty^{-1}$. These values for the probability of frequency response failure and the degree of stability are 0.089 and 0.4815, respectively. In other words, optimum design point A represents 8.9% probability of frequency response failure and its minimum distance to the critical point -1+0j in the Nyquist diagram is 0.4815, representing its degree of stability in 250 Monte Carlo evaluations. Alternatively, the best value of obtained $S_\infty^{-1}$ is that of point C which corresponds to the worst value of $Pr_{resp}$ which are 0.1381 and 0.9798, respectively. In other words, optimum design point C represents 13.81% probability of frequency response failure while its minimum distance to the critical point -1+0j in the Nyquist diagram is 0.9788 representing its improved degree of stability.

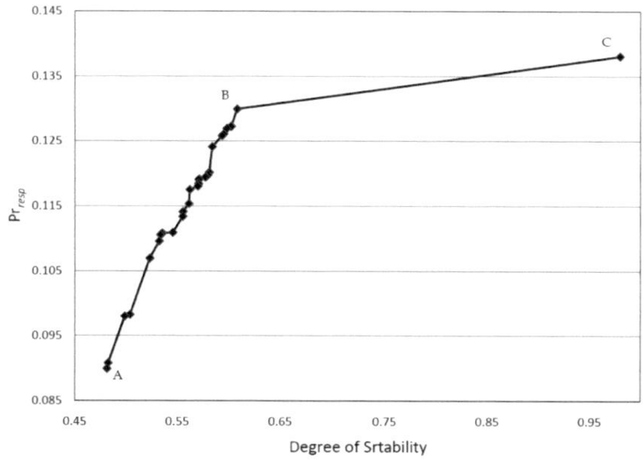


Fig. 11. Pareto fronts of $Pr_{resp}$ and degree of stability ( $S_{\infty}^{-1}$ )

Figure (12) shows the corresponding 1, 10, 30, 50, 70, 90, 99 percentiles of step response of a non-dominated optimum design points B which demonstrate the stochastic behavior of the corresponding PID controllers in 250 Monte Carlo simulations of the plant subjected to the assumed probabilistic uncertainties in the plant. Figure (13) shows the Nyquist diagram for the design point B.

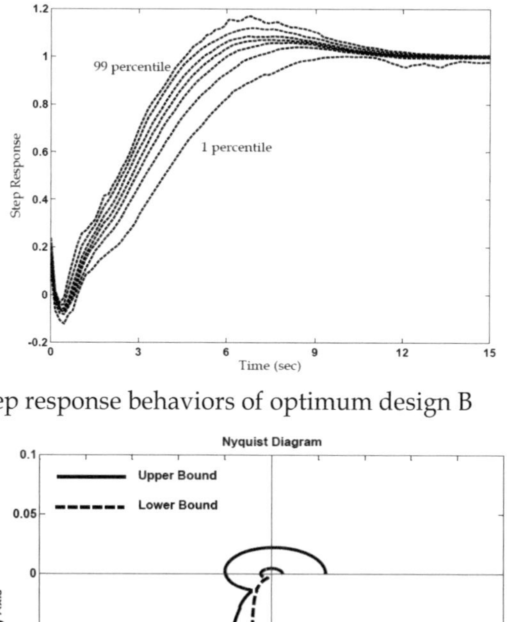


Fig. 12. Probabilistic step response behaviors of optimum design B

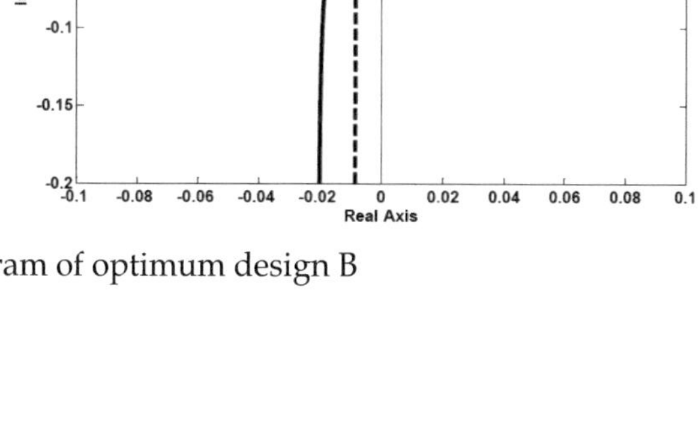


Fig. 13. Nyquist diagram of optimum design B

The robust stability margins of all optimum points have been also shown in figure (14). In this figure, the cumulative distribution functions (CDF) have been shown for all design points.

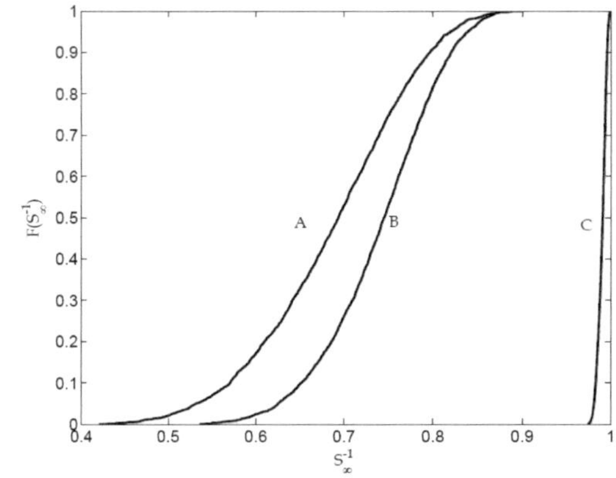


Fig. 14. CDFs for robust stability margins of different optimum designs

Table 2 summarizes the values of those objectives together with the corresponding values of PID controller gains for three optimum design points A, B, and C shown in figure (11).

| Design points | $K_p$ | $K_i$ | $K_d$ | $Pr_{ins}$ | $Pr_{resp}$ | $S_{\infty}^{-1}$ |
|---|---|---|---|---|---|---|
| A | 0.2132 | 0.4035 | 0.0572 | 0 | 0.0899 | 0.4815 |
| B | 0.2210 | 0.3879 | 0.2185 | 0 | 0.1299 | 0.6084 |
| C | 0.0130 | 0.0129 | 0.0119 | 0 | 0.1381 | 0.9798 |

Table 2. Optimum values of objective functions and their gains for the PID controller obtained from 250 Monte Carlo simulations

## 7. Conclusion

A multi-objective genetic algorithm with a recently developed diversity preserving mechanism was used to optimally design PI/PID controllers from a reliability-based point of view in a probabilistic approach. The objective functions which often conflict with each other were appropriately defined using some probabilistic metrics in time and frequency domain. The multi-objective optimization of robust PID controllers led to the discovering some important trade-offs among those objective functions. The framework of such hybrid application of multi-objective GAs and Monte Carlo Simulation of this work for the Pareto optimization of both robust and reliability-based approach using some non-commensurable stochastic objective functions is very promising and can be generally used in the optimum design of real-world complex control systems with probabilistic uncertainties

### 8. References

Atashkari, K.; Nariman-zadeh, N.; Jamali A.& Pilechi A. (2005).Thermodynamic Pareto optimization of turbojet using multi-objective genetic algorithm, *International Journal of Thermal Science*, Vol. 44, No. 11, 1061-1071, Elsevier

Baeyens, E. & Khargonekar, P. (1994). Some examples in mixed $H_2 / H_\infty$ Control, *Proceeding of American Control Conference*, pp. 1608-1612, USA

Coello Coello, C. A. & Becerra, R. L. (2003). Evolutionary Multi-objective Optimization using a Cultural Algorithm, *IEEE Swarm Intelligence Systems.*, pp. 6-13, USA

Coello Coello, C. A., & Christiansen, A. D. (2000). Multiobjective optimization of trusses using genetic algorithms, *Computers & Structures*, Vol. 75, 647-660

Coello Coello, C. A.; Van Veldhuizen, D. A. & Lamont, G. B. (2002). Evolutionary Algorithms for Solving Multi-objective problems, *Kluwer Academic Publishers*, New York

Crespo, L.G. & Kenny, S.P. (2005). Robust Control Deign for systems with probabilistic Uncertainty, *NASA report*, March 2005, TP-2005-213531

Crespo, L.G. (2003). Optimal performance, robustness and reliability base designs of systems with structured uncertainty, *Proceeding of American Control Conference*, pp. 4219-4224, USA, Denver, Colorado,

Deb, K.; Agrawal, S.; Pratap, A. & Meyarivan, T. (2002). A fast and elitist multi-objective genetic algorithm: NSGA-II, *IEEE Transaction on Evolutionary Computation*, Vol. 6, No. 2, 182-197

Diwekar, U.M. & Kalagnaman, J.R. (1997). Efficient sampling technique for optimization under uncertainty, *American Institute of Chemical Engineering Journal*, Vol. 43, No.2, 440-447

Fleming, P.J. & Purshous, R.C. (2002). Evolutionary algorithms in control systems engineering; a survey, *Control Engineering Practice*, 1223-1241

Ge, M.; Chiu, M. & Wang, Q. (2002). Robust PID controller design via LMI approach, *Journal of Process Control*, Vol. 12, 3-13

Goldberg, D.E. (1989). *Genetic Algorithms in Search, Optimization, and Machine Learning*, Addison-Wesley

Hajiloo, A.; Nariman-zadeh, N.; Jamali, A.; Bagheri, A. & Alasti, A. (2007). Pareto Optimum Design of Robust PI Controllers for Systems with Parametric Uncertainty. *International Review of Mechanical Engineering* (IREME), November 2007 Vol. 1, No. 6, 628-640, ISSN 1970-8734

Herreros, A.; Baeyens E. & Persan, J.R. (2002). MRCD: a genetic algorithm for multi objective robust control design, *Engineering Application of Artificial Intelligence*, Vol. 15, 285-301

Jamali, A., Nariman-zadeh, N., Atashkari,K., (2008). Multi-objective Uniform-diversity Genetic Algorithm (MUGA), in *Advances in Evolutionary Algorithms*, Kordic, V., (Ed.), I-Tech Education and Publishing, ISBN 978-3-902613-32-5, Vienna, Austria (in press)

Kalos, M.H. & Whitlock, P.A. (1986). Monte Carlo Methods, Wiley, New York

Kang, Z. (2005). Robust design of structures under uncertainties, PhD. Thesis, University of Stuttgart

Kristiansson, B. & Lennartson, B. (2006). Evaluation and simple tuning of PID controllers with high-frequency robustness, *Journal of Process Control*, Vol. 16, 91-102

Lim, D.; Ong, Y.s. & Lee, B.S. (2005).Inverse multi-objective robust evolutionary design optimization in the presence of uncertainty, *GECCO' 05*, Washington, USA, pp.55-62

Nariman-Zadeh, N.; Atashkari, K.; Jamali, A.; Pilechi, A. & Yao, X. (2005). Inverse modeling of multi-objective thermodynamically optimized turbojet engine using GMDH-type neural networks and evolutionary algorithms, *Engineering Optimization*, Vol. 37, No. 26, 2005, 437-462

Nariman-zadeh, N.; Darvizeh, A.; Jamali, A. & Moeini, A. (2005). Evolutionary Design of Generalized Polynomial Neural Networks for Modeling and Prediction of

Explosive Forming Process, *Journal of Material Processing and Technology*, Vol. 164-165, 1561-1571, Elsevier

Nariman-zadeh, N.; Jamali, A. & Hajiloo, A. (2007). Frequency-based reliability Pareto optimum design of proportional-integral-derivative controllers for systems with probabilistic uncertainty. *Journal of Systems and Control Engineering* (IMECHE), November 2007 Vol. 22, No. 18,1061-1066, ISSN 0959-6518

Nise, N.S. (2004). *Control System Engineering*, John Wiley & Sons, Inc., ISBN 0-471-44577-0, USA

Osyezka, A. (1985). Multicriteria optimization for engineering design, *Design Optimization*, Academic Press, 193-227, New York

Papadrakakis, M.; Lagaros, N. D. & Plevris, V. (2004). Structural optimization considering the probabilistic system response, *Theoretical Applied Mechanics*, Vol. 31, No. 3-4, 361-393, Belgrade

Pareto, V. (1986). Cours d'economic ploitique, Lausanne, Switzerland, Rouge

Porter, B. & Jones, A.H. (1992). Genetic tuning of digital PID controllers, *Electronic Letters*, Vol. 28, No. 9, 843-844

Porter, B. & Nariman-Zadeh, N. (1995). Genetic design of computed-torque fuzzy logic controllers for robotic manipulators, *Proceedings of International Conference on Intelligent Control*, Montgomery, California, USA

Porter, B. & Nariman-Zadeh, N. (1997). Evolutionary design of fuzzy-logic controllers for manufacturing systems, *Annals of the CIRP*, Vol. 46, No. 1

Porter, B.; Sangolola A. & Nariman-Zadeh, N. (1994). Genetic design of computed torque controllers for robotic manipulators, *Proceedings of International Conference on System and Control IASTED*, Lugano, Switzerland

Ray, L. & Stengel, R. F. (1993). A Monte Carlo Approach to the Analysis of Control System Robustness, *Automatica*, Vol. 29, No.1,229-236

Savkin, A.; Peterson, I.R. & Ugronovskii, V.A. (2000). Robust Control Design Using H-infinity methods, pp.1-29, Springer-Verlag, London

Smith, B.A.; Kenny S.P & Crespo, L.G. (2005). Probabilistic Parameter Uncertainty Analysis of Single nput Single Output Control Systems, *NASA report*, March 2005TM-2005-213280

Sree, R.P.; Sriniras, M.N. & Chidambaram, N. (2004).A simple method of tuning PID controller for stable and unstable FOPTD systems, *Computes and Chemical Engineering*, Vol. 28, 2201-2218

Stengel, R.F. & Ryan, L.E. (1989). Stochastic robustness of linear control systems, *Proceeding of Information science and systems Conference*, pp. 556-561

Stengel, R.F. (1986). Stochastic Optimal Control: theory and application, New York, Wiley

Toffolo, A. & Benini, E. (2003). Genetic Diversity as an Objective in Multi-objective evolutionary Algorithms, *Evolutionary Computation*, Vol. 11, No. 2, 151-167, MIT Press

Toscana, R. (2005). A simple robust PI/PID controller design via numerical optimization approach, *Journal of Process Control*, Vol. 15, 81-88

Wang, Q. & Stengel, R.F. (2001).Searching for Robust Minimal-order Compensators, *Journal of Dynamic Systems, Measurement, and Control*, Vol. 123, June 2001,223-236

Wang, Q. & Stengel, R.F. (2002).Robust control of nonlinear systems with parametric uncertainty, *Automatica*, Vol. 38, 1591–1599

Wolovich, W.A. (1994).Automatic Control Systems, Saunders College Publishing, Harcourt Brace College Pub., Orlando, USA

# 13

# Genetic Reinforcement Learning Algorithms for On-line Fuzzy Inference System Tuning "Application to Mobile Robotic"

Abdelkrim Nemra and Hacene Rezine
*Unit of Control, Robotic and Productic Laboratory*
*Polytechnical Military School*
*Algeria*

## 1. Introduction

In the last decade, fuzzy logic has supplanted conventional technologies in some scientific applications and engineering systems especially in control systems, particularly the control of the mobile robots evolving (moving) in completely unknown environments. Fuzzy logic has the ability to express the ambiguity of human thinking and translate expert knowledge into computable numerical data. Also, for real-time applications, its relatively low computational complexity makes it a good candidate. A fuzzy system consists of a set of fuzzy if-then rules. Conventionally, the selection of fuzzy if-then rules often relies on a substantial amount of heuristic observation to express the knowledge of proper strategies.
Recently, many authors proved that it is possible to reproduce the operation of any standard continuous controller using fuzzy controller L. Jouffe, C. Watkins, P. Dayan Dongbing Gu, Huosheng Hu, Libor Spacek . However it is difficult for human experts to examine complex systems, then it isn't easy to design an optimized fuzzy controller.
Generally the performances of Fuzzy inference system (FIS) depend on the formulation of the rules, but also the numerical specification of all the linguistic terms used and an important number of choices is *given a priori,* also it is not always easy or possible to extract these data using human expert. These choices are carried with empirical methods, and then the design of the FIS can prove to be long and delicate vis-à-vis the important number of parameters to determine, and can lead then to a solution with poor performance. To cope with this difficulty, many researchers have been working to find learning algorithms for fuzzy system design. These automatic methods enable to extract information when the experts' *priori* knowledge is not available.
The most popular approach to design Fuzzy Logic Controller (FLC) may be a kind of supervised learning where the training data is available. However, in real applications, extraction of training data is not always easy and become impossible when the cost to obtain training data is expensive. For these problems, reinforcement learning is more suitable than supervised learning. In reinforcement learning, an agent receives from its environment a critic, called reinforcement, which can be thought of as a reward or a punishment. The objective then is to generate a policy maximizing on average the sum of the rewards in the course of time, starting from experiments (state, action, reward).

In this chapter, we used the algorithm of reinforcement learning, Fuzzy Q-Learning (FQL) L. Jouffe, A. Souissi which allows the adaptation of apprentices of the type FIS (continuous states and actions), fuzzy Q-learning is applied to select the consequent action values of a fuzzy inference system. For these methods, the consequent value is selected from a predefined values set which is kept unchanged during learning, and if an improper value set is assigned, then the algorithm may fail. Also, the approach suggested called Fuzzy-Q-Learning Genetic Algorithm (FQLGA), is a hybrid method of Reinforcement Genetic combining FQL and genetic algorithms for on line optimization of the parametric characteristics of a FIS. In FQLGA we will tune free parameters (precondition and consequent part) by genetic algorithms (GAs) which is able to explore the space of solutions effectively. However, many times the priory knowledge about the FIS structure is not available, as a solution, the suggested approach called Dynamic Fuzzy Q-Learning Genetic Algorithm (DFQLGA), which is a hybrid learning method, this method combines the Dynamic Fuzzy Q-Learning algorithm (DFQL) Meng Joo Er, Chang Deng and the genetic algorithms to optimize the structural and parametric characteristics of the FIS, with out any priori knowledge, the interest of the latter (GA) is to explore the space of solutions effectively and permits the optimization of conclusions starting from a random initialization of the parameters.
This chapter is organized as follows. In Section II, overviews of Reinforcement learning, implementation and the limits of the Fuzzy-Q-Learning algorithm is described. The implementation and the limits of the Fuzzy-Q-Learning algorithm are introduced in Section III. Section IV describes the combination of Reinforcement Learning (RL) and genetic algorithm (GA) and the architecture of the proposed algorithm called Fuzzy-Q-Learning Genetic Algorithm (FQLGA). In section V we present the DFQL algorithm, followed by the DFQLGA algorithm in section VI. Section VII shows simulation and experimentation results of the proposed algorithms, on line learning of two elementary behaviours of mobile robot reactive navigation, "Go to Goal" and "Obstacles Avoidance" is presented with discussion. Finally, conclusions and prospects are drawn in Section VIII.

## 2. Reinforcement learning

As previously mentioned, there are two ways to learn either you are told what to do in different situations or you get credit or blame for doing good respectively bad things. The former is called supervised learning and the latter is called learning with a critic, of which reinforcement learning (RL) is the most prominent representative. The basic idea of RL is that agents learn behaviour through trial-and-error, and receive rewards for behaving in such a way that a goal is fulfilled.
Reinforcement signal, measures the utility of the exits suggested relative with the task to be achieved, the received reinforcement is the sanction (positive, negative or neutral) of behaviour: this signal states that it should be done without saying how to do it. The goal of reinforcement learning, is to find the behaviour most effective, i.e. to know, in each possible situation, which action is achieved to maximize the cumulated future rewards. Unfortunately the sum of rewards could be infinite for any policy. To solve this problem a discount factor is introduced.

$$R=\sum_{k\neq 0}^{\infty} \gamma^{k} r_{k} \tag{1}$$

Where $0 \leq \gamma \leq 1$ is the *discount factor.*

The idea of RL can be generalized into a model, in which there are two components: an agent that makes decisions and an environment in which the agent acts. For every time step, the agent perceives information from the environment about the current state, **s**. The information perceived could be, for example, the positions of a physical agent, to simplify say the ***x*** and ***y*** coordinates. In every state, the agent takes an action $u_t$, which transits the agent to a new state. As mentioned before, when taking that action the agent receives a reward.
Formally the model can be written as follows; for every time step ***t*** the agent is in a state $s_t \in S$ where $S$ is the set of all possible states, and in that state the agent can take an action $a_t \in (At)$, where $(At)$ is the set of all possible actions in the state $s_t$. As the agent transits to a new state $s_{t+1}$ at time $t + 1$ it receives a numerical reward $r_{t+1}$. It up to date then its estimate of the function of evaluation of the action using the immediate reinforcement, $r_{t+1}$, and the estimated value of the following state, $Vt\ (S_{t+1})$, which is defined by:

$$V_t\left(s_{t+1}\right) = \max_{u_t \in U_{t+1}} Q\left(s_{t+1}, u_t\right) \tag{2}$$

the Q-value of each state/action pair is updated by

$$Q\left(s_{t+1t}, u_t\right) = Q\left(s_t, u_t\right) + \beta\left\{r_{t+1} + \gamma V_t\left(s_{t+1}\right) - Q\left(s_t, u_t\right)\right\} \tag{3}$$

Where $r_{t+1} + \gamma V_t\left(s_{t+1}\right) - Q\left(s_t, u_t\right)$ the TD error and β is is the learning rate.
This algorithm is called Q-Learning. It shows several interesting characteristics. The estimates of the function Q, also called the Q-values, are independent of the policy pursued by the agent. To calculate the function of evaluation of a state, it is not necessary to test all the possible actions in this state but only to take the maximum Q-value in the new state (eq.4). However, the too fast choice of the action having the greatest Q-value:

$$u_t = \arg \max_{u_t \in U_{t+1}} Q\left(s_{t+1}, u_t\right) \tag{4}$$

can lead to local minima. To obtain a useful estimate of Q, it is necessary to sweep and evaluate the whole of the possible actions for all the states: it is what one calls the phase of exploration L. Jouffe, A. Souissi. In the preceding algorithm, called TD (0), we use only the state which follows the robot evolution, moreover only the running state is concerned. Sutton A. Souissi extended the evaluation in all the states, according to their eligibility traces that memorise the previously visited state action pairs in our case. Eligibility traces can be defined in several ways L. Jouffe, C. Watkins, P. Dayan, A. Souissi. Accumulating eligibility is defined by:

$$e_t(s) = \begin{cases} 1 + \gamma\lambda e_{t-1}(s) & si \quad s = s_t \\ \gamma\lambda e_{t-1}(s) & \text{else} \end{cases} \tag{5}$$

The algorithm Q (λ) is a generalization of Q-Learning which uses the eligibilities traces A. Souissi: *Q-Learning* is then a particular case of Q (λ) when λ=0.

$$Q\left(s_{t+1}, u_t\right) = Q\left(s_t, u_t\right) + \beta\left\{r_{t+1} + \gamma V_t\left(s_{t+1}\right) - Q\left(s_t, u_t\right)\right\} . e_t(s) \tag{6}$$

## 3. Fuzzy Q-learning algorithm

In mobile robotics, input and output variables given by the sensors and effectors are seldom discrete, or, if they are, the number of state is very large. However reinforcement learning such as we described it uses a discrete space of states and actions which must be have reasonable size to enable the algorithms to converge in an acceptable time in practice.
The principle of the Fuzzy Q-Learning algorithm consists in choosing for each rule $R_i$ a conclusion among a whole of actions available for this rule. Then it implements a process of competition between actions. With this intention, the actions corresponding to each rule $R_i$ have a quality $q^i$ which then determines the probability to choose the action. The output action (*discrete* or *continues*) is then the result of the inference between various actions locally elected. The matrix $q$ enables to implement not only the local policies of rules, but also to represent the function of evaluation of the overall t-optimal policy.
For every time step $t$ the agent is in a state $st \in S$ where $S$ is the set of all possible states, and in that state the agent can take an action $at \in (At)$, where $(At)$ is the set of all possible actions in the state $st$. As the agent receives a numerical reward $rt+1 \in R$ at time $t + 1$, and it transits to a new state $st+1$. It then perceives this state by the means of its activated degree of its rules. The algorithm FQL uses a Policy of Exploration/Exploitation (PEE) L. Jouffe, A. Souissi, combining a random exploration part $\rho^i(U)$ and determinist guided part $\eta^i(U)$.
The steps are summarized as follows:

1. Calculate of the evaluation function :

$$Q_t^*(S_{t+1}) = \sum_{R_i \in A_t} (\underset{U \in U}{Max}(q_t^i(U))\alpha_{R_i}(S_{t+1}),$$

2. Calculate the TD error

$$\tilde{\varepsilon}_{t+1} = r_{t+1} + \gamma Q_t^*(S_{t+1}) - Q_t(S_t, U_t)$$

3. Update the matrix of $Q$ values

$$q_{t+1}^i = q_t^i + \beta.\tilde{\varepsilon}_{t+1} e_t^{iT}, \forall R_i$$

4. Election of the action to be applied

$$U_{t+1}(S_{t+1}) = \sum_{R_i \in A_{t+1}} Election_U(q_{t+1}^i)\alpha_{R_i}(S_{t+1}), \forall U \in U,$$

Where Election is defined by:

$$Election_U(q_{t+1}^i) = ArgMax_{U \in U}(q_{t+1}^i(U) + \eta^i(U) + \rho^i(U)),$$

5. Update of the eligibility traces

$$e_{t+1}^i(U^i) = \begin{cases} \gamma\lambda e_t^i(U^i) + \phi_{t+1}^i, (U^i = U_{t+1}^i) \\ \gamma\lambda e_t^i(U^i), \quad \text{sinon} \end{cases}$$

$$Q_{t+1}(S_{t+1}, U_{t+1}) = \sum_{R_i \in A_{t+1}} q^i_{t+1}(U^i_{t+1})\alpha_{R_i}(S_{t+1}),$$

This value will be used to calculate the TD error in the next step time. However the performances of the controller are closely dependent on the correct choice of the discrete actions set, witch is determined using *a priori* knowledge about system, for complex systems like robots, *priori* knowledge are not available, then it becomes difficult to determine a set of correct actions in which figure the optimal action for each fuzzy rule. To solve this problem and to improve the performances of the reinforcement learning, the genetic algorithms will explore the broadest space of solutions to find the solution optimal Dongbing Gu, Huosheng Hu, Libor Spacek Chia-Feng Juang Min-Soeng Kim and Kim and Ju-Jang Lee Chia-Feng Juang and Chun-Feng Lu, and that without any priori knowledge.

## 4. Genetique reinforcement algorithmes

Genetic Algorithms (GA) are stochastic optimization algorithms, founded on species evolution mechanisms David E Goldberg. In GA, a candidate solution for a specific problem is called an individual or a chromosome and consists of a linear list of genes. Each individual represents a point in the search space and, hence, a possible solution to the problem. A population consists of a finite number of individuals. Each individual is decided by an evaluating mechanism to obtain its fitness value. Based on this fitness value and undergoing genetic operators, a new population is generated iteratively, with each successive population referred to as a generation.
Genetic Reinforcement Learning enable to determine the best set of parameters of (antecedents / consequences) starting from a random initialization of these parameters and in each iteration, only one action is applied on the system to the difference of a traditional genetic algorithm GA use three basic operators (the selection, crossover, and mutation) to manipulate the genetic composition of a population:

- *Reproduction*: Individuals are copied according to their fitness values. The individuals with higher fitness values have more offspring than those with lower fitness values.
- *Crossover*: The crossover will happen for two parents that have high fitness values with the crossover probability *pc*. One point crossover is used to exchange the genes.
- *Mutation*: The real value mutation is done by adding a certain amount of noise (Gaussian in this paper) to new individuals to produce the offspring with the mutation probability ***pm***. For the ***ith*** variable in ***jth*** individual, it can be expressed as:

$$a_{t+1} = a_t + \beta(i).N(0, \sigma) \tag{7}$$

Where **N** denote a Gaussian noise, and $\beta(i) = \exp(-i)$ for the $i^{th}$generation.

### A. FQLGA Algorithm

Because of its simplicity, a Takagi-Sugeno Fuzzy inference system is considered, with triangular membership function. The structure (partition of its input space, the determination of the number of *IF-THEN* rules) of the FIS is predetermined.

$$Rule\ R_i : if\ S_1\ is\ L^i_1\ and \ldots\ldots and\ S_N\ is\ L^i_N\ then\ Y\ is\ a^i_j\ with\ q^i_j$$

$a^i_j$ is a vector representing the discrete set of $K$ conclusions generated randomly for the rule $R_i$ with which is associated a vector $q^i_j$ representing the quality of each action ($i= 1 \sim N$ and $j= 1 \sim K$).

The principle of the approach is to use a population of $K$ (FIS) and to treat the output of each one of them as a possible action to apply on the system. FQL algorithm exploits local quality function $q$ witch is associated with each action of a fuzzy rule (équat.6) whereas FQLGA algorithm uses as function fitness the sum of local qualities $q$ given by:

$$f(Ind_j) = Q(S_{t+1}, SIF_{t+1}) = \sum_{i=1}^{N} q^i_j \tag{8}$$

To reduce the training time, the quality matrix Q is not initialized after every iteration, but undergoes the same genetic operations as those applied to the set of the individuals (selection, crossing).

**B. Optimization of the consequent part of a FIS**

A population of $K$ individuals of a predetermined structure is adopted. The size of an individual is equal to number $N$ of the FIS's rules. The architecture of FQLGA algorithm proposed for the optimization of the conclusions is represented on the figure (1).

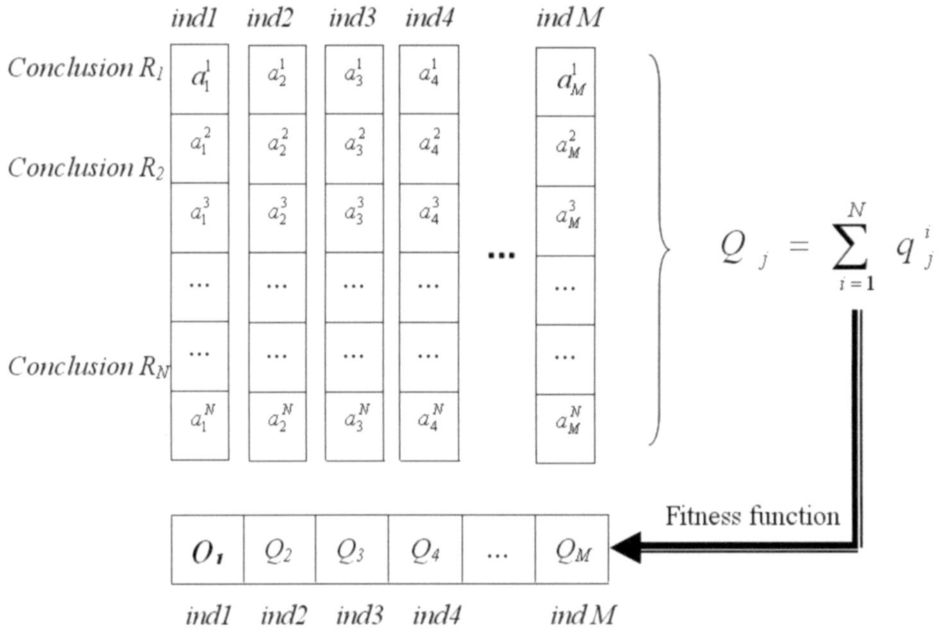


Fig. 1. Representation of the individuals and qualities of the actions in FQLGA algorithm

**C. Optimization of the antecedent part of a FIS**

To find the best set of premises generated by GA, a population made up of $M$ FIS is created. Each individual (FIS) of the population encode the parameters of the antecedents i.e. the

modal points of the FIS and his performance is evaluated by the fitness function of $Q$ (global quality).
The conclusion part of each individual FIS remains fixed and corresponds to the values determined previously. The coding of the membership functions of the antecedent part of a FIS (individual) is done according to the figure (2). To keep the legibility of the FIS, we impose constraints during the evolution of the FIS to ensure the interpretability of the FIS.

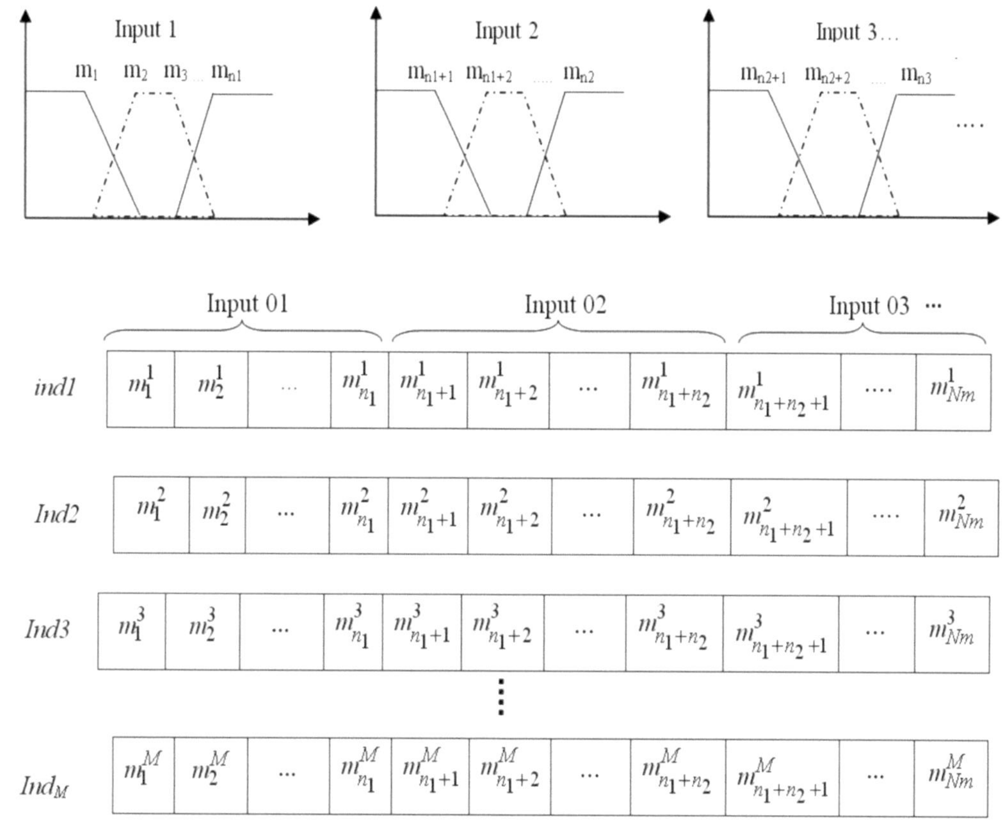


Fig. 2. Coding of the parameters of the antecedent part

Input N: $m^{1}_{N_m - n_N + 1} < .... < m^{1}_{Nm\text{-}2} < m^{1}_{Nm}$

The fitness function used in by genetic algorithm for the optimization of the antecedent part is the global quality of the FIS which uses the degree of activation of the fuzzy rules; this fitness function is given by the following equation:

$$f(ind_i) = Q(S(t), SIF_i) = \frac{\sum_{R_i \in R_A} \alpha_{Ri}(S(t)).q_i^{\hat{i}}(t)}{\sum_{R_i \in R_A} \alpha_{R_i}(S(t))} \tag{9}$$

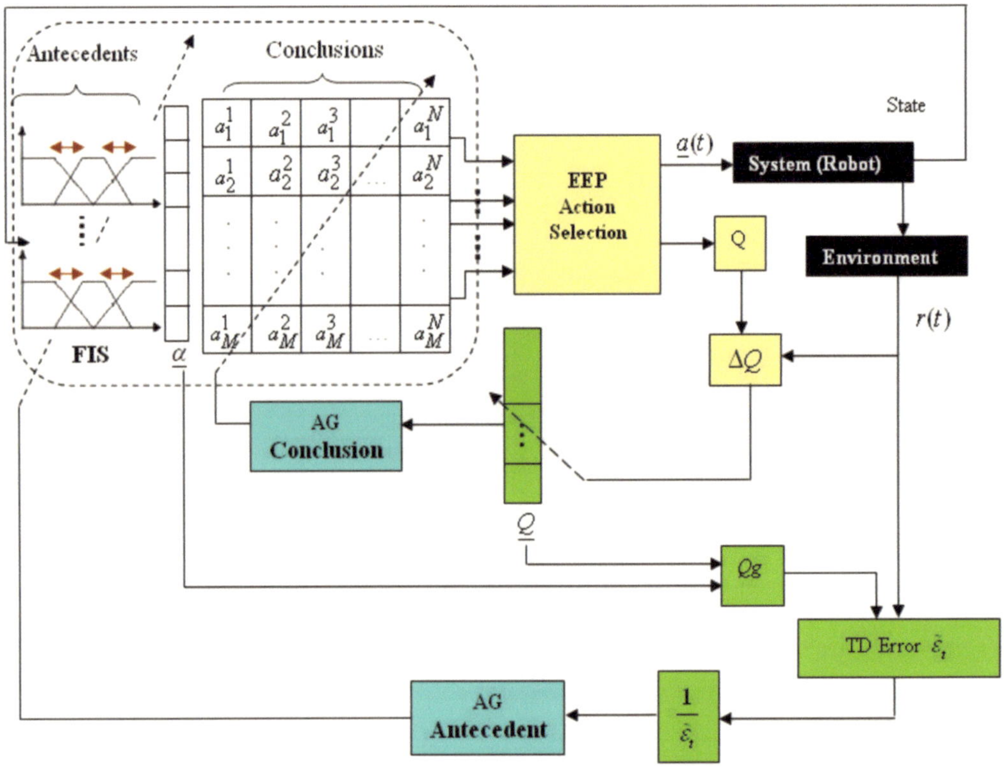


Fig. 3. General Architecture of FQLGA algorithm

**D. Optimization of the Consequent and antecedent part of FIS:**

A FIS consists of a set of fuzzy rules each one of it is made of an antecedent and a consequent part. Optimize the antecedent and the consequent part of the fuzzy controller at the same time is a complex problem which can admit several solutions which are not acceptable in reality, the great number of possible solutions makes the algorithm very heavy and can lead to risks of instability.

FQLGA algorithm proposed in (fig.3) for the optimization of the premises and the conclusions allows the total parametric optimization of the FIS in three stages represented in the flow chart (fig.4).

- At the beginning of the learning process, the quality matrix is initialized at zero, and then traditional algorithm FQL evaluates each action using an exploration policy. This step finishes when a number of negative reinforcements is received.
- After the evaluation of the individuals, the genetic algorithm for the optimization of the consequent part of the fuzzy rules creates a new better adapted generation. This stage is repeated until obtaining convergence of the conclusions or after having reached a certain number of generations. The algorithm passes then to the third stage:
- Once the conclusions of the FIS are optimized, the second genetic algorithm for the optimization of the antecedent part is carried out to adjust the positions of the input membership functions of the controller which are initially equidistant on their universe of discourse.

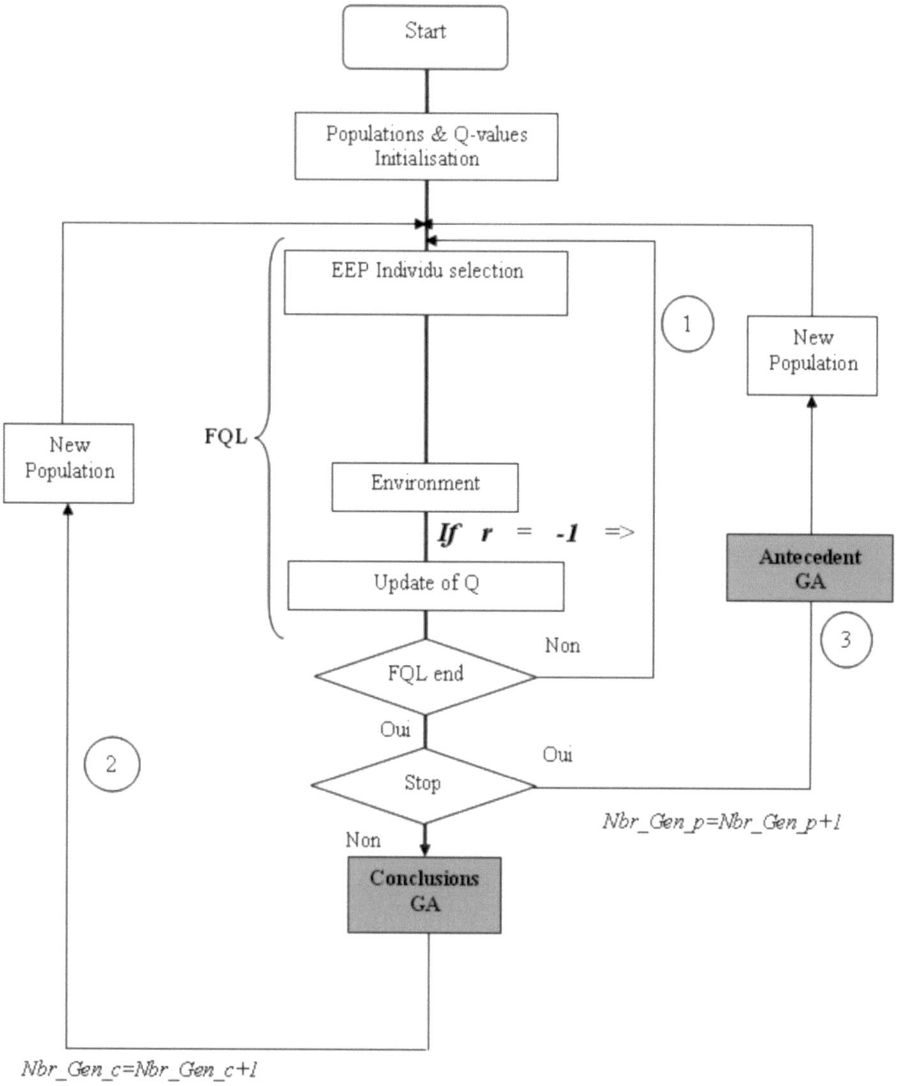


Fig. 4. Flow chart of FQLGA algorithm

## 5. FQLGA simulation & experimental results

To verify the performance of FQLGA two elementary behaviours of reactive navigation of a mobile robot: "Go to *Goal*" and "*Obstacles Avoidance*" are presented in this section; the algorithm was adopted in the experiments for both simulations (Saphira simulator) and real robots (Pioneer II).

### A. «Go to Goal» behaviour:

The two input variables are: the angle"$\theta_{Rb}$" between the robot velocity vector and the robot-goal vector, and the distance robot-goal "$\rho_b$".They are respectively defined by three (*Negative, Zeros, Positive*) and two (*Near, Loin*) fuzzy subsets (fig.5). The two output variables are the rotation speed, *Vrot_CB* and the translation speed *Vtran_CB* each output is represented by nine actions initialised randomly.

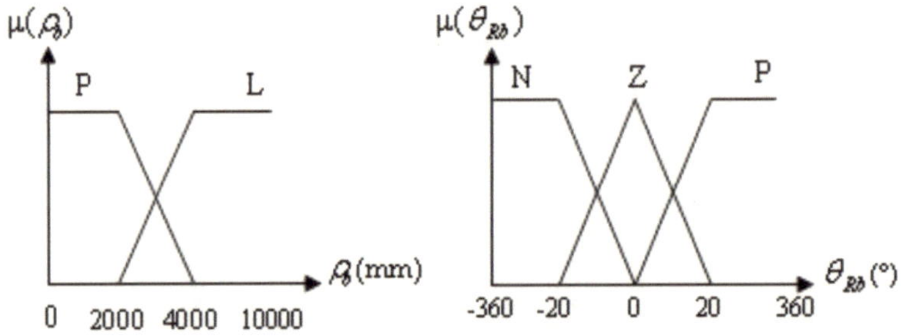


Fig. 5. Membership functions of the input space

The reinforcement functions adopted for the two outputs are respectively given by:

$$r_{Vrot\_CB}(t)=\begin{cases}1 & Si \quad (\theta\cdot\dot{\theta}<0 \;\; ou \;\; -1°<\theta<+1°)\\ 0 & Si \quad (\theta\cdot\dot{\theta}=0)\\ -1 & Else\end{cases} \tag{10}$$

$$r_{Vtran\_CB}(t)=\begin{cases}1 & Si \quad Vtrans\leq 0.15\rho_0 \text{ et } \rho_0\geq 800\\ 1 & Si \quad Vtrans\geq 220 \text{ et } \rho_0\geq 800 \text{ et } abs(\theta)<5\\ -1 & Else\end{cases}$$

The parameters of FQL algorithm and the genetic algorithms are as follows:

| $\gamma$ | $\lambda$ | $L_p$ | $L_c$ | $N_p$ | $N_c$ | $p_m$ |
|---|---|---|---|---|---|---|
| 0.9 | 0.7 | 5 | 6 | 10 | 09 | 0.2 |

*Lp* and *Lc* respectively indicate the sizes of the chromosomes for the antecedents and the conclusions part, *Np*, *Nc* respectively represent the size of the population of the parameters of antecedent and the conclusions and *Pm* the probability of mutation.

The simulation results of the first behaviour "Go to Goal" are presented in the figure (6), 28 generations were sufficient to find the good actions.

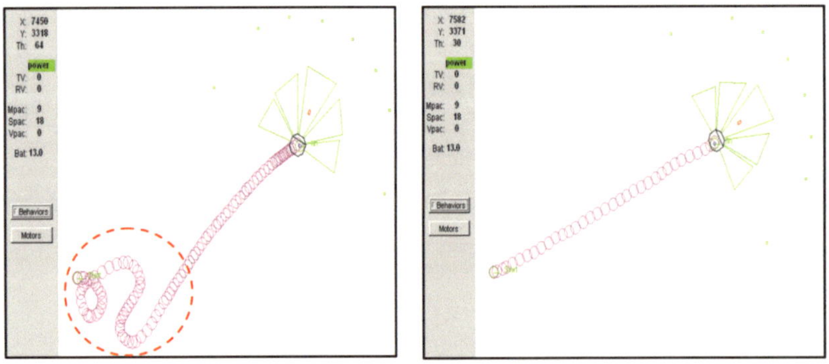

Fig. 6. Go to goal: Training/Validation

Figure (7) shows the convergence of the fitness values of the genetic algorithms for the two output variables *Vrot_CB* and *Vtran_CB* obtained during experimental test.

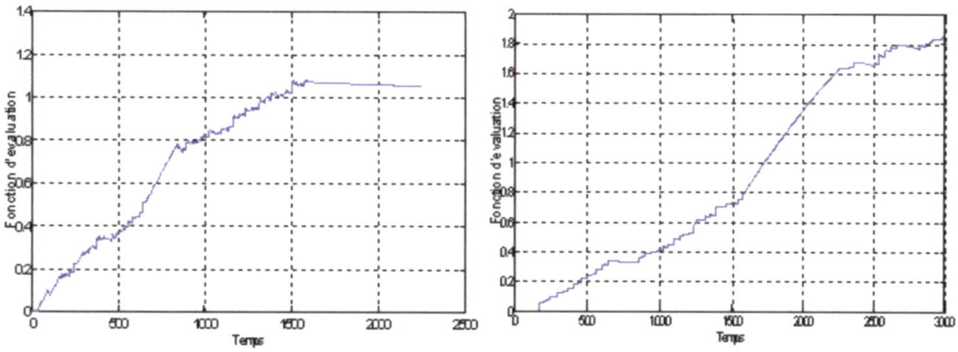


Fig. 7. fitness functions for the two output variables

### B. « Obstacles Avoidance » behaviour

The Inputs of the FIS are the distances provided by the ultrasounds sensors from the three directions (Frontal, Left and Right) and defined by the three fuzzy subsets: near (N), means (M), and far (F) (fig.8).

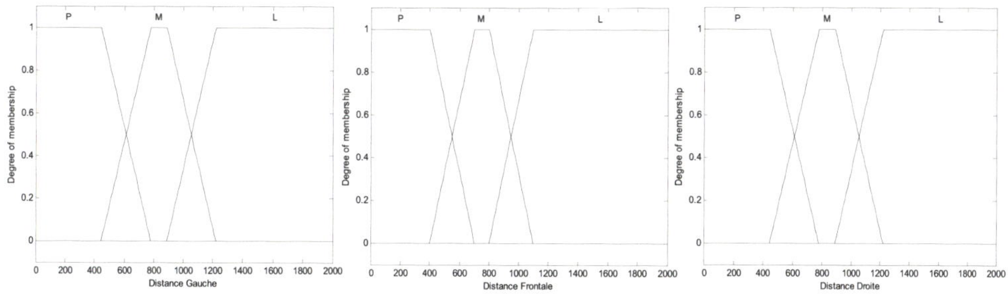


Fig. 8. Memberships Functions of the inputs variables

The conclusions (rotation speeds) are initialized randomly. The translation speed of *Vtran_EO* is given analytically; it is linearly proportional to the frontal distance:

$$Vtran\_EO = \frac{V\max}{D\max}.(Dis\_F - Ds) \tag{11}$$

$V\max$ is the maximum speed of the robot equal to 350mm/s.
$D\max$ is the maximum value measured by the frontal sensors and estimated to 2000mm and $Ds$ is a safe distance witch is fixed at 250mm.
The function of reinforcement is defined as follows [5]:

$$r_{Vrot\_CB}(t) = \begin{cases} 1\times Signe\ (Vrot\_EO) & if\ Dis\_R < Dis\_L\ or\ Dis\_F < Dis\_L \\ & and\ \text{Dmin} < 800 \\ \text{-}1\times Signe\ (Vrot\_EO) & and\ Dis\_L < Dis\_R\ or\ Dis\_F < Dis\_R \\ & and\ \text{Dmin} < 800 \\ 0 & elsewhere \end{cases} \tag{12}$$

with *Dmin=min (Dis_L, Dis_F, Dis_R)*
The parameters of FQL algorithm and the genetic algorithms are identical to the preceding behaviour except for the sizes of the chromosomes *Lp*=12 and *Lc*=27.
Figure (9) represents the trajectories of the robot during the learning phase with FQL algorithm and a random initialization of the consequents part for the 27 fuzzy rules. Several situations of failures are observed, this example show the limits of traditional FQL algorithm when priory knowledge about the system are not available.

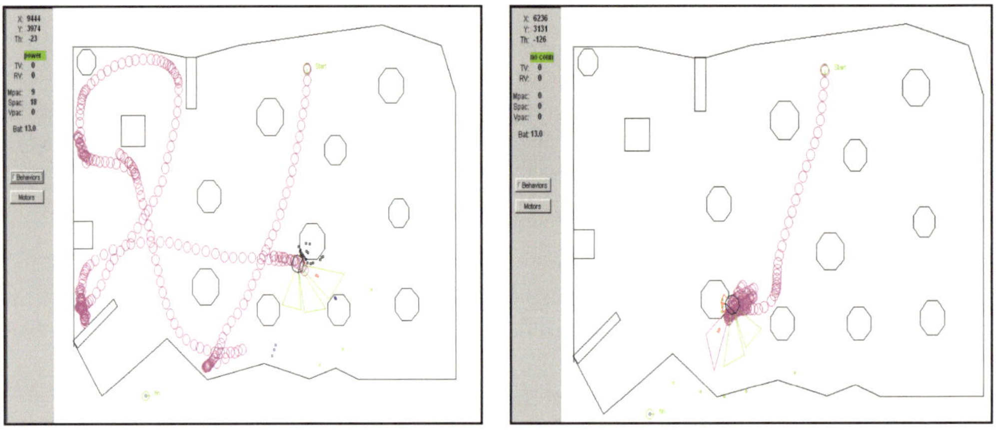

Fig. 9. Trajectories of the robot obtained by FQL algorithm using a random initialization of parameters

The trajectories of figure (10) show the effectiveness of the association of the reinforcement learning FQL and the genetic algorithm as stochastic tool for exploration. FQLGA Algorithm enables to find an optimal FIS for the desired behaviour (obstacles avoidance). The duration of learning depend to the genetic algorithm parameters and the obstruction density of the environment. We observe that after each generation the quality of the FIS (sum of local qualities) increases, which give more chance to the best individuals to be in the next generations.

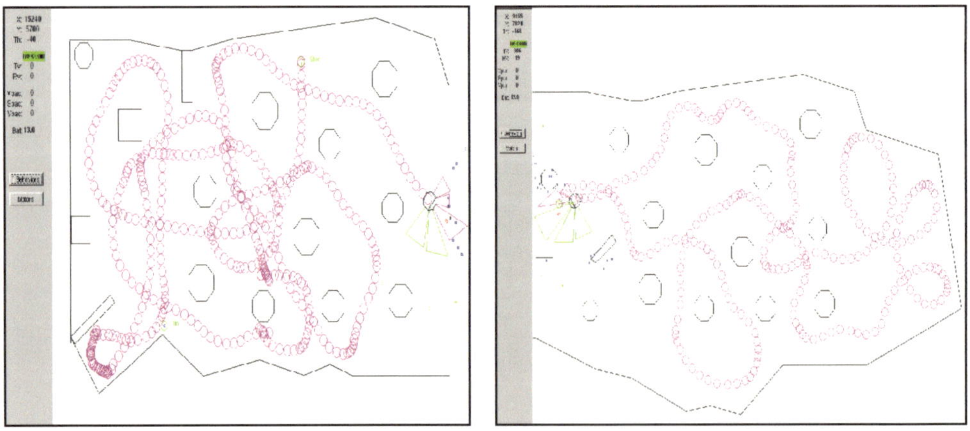

Fig. 10. Learning/Validation Trajectories of the robot with FQLGA algorithm for various environments

Figure (11) shows the performances of FQLGA algorithm compared to the FQL algorithm which can be blocked in a local minimum when the optimal solution is not present in the randomly generated set of actions. On the other hand FQLGA algorithm converges towards the optimal solution independently of the initialized values.

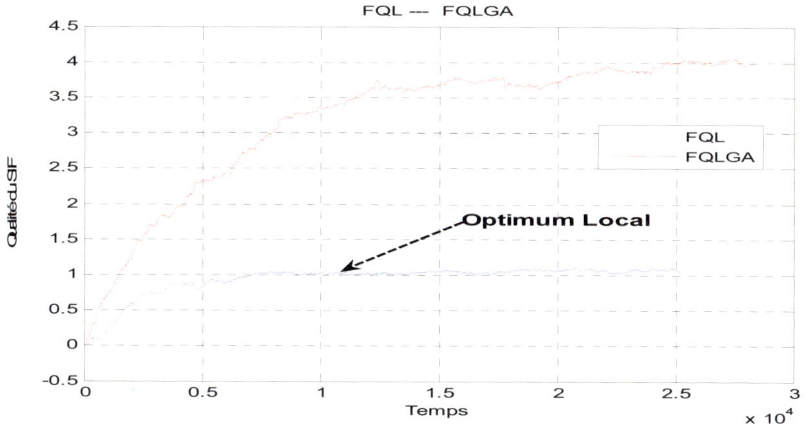


Fig. 11. Evolution of the quality of the Fuzzy controller with FQL and FQLGA algorithms

**C. Experimental results with the real robot Pioneer II**

Figure (12) represents the results of the on line learning of the robot Pioneer II for the behaviour "Go to goal". During the learning phase, the robot does not follow a rectilinear trajectory (represented in green) between the starting point and the goal point because several actions are tested (exploration). Finally the algorithm could find the good actions, and the robot converges towards the goal marked in red colour, the necessary time to find these good actions is estimated at 2mn. Each generation is generated after having noted twenty (20) failures. The learning process requires respectively 32 and 38 generations for GA to determine rotation and translation speed.

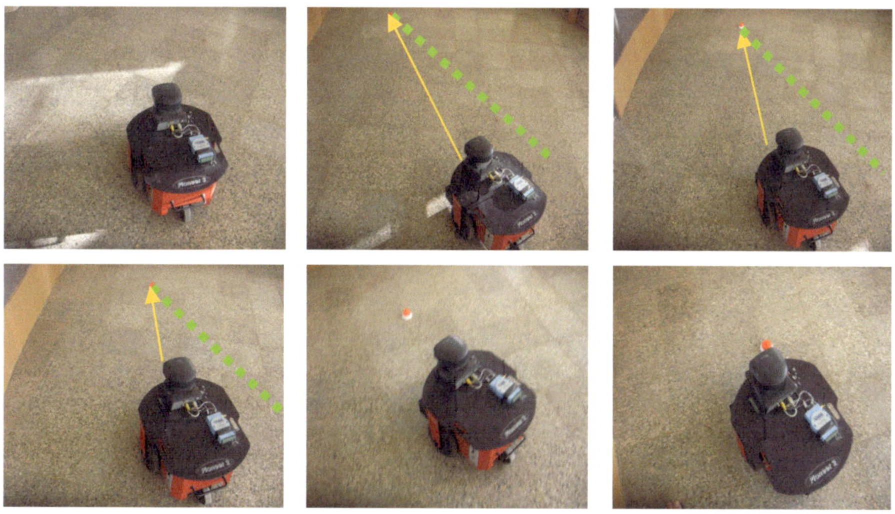

Fig.12. On line learning of the real robot Pioneer II, "Go to goal" behaviour

Figure (13) represents the results of the on line learning of the "Obstacles Avoidance" robot behaviour. For reasons of safety, we consider that the minimal distances detected by the frontal sonars and of left/right are respectively 408 mm and of 345 mm a lower value than these distances is considered then as a failure and involves a retreat (represented in green) of the mobile robot. A generation is created after 50 failures. The genetic algorithms require 14 generations to optimize the conclusions and 24 generations to optimize the parameters of the antecedents. The duration of on line learning is estimated at 20 min, this time is acceptable vis-à-vis the heaviness of the traditional genetic algorithms.

Fig. 13. On line learning of the real robot Pioneer II, behaviour "Obstacle Avoidance»

Figure (14) represents the evolution of the fitness function obtained during this experimentation.

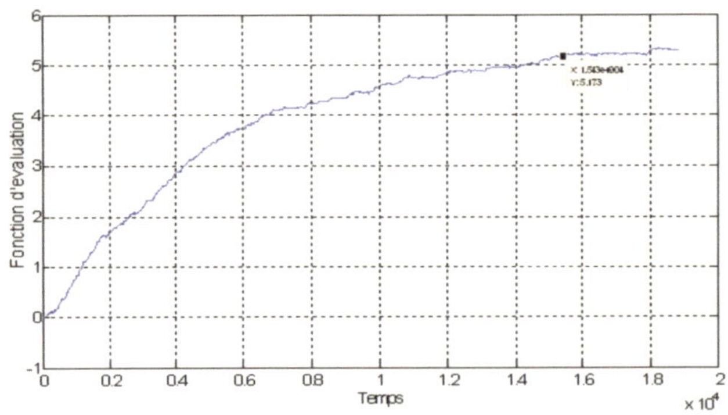


Fig.14 Fitness function evolution, "Obstacles avoidance" robot behaviour

## 6. Dynamic Fuzzy Q-Learning Genetic Algorithm (DFQLGA)

### A. Algorithm DFQL

The algorithm Dynamic Fuzzy Q-Learning (DFQL) proposed by Meng Joo Er, et Chang Deng is an extension of the original Q-learning method into a fuzzy environment. State-space coding is realized by the input variable fuzzy sets. A state described by a vector of fuzzy variables is called a fuzzy state. A learner may partially visit a fuzzy state, in the sense that real-valued descriptions of the state of the system may be matched by a fuzzy state description with a degree less than one. Since more than one fuzzy state may be visited at the same time, possibly with different degrees, we have a smooth transition between a state and its neighbours, and, consequently, smooth changes of actions done in the different states. Both the actions and the Q-function are represented by a FIS whose fuzzy logic rules can be self-constructed based on the -completeness of fuzzy rules and the TD error.

**1) $\varepsilon$ - *Completeness* criteria for Rule Generation**

Definition: -Completeness of fuzzy rules C. C. Lee: For any input in the operating range, there exists at least one fuzzy rule so that the match degree (or firing strength) is no less than $\varepsilon$. In fuzzy applications, the minimum value of $\varepsilon$ is usually selected as $\varepsilon = 0{,}5$.
From the viewpoint of fuzzy rules, a fuzzy rule is a local representation over a region defined in the input space. If a new pattern satisfies $\varepsilon$ - *Completeness*, the DFQL will not generate a new rule but accommodate the new sample by updating the parameters of existing rules.

**2) TD Error Criterion for Rule Generation**

It is not sufficient to consider -completeness of fuzzy rules as the criterion of rule generation only. New rules need to be generated in regions of the input fuzzy subspace where the approximation performance of the DFQL is unsatisfactory. Here, we introduce a separate performance index $\xi^i$, for each fuzzy subspace which enables the discovery of "problematic" regions in the input space. The performance index is updated as follows Meng Joo Er,et Chang Deng:

$$\xi_{t+1}^i = [(K-\alpha_t^i)\xi_t^i + \alpha_t^i(\tilde{\varepsilon}_{t+1})^2]/K \quad \text{avec} \quad K>0$$

By developing the preceding equation, the errors TD coefficients are expressed as follows:

$$\begin{cases} \xi_1^i = \dfrac{\alpha_0^i}{K}\varepsilon_1^2 \\ \xi_2^i = (\dfrac{K-\alpha_1^i}{K})\cdot(\dfrac{\alpha_0^i}{K})\varepsilon_1^2 + (\dfrac{\alpha_1^i}{K})\varepsilon_2^2 \\ \xi_3^i = (\dfrac{K-\alpha_2^i}{K})\cdot(\dfrac{K-\alpha_1^i}{K})\cdot(\dfrac{\alpha_0^i}{K})\varepsilon_1^2 + (\dfrac{K-\alpha_2^i}{K})\cdot(\dfrac{\alpha_1^i}{K})\varepsilon_2^2 + (\dfrac{\alpha_2^i}{K})\varepsilon_3^2 \\ \vdots \end{cases} \tag{13}$$

After each iteration, all preceding errors TD are balanced by the term $\dfrac{K-\alpha_{t-1}^i}{K} < 1$.

Using the squared TD error as the criterion, the rule firing strength $\alpha^i$ determines how much the fuzzy rule $R_i$ affects the TD error. It should be noted that (13) acts as a digital low-pass filter. In this way, TD errors in the past are gradually "forgotten" as time passes, but are never completely lost. The more recent the TD error is received, the more it affects the value of $\xi$. The initial value of $\xi^i$ is set to zero. The parameter $K$ which controls the overall behaviour of $\xi$ is usually selected between ten and 100. A small value of $K$ makes $\xi$ adapt very rapidly and a large value makes $\xi$ more stable in a noisy environment. Thus, if $\xi^i$ is lower than a certain threshold $k_e$, further segmentations should be considered for this fuzzy subspace at least. The figure (15) represents the global flowchart of the DFQL algorithm Meng Joo Er,et Chang Deng.

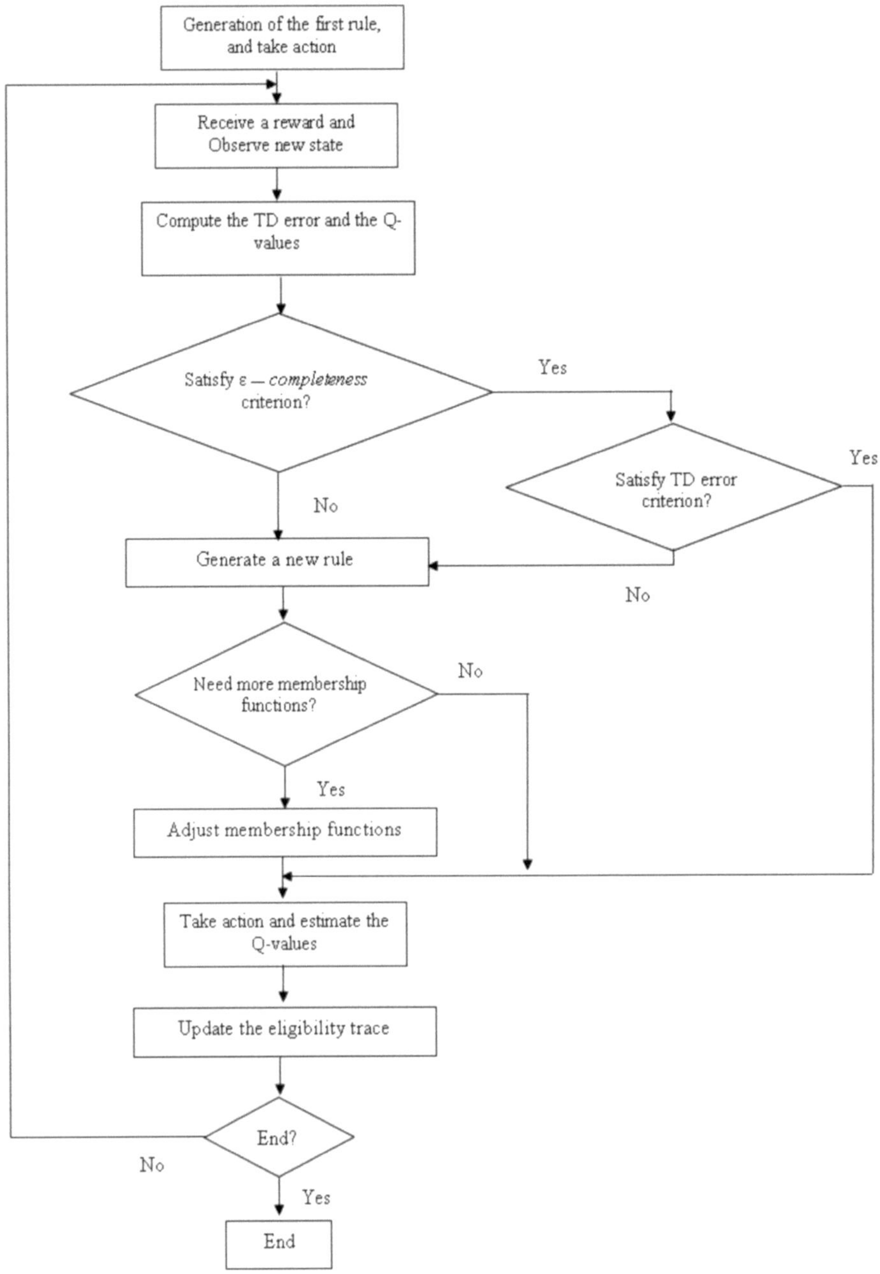


Fig. 15. DFQL algorithm flowchart

**B. Dynamic Fuzzy Q-Learning Genetic Algorithm (DFQLGA)**

The structural and parametric optimization of a fuzzy inference system without any *priori* knowledge, is the objective of several research, several approaches suggested in the literature are based on the reinforcement genetic method.

The Algorithm FQLGA, presented in A. Nemra H. Rezine, ensures a parametric optimization; however the optimization of the conclusions and premises separately doesn't exploit effectively the power of the genetic algorithm. The SEFC Algorithm proposed by Chia-Feng Juang, Jiann-Yow Lin, and Chin-Teng Lin, permit to solve this problem. Using the Michigan approach, when a fuzzy rule (premises & conclusions) is regarded as an individual and the FIS is built by the combination of several individuals.

However for these two algorithms the structure of the FIS (the number of fuzzy rules and membership function) is predefined by the human expert, which is not always easy, especially in the case of the complexes systems. For the latter an automatic design of the FIS becomes necessary, which is the objective of the learning algorithm DFQL which generates the rules and the membership functions automatically during the learning according to the two criterias mentioned before. The results obtained by algorithm DFQL are considered to be satisfactory if the predefined set of actions is correct and suitable for the desired task, nevertheless, if it is not the case (no priori knowledge) then the apprentice is likely to fail to do its task appropriately. Then as a solution for this problem we develop a new algorithm the DFQLGA.

**C. Principle of the DFQLGA algorithm**

The new learning algorithm suggested is called Dynamic Fuzzy Q-Learning Genetic Algorithm (DFQLGA) is a combination of the algorithm DFQL and the genetic algorithm (GA) developed previously. With this algorithm, the fuzzy rules and the memberships functions of the premise part are generated automatically by the DFQL algorithm Meng Joo Er, Chang Deng according to *ε-Completeness* and *error TD* criteria. The conclusion part of the fuzzy rules, as it is randomly initialized (a random set of actions for each fuzzy rule), then the genetic algorithm is used to determine the best conclusions of the rules previously generated.

**D. Optimization of the conclusions by GA**

The considered Fuzzy Inference System (FIS) is a zero order Takagi-Sugeno type, composed by one fuzzy rule only, initialized randomly as follow:

*Rule Ri : if S1 is* $L_1^i$ *and........and* $S_{Ni}$ *is* $L_{Ni}^i$ *then Y is* $a_j^i$ *with* $q_j^i$ $a_j^i$ is a vector representing the discrete set of $K$ conclusions generated randomly for the output variable $Y$ of the rule $R_i$ with which a vector representing the quality $q_j^i$ is associated with each action. The initialization of the fuzzy rule is random for the conclusion part, the membership functions of the premises are of type Gaussian and has unspecified parameters (Means, Standard deviation).

The update of local quality $q_j^i$ is given by the equation (eq.6), the algorithm DFQLGA uses as fitness function the sum of local qualities Q given by (fig.16):

$$f(Ind_j) = Q(S_{t+1}, SIF_{t+1}) = \sum_{i=1}^{N} q_j^i \tag{14}$$

To accelerate the time of learning, the quality matrix Q is not initialized at any iteration, but undergoes the same genetic operations as those applied on the population of the individuals (selection, crossing). A new generation is created not at any iteration but only after having received a specified number of failures (negative rewards) Chia-Feng Juang, Min-Soeng Kim and Kim and Ju-Jang Lee. Between two generations, the evolution of the quality matrix is done by the FQL algorithm according to the equation (6). An exploration/exploitation policy (EEP) is implemented at the beginning of the FQL algorithm; thereafter it is ensured by the mutation operator of the genetic algorithm.

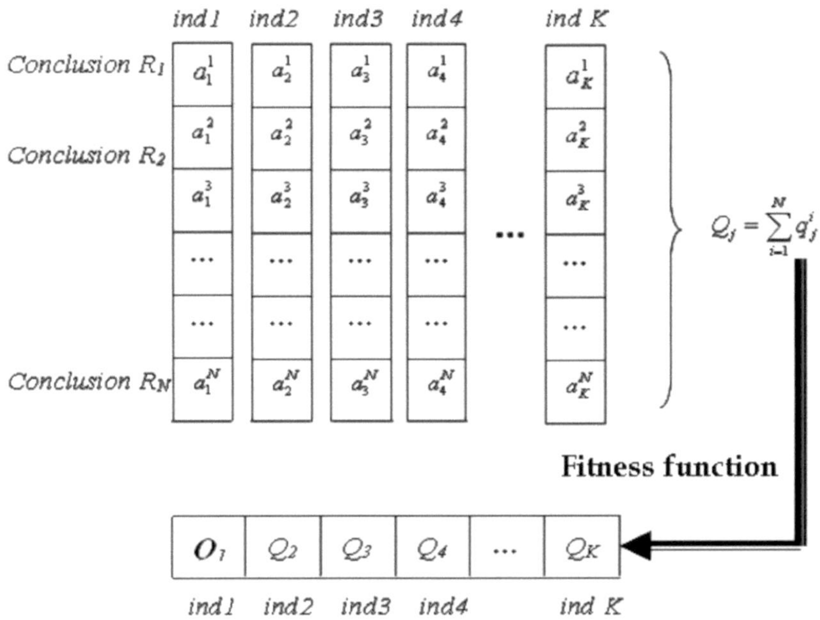


Fig. 16. Evaluation of the individuals

The steps of the algorithm DFQLGA are summarized as follows (fig.17):

1. Initialization with only one fuzzy rule with random parameters.
2. Generation of a new fuzzy rule by using the two criteria ε-*Completeness* and *error TD*.
3. Once a number $N_s$ of fuzzy rules is reached the genetic algorithm is carried out to explore the space of solutions, (creation of a generation after any $N_{ech}$ failure) using the sum of local qualities as fitness function (eq.14).
4. Repeat stage 1 and 2 until the criterion of stop is checked (convergence of the fitness function).

## 7. DFQLGA simulation & experimental results

We study two elementary behaviours of reactive navigation of a mobile robot: "*Go to a goal*" and "*Obstacles Avoidance*" carried out by a fuzzy inference system optimized (structural and parametric characteristics optimization) by the algorithm DFQLGA. We have the results obtained on the Saphira simulator of the robot Pioneer II, then on the real robot.

### A. "Go to Goal" behaviour

The two inputs variables are «$\rho_g$» and «$\theta_{RG}$» represent the angle between the velocity vector of the robot and the vector robot-goal and the distance robot-goal respectively. They are initialized with one fuzzy sub-set only for each fuzzy rule (Fig.18). The two outputs variables are the rotation velocity *Vrot_GG* and the translation, velocity *Vtran_GG*.

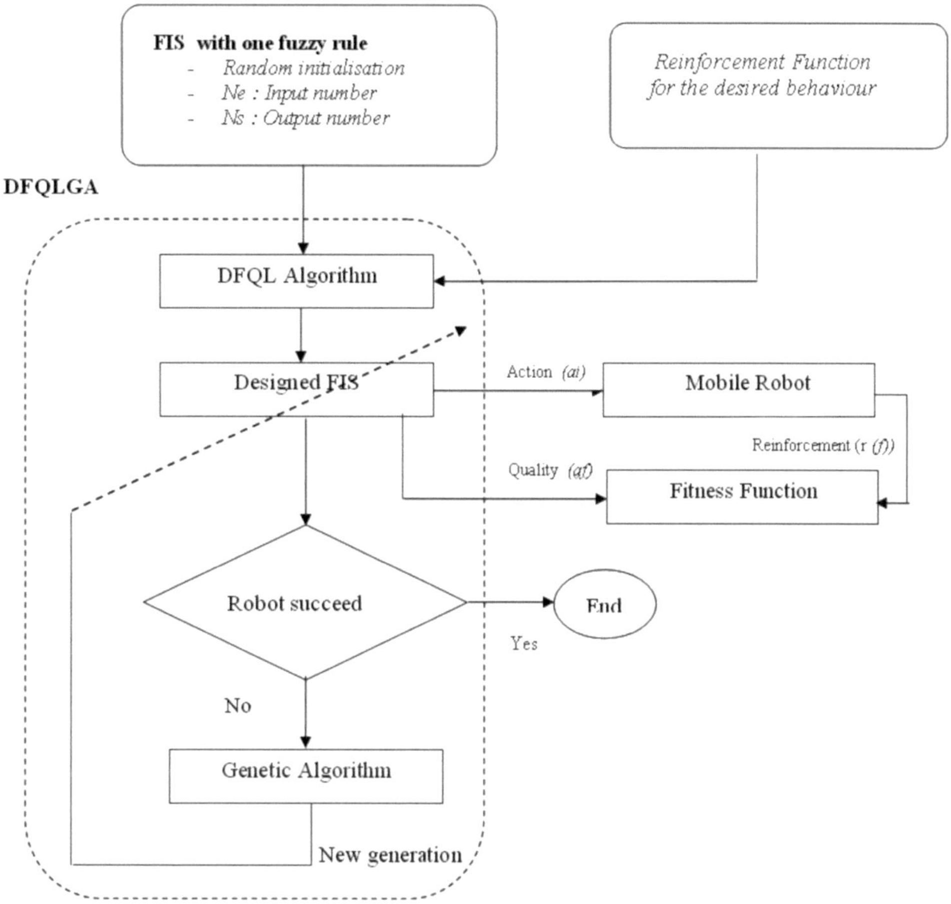


Fig. 17. Flow chart of the Algorithm DFQLGA

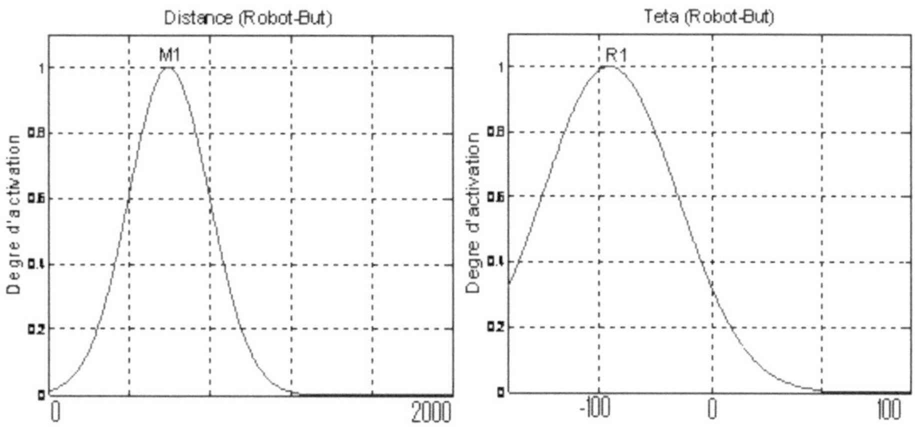


Fig. 18. the initialized membership functions for the two input of the FIS

The adopted reinforcement functions are respectively given by:

$$r_{Vrot\_GG} = \begin{cases} 1 & if \quad (\theta_{RG}\dot{\theta}_{RG} < 0 \;\; or \;\; -1^0 < \theta_{RG} < +1^0) \\ 0 & if \quad (\theta_{RG}\dot{\theta}_{RG} = 0) \\ -1 & else \end{cases}$$

$$r_{Vtrans\_GG} = \begin{cases} 1 & if \;\; Vtrans < 0.15\rho_g \;\; and \;\; \rho_g > 800\ mm \\ 1 & if \;\; Vtrans > 220mm \;\; \& \;\; \rho_g \geq 800\ mm \;\; \& \;\; |\theta_{RG}| < 5^0 \\ -1 & else \end{cases} \qquad (15)$$

The parameters of the algorithm DFQLGA are given as follows:

| DFQL | | | | | | AG | | |
|---|---|---|---|---|---|---|---|---|
| $\varepsilon$ | $k_d$ | $k_{mf1}$ | $k_{mf2}$ | $K$ | $k_e$ | $Ns$ | $Np$ | $Pm$ |
| 0.5 | 0.83 | 1000 | 60 | 20 | 0.7 | 06 | 10 | 02 |

The genetic algorithm run only, once a number of fuzzy rules *Ns* is reached by algorithm DFQL. *Np, Pm* respectively represents the size of the population and the probability mutation.
The figure (19) represents the results at the beginning of the learning, after the generation of four fuzzy rules; the genetic algorithm was not run yet. The membership functions determined by the DFQL algorithm are represented on the figure (20).

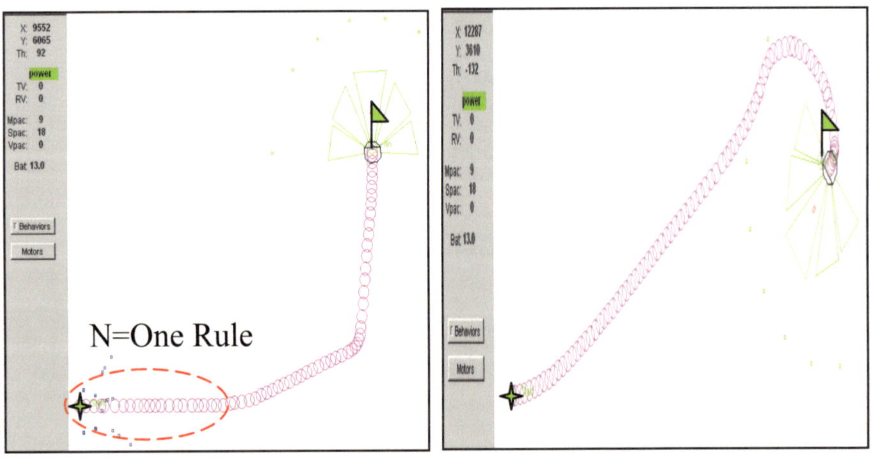


Fig. 19. Go to Goal (Learning)

Progressively with its evolution, the robot discovers new situations, and new fuzzy subsets are then generated (fig.21) and other fuzzy rules are then added and the behaviour of the robot improves (fig.22). At the end of the learning, eight (08) fuzzy rules are built and the figure (21) represents the final membership functions generated for the two inputs.

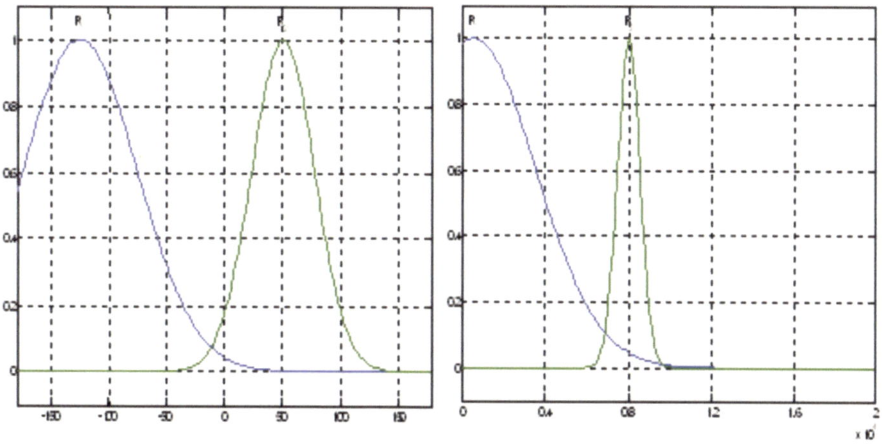

a) Angle $\theta_{Rb}$ (Robot-Goal) b) Distance $\rho$ (Robot-Goal)

Fig. 20. Determined membership functions after the generation of four fuzzy rules

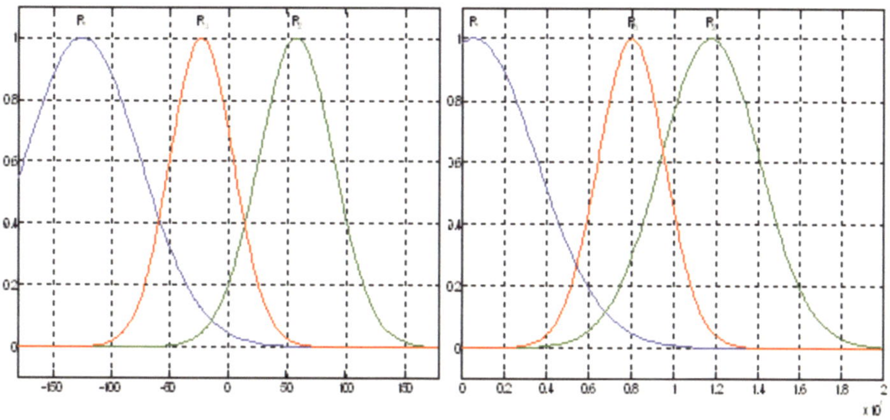

a) Angle $\theta_{RB}$ b) Distance $\rho$

Fig. 21. final membership functions after the generation of 08 fuzzy rules

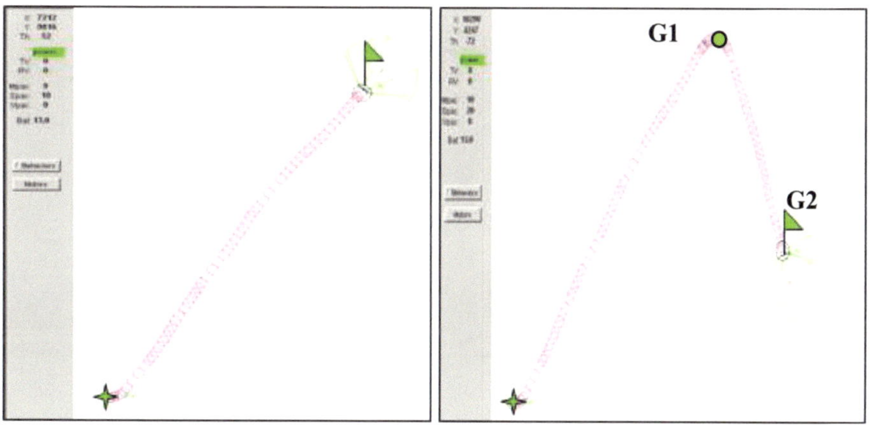


Fig. 22. Validation of the behaviour « Go to Goal »

**B. "Obstacles Avoidance" behaviour**

The algorithm DFQLGA is also validated with the behaviour of obstacles avoidance, the inputs of the FIS are the distances read by the ultrasound sensors, we test two types of FIS (with three and four inputs) and the rotation velocity *Vrot_EO* as output variable, the translation velocity *Vtran_EO* is calculated analytically, as a linear function of the frontal distance.

$$Vtran\_EO = \frac{V\max}{D\max}.(Dis\_F - Ds) \tag{16}$$

$V_{max}$ is the maximum velocity of the robot equal to 350m/s. $D_{max}$ is the maximum reading allowed the frontal sensors equal to 2000mm and $D_s$is a safety distance fixed at 250mm. The parameters of the algorithm are given as follows:

| $\varepsilon$ | $k_d$ | $k_{mf}$ | $K$ | $k_e$ |
|---|---|---|---|---|
| 0.5 | 0.83 | 400 | 20 | 0.7 |

For the GA we will take *Ns=08*, and *Pm=0.1*

**FIS with three inputs:**

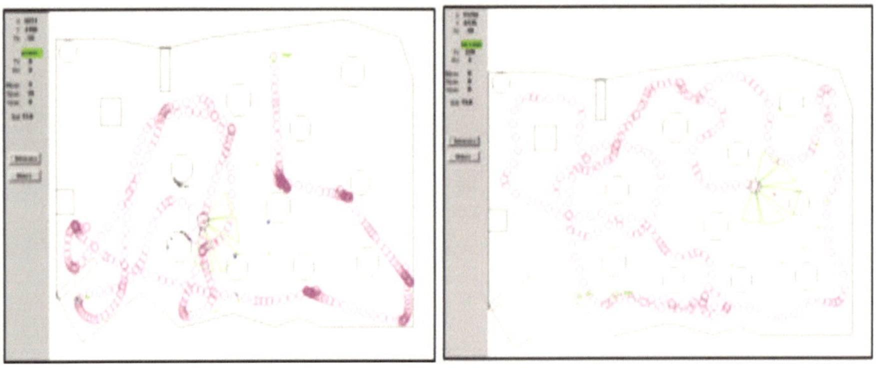

Fig. 23. Trajectories of the robot, Learning & Validation with the algorithm DFQLGA for various environments.

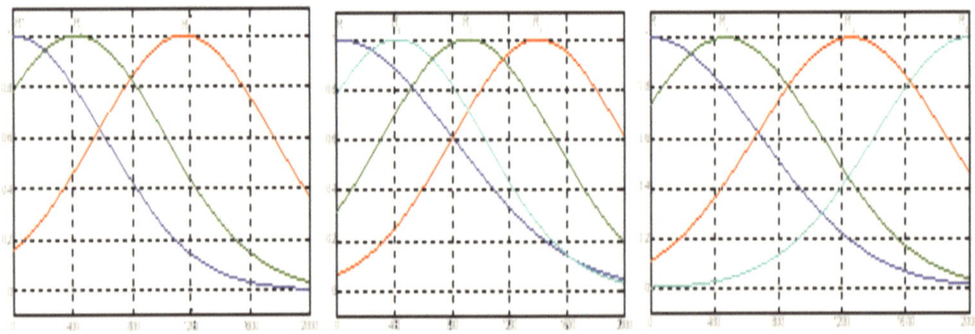

a b c

Fig. 24. Finals membership functions of the inputs of the FIS determined by the DFQLGA algorithm a) Left Distance, b) Right Distance, c) Frontal Distance

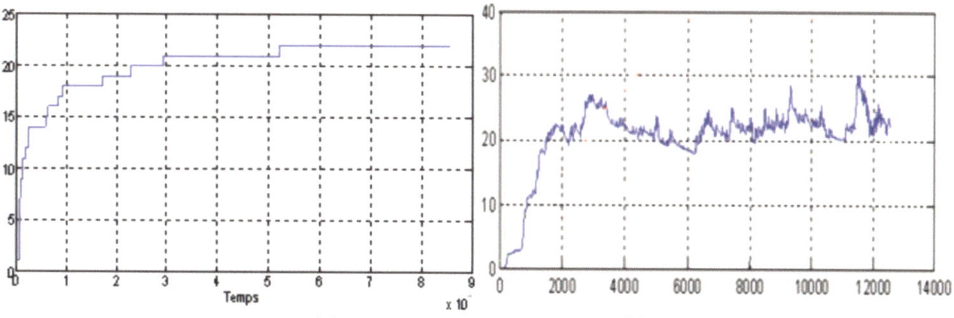


(a) (b)

Fig. 25. (a) Number of fuzzy rules generated by the DFQLGA algorithm (b) Fitness function evolution

**FIS with four inputs:**

The inputs of the fuzzy controller are the minimal distances provided by the four sets of sonars {D1=min (d1, d2), D2=min (d3, d4), G1=min (g1, g2), G2=min (g3, g4)}. The translation velocity of the robot is given as previously (eq.16) with a light modification: it is linearly proportional to the Frontal Distance.

$$Vtrans = \frac{V\max}{D\max}.(\min(g4, d4) - Ds) \tag{17}$$

At the end of the learning process, the final FIS is composed by forty five (45) fuzzy rules and the generated membership functions are represented on the figure (26).

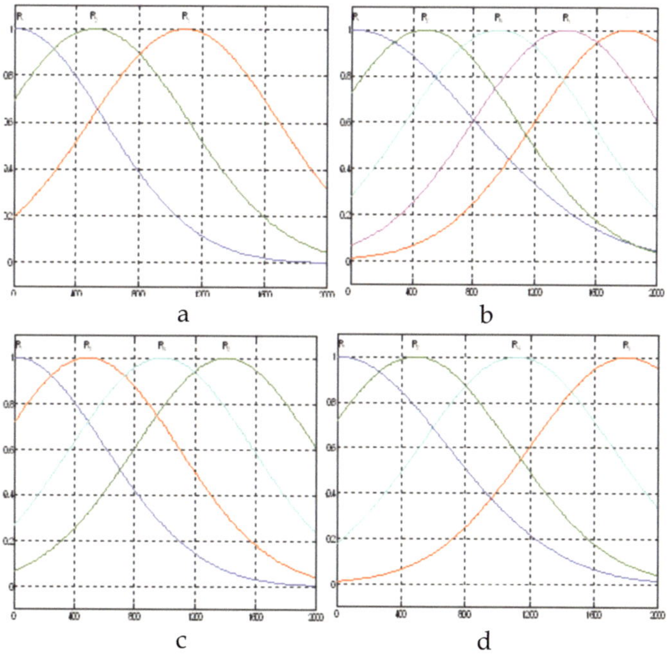

Fig. 26. Final membership functions of the inputs of the FIS determined by the DFQLGA algorithm a) Right Distance, b) Right-Frontal Distance c) Frontal-Left Distance d) Left Distance.

The figure (28) represents the evolution of the number of generations, the execution of the genetic algorithm starts after the generation of eight (08) fuzzy rules. A new generation of the conclusions is carried out after reception of certain number of failures (20 failures in our case).
The figure (27) shows the evolution of the fitness function (sum of the locals qualities) used by the genetic algorithm.

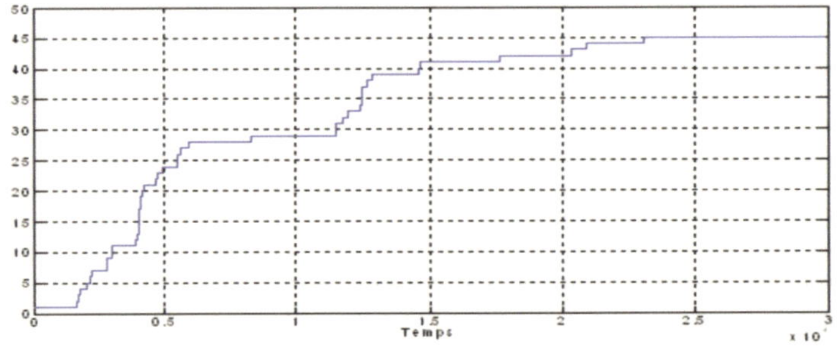


Fig. 27. Evolution of the Number of fuzzy rules generated by the algorithm DFQLGA

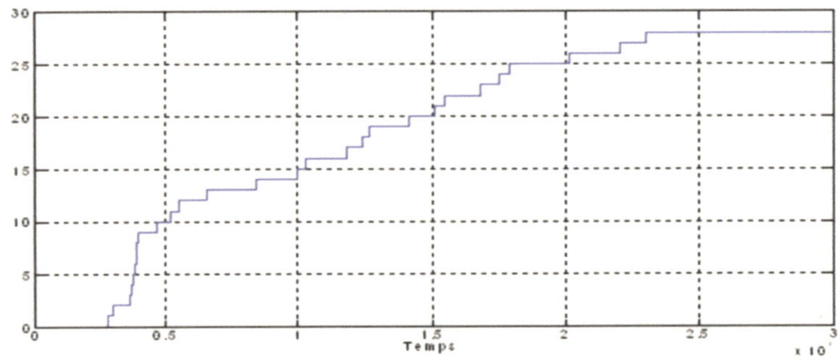


Fig. 28. Number of generation generated by the algorithm DFQLGA

The figures (fig.30) and (fig.31) respectively represent the trajectories of the mobile robot during and after the learning. The learning phase proceeds in three (03) stages (fig.30):

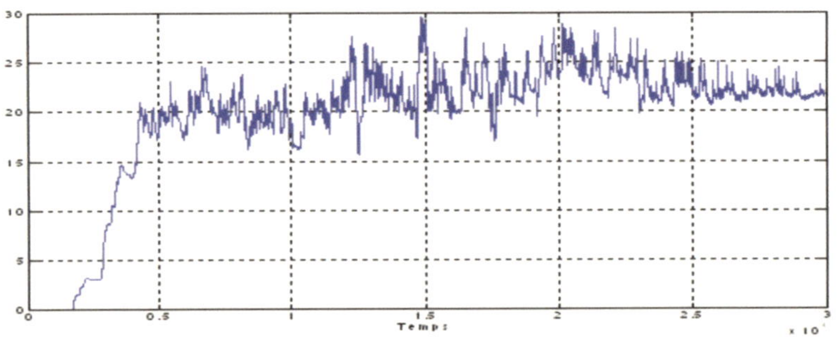


Fig. 29. Evolution of the fitness function

1. In the beginning the robot is almost blind and it moves according to straight line until entering in collision with the obstacles. In this stage the FIS is composed by one fuzzy rule only initialized randomly, this stage contains the first four failures. Once the $\varepsilon$ - *Completeness* and the *Error_TD* criteria are satisfied, then new rules are generated.

2. In a second stage the structure of the found FIS permits the mobile robot to learn the obstacles avoidance behaviour if the correct action is among the set of the suggested actions, otherwise robot affects other failures (failure 5 and failure 6). In this case the genetic algorithm will create new generations of the actions more adapted based on the fitness function.
3. In this third stage, the robot is able to carry out the desired behaviour successfully (fig.31). The necessary number of fuzzy rules and membership functions is generated by the algorithm DFQL when the values of the conclusions are optimized by the genetic algorithm.

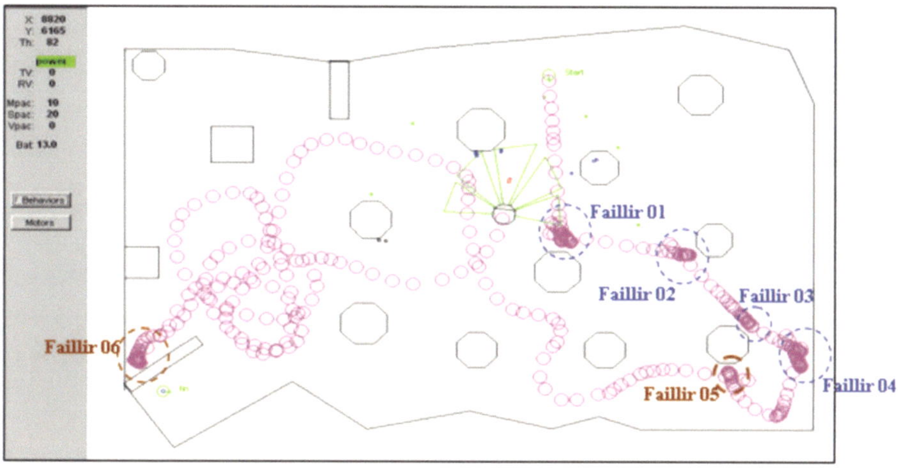


Fig. 30. Trajectories of the robot during the learning of the SIF by algorithm DFQLGA (the first two stages)

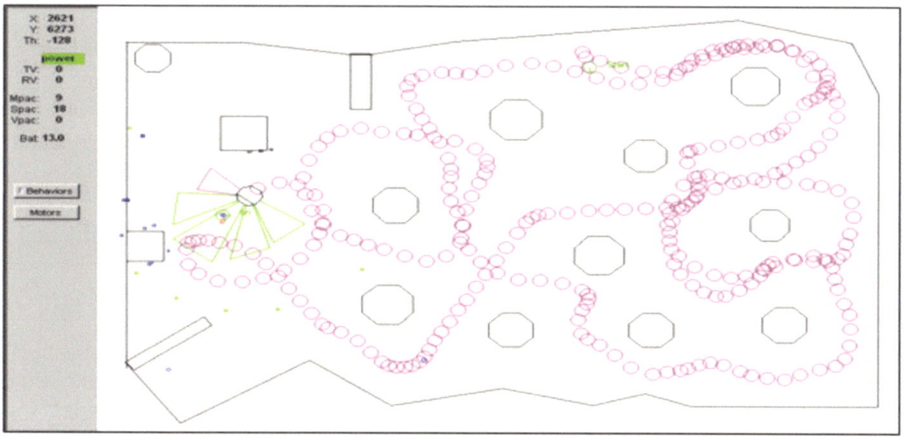

Fig. 31. Trajectories of the robot after the learning of the FIS by algorithm DFQLGA (the third stage)

### C. experimental Results on the robot real Pioneer II

The figure (32) represents the results of validation on the mobile robot Pioneer II for the behaviour "Go to goal" by choosing two goals of coordinate, B1 (2500 mm, - 1000 mm) and

B2 (3250 mm, 1250 mm) the figure (32) shows that the robot succeeds in achieving respectively the two goals B1 then B2.

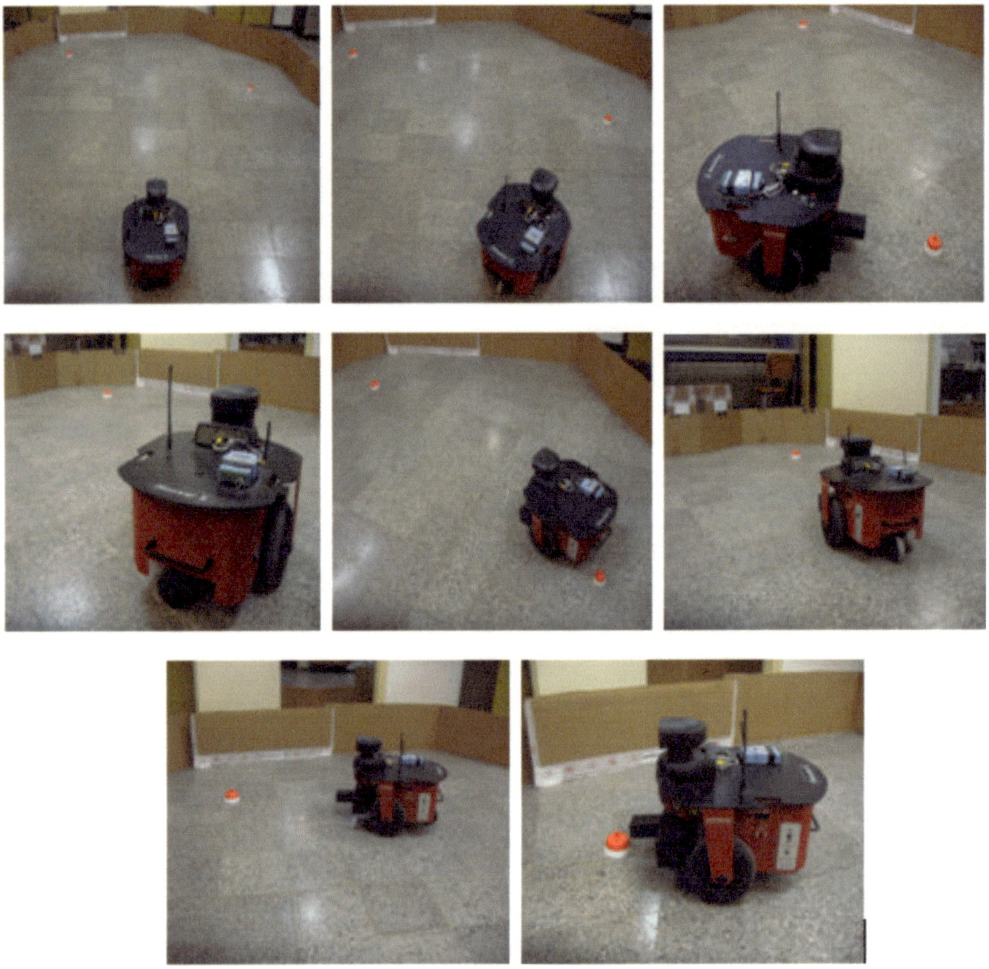

Fig. 32. Validation of the FIS obtained in simulation by algorithm DFQLGA on the real robot Pioneer II for the "Go to goal" behaviour

The figure (33) represents the results of learning (a) and validation (b) in real time for the robot Pioneer II for the "go to goal" behaviour.
The figure (34) represents the results of the on line learning of the robot for the behaviour "Obstacles Avoidance". For reasons of safety, we consider that the minimal distance detected by the frontal sonar is 408 mm and the right/left sonar is 345 mm, if the detected value is lower than these thresholds then the behaviour is regarded as a failure and involves a retreat (represented in green) of the mobile robot.
The figure (34-a) represents the behaviour of the robot during the real time learning of the robot with the behaviour "obstacles avoidance". The figure (34-b) represents the evolution of the fitness function, according to this figure we can distinguish two phases, the first

corresponds to the exploration stage, where the robot carries out several tests/failures, then in a second phase the robot begin to choose the suitables actions at the appropriate state, it is the end of learning. The algorithm DFQLGA generates fourteen (14) fuzzy rules (fig.34.c) after 21 generations; the time of real time learning is estimated to 19 minutes.

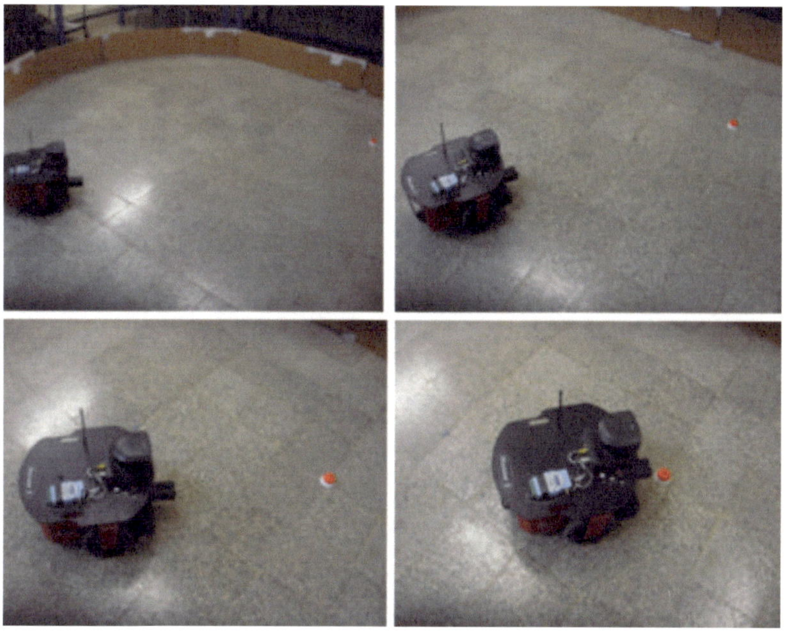

a) Real time learning

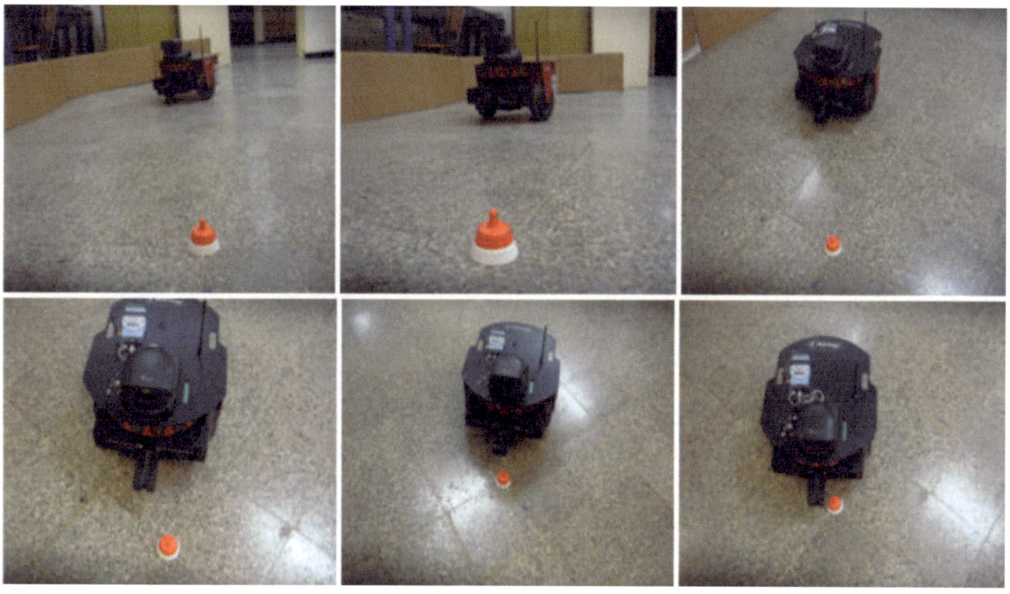

b) Validation

Fig. 33. Learning/Validation of the FIS conceived in real time on the robot pioneer II

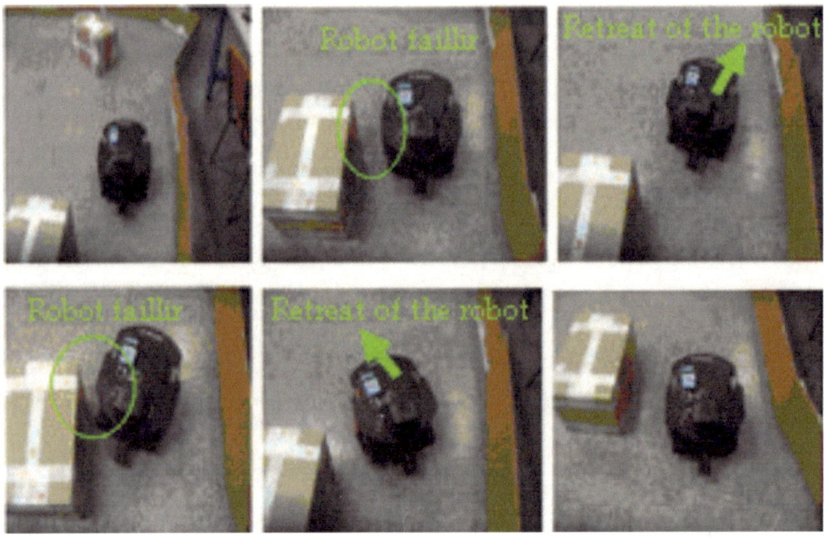


a. Robot Behaviour

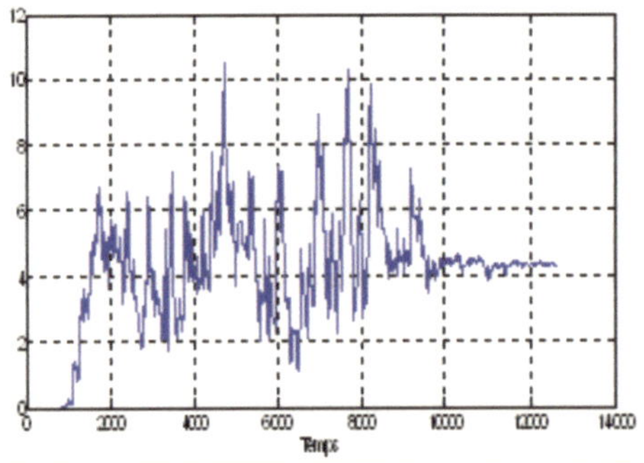


b. Fitness Function

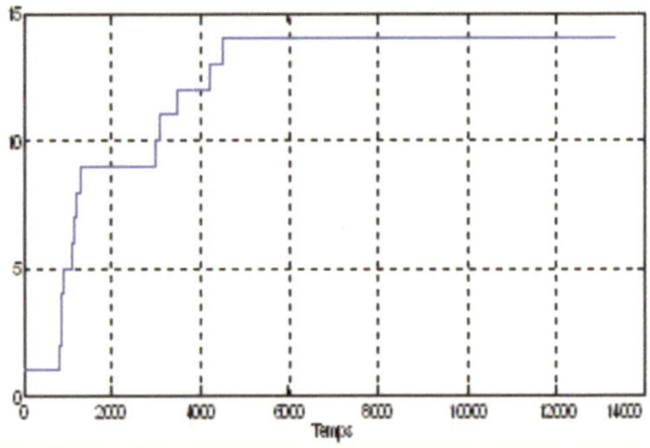


c. Number of the fuzzy rules

Fig. 34. Real time learning by the algorithm DFQLGA of the robot Pioneer II, behaviour "Obstacles Avoidance" (FIS with 03 inputs)

Figure.35 represents the validation results of a FIS conceived in real time, for the behaviour "Obstacles Avoidance".

## 8. Conclusion

The combination of the reinforcement Q-Learning algorithm and genetics Algorithms give a new type of hybrid algorithms (FQLGA) which is more powerful than traditional learning algorithms. FQLGA proved its effectiveness when no *priori* knowledge about system is available. Indeed, starting from a random initialization of the conclusions values and equidistant distribution for the membership functions for antecedent part the genetic algorithm enables to find the best individual for the task indicated using only the received reinforcement signal. The controller optimized by FQLGA algorithm was validated on a real robot and satisfactory results were obtained. The next stage of this work is the on line optimization of the structure of the Fuzzy controller.

Fig. 35. Validation of a FIS conceived on real time by algorithm DFQLGA

The algorithm DFQLGA is a combination of the algorithm DFQL and the genetic algorithm GA. The fuzzy rules and the fuzzy subsets (membership functions) are generated automatically by the algorithm DFQL according to the *ε-Completeness* and the *error TD* criteria. The conclusions part of the fuzzy rules is initialized randomly (a random set of actions is proposed for each rule); the genetic algorithm permits to determine the best conclusions of the rules previously generated. The algorithm DFQLGA showed its effectiveness to optimize on line and in an automatic way the structural and parametric characteristics of a fuzzy controller without any priori knowledge on the system to be controlled. Satisfactory results are obtained in simulation then in practice on the mobile robot Pioneer II, for the behaviour "Go to goal" and "Obstacles avoidance". The continuation of this work consists in reducing the duration of learning.

## 9. Acknowledgment

Abdelkrim Nemra: was born in August, 7th, 1980 in Tenes (Algeria). He received the Engineering and the Master degree in 2004, 2006 respectively from the Military Polytechnic School (Algeria). Currently, he is PhD student at Cranfield University, Department Aerospace, Power and Sensors (UK). His current research areas include Robotic, Control of System, Unmanned Aerial Vehicle (UAV) and Visual SLAM.
Hacène Rezine: was born in 1955 in El-Kala, Algeria. He received an engineering degree from the Ecole Nationale d'Ingénieurs et de Techniciens, Alger in 1979, the Engineer Degree from the Ecole Nationale Supérieur d'Electrotechnique, d'Electronique, d'Informatique, et d'Hydrolique, Toulouse, France in 1980 and the Doctor-Engineer Degree from the Institut Nationale Polytechnique of Toulouse in 1983. He is currently the Head of Unit of teaching

and control research of the Algerian Polytechnic Military School. His current research areas include Electrical drives, Mobile robotic, Fuzzy control and Artificial Intelligent Control.

## 10. References

L. Jouffe: "Actor-Critic Learning Based on Fuzzy Inference System", Proc of the IEEE International Conference on Systems, Man and Cybernetics. Beijing, China, pp.339-344, 1996.

C. Watkins, P. Dayan: "Q-Learning Machine Learning", pp.279-292, 1992.

P. Y. Glorennec and L. Jouffe, "Fuzzy Q-learning," Proc. Of IEEE Int. Con On Fuzzy Systems, pp. 659-662, 1997

Dongbing Gu, Huosheng Hu, Libor Spacek " Learning Fuzzy Logic Controller for Reactive Robot Behaviours" Proc. of IEEE/ASME International Conference on Advanced Japan, pp.20-24 July, 2003.

A. Souissi : "Apprentissage par Renforcement des Systèmes d'inférence Floue par des Méthodes par renforcement application à la navigation réactive d'un robot mobile", Thèse de Magistère, Ecole Militaire Polytechnique, Janvier 2005.

Meng Joo Er,et Chang Deng " Online Tuning of Fuzzy Inference Systems Using Dynamic Fuzzy Q-Learning "IEEE transactions on systems, man, and cybernetics, part B: cybernetics, Vol. 34, N°. 3, pp.1478-1489, June 2004

Chia-Feng Juang " Combination of online Clustering and Q-Value Based GA for Reinforcement Fuzzy System Design" IEEE Transactions On Systems, Man, And Cybernetics Vol. 13, N°. 3, pp.289-302, June 2005

Min-Soeng Kim and Kim and Ju-Jang Lee. "*Constructing a Fuzzy Logic ControllerUsing Evolutionary Q-Learning*" IEEE. pp.1785-1790, 2000

Chia-Feng Juang and Chun-Feng Lu. Combination of On-line Clustering and Q-value Based Genetic Reinforcement Learning For Fuzzy Network Design 0-7803-7898-9/03/$17.00 0 2003 IEEE.

David E Goldberg "*Algorithme génétique Exploration, optimisation et apprentissage automatique*"» Edition Addison-Wesley, France 1994.

C. C. Lee, "Fuzzy logic in control system: Fuzzy logic controller," IEEE Trans. Syst., Man, Cybern., vol. 20, pp. 404–435, Mar./Apr. 1990.

A. Nemra H. Rezine : « Genetic reinforcement learning of fuzzy inference system Application to mobile robotic », ICINCO, may 2007, P 206-2130.

Chia-Feng Juang, Jiann-Yow Lin, and Chin-Teng Lin "Genetic Reinforcement Learning through Symbiotic Evolution for Fuzzy Controller Design" IEEE transactions on systems, man, and cybernetics-part b: cybernetics, vol. 30, no. 2, april 2000.

Chia-Feng Juang " Combination of online Clustering and Q-Value Based GA for Reinforcement Fuzzy System Design" IEEE Transactions On Systems, Man, And Cybernetics Vol. 13, N°. 3, pp.289-302, June 2005.

Min-Soeng Kim and Kim and Ju-Jang Lee. "Constructing a Fuzzy Logic ControllerUsing Evolutionary Q-Learning" IEEE. pp.1785-1790, 2000.

# 14

# Control of Redundant Submarine Robot Arms under Holonomic Constraints

E. Olguín-Díaz, V. Parra-Vega and D. Navarro-Alarcón
*Robotics and Advanced Manufacturing Division*
*Research Center for Advanced Studies-CINVESTAV*
*Saltillo, México*

## 1. Introduction

AUV has undergone a major leap as technology allows higher integration, while faster sensors and actuators and new modelling techniques and its corresponding control algorithms are available. One of the new improvements of AUV technology is the capability to produce dexterous motion using robot manipulators as its end effector. This robot manipulator behaves as a multi-degrees of freedom active tool, such that the AUV stands as the free-floating base of the robot manipulator. In this case, the AUV navigates to drive the RA to its working environment, with two independent controllers, one for the AUV and another one for the RA. However, when the RA is working out its task, it is convenient to automatically control the whole AUV+RA, coined in this paper as Submarine Robot Arm or SRA for short, as a whole and unique system so as to take advantage of its redundancy and achieve better accomplishment in comparison to control the AUV and the RA independently.

When the RA is in contact to a rigid object, a constrained SRA appears and the control system now must control additionally the contact forces. This sort of systems have become a new area in AUV technologies, however there is no available and proved control system for constrained SRA, which posse a complex problem because there appears a tightly coupled hyper-redundant nonlinear system subject to holonomic constraint, which produces all together a set of nonlinear algebraic differential equations of index 2.

Constrained SRA deals simultaneously with navigation of its non-inertial base while controlling the pose and contact force of its RA. For the general case, we would have a free-floating hyper-redundant constrained RA. Additionally, the holonomic constraint must be satisfied all the time to maintain stable contact to a rigid underwater object, thus an efficient force controller is required to achieve stable contact while exerting a given desired contact force on this object. This rather new problem deserves a separate attention in AUV technologies, due to the subtle complexities of constrained SRA in its own right.

### 1.1 Contribution

After a brief discussion in Section 2 on the nature of the control problem of constrained SRA, which deserves a particular treatment apart to the AUVs control problem, we go through the full dynamic model of an SRA in Section 3. Then, Section 4 shows the key design of the open-loop error system. In Section 5 a quite simple force/posture model-free decentralized control structure is proposed, which guarantees robust tracking of time-varying contact force and posture, without any knowledge of SRA dynamics, which manage redundancy to introduce primary and secondary tasks. Closed-loop stability properties are obtained in the sense of Lyapunov and Variable Structure Systems arguments deliver second order sliding modes to obtain very fast tracking, while satisfying both tasks. A representative set of simulations for a 12 DoF SRA system are presented and discussed in Section 6. Some remarks are presented in Section 7, and final conclusions are given in Section 8.

## 2. The motion control problem of redundant constrained SRA

Controlling redundant constrained SRA subject to holonomic constraints requires the latest scientific knowledge and technological achievements for AUV and RA, from a simple torpedo to modern AUV and RA subject to hydrodynamic forces. Those vehicles pose at the same time tantamount scientific and technological challenges in robotics, control, man-machine interfaces, submarine telecommunications and mechatronics. This kind of SRA systems provide dexterity to a level yet unknown in AUV, which some day could superpose the limited capabilities of deep water divers, where bulky equipment is required to survive, thus less dexterity is exhibited at this waters by human divers. Despite these evident benefits, little study has been published on the automatic control of this systems. So far, must of the contributions point out on how to provide an acceptable level of (perhaps autonomous or automatic) navigation capabilities of the main body of the AUV, without compromising security, rather than in the manipulation capabilities of its tools, perhaps a RA with few DoF. Therefore, we bring the attention of a new breed of AUV whose main task is manipulation, perhaps with more than one robot arm, where the underlined issue is that the main body, the AUV, is considered as a free-floating fully actuated base. Notice that a common assumptions on this problem is that AUV is several times heavier than the RA so as to provide inertial decoupling between the AUV and the robot arm. For this case, the control system is designed so as to controlling independently the base and the arm, while a coupling endogenous disturbance is presented. In this case, we have $n$ trusters to drive the AUV and $m$ actuators to drive the SRA. However, with lighter materials and batteries and powerful trusters, this assumption may fail, since the mass and inertial might be quite similar and singular perturbation theory hardly applies anymore. As a consequence, technological improvements bring coupling and thus more elaborated control schemes are required to deal with the whole nonlinear highly coupled dynamics, from the base AUV to the end-effector RA, as one integrated free-floating constrained system.

### 2.1 The free-motion SRA problem

Pioneering efforts on SRA were focused on motion control with simple PD regulators in unconstrained (free-motion) motion, similar to the case of fixed-base robots in our labs. Acceptable performance for tracking has been proposed using more complicated (saturated or nonlinear) PID schemes and few model-based controllers have been proposed for tracking, under lab conditions (Spong; Villani et al.). In (Smallwood & Whitcomb), some heuristical simplifications are propose to deal with simple control techniques, however formal results are not provided, which may become potentially unstable under several possible working conditions. Though the Euler Lagrange dynamics, coming from the

Kirchhoff formulation and its equivalence in Euler-Lagrange formalisms shows passivity and nice energetic properties, and although it seems plausible that several passivity-based techniques could be extended to the realm of SRA, few publications address this problem in comparison to the wide variety of available passivity-based techniques for fixed-based RA. Automatic unconstrained motion control problem of SRA, though formally studied during the last two decades, is still in its infancy basically due to the fact that the full nonlinear model is quite complex, besides that the estimation of physical parameters are really difficult to obtain, so model-based controllers are difficult to implement. This is one reason why human operators are still the preferred controllers at risky missions. Since the model is hardly available, soft-computing techniques may be an option to approximate the inverse dynamics and implement controllers with implicit knowledge of the full system, however, no formal publications are known by the authors in this area for the full nonlinear model with formal stability results.

When the task is to achieve contact between the end-effector and a rigid object clamped on the sea floor or in a submarine structure, contact wrenches are propagated all over the RA and along the AUV through its rigid structure. Powerful and quick trusters are required to establishing, maintain and achieve stable contact, and finally to exert forces while moving along the surface of the object. This is known as the constrained SRA problem.

### 2.2 The constrained SRA problem

Stable contact for SRA is a more complex problem in comparison to the typical force/position control problem of robot manipulators fixed to ground in our laboratory because not only complementary complex dynamics are presented in SRA, such as buoyancy and added masses, as well as complex hydrodynamic effects, but to the fact that the vehicle reference frame is not longer inertial, see (Schjølberg & Fossenl; E. Olguín Díaz), thus there is not a fixed reaction point to react to. Thus, in this case, the truster of the AUV must react accordingly to hold still or accommodate these forces while still achieving simultaneously tracking not only for the UAV but also for the RA.

However, more interesting submarine tasks involves the more challenging problem of establishing stable contact while moving along the contact surface, like pushing itself against a wall or polishing a sunken surface vessel surface or manipulating tools on submarine pipe lines. In all these cases, contact forces are presented, and little is known about the structural properties of these contact forces, let alone exploit them either for design or control. This problem leads us to study the simultaneous force and pose (position and orientation) control of free-floating SRA under realistic conditions. By realistic we mean that the full nonlinear coupled dynamics are considered subject to holonomic constrains.

In the sequel, we assume that the dynamical model, and its parameters, are hardly known in practice, though the the full state is available as well as the geometric description of the contact surface or object.

There are two main general reasons that help us to explain why that force/posture problem remains rather an open problem. One reason is that we really know little about how to exploit its apparent generously well-behaved and slow dynamics. On one hand, how to model and control properly a fully immersed vehicle with a RA constrained by rigid object is an open issue in terms of exploitation of its passivity properties, when the model is subject to holonomic constrains. The second reason is rather technological and economical. On one hand, present day commercial submarine force control technology lies behind today system

requirements, such as very fast sampling and uniform rates of sensors and actuators, as well as low power consumption, even when the bandwidth of the submarine robot is very low.
Despite brilliant -for the simplicity of this complex problem- control schemes for free motion submarine robots published in the past few years, in particular those of (Yoerger & Slotine; Smallwood & Whitcomb 2001; Smallwood & Whitcomb 2004) does not formally guarantee convergence of tracking errors, let alone simultaneous convergence of posture and contac force tracking errors. There are several results that suggest empirically that a simple PD control structure behaves as stiffness control for submarine robots to produce acceptable low performance contact tasks. However, for more precise and fast tasks, the fast simultaneous convergence of timevarying contact forces and posture remains an open problem.
Since SRA dynamics are very hard to known exactly in practice, the dynamic model and its dynamic parameters should be considered uncertain, or at least parametrically unknown. Recently, some efforts have focused on how to obtain simple control structures to control the time-varying pose of the AUV under the assumption that the relative velocities are low (Smallwood & Whitcomb 2004; Perrier & Canudas de Wit). For force control of SRA, when dynamics are unknown, virtually none complete and formal control system is known. We believe that to obtain better performance in contact tasks a better understanding of the structural properties of submarine robots in stable contact to rigid objects are required. To this end, we assume that the rigid body dynamics of SRA is subject the now well-known holonomic constraints and thus we can extend some schemes to the case of SRA. Notice that during rigid contact the system exhibits similar structural properties of fixed-base constrained robots, under the full formulation of the Kirchhoff dynamics. Thus, in this paper we have chosen the Orthogonalization Principle (Parra-Vega & Arimoto) to extend from fix base to free-floating base to redundant SRA to propose a simple, yet high performance, controller with advanced tracking stability properties.

### 2.3 The constrained redundant SRA problem

When the SRA is redundant, there is some degrees of freedom available that can be used to satisfy a secondary task, being the primary task convergence of tracking errors. For instance, the AUV could be relocated dynamically all over the time at the pose of minimum power consumption such that the RA carries out the main task with greater manipulability index or avoiding joint limits or avoiding obstacles or with less energy consumption or keeping the AUV in a still position while the RA moves around.
If we could pursue primary and secondary tasks fulfillment, then some sort of Cost Index should be penalized, similar to the case of free motion grounded robots. That is, the cost function may constraint the motion of the SRA's base within an envelope to achieve better manipulability index or to minimize control effort/energy without compromising maneuverability, while tracking desired force/posture trajectories. In any case, besides the simultaneous tracking control problem of position and force, for redundant SRA, an optimal control problem is involved, which could be treated at the kinematic or dynamic level to take full advantage of the capability to carry out simultaneously a primary and secondary task. In this way, a certain degree of dexterity can be introduced when solving online the redundancy of the SRA.

## 3. The full nonlinear coupled model of the SRA

For completness, the AUV dynamic model is presented firstly, including under contact. That is a six DOF AUV is derived, including contact wrench. Then, the RA is attached to the AUV to build the SRA and the full expresions are presented.

### 3.1 The AUV model

The model of a submarine can be obtained with the momentum conservation theory and Newton's second law for rigid objects in free space via the Kirchhoff formulation (Fossen), the inclusion of hydrodynamic effects such as added mass, friction and buoyancy and the account of external forces/torques like contact effects (Olguín Díaz). The model is then expressed by the next set of equations:

$$M_v\dot{v}_v + C_v(v_v)v_v + D_v(v_v,t)v_v + g(q_v) = u_v + F_c^{(v)} + \eta_v(v_v,t), \quad (1)$$

$$v_v = J_v(q_v)\dot{q}_v \quad (2)$$

From this set, (1) is called the dynamic equation while (2) is called the kinematic equation. The generalized coordinates vector $q_v \in \Re^6$ is given on one hand by the 3D Cartesian position $d_v = (x_v, y_v, z_v)^T$ of the origin of the submarine frame ($\Sigma_v$) with respect to a inertial frame ($\Sigma_0$), and on the other hand by any set of attitude parameters that represent the rotation of the vehicle's frame with respect to the inertial one. Most common sets of attitude representation such a Euler angles, in particular roll-pitch-yaw ($\phi$, $\theta$, $\psi$), use only 3 variables (which is the minimal number of orientation variables). Then, for a submarine, the generalized coordinates represents its 6 degrees of freedom:

$$q_v = \begin{pmatrix} d_v \\ \vartheta_v \end{pmatrix} \quad (3)$$

where $\vartheta_v = (\phi_v, \theta_v, \psi_v)^T$ stands for the attitude parameter vector.

The vehicle velocity $\nu_v \in \Re^6$ is the velocity wrench (vector representing both linear and angular velocity) of the submarine in the vehicle's frame. This vector is then defined as $v_v = (\mathrm{v}_v^{(v)^T}, \omega_v^{(v)^T})^T$ The relationship between this vector and the generalized coordinates is given by the kinematic equation. The linear operator $J_v\ (q) \in \Re^{6\times 6}$ in (2), is built by the concatenation of two transformations. The first is $J_q(q_v)\ \in \Re^{6\times 6}$ which converts time derivatives of generalized coordinates to velocity wrench in the inertial frame. This operator is necessary because the angular velocity of a body ($\omega$) is not given by the time derivative of its angular parameters ($\dot{\vartheta} \neq \omega$). However, there is always a transformation operator given by the very specific type of chosen orientation parameters:

$$\omega_v = J_\theta(\vartheta)\dot{\vartheta}_v \quad (4)$$

Then the operator $J_q(q_v)$ is defined as:

$$J_q(q_v) \triangleq \begin{bmatrix} I & 0 \\ 0 & J_\theta(q_v) \end{bmatrix} \quad (5)$$

The second operator is

$$J_{R_v}(q_v) = \begin{bmatrix} R_0^v & 0 \\ 0 & R_0^v \end{bmatrix} \in \Re^{6\times 6} \tag{6}$$

which transforms a 6 dimension tensor from the inertial frame to vehicle's frame. The matrix $R_0^v(\vartheta_v) \in SO_3$ is the rotation matrix of the vehicle. Thus, the linear operator is defined as

$$J_\nu(q) \triangleq J_R^T(q) J_q(q)$$

A detailed discussion on the terms of (1) can be found in (Olguín Díaz & Parra-Vega).

In the dynamic equation (1), matrices $M_v, C_v(\nu), D_v(\cdot) \in \Re^{6\times 6}$ are Inertia matrix, Coriolis matrix and Damping matrix. $M_v$ includes the terms of classical inertia plus the hydrodynamic coefficients of the added mass effect (due to the amount of extra energy needed to displace the surrounding water when the submarine is moving). The Inertia matrix is constant, definite positive and symmetric only when the submarine is complete immersed and the relative water incidence velocity is small (Fossen). This condition is met for a great amounts of activities. The Coriolis vector $C_v(\nu)\nu$ represents the Coriolis and gyroscopic terms, plus the velocity quadratic terms induced by the added mass. The Coriolis matrix in this representation does not depend on the position but only on the velocity, in contrast to the same expression for a Robot Manipulator. It is indeed skew symmetric and fulfills the classic relationship for Lagrangian systems: $\dot{M}_v - 2C_v(\nu) = Q$; $Q + Q^T = 0$. The Damping matrix represents all the hydrodynamic effects of energy dissipation. For that reason it is a strictly positive definite matrix, $D_v(q, \nu, t) > 0$. Its arguments are commonly the vehicle's orientation $\vartheta_v$, the generalized velocity $\nu$, and the velocity of the surrounding water $\zeta(t)$. The diagonal components represents the drag forces while the off-diagonal components represent the lift forces. Vectors $g_v(q)$, $u$, $F_c^{(v)} \in \Re^6$ are all force wrenches (force-torque vector) in the vehicle's frame. They represent respectively: gravity, input control and the contact force. Gravity vector includes buoyancy effects and it does not depend on velocity but on the orientation (attitude) of the submarine with respect to the inertial frame. The contact force wrench is the one applied by the environment *to* the submarine. The input control are the forces/torques induced by the submarine thrusters in the vehicle frame.

The disturbance $\eta_v(\nu, \zeta(t), \dot{\zeta}(t))$of the surrounding fluid depends mainly in the incidence velocity, i.e. the relative velocity of the vehicle velocity and the fluid velocity. The last is a non-autonomous function, but an external perturbation. This disturbance has the property of

$$\eta_v(\nu, 0, 0) = 0. \tag{7}$$

That is that all the disturbances are null when the fluid velocity and acceleration are null.

The dynamic model (1)-(2) can be rearranged by replacing (2) and its time derivative into (1). The result is one single equation model:

$$M_q(q_v)\ddot{q}_v + C_q(q_v, \dot{q}_v)\dot{q}_v + D_q(\cdot)\dot{q}_v + g_q(q_v) = u_{q_v} + \tau_c + \eta_q(\dot{q}_v, \zeta(t), \dot{\zeta}(t)); \tag{8}$$

which, whenever $\zeta(t) = \dot{\zeta}(t) = 0$, i.e. $\eta_q(\cdot) = 0$, has the form of any Lagrangian system. Its components fulfills all properties of such systems i.e. definite positiveness of inertia and damping matrices, skew symmetry of Coriolis matrix and appropriate bound of all components (Sagatun & Fossen). The control input in this equation is obtained by a linear transformation of the real input using the linear operator given by the kinematic equation:

$$u_{q_v} = J_\nu^T(q_v)u_v \tag{9}$$

The contact effect is also obtained by the same transformation. However it can be expressed directly from the contact wrench in the inertial frame ($\Sigma_0$) by the relationship

$$\tau_c = J_\nu^T(q_v)F_c^{(v)} = J_q^T(q_v)F_c^{(0)}, \tag{10}$$

where the contact force $F_c^{(0)}$ is the one expressed in the inertial frame. By simplicity it will be noted as $F_c$ from this point further. The relationship with the one expressed in the vehicle's frame is given by $F_c = J_R^T(q)F_c^{(v)}$.This wrench represents the contact forces/torques exerted by the environment to the submarine as if measured in a non moving frame. These forces/torques are given by the normal force of an holonomic constraint when in contact and the friction due to the same contact. For simplicity in this work, tangential friction is not considered. The equivalent of the disturbance is obtained also with the linear operator given as:

$$\eta_q(\cdot) = J_\nu^T(q)\eta_v(\cdot). \tag{11}$$

### 3.2 Contact force due to an holonomic constraint

A holonomic constraint (or infinitely rigid contact object) can be expressed as a function of the generalized coordinates of the submarine as

$$\varphi(q_v) = 0, \tag{12}$$

with $\varphi(q_v) \in \Re^r$, where $r$ stands for the number of independent contact points between the SRA and the motionless rigid object. Equation (12) means that stable contact appears while the SRA submarine does not deattach from the object $\varphi(q_v) = 0$. Evidently all time derivatives of (12) are zero, which for $r = 1$

$$J_\varphi(q_v)\dot{q}_v = 0 \tag{13}$$

where $J_\varphi(q_v) = \frac{\partial \varphi(q_v)}{\partial q_v} \in \Re^{r\times 6}$ is the constraint jacobian. Last equation means that velocities of the submarine in the directions of constraint jacobian are restricted to be zero. This directions are then normal to the constraint surface $\varphi(q_v)$ at the contact point. As a consequence, the normal component of the contact force has exactly the same direction as those defined by $J_\varphi(q_v)$, consequently, the contact force wrench can be expressed as

$$F_c = J_{\varphi+}^T(q_v)\lambda \tag{14}$$

where $J_{\varphi+}(q_v) \equiv \frac{J_\varphi}{\|J_\varphi\|}$ is a normalized version of the constraint jacobian; $\lambda \in \Re^r$ is the magnitude of the normal contact force at the origin of vehicle frame: $\lambda = \|F_c\|$. The free moving model expressed by (1)-(2), when no fluid disturbance and in contact with the holonomic constraint can be rewritten as:

$$M_v \dot{\nu} + h_v(q_v, \nu, t) = u_v + J_R^T(q_v) J_{\varphi+}^T(q_v)\lambda, \tag{15}$$

$$\nu = J_\nu(q_v)\dot{q}_v, \tag{16}$$

$$\varphi(q_v) = 0, \tag{17}$$

where $h_v(q_v, \nu, t) = C_v(\nu)\nu + D_v(q_v, \nu, t)\nu + g_v(q_v)$. Equivalently, the model (8) is also expressed as

$$M_q(q_v)\ddot{q}_v + h_q(q_v, \dot{q}_v, t) = u_{q_v} + J_{\bar{\varphi}}^T(q_v)\lambda, \tag{18}$$

$$\varphi(q_v) = 0, \tag{19}$$

with $h_q(q_v, \dot{q}_v, t) = C_q(q_v, \dot{q}_v)\dot{q}_v + D_q(q_v, \dot{q}_v, t)\dot{q}_v + g_q(q_v)$ and $J_{\bar{\varphi}}(q_v) = J_{\varphi+}(q_v)J_q(q_v)$. Equations (18)- (19) are a set of Differential Algebraic Equations index 2 (DAE-2). To solve them numerically, a DAE solver is required. This last representation has the same structure and properties as those reported in (Parra-Vega).

### 3.3 The robot arm

This section formulates the problem of a manipulator having free mobility on its base. That means, when the base of the robot arm is no longer inertial and thus does not fulfils Newton's laws unless all its dynamic is at new, expressed in a inertial frame.

In order to include the movent of the base of the robot arm, it is necessary to introduce some extra elements which do not appear in the classical fixed-base model. For this case, the free moving base, the inertial frame shall be chosen in the same way it is chosen for the vehicle's: at some point attached to the earth. It is evident that this two references can be identical for the fixed base case, but should certainly be different for the free moving base case. Lets use the inertial reference $\Sigma_0$ used for the submarine and define as $\Sigma_b$ the base frame of the arm when its base is no moving, known as the fixed-base condition.

As a result there are two new homogeneous transformations in the kinematic chain: $H_0^v(q_v)$ from inertial frame $\Sigma_0$ to the vehicle's frame $\Sigma_v$ and $H_v^b$ from $\Sigma_v$ to the fixed-base first reference frame $\Sigma_b$ from which all the modelling is obtained.

The homogeneous transformation from inertial frame $\Sigma_0$ to the vehicle frame $\Sigma_v$ is then given by:

$$H_0^v = \begin{bmatrix} R_0^v(\vartheta_v) & d_v \\ 0 & 1 \end{bmatrix} \tag{20}$$

where $d_v$ is the inertial position of the vehicle and $R_0^v \in SO_3$ represents its orientation. Recall that the generalized coordinates of the vehicle are given in (3).

The homogeneous transformation from $\Sigma_v$ to $\Sigma_b$ is then given by:

$$H_v^b = \begin{bmatrix} R_v^b & \boldsymbol{d}_{b/v} \\ 0 & 1 \end{bmatrix} \tag{21}$$

where $d_{b/v} \in \Re^3$ is the position vector of $b$ wrt vehicle's frame (expressed in $\Sigma_v$) and $R_v^b \in SO_3$ represents the orientation that the arm is attached to the vehicle. With the reasonable assumption that the vehicle is a rigid body, and that the assembling is as well, this transformation is constant.

The forward kinematics of the free-base robot arm is given by the concatenation of the proper homogeneous transformations. For instance, the forward kinematic of the end-effector $x_e$ is given by:

$$H_0^e = H_0^v(q_v)H_v^bH_b^e(q_m) = \begin{bmatrix} R_0^e(q_v, q_m) & d_0^e(q_v, q_m) \\ 0 & 1 \end{bmatrix}$$

where $H_b^e(q_m)$ stands for the homogeneous transformation of the manipulator , and $q_m$ are the generalized coordinates of the arm chain, both under the fixed-base conditions.

From here on, it is evident that the generalized coordinates for the free-base manipulator shall be extended to include the vehicle configuration as

$$q \triangleq \begin{pmatrix} q_v \\ q_m \end{pmatrix} \in \Re^{6+n} \tag{22}$$

An acceptable interpretation is that the vehicle is an extra link in the manipulator's chain that has a six degree-of-freedom articulation.

The forward kinematics of the end-effector are given by:

$$d_e = f(q) \quad = \quad d_v + R_0^v(\vartheta_v)\left(r_{e/v}^{(v)}(q_m)\right) \tag{23}$$

$$= d_v + R_0^v(\vartheta_v)\left(r_{b/v}^{(v)} + R_v^b d_{e/b}^{(b)}(q_m)\right) \tag{24}$$

where $d_{e/b}^{(b)}(q_m) \in \Re^3$ is the forward kinematic equation on the the fixed-base condition, and $r_{e/v}^{(v)}$ is the position vector of the end-effector from the origin of the vehicle's frame, expressed in that same frame $\Sigma_v$.

From eqs. (23)-(24) the linear inertial velocity of the end-effector is:

$$\dot{d}_e = \dot{d}_v + \omega_v \times R_0^v(\vartheta_v)r_{e/v}^{(v)}(q_m) + R_0^v(\vartheta_v)R_v^b\dot{d}_{e/b}^{(b)}(q_m) \tag{25}$$

The linear velocity $\dot{d}_{e/b}^{(b)}(q_m)$ is the fixed-based condition's linear velocity of the end-effector and can be calculated via the linear velocity jacobian: $v_e = J_{vel_{fb}}(q_m)\dot{q}_m$. Then last relationship can also be expressed as

$$\dot{d}_e = \dot{d}_v - \left(R_0^v(\vartheta_v)r_{e/v}^{(v)}(q_m)\right) \times \omega_v + R_0^b(\vartheta_v)J_{vel_{fb}}(q_m)\dot{q}_m \tag{26}$$

The angular velocity of the end-effector is the sum of the vehicle's angular velocity $\omega_v$ plus the relative angular velocity of the end-effector respect to the base $\omega_{e/b}$ expressed in the inertial frame. Then, the angular velocity of the end-effector is:

$$\omega_e = \omega_v + R_0^b(\vartheta_v) J_{\omega_{fb}}(q_m)\dot{q}_m \tag{27}$$

where $J_{\omega_{fb}}$ is the angular velocity jacobian for the fixed-base condition. By replacing equation (4) in equations (26) and (27), the end-effector velocity wrench can be written as a function of the extended generalized coordinates and its time derivative as:

$$\nu_{e0} \triangleq \nu_e^{(0)} = \begin{pmatrix} \dot{d}_e \\ \omega_e \end{pmatrix} = \begin{pmatrix} \dot{d}_v - \left[ \left( R_0^v(\vartheta_v) r_{e/v}^{(v)}(q_m) \right) \times \right] J_\theta(\vartheta_v)\dot{\vartheta}_v + R_0^b(\vartheta_v) J_{vel_{fb}}(q_m)\dot{q}_m \\ J_\theta(\vartheta_v)\dot{\vartheta}_v + R_0^b(\vartheta_v) J_{\omega_{fb}}(q_m)\dot{q}_m \end{pmatrix} \tag{28}$$

Last equation can also be written in block matrices in the next way:

$$\begin{aligned} \nu_{e0} &= \begin{bmatrix} I_3 & -\left[ \left( R_0^v(\vartheta_v) r_{e/v}^{(v)}(q_m) \right) \times \right] J_\theta(\vartheta_v) & R_0^b(\vartheta_v) J_{vel_{fb}}(q_m) \\ 0 & J_\theta(\vartheta_v) & R_0^b(\vartheta_v) J_{\omega_{fb}}(q_m) \end{bmatrix} \begin{pmatrix} \dot{d}_v \\ \dot{\vartheta}_v \\ \dot{q}_m \end{pmatrix} && (29) \\ &= J_v(q)\dot{q}_v + J_m(q)\dot{q}_m && (30) \\ &= J\dot{q} && (31) \end{aligned}$$

where the *vehicle Jacobian* $J_v \in \Re^{6\times 6}$ is defined as:

$$J_v(q) \triangleq \begin{bmatrix} I_3 & -[R_0^v(\vartheta_v) r_{e/v}^{(v)}(q_m)\times] J_\theta(\vartheta_v) \\ 0 & J_\theta(\vartheta_v) \end{bmatrix} \tag{32}$$

and the *manipulator Jacobian* $J_m \in \Re^{6\times n}$ is defined as:

$$J_m(q) \triangleq J_{R_b}(\vartheta_v) J_{fb}(q_m) \tag{33}$$

In the above definitions, the term $[a\times]$ stands for the skew symmetric matrix representation of the cross product of a vector (Spong & Vidyasagar), $J_{R_b} \in SO_6$ is defined as (6) and $J_{fb}$ is the manipulator fixed-base geometric Jacobian. The geometric version of the vehicle Jacobian $J_{v_g}(\vartheta_v, q_m)$ and the manipulator Jacobian $J_m(\vartheta_v, q_m)$ make up the Mobile Manipulator Jacobian defined in (Hootsman & Dubowsky). However, in this work we prefer to use this geometric jacobian because it maps the generalized velocities $\dot{q}$ in linear an angular velocities at any point in the vehicle/ manipulator system.

$$J = \begin{bmatrix} J_v(\vartheta_v, q_m) & J_m(\vartheta_v, q_m) \end{bmatrix} \in \Re^{6\times(6+n)} \tag{34}$$

The dynamics of the free base manipulator can be obtained using the expressions of the kinetic and potential energy of any mass, and using expressions (23) and (29). Because the

generalized coordinates vector has a 6 + $n$ dimension, there must be an inertia matrix $\bar{H}(q)$ of the size (6 + $n$) × (6 + $n$) and the vector of generalized forces $\tau$ should also have a 6 + $n$ dimension.
The kinetic energy of a free-base manipulator is given by:

$$K_m = \sum_{i=0}^{n} \frac{1}{2}\left(\dot{r}_{ci}^T(m_i)\dot{r}_{ci} + {\omega_i}^T R^i(p) I_i {R^i}^T(p)\omega_i\right) \quad (35)$$

$$= \frac{1}{2}\dot{q}^T \bar{H}(q)\dot{q} \quad (36)$$

where the body 0 is the base, that has no movement in the fixed-base conditions. The linear velocity $\dot{d}_{ci}$ and $\omega_i$ are given by equations (26) and (27), respectively, but calculating the distance to the center of mass of the corresponding link.
The resulting solution for this extended inertia matrix can be written as follows:

$$\bar{H}(q) = \begin{bmatrix} M_m(q_v, q_m) & H_A(q_v, q_m) \\ H_A^T(q_v, q_m) & H_{fb}(q_m) \end{bmatrix} = \bar{H}^T(q) > 0 \quad (37)$$

which by definition is symmetric and definite positive.

$$M_m \triangleq \begin{bmatrix} \left(\sum_{i=0}^{n} m_i\right) I_3 & -\sum_{i=0}^{n}\left(m_i[d_{ci/v}^{(0)}(q)\times]\right) \\ \sum_{i=0}^{n}\left(m_i[d_{ci/v}^{(0)}(q)\times]\right) & \sum_{i=0}^{n}\left(R_0^b(R_b^i I_i {R_b^i}^T){R_0^b}^T - m_i[d_{ci/v}^{(0)}(q)\times]^2\right) \end{bmatrix} \in \Re^{6\times 6} \quad (38)$$

$$H_A \triangleq \begin{bmatrix} R_0^b(q_v)\sum_{i=1}^{n} m_i J_{vel_{fb_{ci}}}(q_m) \\ \sum_{i=0}^{n}\left(m_i[d_{ci/v}^{(0)}]R_0^b(q_v)J_{vel_{fb_{ci}}}(q_m) + R^0(R_0^i I_i {R_0^i}^T)J_{\omega_{bf_i}}\right) \end{bmatrix} \in \Re^{6\times n} \quad (39)$$

$$H_{fb} \triangleq \sum_{i=1}^{n}\left(m_i J_{vel_{fb_{ci}}}(q_m)^T J_{vel_{fb_{ci}}}(q_m) + {J_{\omega_{bf_i}}}^T R_0^i I_i {R_0^i}^T J_{\omega_{fb_i}}\right) \in \Re^{n\times n} \quad (40)$$

Note that Matrix $H_{fb}$ is the inertial matrix of the same robot arm for the fixed-base condition and it depends only in the manipulator coordinates $q_m$.
On the other hand, being the potential energy, gravitational and buoyant is also function of the vehicle positions, it can be written as a function of the generalized coordinates:

$$V_m = V_m(q) = V_m(q_v, q_m) \quad (41)$$

Then the dynamic equation can be obtained by solving the Euler-Lagrange equation. The resulting model would be of the form:

$$\bar{H}(q)\ddot{q} + \bar{C}(q,\dot{q})\dot{q} + \bar{g}(q) = \bar{\tau}_q + \bar{\tau}_{hydro} \quad (42)$$

where $\bar{C}(q,\dot{q}) \in \Re^{6+n\times 6+n}$ is the Coriolis matrix which has the same properties that for the fixed-base case (i.e. $\frac{1}{2}\dot{\bar{H}} - \bar{C} = Q$ ;$Q - Q^T = 0$ always true), $\bar{g}(\bar{q}) = \frac{\partial V_m}{\partial \bar{q}}$ is the

gravitational vector of the manipulator and its influence over the vehicle's coordinates and includes the restoring forces due to the floatability of each link, $\bar{\tau}_q$ is the generalized coordinates force vector, and $\bar{\tau}_{hydro}$ are the generalized forces due to the hydrodynamic effects.
This last term is somehow complicated to determine. However, a good approximation is to compute these forces over each link and to translated them to the generalized coordinates, using the virtual work principle (Spong & Vidyasagar), by means of the Mobile Manipulator Jacobian (34) of the geometric center of each link. The resulting vector shall have the next structure (Olguín Díaz):

$$\bar{\tau}_{hydro} = -\bar{D}_m(\cdot)\dot{q} + \bar{\eta}_m(\dot{q}, \zeta(t), \dot{\zeta}(t))$$

where the damping matrix $\bar{D}_m > 0$ is positive definite, due to the fact that the hydrodynamic effects are dissipative, and the hydrodynamical perturbation forces $\bar{\eta}_m$ becomes null when the current is steady ($\zeta(t) = \dot{\zeta} = 0$). The Damping matrix $\bar{D}_m(\cdot)$ can be also be written in block submatrices as:

$$\bar{D}_m(\cdot) = \begin{bmatrix} D_{vv} & D_{vm} \\ D_{mv} & D_{mm} \end{bmatrix}$$

Coriolis, gravitational terms, Hydrodynamic damping and current perturbations are highly nonlinear so it is very common to write then together as the non-linear vector $\bar{h}(q,\dot{q}) = \bar{C}(q,\dot{q})\dot{q} + \bar{g}(q) + \bar{D}_m(\cdot)\dot{q} - \bar{\eta}_m(\dot{q}, \zeta(t), \dot{\zeta}(t))$.Then model (42) can be presented as a function of vehicle's and arm's coordinates $q_v$ and $q_m$:

$$\begin{bmatrix} M_m(q) & H_A(q) \\ H_A^T(q) & H_{fb}(q_m) \end{bmatrix} \begin{pmatrix} \ddot{q}_v \\ \ddot{q}_m \end{pmatrix} + \begin{pmatrix} h_{m/v}(q_v, q_m, \dot{q}_v, \dot{q}_m) \\ h_m(q_v, q_m, \dot{q}_v, \dot{q}_m) \end{pmatrix} = \begin{pmatrix} \tau_{m/v} \\ \tau_m \end{pmatrix} \quad (43)$$

Or else, it can be written as two coupled equations as:

$$M_m(q)\ddot{q}_v + H_A(q)\ddot{q}_m + h_{m/v}(q_v, \dot{q}_v, q_m, \dot{q}_m) = \tau_{m/v} \in \Re^6 \quad (44)$$

$$H_A^T(q)\ddot{q}_v + H_{fb}(q_m)\ddot{q}_m + h_m(q_v, \dot{q}_v, q_m, \dot{q}_m) = \tau_q \in \Re^n \quad (45)$$

### 3.4 The submarine AUV+Robot Arm=SRA

The interaction between the models of the vehicle and the free-base manipulator are forces-torques at the attaching point. So if the original assumption where this attachment is rigid, i.e. it does not nave elastic deformation behaviour, this force wrench shall appeared in both models with opposite direction (due to Newton's 3rd law).
On one hand, this interaction wrench is given in the vehicle dynamics as an external perturbation wrench. This can be seen in the dynamic equation (8) as follows:

$$M_q(q_v)\ddot{q}_v + h_q(q_v, \dot{q}_v) = u_{q_v} + \tau_{arm} \quad (46)$$

where $h_q(q_v, \dot{q}_v) = C_q(q_v, \dot{q}_v)\dot{q}_v + D_q(\cdot)\dot{q}_v + g_q(q_v)) - \eta_q(\dot{q}_v, \zeta(t), \dot{\zeta}(t))$ is the non-linear vector term and $\tau_{arm}$ is the perturbation produced by the manipulators movements interaction.

On the other hand, the interaction between the vehicle and the arm, *seen* from the manipulator is the component $\tau_{m/v}$ on either model (43) or (44). By Newton's $3_{rd}$ law it can be seen that the force wrench $\tau_{m/v}$ on the manipulator model is the same but with opposite direction of the perturbation $\tau_{arm}$ on the vehicle's model.

$$\tau_{m/v} = -\tau_{arm} \tag{47}$$

Then by using last equality, a single expression for both model (44) and (46) is found to be:

$$\Big[M_q(q_v) + M_m(q_v, q_m)\Big]\ddot{q}_v + H_A(q_v, q_m)\ddot{q}_m + h_q(q_v, \dot{q}_v) + h_{m/v}(q_v, \dot{q}_v, q_m, \dot{q}_m) = u_{q_v} \tag{48}$$

Then equations (45) and (48) can be represented by a single whole-system differential equation in a compact form by a coupled pair of differential equations:

$$H(q)\ddot{q} + C(q,\dot{q})\dot{q} + D(\cdot)\dot{q} + g(q) = \tau + \eta(\dot{q}, \zeta(t), \dot{\zeta}(t)) \tag{49}$$

where the nonlinear terms can also be written in a compact form as $h(q,\dot{q}) = C(q,\dot{q})\dot{q} + D(\cdot)\dot{q} + g(q) - \eta(\dot{q}, \zeta(t), \dot{\zeta}(t))$ and the overall terms are given by the next set of relationships [6]:

$$H(q) = \begin{bmatrix} M_v(q_v) + M_m(q_v, q_m) & H_A(q_v, q_m) \\ H_A^T(q_v, q_m) & H_{fb}(q_m) \end{bmatrix} \in \Re^{6+n \times 6+n} \tag{50}$$

$$C(q,\dot{q}) = \begin{bmatrix} C_q(q_v, \dot{q}_v) + C_{vv}(q_v, q_m, \dot{q}_v, \dot{q}_m) & C_{vm}(q_v, q_m, \dot{q}_v, \dot{q}_m) \\ C_{mv}(q_v, q_m, \dot{q}_v, \dot{q}_m) & C_{mm}(q_v, q_m, \dot{q}_v, \dot{q}_m) \end{bmatrix} \tag{51}$$

$$D(\dot{q}, \zeta(t), \dot{\zeta}(t)) = \begin{bmatrix} D_q(\dot{q}_v, \zeta(t), \dot{\zeta}(t)) + D_{vv}(\dot{q}_v, \dot{q}_m, \zeta(t), \dot{\zeta}(t)) & D_{vm}(\dot{q}_v, \dot{q}_m, \zeta(t), \dot{\zeta}(t)) \\ D_{mv}(\dot{q}_v, \dot{q}_m, \zeta(t), \dot{\zeta}(t)) & D_{mm}(\dot{q}_v, \dot{q}_m, \zeta(t), \dot{\zeta}(t)) \end{bmatrix} \tag{52}$$

$$g(q) = \begin{pmatrix} g_q(q_v) + g_{m/v}(q_v, q_m) \\ g_m(q_v, q_m) \end{pmatrix} \tag{53}$$

$$\tau = \begin{pmatrix} u_{q_v} \\ u_{q_m} \end{pmatrix} \in \Re^{6+n} \tag{54}$$

$$\eta(\dot{q}, \zeta(t), \dot{\zeta}(t)) = \begin{pmatrix} \eta_q(\dot{q}_v, \zeta(t), \dot{\zeta}(t)) + \eta_{m/v}(\dot{q}_v, \dot{q}_m, \zeta(t), \dot{\zeta}(t)) \\ \eta_m(\dot{q}_v, \dot{q}_m, \zeta(t), \dot{\zeta}(t)) \end{pmatrix} \tag{55}$$

As well as inthe case of the AUV alone, whenever $\zeta(t) = \dot{\zeta}(t) = 0$, then $)(\cdot) = 0$,and the dynamic equation (49) has the form of a Lagrangian system. Thus, its components fulfills all properties of such systems i.e. definite positiveness of inertia and damping matrices, skew symmetry of Coriolis matrix and appropriate bound of all components.

### 3.5 SRA in contact

When the end-effector of the SRA gets in contact with the environment, external forces and torques appear in the dynamics that was not taken into account when the dynamics equations was obtained. Let $F_e = \left(f_e^T, n_e^T\right)^T \in \Re^6$ express the force wrench due to contact forces and torques at the end-effector. From the virtual work principle, this contact force $F_e$ will modify the dynamics of the system through the transpose of the Mobile-Manipulator-Jacobian given by equation (34) as

$$\tau_c = J^T F_e = \begin{bmatrix} J_v^T \\ J_m^T \end{bmatrix} F_e \tag{56}$$

Which means that the contact forces are translated to the vehicle coordinates by the vehicle's Jacobian and to the joints by the manipulator's Jacobian.
Then equation (49) is modified by adding $\tau_c$ as an external force to the righthand side yielding to

$$H(q)\ddot{q} + h(q,\dot{q}) = \tau + J^T F_e \tag{57}$$

In this work, this contact force is also modelled in the same manner as treated in section 3.2, as

$$\tau_c = J^T F_e = J_\varphi^T(q)\lambda \tag{58}$$

Where the $J_\varphi(q)=J_{\varphi+}(q)J(q)$ is the jacobian of the holonomic restriction and $\lambda$ is the magnitude of the contact force.

## 4. Open-loop error equation

The introduction of a so called Orthogonalization Principle has been a key in solving, in a wide sense, the force control problem of a robot manipulators with fix base. This physical-based principle states that the orthogonal projection of contact torques and joint generalized velocities are complementary, and thus its dot product is zero, carrying no power and no work is done. Relying on this fundamental observation, passivity arises from torque input to generalized velocities, in open-loop. To preserve passivity in closed-loop, then, the closed-loop system must satisfy the passivity inequality for a given error velocity function. This is true for robot manipulators with fixed frame, and here we extend this approach for robots whose reference frame is not inertial, like SRA. Additionally, we present here the developments that this holds true also for redundant SRA.

### 4.1 Orthogonalization principle and linear parametrization

Similar to (Liu et al.), the orthogonal projection of $J_\varphi(q)$, which arises onto the tangent space at the contact point, is given by the following operator

$$Q(q) \equiv I_n - J_\varphi^\dagger(q)J_\varphi(q) \in \Re^{(6+n)x(6+n)} \tag{59}$$

where $I_{6+n} \in \Re^{(6+n)x(6+n)}$ is the identity matrix and $J_\varphi^\dagger(q) = J_\varphi^T(q)(J_\varphi(q)J_\varphi^T(q))^{-1}$, which always exists since rank{$J_\varphi(q)$} = r. Notice that $rank\{Q(q)\} = (6+n)-r$ and $Q\dot{q} = \dot{q}$, then $Q(q)\, J_\varphi^t(q)$=0. Therefore, according to the Orthogonalization Principle, the integral of $(\tau, \dot{q})$ is upper bounded by $-\mathcal{H}(t_0)$, for $\mathcal{H}(t)$ = K + P whenever $\eta(q,\dot{q},\zeta(t),\dot{\zeta}(t)) = 0$ because $\dot{q}^T J_\varphi^T(q)\lambda = 0$

Then passivity arise for the full constrained SRA, under no fluid disturbances. This conclusion gives a very useful and promising theoretical framework, similar to the approach of passivity-based control for fix-base robot arms. On the other hand, it is known that the dynamic equation (49) with no fluid perturbation can be linearly parameterized as follows

$$H(q)\ddot{q} + C(q,\dot{q})\dot{q} + D(\cdot)\dot{q} + g(q) = Y(q,\dot{q},\ddot{q})\Theta, \tag{60}$$

where the regressor $Y(q,\dot{q},\ddot{q}) \in \Re^{n\times p}$ is composed of known nonlinear functions and $\Theta \in \Re^{p}$ by $p$ unknown but constant parameters. This is useful to obtain the fundamental change of coordinates of the SRA into the controlled error system, the system expressed in error coordinates, wherein we want to control the system in the trivial equilibria.

### 4.2 Change of coordinates

In order to design the controller, we need to work out the open loop error equation using (60), in terms of nominal references $\dot{q}_r$, as follows. Consider

$$H(q)\ddot{q}_r + [C(q,\dot{q}) + D(\cdot)]\,\dot{q}_r + g(q) \;=\; Y_r(q,\dot{q},\dot{q}_r,\ddot{q}_r)\Theta, \tag{61}$$

where $\ddot{q}_r$ is the time derivative of $\dot{q}_r$, *to be defined.* Then the open loop (49) can be written by adding and subtracting (61) as

$$H(q)\dot{s} \;=\; -\,[C(q,\dot{q}) + D(\cdot)]\,s - Y_r(q,\dot{q},\dot{q}_r,\ddot{q}_r)\Theta + J_\varphi^T(q)\lambda + u_q, \tag{62}$$

where $s \equiv \dot{q} - \dot{q}_r$ is called the extended error. The problem of designing a controller for the open loop error equation (62) is to find $u_q$ such that $s(*)$ exponentially converges when $Y_r\Theta$ is not available.

### 4.3 Kinematic redundancy

Notice that

$$X = f(q) \rightarrow \dot{X} = J(q)\dot{q} \tag{63}$$

Since dimensions of $X \in \Re^m$ and $q$ are not the same, jacobian $J(q) \in \Re^{m\times(6+n)}$, then its inverse does not exists, then to obtain the inverse mapping of (63), we use the pseudoinverse of Penrouse to get

$$\dot{q} = J^{\dagger}\dot{X} + Q_k v \tag{64}$$

where matrix $Q_k = \left(I_{6+n} - J^{+}(q)J(q)\right) \in \Re^{(6+n)\times(6+n)}$ stands for the orthogonal projection of $J(q)$ and spans the $6 + n - m$ kernel of $J(q)$, that is $J(q)$ and $Q_k$ are orthogonal complements and its dot product is zero. Now, let consider that $Q_k$ maps any arbitrary vector $v \in \Re^{(6+n)}$ into the null space of $J(q)$. Consider, let

$$\dot{z} = Q_k v \tag{65}$$

be a vector which belongs to the null space of $J(q)$. This vector yields

$$\dot{z} = Q_k \dot{z} \tag{66}$$

which means that (64) can be written as

$$\dot{q} = J^{\dagger}\dot{X} + \dot{z} \tag{67}$$

That is, given $m$ values of $X$, we can complete the remaining $6 + n - m$ values of $q \in \Re^{6+n}$ by designing $\dot{z}$ under a given criteria.

### 4.4 Orthogonal nominal reference

Since $\dot{q} = Q\dot{q}$, and considering the decomposition (67) to design the extended error $s = \dot{q} - \dot{q}_r \equiv Q\dot{q} - \dot{q}_r$, and aiming at preserving passivity in closed loop, it is natural to consider a structure for $\dot{q}_r$ similar to $\dot{q}$, that is the nominal reference $\dot{q}_r$ at the velocity level takes de following form

$$\dot{q}_r = Q\left(J^{\dagger}\dot{X}_r + \dot{z}_r\right) + \beta J_{\varphi}^T \left(s_F - s_{dF} + \gamma_2 \int_{t_0}^{t} sgn\{s_{qF}(t)\}dt\right), \tag{68}$$

with

$$\dot{X}_r = \dot{X}_d - \sigma\tilde{X} + s_{dp} - \gamma_1 \int_{t_0}^{t} sgn\{s_{qp}(t)\}dt \tag{69}$$

where $\widetilde{X} \triangleq X(t) - X_d(t)$, $X_d(t)$ and $\lambda_d(t)$ are the desired smooth trajectories of position and contact force $\tilde{\lambda} \triangleq \lambda(t) - \lambda_d(t)$ as the position and force tracking errors, respectively. Parameters $\beta$, $\sigma$, $\gamma_1$ and $\gamma_2$ are constant matrices of appropriate dimensions; and $sgn(y)$ stands for the entrywise signum function of vector $y$, and

$$s_p = \dot{\tilde{X}} + \sigma\tilde{X}, \tag{70}$$

$$s_{dp} = s_p(t_0)e^{-\alpha(t-t_0)}, \tag{71}$$

$$s_{qp} = s_p - s_{dp}, \tag{72}$$

$$s_{vp} = s_{qp} + \gamma_1 \int sgn(s_{qp}(\varsigma))d\varsigma, \tag{73}$$

$$s_F = \int_{t_0}^{t} \tilde{\lambda}dt, \tag{74}$$

$$s_{dF} = s_F(t_0)e^{-\eta(t-t0)}, \tag{75}$$

$$s_{qF} = s_F - s_{dF}, \tag{76}$$

$$s_{vF} = s_{qF} + \gamma_2 \int sgn(s_{qF}(\varsigma))d\varsigma \tag{77}$$

for $\alpha > 0$, $\eta > 0$. Finally, the reference for $\dot{z}$, that is $\dot{z}_r$ introduces a reconfigurable error, such that tracking errors in the null space will also converge to its desired value and full control on the redundancy is introduced. To this end, consider

$$\dot{z}_r = \dot{z}_d - \alpha_z \Delta z + s_{dz} - \gamma_z \int \mathrm{sgn}(s_{qz}) \tag{78}$$

where $\dot{z}_r$ fulfills $Q_k \dot{z}_r = \dot{z}_r$ in such a way that

$$s_z = \Delta \dot{z} + \alpha_z \Delta z = (\dot{z} - \dot{z}_d) + \alpha_z (z - z_d) \tag{79}$$

$$s_{dz} = \beta_z s_z(t_0) e^{-\kappa t} \tag{80}$$

$$s_{qz} = s_z - s_{dz} \tag{81}$$

$$s_{vz} = s_{qz} + \gamma_z \int \mathrm{sgn}(s_{qz}) \tag{82}$$

for positive definite feedback gains $\alpha_z, \beta_z, \gamma_z$. To complete the definitions, consider

$$\dot{z}_d = Q_k v_d \tag{83}$$

where $v_d = k \frac{\partial}{\partial q} \Omega$ stands for the gradient of a given cost function $\Omega$ to be optimized. According to this cost function, the redundant degrees of freedom of the full open kinematic chain tracks $\dot{z}_d$ and $z_d$, as it will be proved in the following, where

$$z_d = \int \dot{z}_d + z_d(t_0) \tag{84}$$

for without loss of generality it is assumed that $z_d(t_0) = q(t_0)$. Finally, owing to the fact that $\dot{q} = Q(J^\dagger \dot{X} + \dot{z})$ and that $s_r = \dot{q} - \dot{q}_r$, we obtain then

$$s = Q\left\{J^\dagger(\dot{X} - \dot{X}_r) + (\dot{z} - \dot{z}_r)\right\} - \beta J_\varphi^T s_{vf} \tag{85}$$

$$= Q\left\{J^\dagger s_{vp} + s_{vz}\right\} - \beta J_\varphi^T s_{vf} \tag{86}$$

where $s_{vp}$, $s_{vF}$, and $s_{vz}$ respectively, are given by (73), (77) and (82). Notice that $J^\dagger s_{vp}$ and $s_{vz}$ are orthogonal complements $(J^\dagger s_{vp})^T s_{vz} = 0$ and so does $Q(*)$ and $J_\varphi^T$. Notice that although the time derivative of $\dot{q}_r$ is discontinuous, that is not of any concern because it is not used in the controller.

## 5. Model-free second order sliding mode controller

Consider the following nominal continuous control law:

$$u_q = -K_d s + J_{\varphi+}^T(q)\left(-\lambda_d + \dot{s}_{dF} + \gamma_2 \tanh(\mu s_{qF}) + \mu s_{vF}\right) \tag{87}$$

with $\mu > 0$ and $K_d = K_d^T > 0, \in \Re^{(6+n)x(6+n)}$. This nominal control, designed in the q-space can be mapped to the original coordinates of the vehicle model, expressed by the set (1)-(2), using the next relationship $u_v = J_\nu^{-T}(q)u_{q_v}$. Thus, the physical controller in the vehicle $u_v$ is implemented in terms of a key inverse mapping $J_v^{-T}$.

### 5.1 Closed-loop system

The open loop system (62) under the continuous model-free second order sliding mode control (87) yields to

$$H\dot{s} = -[C + D + K_d]s - Y_r\Theta + J_{\varphi+}^T(\dot{s}_{vF} + \eta s_{vF}) + \gamma_2 J_\varphi^T Z + \tau^* \tag{88}$$

where $Z = tanh(\mu s_{qF}) - sgn(s_{qF})$, and $\tau^* \equiv 0$ is useful for the passivity analysis.

### 5.2 Stability analysis

**Theorem** Consider a constrained SRA (57)under the continuous model-free second order sliding mode control (87). The Underwater system yields a second order sliding mode regime with local exponential convergence for the position, and force tracking errors.

**Proof.** A passivity analysis $\langle S, \tau^*\rangle$indicates that the following candidate Lyapunov function $V$ qualifies as a Lyapunov function

$$V = \frac{1}{2}(s^T H s + \beta s_{vF}^T s_{vF}), \tag{89}$$

for a scalar $\beta > 0$. The time derivative of the Lyapunov candidate equation immediately leads to

$$\begin{aligned} \dot{V} &= -s^T(K_d + D_q)s - \beta\eta s_{vF}^T s_{vF} - s^T Y_r\Theta + s^T\gamma_2 J_\varphi^T Z \\ &\leq -s^T K_d s - \beta\eta s_{vF}^T s_{vF} + \|s\|\|Y_r\Theta\| + \|s\|\|\gamma_2\|\|J_\varphi\|\|Z\| \\ &\leq -s^T K_d s - \beta\eta s_{vF}^T s_{vF} + \|s\|\|\epsilon\|, \end{aligned} \tag{90}$$

where it has been used the skew symmetric property of $\dot{H} - 2C(q, \dot{q})$, the boundedness of both the feedback gains and submarine dynamic equation (there exists upper bounds for $H, C(q, \dot{q}), g(q), \dot{q}_r, \ddot{q}_r$), the smoothness of $\varphi(q)$ (so there exists upper bounds for $J_\varphi$ and $Q(q)$), and finally the boundedness of $Z$. All these arguments establish the existence of the functional $\epsilon$. Then, if $K_d$, $\beta$ and $\eta$ are large enough such that $s$ converges into a neighborhood defined by $\epsilon$ centered in the equilibrium $s = 0$, namely as $t \to \infty$, one obtains

$$s \to \epsilon \tag{91}$$

This result stands for local stability of $s$ provided that the state is near the desired trajectories for any initial condition. This boundedness in the $\mathcal{L}_\infty$ sense, leads to the existence of the constants $\epsilon_3 > 0$, $\epsilon_4 > 0$ and $\epsilon_5 > 0$ such that

$$\|\dot{s}_{vp}\|_{\mathcal{L}_\infty} < \epsilon_3, \tag{92}$$

$$\|\dot{s}_{vz}\|_{\mathcal{L}_\infty} < \epsilon_4, \quad (93)$$
$$\|\dot{s}_{vF}\|_{\mathcal{L}_\infty} < \epsilon_5. \quad (94)$$

An sketch of the local convergence of $s_{vp}$ is as follows[1]. Boundedness of $s$ means boundedness of the projected vectors $Q_k s_z$ and $J^{\dagger}(q)s_{vp}$, then if we multiply them we get zero, then we can analyze independently each projected vector. Locally, in the $n - r$ dimensional image of Q, we have that $s^*_{vp} = Qs_{vp} \in \Re^n$. Consider now that under an abuse of notation that $s_{vp} = s^*_{vp}$, such that for small initial conditions, if we multiply the derivative of $s_{vp}$ in (73) by $s^T_{qp}$, we obtain

$$s^T_{qp}\dot{s}_{qp} = -\gamma_1 \|s_{qp}\| + s^T_{qp}\dot{s}_{vp} \leq -\gamma_1 \|s_{qp}\| + \|s_{qp}\| \|\dot{s}_{vp}\| \leq -(\gamma_1 - \varepsilon_3) \|s_{qp}\| \quad (95)$$

which have used (92), and $\gamma_1 > \epsilon_3$, to guarantee the existence of a sliding mode at $s_{qp}(t) = 0$ at time $t \leq \|s_{qp}(t_0)\|/(\gamma_1 - \epsilon_3)$, and according to the definition of $s_{qp}$ (72), $s_{qp}(t_0) = 0$, which simply means that $s_{qp}(t) = 0$ for all time.
Similar than for $s_{vp}$, if we multiply the derivative of $s_{vz}$ in (79) by $s^T_{qz}$, we obtain

$$s^T_{qz}\dot{s}_{qz} = -\gamma_z \|s_{qz}\| + s^T_{qz}\dot{s}_{vz} \leq -\gamma_z \|s_{qz}\| + \|s_{qz}\| \|\dot{s}_{vz}\| \leq -(\gamma_z - \varepsilon_5) \|s_{qz}\| \quad (96)$$

which have used (93), and $\gamma_z > \epsilon_4$, to guarantee the existence of a sliding mode at $s_{qz}(t) = 0$ at time $t \leq \|s_{qz}(t_0)\|/(\gamma_z - \epsilon_4)$, and according to the definition of $s_{qz}$ in (79)-(82), $s_{qz}(t_0) = 0$, which simply means that $s_{qz}(t) = 0$ for all time.
Finally, we see that if we multiply the derivative of (77) by $s^T_{vf}$, we obtain

$$s^T_{qF}\dot{s}_{qF} = -\gamma_2 \|s_{qF}\| + s^T_{qF}\dot{s}_{vF} \leq -\gamma_2 \|s_{qF}\| + \|s_{qF}\| \|\dot{s}_{vF}\| \leq -(\gamma_2 - \varepsilon_4) \|s_{qF}\| \quad (97)$$

which have used (94), and $\gamma_2 > \epsilon_5$, to guarantee the existence of a sliding mode at $s_{qp}(t) = 0$ at time $t \leq \|s_{qF}(t_0)\|/(\gamma_2 - \epsilon_5)$ and,according to (77), $s_{qF}(t_0) = 0$, which simply means that $s_{qf}(t) = 0$ for all time, which simply implies that $\lambda \to \lambda_d$ exponentially fast.

## 6. Simulation results

Simulations has been made on a simplified and idealized platform of a 6 dof submarine with a mass of 100 kg, where the floatability point it is assumed to be in the geometrical centre of the vehicle. It is also assumed that the vehicle has neutral floatability i.e. the buoyancy forces equals the gravity ones. The submarine vehicle was modeled as a regular cube, so its principal moments of inertia are represented by a constant diagonal matrix. In order to get a simple but reliant dynamic model, we considered the base of the robot arm to be in the geometrical center of the vehicle. This robot arm was based on the 6 DoF Mitsubishi PA-10 with a mass of 30 kg. We also assumed that the spherical wrist only contributes in orientation, that is, the last link has length zero. The robot arm principal moments of inertia are as well assumed to be diagonal matrices.

---

[1] The strict analysis follows (*Liu, et. al.*)

Now even when our simulator is simple, it contains most of the coupled dynamics from the vehicle and the robot arm, which still makes the control problem quite a challenge.
Simulations were realized using the simulation software 20Sim 4.0, on a 64-bit Windows Vista computer with 4 GB RAM memory at 2.1 GHz Dual-Core processor. Simulator presents tridimensional results (x, y, z, roll, pitch, yaw), so the generalized coordinates for the vehicle is given by:

$$q_v = \begin{pmatrix} x_v & y_v & z_v & \phi_v & \theta_v & \psi_v \end{pmatrix}^T \tag{98}$$

which represents the cartesian position and orientation of the vehicle. The generalized coordinate for the robot arm is given by:

$$q_m = \begin{pmatrix} q_1 & q_2 & q_3 & q_4 & q_5 & q_6 \end{pmatrix}^T \tag{99}$$

which represents the manipulator joint positions. The generalized coordinates for the SRA system is composed by the vehicle and robot arm generalized coordinates, and is given by:

$$q = \begin{pmatrix} q_v \\ q_m \end{pmatrix} \tag{100}$$

The holonomic constraint is given by a vertical surface located at 1 meter from the origin, who's expression is

$$\varphi(q) \equiv x - 1 = 0 \tag{101}$$

It is assumed that the SRA end-effector remains in contact with the environment for all the time. In order to satisfy the constraint equation, the initial conditions were chosen as:

$$q_v(t_0) = \begin{pmatrix} 0 & 0 & 0 & 0 & 0 & 0 \end{pmatrix}^T \tag{102}$$

$$q_m(t_0) = \begin{pmatrix} 0 & 0 & -\pi/2 & 0 & -\pi/4 & 0 \end{pmatrix}^T \tag{103}$$

which leads to the following end-effector initial pose:

$$X(t_0) = \begin{pmatrix} 1 & 0 & 1 & 0 & -3\pi/4 & 0 \end{pmatrix}^T \tag{104}$$

Now, the desired end effector position and orientation trajectory task are defined as:

$$X_d = \begin{pmatrix} x_d & y_d & z_d & \phi_d & \theta_d & \psi_d \end{pmatrix}^T \tag{105}$$

$$X_d = \begin{pmatrix} 1 & 1-\cos(t/10) & 0.9+0.1\cos(t/2) & 0 & 3\pi/4 & 0 \end{pmatrix}^T \tag{106}$$

On the other hand, the desired force profile applied by the SRA end effector on the contact surface (holonomic constraint) is represented by:

$$\lambda_d = 20\sin(t/2) + 100 \tag{107}$$

The model-free control parameters are as follows:

| $q_i$ | $K_d$ | $\gamma_1$ | $\sigma$ | $\alpha$ | $\beta$ | $\mu$ |
|---|---|---|---|---|---|---|
| $x_v$ | 400 | $10^{-3}$ | 9 | 20 | 1 | 100 |
| $y_v$ | 400 | $10^{-3}$ | 9 | 20 | 1 | 100 |
| $z_v$ | 400 | $10^{-3}$ | 9 | 20 | 1 | 100 |
| $\phi_v$ | 400 | $50^{-3}$ | 9 | 20 | 1 | 100 |
| $\theta_v$ | 400 | $50^{-3}$ | 9 | 20 | 1 | 100 |
| $\psi_v$ | 400 | $50^{-3}$ | 9 | 20 | 1 | 100 |
| $q_1$ | 80 | $80^{-3}$ | 22 | 70 | 5 | 100 |
| $q_2$ | 110 | $80^{-3}$ | 22 | 70 | 5 | 100 |
| $q_3$ | 100 | $80^{-3}$ | 18 | 70 | 5 | 100 |
| $q_4$ | 28 | $50^{-3}$ | 5 | 70 | 5 | 100 |
| $q_5$ | 25 | $50^{-3}$ | 5 | 70 | 5 | 100 |
| $q_6$ | 20 | $50^{-3}$ | 4 | 70 | 5 | 100 |

And the force scalar gains were defined as $\eta = 2$, $\gamma_2 = 10^{-1}$.

### 6.1 Numerical considerations

To compute the value of $\lambda$, the constrained Lagrangian that fulfils the constrained movement, can be calculated using the second derivative of the holonomic restriction: $\ddot{\varphi}(q) = 0$. Then, by using the dynamic equation (49) and after some algebra its expression becomes either:

$$\lambda = \left[J_\varphi H^{-1} J_\varphi^T\right]^{-1} \left(J_\varphi H^{-1} \left(h(q,\dot{q},t) - u_q\right) - \dot{J}_\varphi \dot{q}\right) \tag{108}$$

The set of eqns. (49)-(108) describes the constrained motion of the SRA when in contact to infinitely rigid surface described by (12). Numerical solutions of these sets can be obtained by simulation, however the numerical solution, using a DAE solver, can take too much effort to converge due to the fact that these sets of equation represent a highly stiff system. In order to minimize this numerical drawback, the holonomic constraint has been treated as a numerically compliant surface which dynamic is represented by

$$\ddot{\varphi}(q) + D\dot{\varphi}(q) + P\varphi(q) = 0. \tag{109}$$

This is known in the force control literature of robot manipulators as constrained stabilization method, which bounds the nonlinear numerical error of integration of the backward integration differentiation formula. With a appropriate choice of parameters $P$ and $D$, the solution of $\varphi(, t) \to 0$ is bounded. This dynamic is chosen to be fast enough to allow the numerical method to work properly. In this way, it is allowed very small deviation on the computation of $\lambda$, typically in the order of $-10^6$ or less, which may produce, according to some experimental comparison, less than 0.001% numerical error. Then, the value of the normal contact force magnitude becomes:

$$\lambda = \left[J_\varphi H^{-1} J_\varphi^T\right]^{-1}\left(J_\varphi H^{-1}\big(h_v(q,\dot{q},t) - u_q\big)\dot{J}_\varphi \dot{q} - DJ_\varphi \dot{q} - P\varphi(q)\right), \tag{110}$$

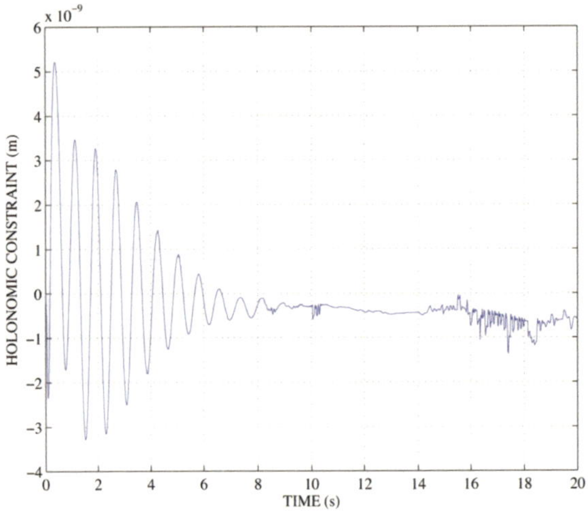


Figure 1. Holonomic constraint using the constraint stabilizer

### 6.2 Model free position-force tracking control

Figure 2 shows the end-effector's cartesian position. It is important to remark that the end-effector $x$ = 1 position, correspond to the holonomic constraint, which is asumued to be satisfied for all the time, that because of the infinitely rigid object assumption given in in (12). On the other hand, the $y$, $z$ positions are the desired end effector trajectories over the contact surface. Note that all the end effector positions are given in the inertial frame.

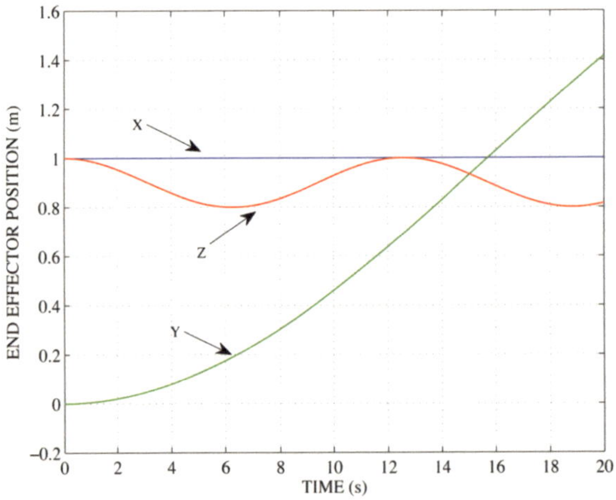


Figure 2. End-effector tracking desired position $y$, $z$ while remaining in contact with the holonomic constraint all the time.

On Figure 3 we find the end-effector orientation angles, which were defined as the roll-pitch-yaw angles around the x-y-z axis of the inertial frame. It can be seen that orientation tracks a desired orientation with all angles at at constant values of 0 for the roll and yaw angles, and $3\pi/4$ for the pitch angle.

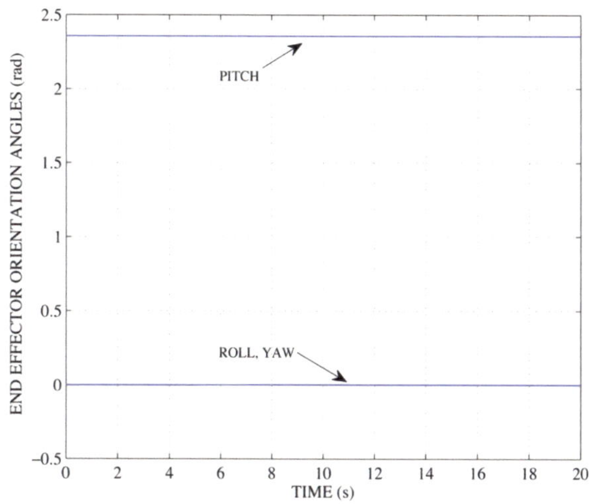


Figure 3. End-effector's orientation angles (roll, pitch, yaw), where we can see that set-point tracking tasks on 0 and $3\pi/4$ rad are being performed by the system.

Figure 4 shows how the end effector linear velocity vector evolves over time. In this Figure, we can observe that x velocity remains on 0, while the y and z velocities tracks $1/10\cos(t/10)$ and $0.05\cos(t/2)$ respectively. This way, both set-point control and tracking are being performed by the system.

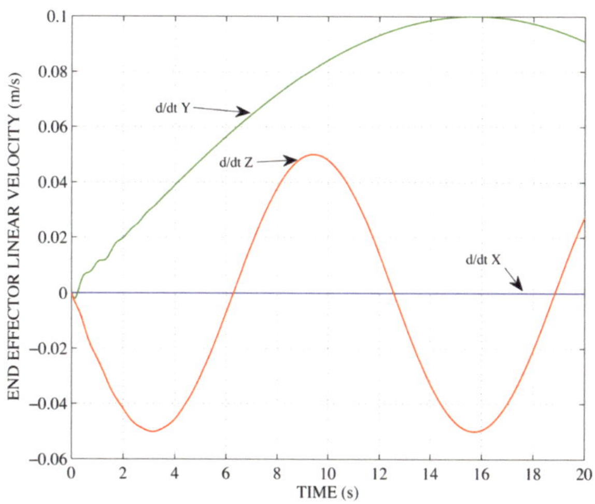


Figure 4. End-effector linear x-y-z velocities, where both, regulation and tracking are performed

Given the parameters used to describe the end-effector orientation, Figure 5 shows the time derivative of these orientation parameters, which in the strictly mathematical sense, are different form the end-effector angular velocity. In this figure, we can see that the orientation derivative, after a small transient finally establish at zero radians, that is, the end effector orientation will remain in the same position for all the time.

Now, let us define $X$ as the 6 dof end-effector pose (position and orientation) vector, and $\tilde{X} = X - X_d$ as the pose error vector. Figure 6 shows the Euclidean norm of the end effector pose error ($\| \tilde{X} \|$) and the norm of the end effector analytic velocity error ($\| \dot{\tilde{X}} \|$), which must not be confused with the geometric velocity error. In this figure, it is very clear that the error manifold are nearly zero, because of the controller inner properties.

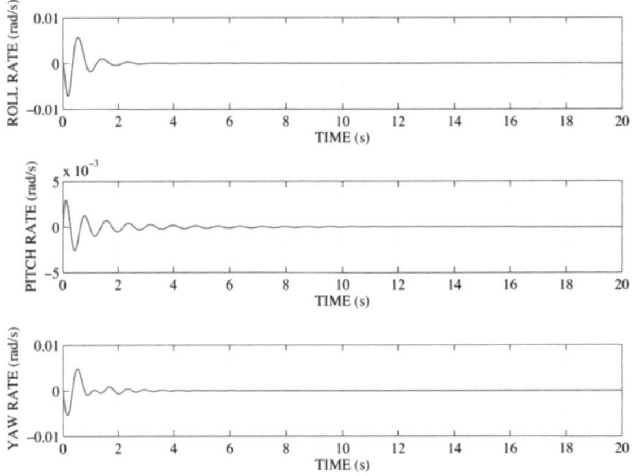


Figure 5. End-effector variation of angle parameters ($\dot{\phi}, \dot{\theta}, \dot{\psi}$) converge to zero after a small transient

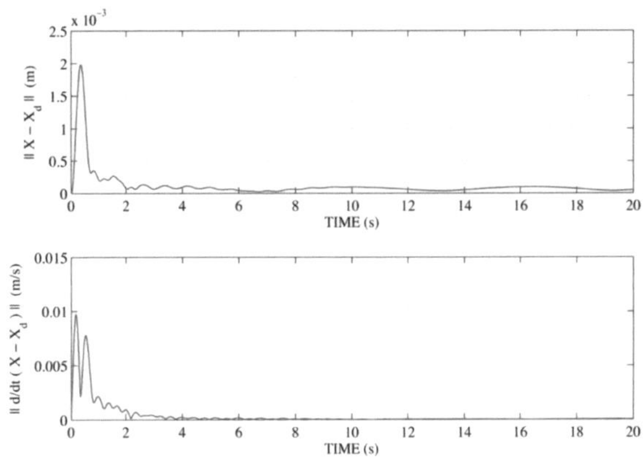


Figure 6. Euclidian norms of the end-effector's pose and its time derivative ($\tilde{X}$ , $\dot{\tilde{X}}$ ), where it is very clear the stability properties of the controller

Similarly as with the position, we defined a force task which must be accomplished by the end effector when moving around the contact surface. This task is independent from the position one, in other words, we can control both position and force simultaneous and independently. Figure 7 presents the end-effector force tracking trajectory ($\lambda$), where a periodic signal mounted on an offset of 100 N is shown. Additionally, Figure 7 shows on the right upper frame, a zoom of how the contact force, tracks the desired force profile. On the right lower frame, we can see the norm of the contact force error which is defined as $\| \tilde{\lambda} \| = \|\lambda - \lambda_d\|$. This error shows how fast the controller makes the contact force to track, with very little error, the desired contact force trajectory.

In relation with the control signals, Figure 8 shows the x-y-z linear forces applied to the vehicle. On this set of figures we can observe that no linear forces present chattering. This characteristic is very representative of the second order sliding mode controllers. Note that the biggest control signal belongs to the x linear force, this because of the shared contribution of the vehicle and the manipulator to achieved the contact force on the holonomic constraint, positioned along the x axis.

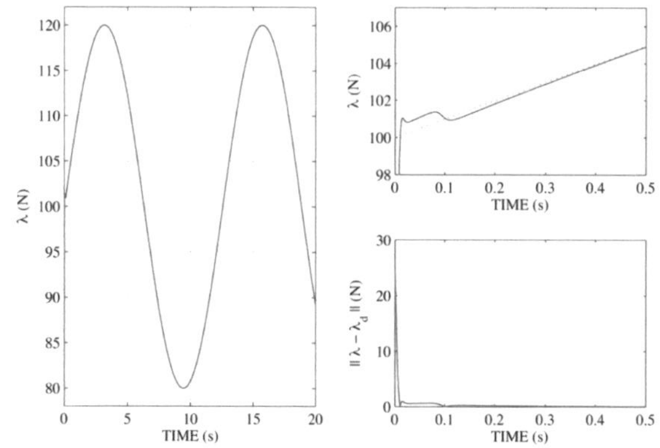


Figure 7. Left frame shows the end-effector contact force trajectory ($\lambda$), right upper frame shows a zoom of this force, and the right lower frame shows the norm of the end effector contact force $\|\tilde{\lambda}\|$

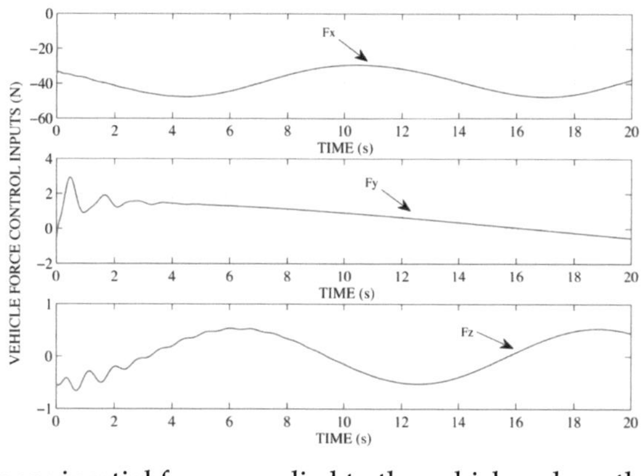


Figure 8. The x-y-z linear inertial forces applied to the vehicle, where the contact force contribution is mainly made through the x linear force

The vehicle torque control inputs can be seen in Figure 9, where, just as in the linear force case, no chattering is present on the signals. In this figure, we can see that x and z torque inputs are nearly cero, contrary to the y torque input, whose value oscillates around -30 Nm. due to the torque created by the distance of the contact point to the vehicle's origin.

Likewise, Figure 10 shows the torques ($\tau_i$) applied to the robot arm joints, where torque inputs $\tau_4$, $\tau_5$, $\tau_6$ are nearly cero, this is because these three joints only contribute on the end effector orientation in other words, the final link connected to them has length zero. Note that joint torque 2 are the one that mostly contribute to the end effector contact force.

Now, Figures 11-14 show the generalized coordinates sliding surfaces, that is, the vehicle and manipulator generalized coordinates surfaces. On these figures it can been shown that the error manifolds converge to a stable bounded neighborhood centered in the origin. We know that even when actual modern actuators have been growing really fast in past few year, it is still imposible to implement a signum function on them, that is the main reason why a saturation function is preferred. With the use of this saturation function we may no get ideal results, instead we get this convergence region where error stabilize.

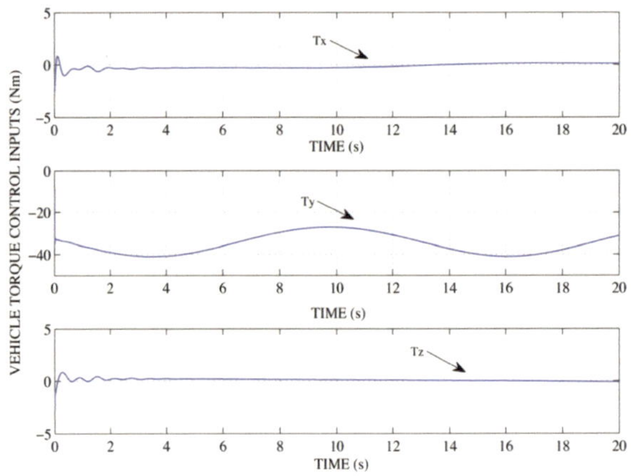


Figure 9. Input torques ($n_x$, $n_y$, $n_z$) applied to the vehicle, where we can see that $n_x$ and $n_z$ are nearly cero

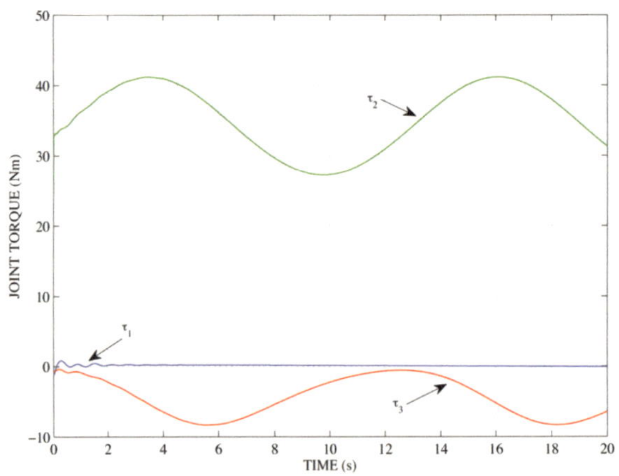


Figure 10. Input control torques ($\tau_i$) applied to the robot-arm, where the contact force contribution is mainly made through joint 2

On the other hand, Figure 15 shows the force error manifold, on which can be seen that the force error converge to a stable bounded neighborhood centered in the origin. It is very clear that this force manifold is the strongest one on the controller.

In consequence of the highly redundancy of the submarine robot arm (12 dof) working in normal operational conditions (6 dof), some degrees of freedom can be used to perform an additional task like minimize energy consumption, maximize dexterity, maximize the distance to the mechanical limits, etc. These additional tasks can be made by projecting a vector into the null space of the jacobian matrix.

Using the very know result of the gradient descendent, we can locally optimize a performance criteria like the ones already mentioned.

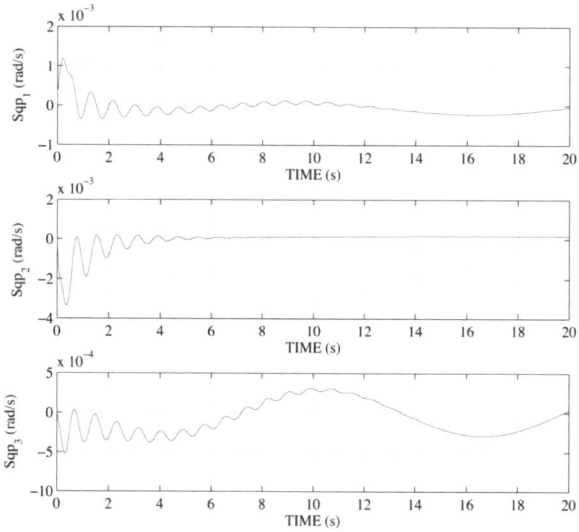


Figure 11. Generalized coordinates 1-3 position sliding surface

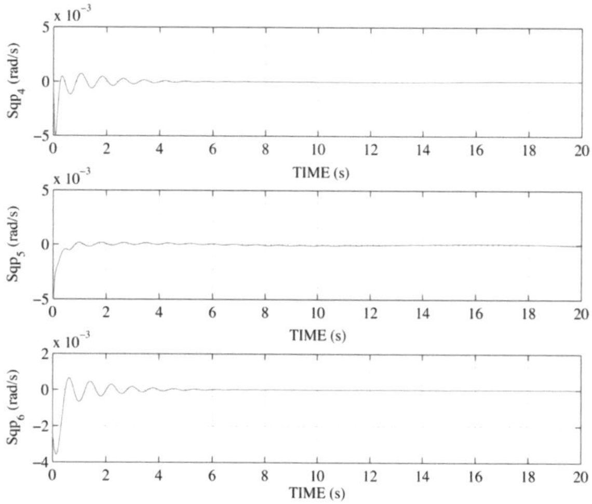


Figure 12. Generalized coordinates 4-6 position sliding surface

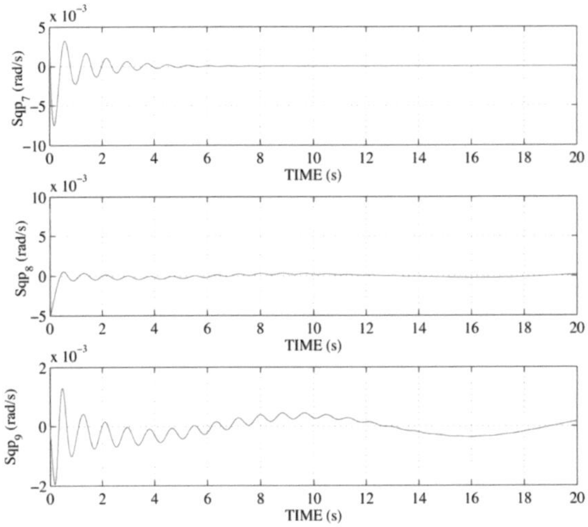


Figure 13. Generalized coordinates 7-9 position sliding surface

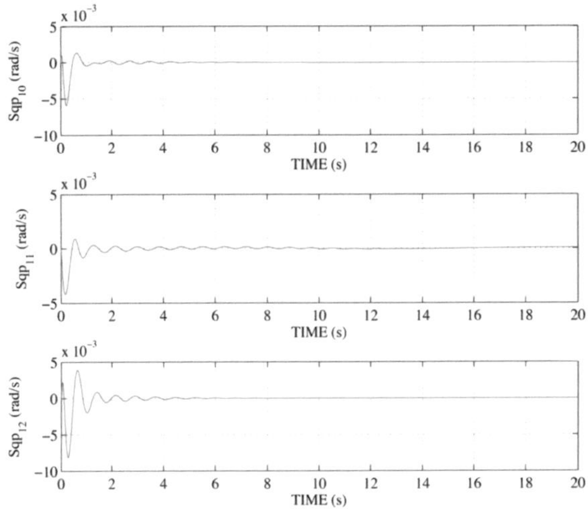


Figure 14. Generalized coordinates 10-12 position sliding surface

Simulations results are presented using the distance to the mechanical limits as the performance criteria, this criteria is given by:

$$\omega = \frac{1}{2n}\sum_{i=1}^{n}\left(\frac{q_i - \bar{q}_i}{q_{iM} - \bar{q}_{im}}\right)^2 \tag{111}$$

where $q_{iM}(q_{im})$ represents the maximum (minimum) limit for the generalized coordinate $q_i$, and $\overline{q}_i$ is the medium point for a coordinate range.

Figure 16 shows the performance criteria locally optimized for the required position-force trajectory, where curve lower minima represents a local optimal configuration.

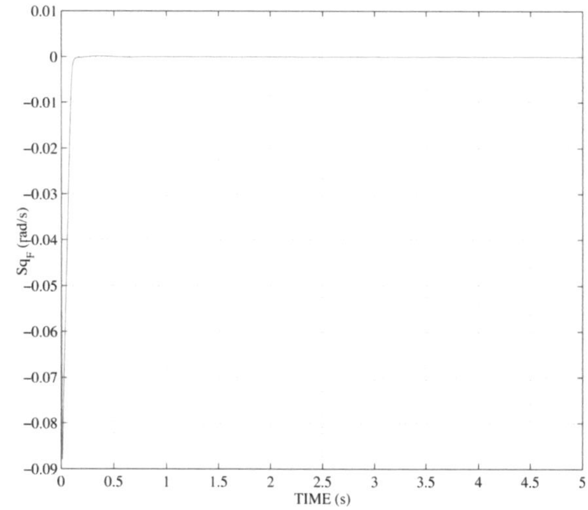


Figure 15. Force sliding surface

## 7. Remarks

There are several discussions that arise in such a complex system. In particular the dynamic model deserves further discussions related to its structural properties. Also, the simple control structure highlights its simplicity in contrast to the complex dynamics. Friction at the contact point is another issue we did not include. Finally redundancy is an additional structural degree of freedom which could be used to successfully tackle security issues too.

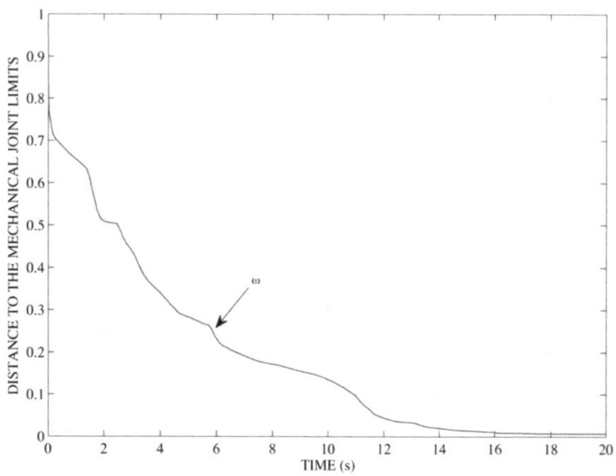


Figure 16. Redundancy resolution performance criteria ($\omega$)

**Properties of the Dynamics** As it was pointed out in section 3, the model of a submarine robot can be expressed in either a *self space* where the inertia matrix is constant for some conditions that in practice are not difficult to met or in a generalized coordinate space in which the inertial matrix is no longer constant but the model is expressed by only one equation likewise the kinematic lagrangian chains. Both representations preserves the all well-known properties of lagrangian systems such skew symmetry for the Coriolis/Inertia matrix, boundedness of all the components and passivity preserved properties for the hydrodynamics added effects, including buoyancy.

**Properties of the Kinematics** In this way, the kinematic equation gives rise to an equivalence of these representations. The latter is a linear operator that maps generalized coordinates time derivative with a generalized physical velocity. This relationship is specially important due to the fact that time derivative of angular representations (such a roll-pitch-yaw) is not the angular velocity, however there is always a correspondence between these vectors. For external forces this mapping is indeed important. It relates a physical force/torque wrench to the generalized coordinates $q = (x_v, y_v, z_v, \phi_v, \theta_v, \psi_v)$ whose last 3 components does not represent a unique physical space. In this work such mapping is given by $J_v$ and appears in the contact force mapping by the either the mobile manipulator Jacobian $J(q)$ or the holonomic restriction Jacobian $J_\varphi$.

**The Controller** Although a PD controller, recently validated by (Smallwood & Whitcomb 2004), has a much simpler structure and acceptable performance for underwater vehicles in position control schemes, it does not deal with the task of contact force control. Notice that the controller exhibits a PD structure plus a nonlinear *I*-tame control action, with nonlinear time-varying feedback gains, for each orthogonal subspace. It is in fact a decentralized PID-like, called "Sliding PD" controller, with two inner control loops and two kernels, one for contact point, where generalized velocities arise, and one for the self motion, evolving in the kernel of the observable output. It is indeed surprising that similar control structures can be implemented seemingly for a robot in the surface or below water, with similar stability properties, attaining simultaneous tracking of contact force and posture. Of course, this is possible under a proper mapping of Jacobians, key to implement this controller.

**Friction at the contact point** When friction at the contact point arises, which is expected to submarine tasks wherein the contact object is expected to exhibits a rough surface, with high friction coefficients, a tangential friction model should be added in the right hand side. Particular care must be placed to map the generalized velocities. Since this forces are nonconservative, it is expected, under proper handling of these dissipative forces in the error coordinates system, that stability will be enhanced.

## 8. Conclusions

Structural properties of the open-loop dynamics of submarine robots are key to design passivity-based controllers. In this paper, an advanced, yet simple, controller is proposed to achieve simultaneously tracking of time-varying contact force and posture, without any knowledge of the dynamics of the SRA. This is a significant feature of this controller, since in particular for submarine robots the dynamics are very difficult to compute, let alone its parameters. The redundancy is resolved at the kernel of the generalized velocities, such as primary, typically tracking errors, and secondary tasks can be satisfied. A simulation study of a hyper-redundant SRA provides additional insight into the closed-loop dynamic properties for set-point control and tracking case. It is proved that redundancy introduces a level of dexterity previously unexposed for SRA, while achieving critical contact force control.

## 9. References

E. Olguín-Díaz, V. Parra-Vega *On the Force/Posture Control of a Constrained Submarine Robot* 4th International Conference on Informatics in Control, Robotics and Automation, Conference Prodeedings. May 2007

Smallwood, D.A.; Whitcomb, L.L. *Model-based dynamic positioning of underwater robotic vehicles: theory and experiment* Oceanic Engineering, IEEE Journal of Volume: 29 Issue: 1 Jan. 2004

V. Parra-Vega, Second Order Sliding Mode Control for Robot Arms with Time Base Generators for Finitetime Tracking, Dynamics and Control, 2001.

Smallwood, D.A.; Whitcomb, L.L. *Toward model based dynamic positioning of underwater robotic vehicles* OCEANS, 2001. MTS/IEEE Conference and Exhibition Volume: 2 2001

Villani, L.; Natale, C.; Siciliano, B.; Canudas de Wit, C.; *An experimental study of adaptive force/position control algorithms for an industrial robot* Control Systems Technology, IEEE Transactions on Volume 8, Issue 5, Sept. 2000

E. Olguín-Díaz, *Modélisation et Commande d'un Système Véhicule/Manipulateur Sous-Marin* PhD Thesis. Laboratoire d'Automatique de Grenoble, January 1999

Y. H. Liu, S. Arimoto, V. Parra-Vega, and K. Kitagaki, *Decentralized Adaptive Control Of Multiple Manipulators in Cooperations*, International Journal of Control, Vol. 67, No. 5, pp. 649-673, 1997.

M. Perrier, C. Canudas de Wit, "Experimental Comparison of PID versus PID plus Nonlinear Controller for Subsea Robots" *Journal of Autonomous Robots, Special issue on Autonomous Underwater Robots, 1996.*

V. Parra-Vega and S. Arimoto, *A Passivity-based Adaptive Sliding Mode Position-Force Control for Robot Manipulators*, International Journal of Adaptive Control and Signal Processing, Vol. 10, pp. 365-377, 1996.

Arimoto, S. *Fundamental problems of robot control* Robotica (1995), volume 13, pp 19-27,111-122, Cambrige University Press

Thor I. Fossen. *Guidance and Control of Ocean Vehicles.* John Wiley and Sons, Chichester 1994

I. Schjølberg, T. I. Fossen. "Modelling and Control of Underwater Vehicle-Manipulator Systems" Proceedings the 3rd Conference on Marine Craft Maneuvering and Control (MCMC'94), Southampton, UK, 1994.

Chiaverini, S.; Sciavicco, L. *The parallel approach to force/position control of robotic manipulators.* IEEE Transactions on Robotics and Automation Volume 9, Issue 4, Aug. 1993

IFREMER, Project VORTEX: *Modélisation et simulation du comportement hidrodynamique d'un véhicule sous-marin asservi en position et vitesse.* 1992

Sagatun, S.I.; Fossen, T.I.; *Lagrangian formulation ofunderwater vehicles' dynamics* Decision Aiding for Complex Systems, Conference Proceedings., 1991 IEEE International Conference.

N. Hootsman & S. Dubowsky. Large Motion Control of Mobile Manipulators Including Vehicle Suspension Characteristics *Procedings of the 1991 IEEE International Conference on Robotics and Automation. Sacramento, California. (April 1991)*

Spong M.W., Vidyasagar M. *Robot dynamics and control*. John Wiley, New York 1989
Yoerger, D.; Slotine, J. *Robust trajectory control of underwater vehicles* Oceanic Engineering, IEEE Journal of Volume: 10 Issue: 4 Oct 1985

15

# Predictive Control with Local Visual Data

Lluís Pacheco, Ningsu Luo and Xavier Cufí
*University of Girona*
*Spain*

## 1. Introduction

Last half century depicted an enormous progress in robotics. However, this field is still in its infancy as compared with the human behaviour. Therefore, the ability of a robot is limited in its movement, understanding complex sensorial inputs or performing higher level reasoning. Important clues to enable further progress to be made are the integration of several different fields of knowledge, such as computing, communications and control sciences, with the aim of enabling robots to use higher level reasoning and decision tools with a strong theory base (Murray et al., 2003). The research work presented here integrates control science and robot vision knowledge in a computer science environment. Nowadays, the local visual perception research, applied to autonomous mobile robots, has succeeded in some important objectives, such as feasible obstacle detection and structure knowledge (Campbell, et al., 2004). In this context, the allowed navigation control signals should achieve the obstacle avoidance as well as the final desired coordinates. Scientific community has developed several studies in this field. Based on the dynamic window approach with available robot speeds, the reactive avoidance collisions, safety stop and goal can be achieved using the dynamic constraints of WMR (Fox, et al., 1997). In other works, it is presented the methodologies for the exact motion planning and control, based on the artificial potential fields where the complete information about the free space and goal are encoded (Rimon and Koditschek, 1992). Some approaches on mobile robots propose the use of potential fields, which satisfy the stability in a Lyapunov sense, in a short prediction horizon (Ögren & Leonard, 2005).
The main contribution of this paper is the use of the visual information as a dynamic window where the collision avoidance and safety stop can be planned. Thus, local visual data and local attraction of artificial potential fields are used in order to achieve the Lyapunov stability. This work relates the on-robot visual perception and odometer system information with the nonlinear mobile robot control system, consisting in a differential driven robot with a free rotating wheel. It is developed an easily portable methodology to plan the goal achievement by using the visual data as an available source of positions. The dynamic interactions of the robotic system arise from the knowledge of a set of experimental robot models that allow the development of model predictive control strategies based on the mobile robot platform PRIM available in the Laboratory of Robotics and Computer Vision. Moreover, the local visual information is used as an occupancy grid that allows planning feasible trajectories towards the desired objective. The knowledge of the objective allows the optimal solution of the local desired coordinates based on the acquired images.

This chapter is organized as follows: Section 1 gives a brief presentation about the aim of the present work. In the Section 2, the platform is introduced as an electromechanical system. This section also describes the experiments to be realized in order to find the parametric models of the robot suitable for designing and implementing MPC methods. Section 3 presents the MPC strategies used for achieving the path following of the reference trajectories. In the Section 4, the use of visual data is presented as a horizon where optimal trajectories can be planned by using MPC strategies. In the Section 5, the experimental results using MPC strategies are depicted. Finally, in Section 6 some conclusions are made with special attention paid into the future research works.

## 2. The Robot and the control system identification

### 2.1 Introduction

This section gives some description on the main robot electromechanical and sensorial systems of the platform tested in this work. Hence, the WMR PRIM, available in our lab, has been used in order to test and orient the research (Pacheco et al., 2008). The modelling methodology is depicted as an experimental indentification. Moreover, the simplified control model and the odometer system are also introduced.

### 2.2 Electromechanical and sensorial system of the robot

Fig. 1.a shows the robot PRIM used in the research work. The mechanical structure of the robot is made of aluminum, with two independent wheels of 16cm diameters actuated by two DC motors. The distance between two wheels is 56.4cm. A third spherical omni-directional wheel is used to guarantee the system stability. The maximum continuous torque of each motor is 131mNm. The proportion of gear reduction for each motor is 86:1 and thus the total force actuating on the robot is near 141N. Shaft encoders with 500 counts/rev are placed at the motor axes, which provide 43000 counts for each turn of the wheel.

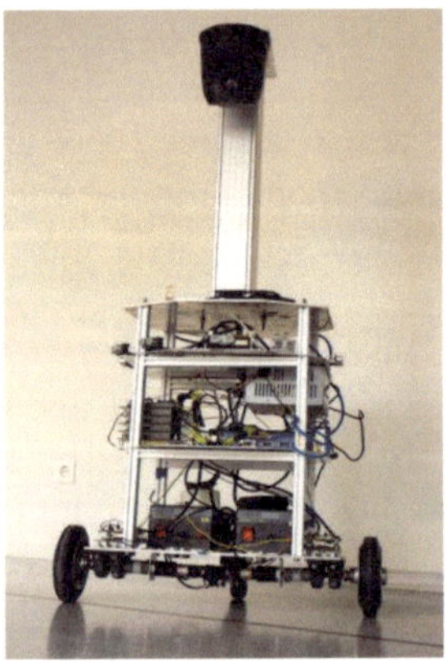

(a)

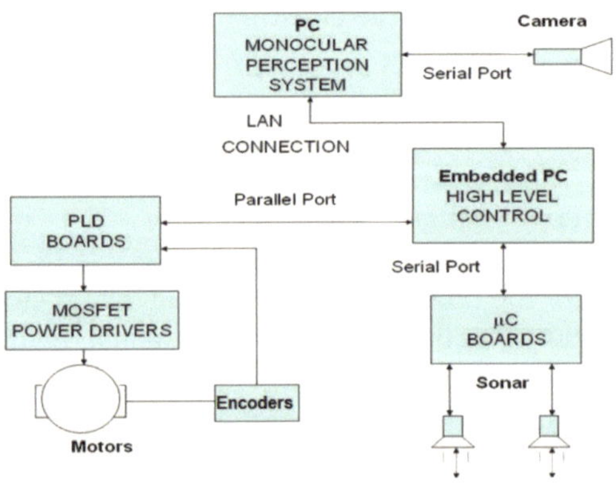


(b)

Fig. 1. (a) The robot PRIM used in this work; (b) The sensorial and electronic system blocs

A set of PLD (programmable logic device) boards is connected to the digital outputs of the shaft encoders. The printed circuits boards (PCB) are used to measure the speed of each motor at every 25ms. An absolute counter provides the counts in order to measure the robot position and speed by the odometer system. Another objective of these boards is to generate a signal of 23khz PWM for each motor. The communication between the central digital computer and the boards is made through the parallel port. The speed is commanded by a byte and thus it can generate from 0 to 127 advancing or reversing speed commands. The maximal speed is near 0.5m/s. A set of microcontroller boards (MCS-51) is used to read the information available from different connected sensors. The rate of communication with these boards is 9600 b/s. Fig. 1.b shows the electronic and sensorial system blocks. The data gathering and the control by digital computer is set to 100ms. The system flexibility is increased with the possibility of connecting with other computer systems through a local LAN. In this research, it is connected to a machine vision system that controls a colour camera EVI-D70P-PAL through the VISCA RS232-C control protocol. For instance, the camera configuration used in this work is of a horizontal field of view of 48° and a vertical field of 37°. The focus, pan and tilt remain fixed under present configuration. The camera pose is set to 109cm from the floor with a tilt angle of 32°. The local desired coordinates, obtained by the machine vision system, can be transmitted to the control unit connecting the USB port to the LAN.

### 2.3 Experimental model and system identification

The parametric identification process is based on black box models (Norton, 1986; Ljung, 1989). Thus, the transfer functions are related to a set of polynomials that allow the use of analytic methods in order to deal with the problem of controller design. The nonholonomic system dealt with in this work is considered initially as a MIMO (multiple input multiple output) system, which is composed of a set of SISO subsystems with coupled dynamic influence between two DC motors. The approach of multiple transfer functions consists in making the experiments with different speeds. In order to find a reduced-order model, several studies and experiments have been done through the system identification and model simplification. The parameter estimation is done by using a PRBS (Pseudo Random Binary Signal) as excitation input signal. It guarantees the correct excitation of all dynamic sensible modes of the system along the spectral range and thus results in an accurate precision of parameter estimation. The experiments to be realized consist in exciting two DC motors in different (low, medium and high) ranges of speed.

The ARX (auto-regressive with external input) structure has been used to identify the parameters of the robot system. The problem consists in finding a model that minimizes the error between the real and estimated data. By expressing the ARX equation as a lineal regression, the estimated output can be written as:

$$\hat{y} = \theta\varphi \tag{1}$$

with $\hat{y}$ being the estimated output vector, $\theta$ the vector of estimated parameters and $\varphi$ the vector of measured input and output variables. By using the coupled system structure, the transfer function of the robot can be expressed as follows:

$$\begin{pmatrix} Y_R \\ Y_L \end{pmatrix} = \begin{pmatrix} G_{RR} & G_{LR} \\ G_{RL} & G_{LL} \end{pmatrix} \begin{pmatrix} U_R \\ U_L \end{pmatrix} \tag{2}$$

where $Y_R$ and $Y_L$ represent the speeds of right and left wheels, and $U_R$ and $U_L$ the corresponding speed commands, respectively. In order to know the dynamics of robot system, the matrix of transfer function should be identified. Fig. 2 shows the speed response of the left wheel corresponding to a left PBRS input signal. The treatment of experimental data is done before the parameter estimation. In concrete, it includes the data filtering, using the average value of five different experiments with the same input signal, the frequency filtering and the tendency suppression. The system is identified by using the identification toolbox "ident" of Matlab for second order models. The following continuous transfer function matrix for medium speed is obtained:

$$\begin{pmatrix} Y_L \\ Y_R \end{pmatrix} = \begin{pmatrix} \frac{0.35s^2 + 4.82 + 4.46}{s^2 + 5.84s + 4.89} & \frac{0.02s^2 + 0.27s + 0.32}{s^2 + 5.84s + 4.89} \\ \frac{0.26s^2 + 3.41s + 0.28}{s^2 + 5.84s + 4.89} & \frac{0.11s^2 + 1.72s + 5.12}{s^2 + 5.84s + 4.89} \end{pmatrix} \begin{pmatrix} U_L \\ U_R \end{pmatrix} \tag{3}$$

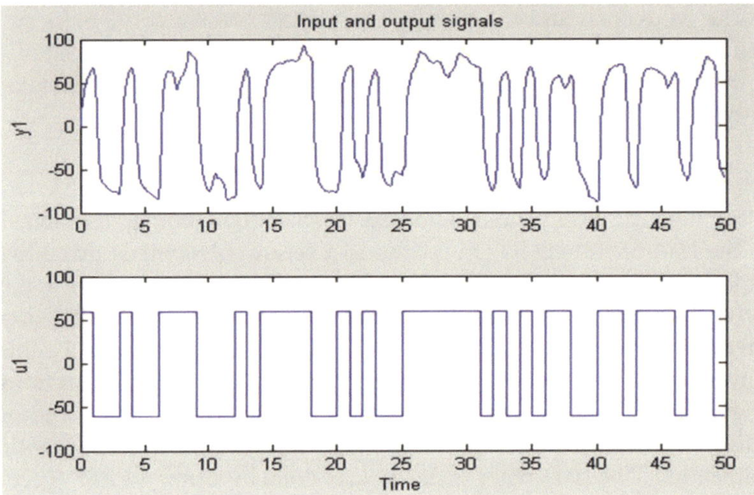


Fig. 2. Left speed output for a left PRBS input signal

It is shown by simulation results that the obtained model fits well to the experimental data.

### 2.4 Simplified control model and odometer system

This section studies the coupling effects and the way for obtaining a reduced-order dynamic model. It is seen from (3) that the dynamics of two DC motors are different and the steady gains of coupling terms are relatively small (less than 20% of the gains of main diagonal terms). Thus, it is reasonable to neglect the coupling dynamics so as to obtain a simplified model. In order to verify it from real results, a set of experiments have been done by sending non-zero speed commands to one motor while commanding zero speed to the other motor. In Fig. 3, it is shown a response obtained on the left wheel, when a medium speed command is sent to the right wheel. The experimental result confirms the above facts. The existence of different gains in steady state is also verified experimentally.

Finally, the order reduction of system model is carried out trough the analysis of pole positions by using the method of root locus. Afterwards, the system models are validated through the experimental data by using the PBRS input signal.

A two dimensional array with three different models for each wheel is obtained. Hence, each model has an interval of validity where the transfer function is considered as linear, and classic PID speed control can be developped.

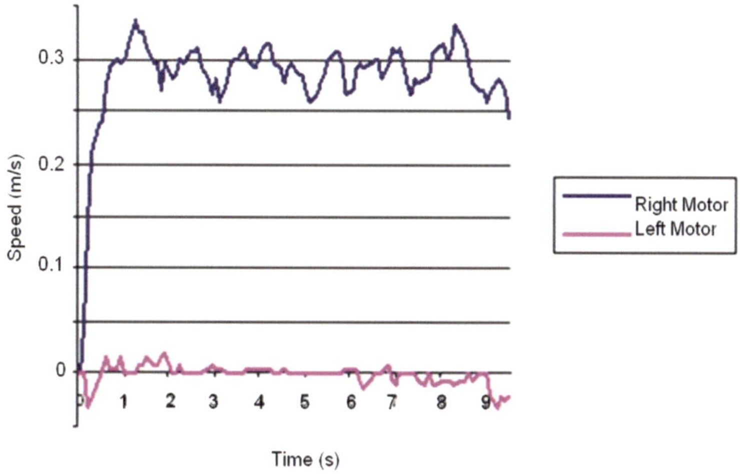


Fig. 3. Coupling effects at the left wheel

The robot speed and position are provided by the odometer system; hence, $(x, y, \theta)$ denote the coordinates of position and orientation, respectively. The Fig. 4 describes the positioning of robot as a function of the radius of left and right wheels $(R_e, R_d)$, and the angular incremental positioning $(\theta_e, \theta_d)$, with $E$ being the distance between two wheels and $dS$ the incremental displacement of the robot. The position and angular incremental displacements are expressed as:

$$dS = \frac{R_d d\theta_d + R_e d\theta_e}{2} \qquad d\theta = \frac{R_d d\theta_d - R_e d\theta_e}{E} \tag{4}$$

The coordinates (x, y, θ) can be expressed as:

$$\begin{aligned} x_n &= x_{n-1} + dS\cos(\theta_{n-1} + d\theta) \\ y_n &= y_{n-1} + dS\sin(\theta_{n-1} + d\theta) \\ \theta_n &= \theta_{n-1} + d\theta \end{aligned} \tag{5}$$

Thus, the incremental position of the robot can be obtained through the odometer system with the available encoder information obtained from (4) and (5).

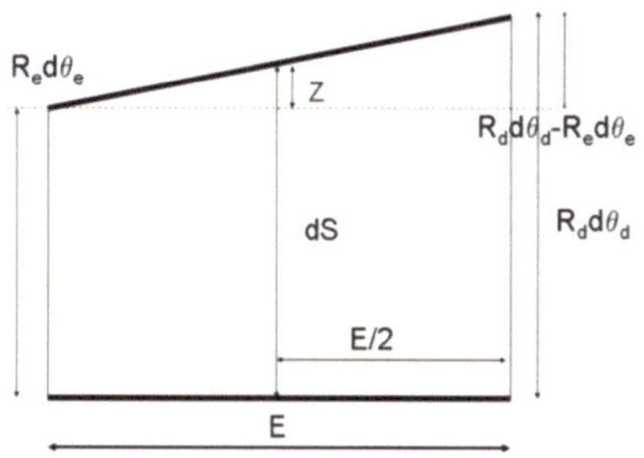


Fig. 4. Positioning of the robot as functions of the angular movement of each wheel

## 3. Model Predictive Control

### 3.1 Introduction

The MPC (model predictive control) MPC has many interesting aspects for its application to mobile robot control. It is the most effective advanced control technique, as compared to the standard PID control, that has made a significant impact to the industrial process control (Maciejowski, 2002). Recently, real time mobile robot MPC implementations have been developed using global vision sensing (Gupta et al., 2005). In (Küne et al., 2005), it was studied the MPC based optimal control useful for the case when nonlinear mobile robots are used under several constraints, as well as the real time implementation possibilities when short prediction horizons are used. In general, the global trajectory planning becomes unfeasible since the sensorial system of some robots is just local. By using a MPC, the idea of the receding horizon can deal with the local sensor information. In this way, it is proposed a LMPC (local model predictive control) in order to use the available visual data in the navigation strategies for the goal achievement. The MPC is based on minimizing a cost function, related to the objectives, through the selection of the optimal inputs. In this case, the cost function can be expressed as follows:

$$J(n,m) = \min_{\left\{U(k+i|k)\right\}_{j=0}^{j=m-1}} \left\{ \begin{array}{l} \left[X(k+n|k)-X_d\right]^T P\left[X(k+n|k)-X_d\right] \\ +\sum_{i=1}^{n-1}\left[X(k+i|k)-X_d\right]^T Q\left[X(k+i|k)-X_d\right] \\ +\sum_{i=0}^{m-1} U^T(k+i|k) R U(k+i|k) \end{array} \right\} \tag{6}$$

where $X_d=(x_d, y_d, \theta_d)$ denotes the desired trajectory coordinates. The first term of (6) is referred to the final desired coordinate achievement, the second term to the trajectory to be followed, and the last one to the input signals minimization. $P$, $Q$ and $R$ are weighting parameters. $X(k+n|k)$ represents the terminal value of the predicted output after the horizon of prediction $n$ and $X(k+i|k)$ represents the predicted output values within the prediction horizon. The system constrains are also considered:

$$\left\{ \begin{array}{ll} \left|U(k+i|k)\right| \le G_1 & \alpha \in [0,1) \\ \left\|\left[x_{k+i}, y_{k+i}\right]-\left[x_o, y_o\right]\right\| \ge G_2 & \\ \left\|\left[x_{k+n}, y_{k+n}\right]-\left[x_d, y_d\right]\right\| \le \alpha\left\|\left[x_k, y_k\right]-\left[x_d, y_d\right]\right\| & \end{array} \right\} \tag{7}$$

The limitation of the input signal is taken into account in the first constraint. The second constraint is related to the obstacle points where the robot should avoid the collision. The last one is just a convergence criterion.

### 3.2 LMPC Algorithms

This section gives the LMPC algorithms by using the basic ideas presented in the MPC introduction. The LMPC algorithm is run in the following steps:

1. To read the actual robot position;
2. To minimize the cost function and to obtain a series of optimal input signals;

3. To choose the first component of the obtained input signals as the command signal;
4. To go back to the step 1 in the next sampling period.

The minimization of the cost function is a nonlinear problem in which the following equation should be verified:

$$f(\alpha x + \beta y) \leq \alpha f(x) + \beta f(y) \tag{8}$$

It is a convex optimization problem caused by the trigonometric functions used in (5), (Boyd & Vandenberghe, 2004). The use of interior point methods can solve the above problem (Nesterov et al., 1994). Among many algorithms that can solve the optimization, the descent methods are used, such as the gradient descent method among others, (Ortega, et al. 2000). The gradient descent algorithm has been implemented in this work. In order to obtain the optimal solution, some constraints over the inputs are taken into account:

- The signal increment is kept fixed within the control horizon
- The input signals remain constant during the remaining interval of time

The input constraints present advantages such like the reduction in the computation time and the smooth behavior of the robot during the prediction horizon. Thus, the set of available input is reduced to one value. In order to reduce the optimal signal value search, the possible input sets are considered as a bidimensional array, as shown in Fig. 5. Then, the array is decomposed into four zones, and the search is just located to analyze the center points of each zone. It is considered the region that offers better optimization, where the algorithm is repeated for each sub-zone, until no sub-interval can be found. Once the algorithm is proposed, several simulations have been carried out in order to verify the effectiveness, and then to make the improvements. Thus, when only the desired coordinates are considered, the robot could not arrive in the final point. Fig. 6.a and Fig. 6.b show the simulated results corresponding to the use of two different horizons (one with a prediction horizon n=5 and a control horizon m=3, and the other with a prediction horizon n=10 and a control horizon m=5), where it is depicted that the inputs can minimize the cost function by shifting the robot position to the left.

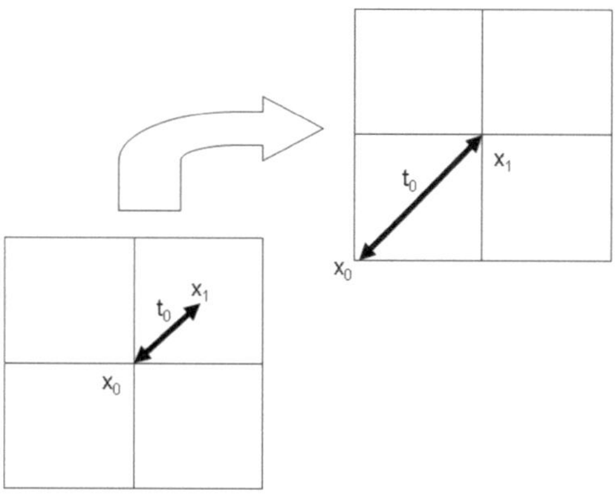


Fig. 5. Optimal interval search.

The reason can be found in (3), where the right motor has more gain than the left. This problem can be easily solved by considering a straight-line trajectory from the actual point of the robot to the final desired point. Thus, the trajectory should be included into the LMPC cost function. The possible coordinates available for prediction, as shown in Fig. 6.a, depict a denser horizon due to the shorter prediction horizon. Therefore, prediction horizons between 0.5s and 1s were proposed and the computation time for each LMPC step was set to less than 100ms, running in an embedded PC of 700MHz. Trajectory tracking and final goal achievement are other interesting aspects to be analyzed. Fig. 7.a shows the simulated results obtained in tracking a straight line of 2m using two different prediction horizons. Fig. 7.b shows the velocities of both wheels with the above strategies. The wide prediction strategy shows a softer behaviour due to the larger control horizon.

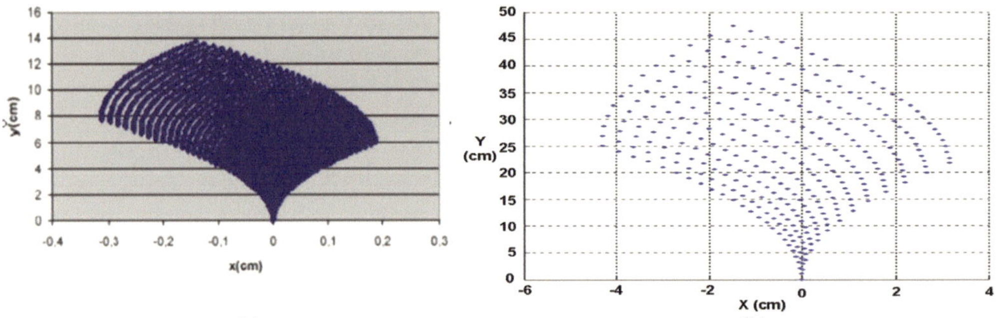


(a) (b)

Fig. 6. Predicted coordinates from speed zero (a) n=5 m=3 (b) n=10 m=5

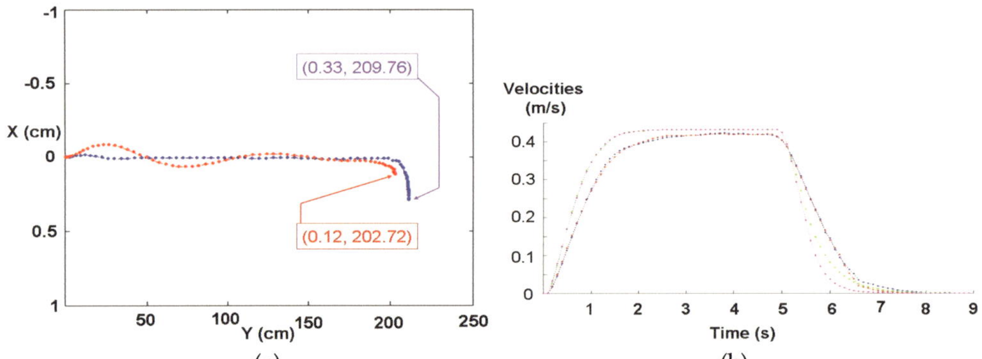


(a) (b)

Fig. 7. (a) Trajectory tracking in red (n=10, m=5) and in blue (n=5, m=3). The larger prediction horizon shows a closer final goal achievement and worse trajectory tracking. (b) Wheel speeds during the 2m straight line tracking. The red and blue dots show the right and left speeds respectively, with n=10 and m=5. The magenta and green dot lines depict the right and left speeds with n=5 and m=3.

## 4. The Horizon of local visual perception

### 4.1 Introduction

The computer vision techniques applied to WMR have solved the problem of obstacle detection by using different methods as stereo vision systems, optical flow or DFF (depth

from focus). Stereo vision systems seem to provide the easiest cues to infer scene depth (Horn, 1998). The optical flow techniques used in WMR result in several applications as i.e. structure knowledge, obstacle avoidance, or visual servoing (Campbell et al., 2004). The DFF methods are also suitable for WMR. For example, three different focused images were used, with almost the same scene, acquired with three different cameras (Nourbakhsh et al., 1997). In this work, it is supposed that available obstacle positions are provided by using computer vision systems. The use of sensor information as a useful source to build 2D environment models consists of a free or occupied grid proposed by (Elfes, 1989). The knowledge of occupancy grids has been used for static indoor mapping with a 2D grid (Thrun, 2002). In other works of multidimensional grids, multi target tracking algorithms are employed by using obstacle state space with Bayesian filtering techniques (Coué et al., 2006). In this work it is proposed the use of the local visual information available from the camera as a local map that has enough information in order to achieve a desired objective. The present research assumes that the occupancy grid is obtained by a machine vision system. It is proposed an algorithm that computes the local optimal desired coordinate as well as the local trajectory to be reached. The research developed assumes indoor environments as well as flat floor constraints. However, it can be also applied in outdoor environments.
This section presents firstly the local map relationships with the camera configuration and poses. Hence, the scene perception coordinates are computed. Then, the optimal control navigation strategy is presented, which uses the available visual data as a horizon of perception. From each frame, it is computed the optimal local coordinates that should be reached in order to achieve the desired objective. Finally, the WMR dynamics constraints and the strategy of navigation drawbacks are also commented.

### 4.2 Scene perception

The local visual data provided by the camera are used in order to plan a feasible trajectory and to avoid the obstacle collision. The scene available coordinates appear as an image, where each pixel coordinates correspond to a 3D scene coordinates. In the case attaining to this work, flat floor surface is assumed. Hence, scene coordinates can be computed using camera setup and pose knowledge, and assuming projective perspective. The Fig. 8 shows the robot configuration studied in this work. The angles $\alpha$, $\beta$ and $\varphi$ are related to the vertical and horizontal field of view, and the tilt camera pose, respectively. The vertical coordinate of the camera is represented by $H$. Using trigonometric relationships, the scene coordinates can be computed:

$$y_j = H\tan\left(\varphi - {}^{\alpha}/_{2} + \Delta\alpha\right) \quad \Delta\alpha = K_j\frac{\alpha}{R} \quad \left(0 \le K_j \le R\right) \tag{9}$$

$$x_{i,j} = \pm\frac{\mathrm{H}}{\cos(\varphi - \alpha/2 + \Delta\alpha)}\tan(\Delta\beta) \quad \Delta\beta = K_i\frac{\beta}{C} \quad (0 \le K_i \le C/2) \tag{10}$$

The $K_i$ and $K_j$ are parameters used to cover the image pixel discrete space. Thus, $R$ and $C$ represent the image resolution through the total number of rows and columns. It should be noted that for each row position, which corresponds to scene coordinates $y_j$, there exist $C$ column coordinates $x_{i,j}$. The above equations provide the available local map coordinates when no obstacle is detected. Thus, considering the experimental setup reported in Section 2, the local on-robot map depicted in Fig. 9 is obtained.

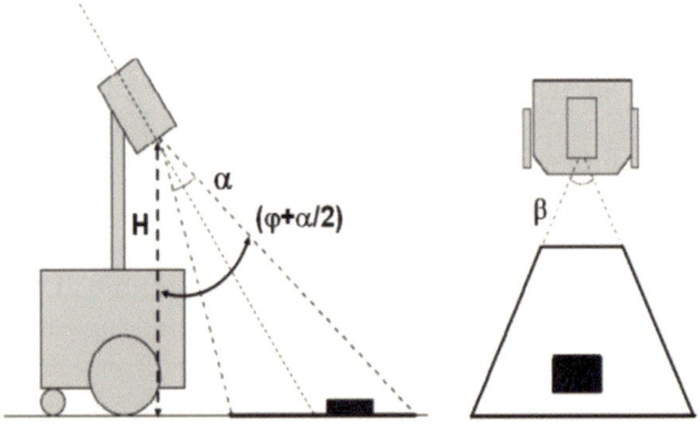


Fig. 8. Fixed camera configuration including vertical and horizontal field of view, and vertical tilt angle

**4.3 Local optimal trajectory, algorithms and constraints**

The available information provided by the camera is considered as a local horizon where the trajectory is planned. Hence, a local map with free obstacle coordinates is provided. In this sense, the available local coordinates are shown in Fig. 9. It is noted that low resolution scene grids are used in order to speed up the computing process.

The minimization of a cost function, which consists of the Euclidean distance between the desired coordinates and the available local scene coordinates, can be optimally solved by finding the local desired coordinates. Hence, the algorithm explores the image pixels, *IMAGE(i,j)*, considering just the free obstacle positions. Once the local desired point is obtained, a trajectory is planned between the robot coordinates, at the instant when the frame was acquired, and the optimal scene coordinates. Thus, the current robot coordinates are related to this trajectory, as well as to the control methods.

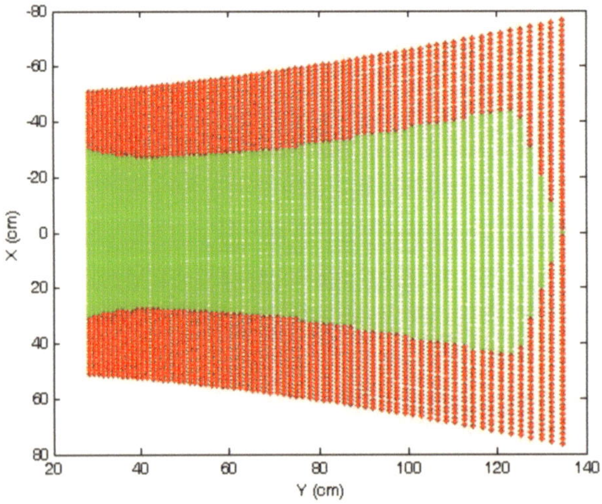


Fig. 9. Local visual perception free of obstacles, under 96x72 low resolution grids. Available local map coordinates (in green), the necessary wide-path (in red).

WMR movements are planned based on the local visual data, and always in advancing sense. The change in the WMR orientation can be done in advancing sense by the use of the trajectory/robot orientation difference as the cost function computed over the available visual data. The proposed algorithm, concerning to obtaining the local visual desired coordinates, consists of two simple steps:

- To obtain the column corresponding to the best optimal coordinates that will be the local desired $X_i$ coordinate.
- To obtain the closer obstacle row, which will be the local desired $Y_j$ coordinate.

The proposed algorithm can be considered as a first order approach, using a gross motion planning over a low resolution grid. The obstacle coordinates are increased in size with the path width of the robot (Schilling, 1990). Consequently, the range of visually available orientations is reduced by the wide-path of WMR. Other important aspects as dynamic reactive distance and safety stop distance should also be considered. The dynamic reactive distance, which should be bigger than the safety stop distance, is related to the robot dynamics and the processing time for each frame. Moreover, the trajectory situated in the visual map should be larger than a dynamic reactive distance. Thus, by using the models corresponding to the WMR PRIM, three different dynamical reactive distances are found. As i.e. considering a vision system that processes 4 frames each second; the table 1 shows these concepts.

The local minimal failures will be produced when a convergence criterion, similar to that used in (7), is not satisfied. Thus, the local visual map cannot provide with closer optimal desired coordinates, because obstacles block the trajectory to the goal. In these situations, obstacle contour tracking is proposed. Hence, local objectives for contour tracking are used, instead of the goal coordinates, as the source for obtaining a path until the feasible goal trajectories are found. The Fig. 10 shows an example with local minimal failures.

| Model | Safety stop distance | Obstacle reactive distance | Robot displacement | Minimal allowable distance |
|---|---|---|---|---|
| Low velocity (0.15 m/s) | 0.085m | 0.125m | 0.038m | 0.248m |
| Medium velocity (0.3 m/s) | 0.178m | 0.125m | 0.075m | 0.378m |
| High velocity (0.45 m/s) | 0.310m | 0.125m | 0.113m | 0.548m |

Table 1. Reactive criterion and minimal allowable distances

It is seen that in A, the optimal trajectory is a straight line between A and E. However, an obstacle is met at B, and local minimal failure is produced at B. When this situation occurs, no trajectory can approach to the desired goal, *(Xd, Yd)*. Then, obstacle contour tracking is proposed between B and C. Once C is attained, local minimization along coordinates Y is found and the trajectory between C and D is planned. From D to E local minimums are reached until the final goal is achieved. It should be noted that once B is reached, the left or right obstacle contour are possible. However, the right direction will bring the robot to an increasing $Y_j$ distance.

The robot follows the desired goals except when the situation of obstacle contour tracking is produced, and then local objectives are just the contour following points. The local minimal failures can be considered as a drawback that should be overcome with more efforts. In this sense, by taking into account the vision navigation strategies (Desouza & Kak, 2002), it is proposed in this work the use of feasible maps or landmarks in order to provide local objective coordinates that can be used for guiding the WMR to reach the final goal coordinates. The use of effective artificial attraction potential fields should also be considered.

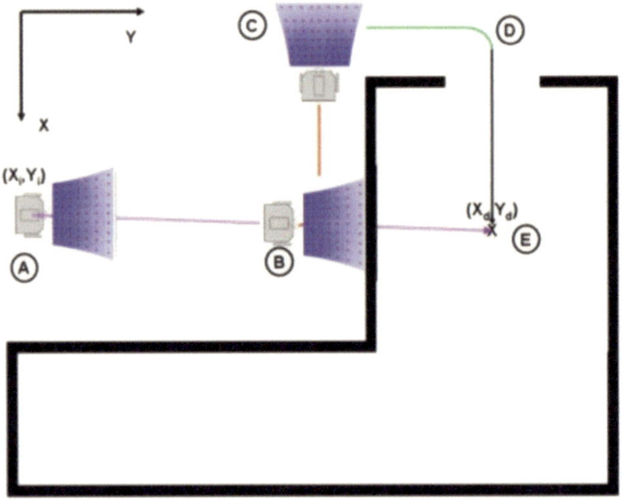


Fig. 10. Example of local minimal failures produced at B with A being the starting point and E the desired goal.

## 5. Testing local visual trajectories using LMPC strategies

### 5.1 Introduction

The minimization of path following error is considered as a challenging subject in mobile robotics. The main objective of highly precise motion tracking consists in minimizing the error between the robot and the desired path. Real-time implementation of MPC in the mobile robotics has been developed using global vision sensing (Gupta et al., 2005). In (Küne et al, 2005), MPC based optimal control was studied, which is useful for cases when nonlinear mobile robots are used under several constraints. In general, real-time implementation is possible when a short prediction horizon is used. By using MPC, the idea of receding horizon can deal with the local sensor information. MPC is based on minimizing a cost function related to the objectives for generating the optimal inputs.

The LMPC (local model predictive control) is proposed to use the available visual data in the navigation strategies for the goal achievement (Pacheco & Luo, 2007). Define the cost function as follows:

$$J(n,m) = \min_{\left\{U(k+i|k)_{i=0}^{i=m-1}\right\}} \left\{ \begin{array}{l} \left[X(k+n|k) - X_{ld}\right]^T P \left[X(k+n|k) - X_{ld}\right] \\ + \sum_{i=1}^{n-1} \left[X(k+i|k) - \overline{X_{ld} X_{l0}}\right]^T Q \left[X(k+i|k) - \overline{X_{ld} X_{l0}}\right] \\ + \sum_{i=0}^{m-1} U^T(k+i|k) R U(k+i|k) \end{array} \right\} \quad (11)$$

The first term of (11) refers to reaching the local desired coordinates, $X_{ld}=(x_d, y_d, \theta_d)$. The second one is related to the distance between the predicted robot positions and the trajectory segment $\overline{X_{ld}X_{l0}}$ given by a straight line between the initial robot coordinates $X_{l0}=(x_{l0}, y_{l0}, \theta_{l0})$ from where the local perception frame was acquired, and the desired local position $X_{ld}=(x_{ld}, y_{ld}, \theta_{ld})$ belonging to the local perception. The last one is related to the input signals denoted as $U$. $P$, $Q$ and $R$ are weighting parameters that express the importance of each term. $X(k+n|k)$ represents the terminal value of the predicted output after the horizon of prediction $n$ and $X(k+i|k)$ represents the predicted output values within the prediction horizon. The system constraints are also considered:

$$\begin{Bmatrix} |U(k+1|k)| \leq G_1 & \alpha \in (0,1] \\ \|[X_{k+n}]-[X_{ld}]\| \leq \alpha \|[X_k]-[X_{ld}]\| \end{Bmatrix} \tag{12}$$

Where $X_{k+n}$ denotes the predicted coordinates and $X_k$ the actual coordinates. The limitation of the input signal is taken into account in the first constraint. The last one is a contractive constraint (Wan, 2007). Contractive constraint arises in a convergence towards the desired trajectory; hence until a new trajectory is not commanded the control system will achieve the objective coordinates. Therefore, path planning consists in a set of points, obtained within the available field of view, and tracked by the LMPC strategy.

### 5.2 Experimental results

The trajectory tracking accuracy and time performance are two important aspects to be considered. In this context, the odometer system performance was analyzed by measuring the accuracy of the system. It was done by commanding long trajectories along lab corridors. After calibrating the odometer, the results showed that a commanded trajectory of 22m provided averaged final distance errors of less than 0.5m, and angular orientation errors of less than 5°. Hence in this research, it is analyzed local trajectories of less than 1.5m accordingly with the narrow visual perception provided. Thus, the odometer system errors can be neglected when local trajectories are considered. Therefore, the odometer system is locally used to compute LMPC trajectory tracking errors. The tested trajectories are obtained from the available set of local map coordinates as shown in Fig. 9. The LMPC results are analyzed when different trajectories tracking are commanded, as it is depicted in Fig. 11.
Denote $E_1$ as the average final error, $E_2$ the maximal average tracking error, $E_3$ the average tracking error, and $E_4$ the standard deviation of average tracking error. Table 2 presents the statistics concerning about the error obtained in cm testing the trajectories shown in Fig. 11.
It can be seen that the accuracy of trajectory tracking, when straight line is commanded, has a deviation error of 0.54cm. However, when a turning action is performed, the error in straight line tracking is bigger as consequence of the robot dynamics when it is moving forward. Hence, the forward movement consists in usually a steering action. Fig. 11 gives a clue about what is happening. Thus, the major turning angle will produce the major deviation distance. Usually, it is very difficult to reduce the approaching distance to zero, due to the control difficulty of dead zone for the WMR and to the fact that the final target is considered in the present work as being reached by the robot when the Euclidean approaching distance is less than 5cm.

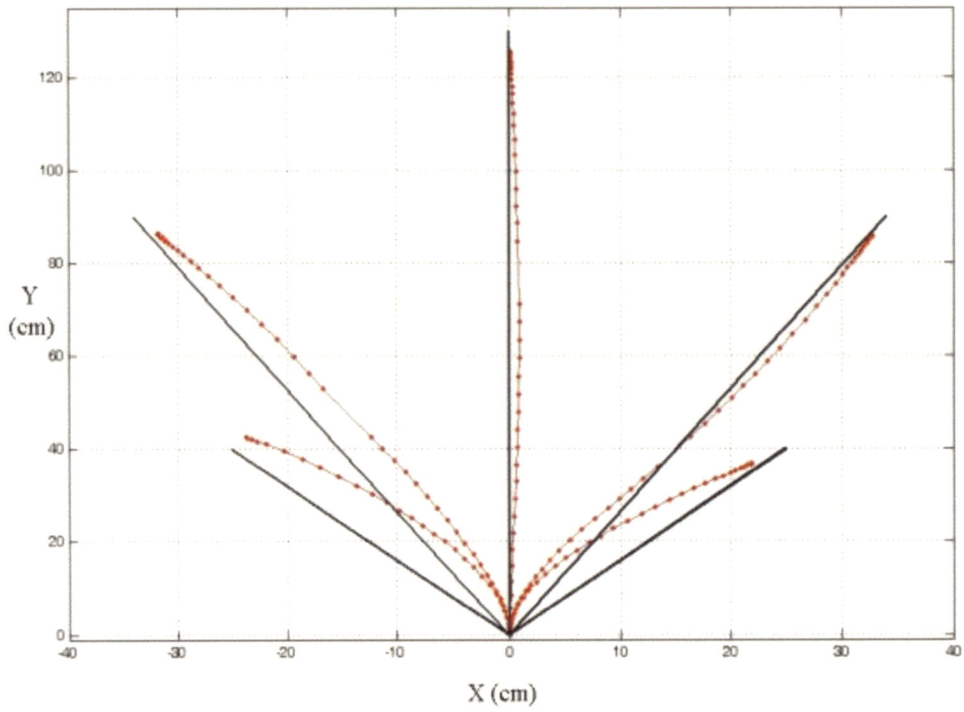


Fig. 11. Trajectory tracking tested from point to point by using the available local map coordinates provided by the monocular perception system.

| Trajectory | $E_1$ | $E_2$ | $E_3$ | $E_4$ |
|---|---|---|---|---|
| From (0,0) to (0,130) | 4.4cm | 0.9cm | 0.54cm | 0.068 |
| From (0,0) to (34,90) | 3.8cm | 3.9cm | 2.3cm | 0.82 |
| From (0,0) To (25,40) | 4.5cm | 5.3cm | 3.9cm | 1.96 |

Table 2. Point to point trajectory tracking statistics

Other interesting results consist in testing the LMPC performance when the trajectory is composed of a set of points to be tracked. In this sense, when it is regarded to the kind of robot used, a pure rotation is possible by commanding the same speed with different sense to each wheel motor. Hence, when a trajectory is composed of many points, two possibilities exist: *continuous movement* in advancing sense, or *discontinuous movement* in which the robot makes the trajectory orientation changes by turning around itself at the beginning of the new straight segment. Fig. 12 shows the tracking performance of the robot by tuning around itself, when the robot follows a trajectory composed of a set of points (0,0), (-25,50), (-25,100), (0, 150) and (0,200). The reported trajectory deviations are less than 5 cm. However, the tracking time may reach up to 25s.

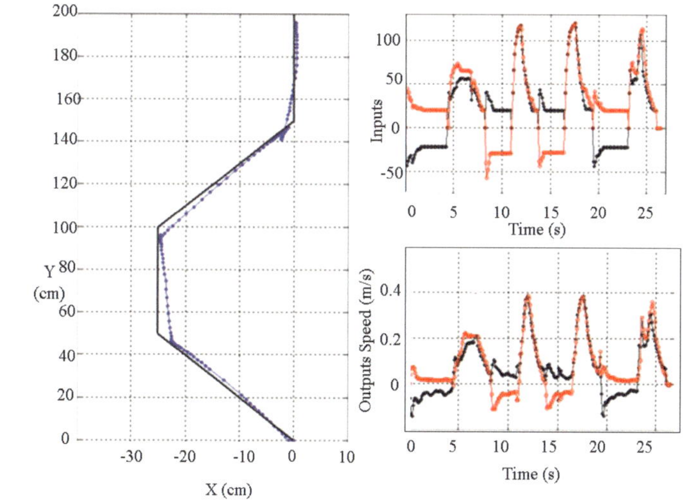


Fig. 12. Trajectory tracking with discontinuous movement

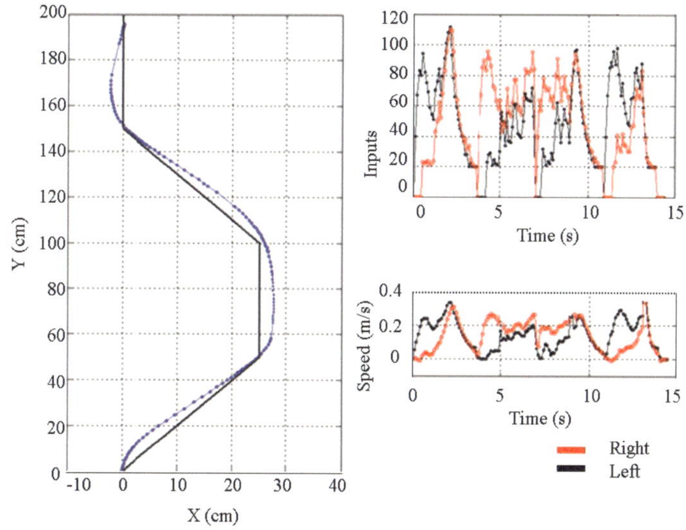


Fig. 13. Trajectory tracking with continuous movement

The trajectory tracking strategy with *continuous movement*, for a set of points (0,0), (25,50), (25,100), (0,150) and (0,200), is represented in Fig. 13. In this case, it is reported a bigger trajectory deviation, due to the WMR's mechanical dynamics. The trajectory tracking is performed much faster (≤15s). Hence, in the continuous moving case, it needs a turning action with a minimum radius, once the direction is attained the robot deviation is very small. Thus, trajectories following straight lines have reported errors less than 1cm. When time performance is analyzed the *continuous movement* presents a better behaviour. Moreover, in the research reported in this work, just *continuous movement* is possible, due to the narrow available field of view showed in Fig. 9.

## 6. Conclusions and future work

This work has integrated the control science and the robot vision knowledge into a computer science environment. Hence, local path planning by using local information is reported. One of the important aspects of the paper has been the simplicity, as well as the easy and direct applicability of the presented approaches. The proposed methodology has been attained by using the on-robot local visual information, acquired by a monocular camera system, and the techniques of LMPC. The use of sensor fusion, especially the odometer system information, is of a great importance. The odometer system uses are not just constrained to the control of the velocity of each wheel. Thus, the absolute robot coordinates have been used for planning a trajectory to the desired global or local objectives. The local trajectory planning has been done using the relative robot coordinates, corresponding to the instant when the frame was acquired. The available local visual data provide a local map, where the feasible local minimal goal is selected, considering obstacle avoidance politics.
Nowadays, the research is focused to implement the presented methods through developing flexible software tools that should allow to test the vision methods and to create locally readable virtual obstacle maps. The use of virtual visual information can be useful for testing the robot under synthetic environments and simulating different camera configurations. Further studies on LMPC should be done in order to analyze improvements such as changing the tracking set-point when the WMR is not close to the desired point or its relative performance with respect to other control laws. The influence of the motor dead zones is also an interesting aspect that should make further efforts to deal with.

## 7. Acknowledgments

This work has been partially funded by the Commission of Science and Technology of Spain (CICYT) through the coordinated project DPI-2007-66796-C03-02, and by the Government of Catalonia through the Network Xartap and the consolidated research group's grant SGR2005-01008.

## 8. References

Boyd, S. & Vandenberghe, L. (2004). *Convex Optimization*, Cambridge University Press, ISBN-13: 9780521833783

Campbell, J.; Sukthankar, R. & Nourbakhsh, I. (2004). Techniques for Evaluating Optical Flow in Extreme Terrain. *Proceedings of Intelligent Robots and Systems*, 3704-3711, ISBN 0-7803-8463-6, Sendai (Japan), September 2004.

Coué, C.; Pradalier, C.; Laugier, C.; Fraichard, T. & Bessière, P. (2006). Bayesian Occupancy Filtering for Multitarget Tracking: An Automotive Application. International Journal of Robotics Research, Vol. 25, No. 1, (Jan 2006) 19-30

DeSouza, G.N. & Kak, A.C. (2002). Vision for Mobile Robot Navigation: a survey. Patter Analysis and Machine Intelligence, Vol. 24, No. 2, (Feb 2002), 237-267

Elfes, A. (1989). Using occupancy grids for mobile robot perception and navigation, IEEE Computer, Vol. 22, No. 2, (Jun 1989) 46-57, ISSN: 0018-9162

Fox, D.; Burgard, W. & Thrun, S. (1997). The dynamic window approach to collision avoidance. *IEEE Robotics & Automation Magazine*, Vol. 4, No. 1, (Mar 1997) 23-33, ISSN 1070-9932

Gupta, G.S.; Messom, C.H. & Demidenko, S. (2005). Real-time identification and predictive control of fast mobile robots using global vision sensor. *IEEE Trans. On Instr. and Measurement*, Vol. 54, No. 1, (February 2005) 200-214, ISSN 1557-9662

Horn, B. K. P. (1998). Robot Vision. *MIT press*, Ed. McGraw-Hill, ISBN 0-262-08159-8, London (England)

Küne, F.; Lages, W. and Da Silva, J. (2005). Point stabilization of mobile robots with nonlinear model predictive control. *Proc. IEEE Int. Conf. On Mech. and Aut.*, 1163-1168, ISBN 0-7803-9044, Ontario (Canada), July 2005

Ljung, L. (1991). Issues in System Identification. *IEEE Control Systems Magazine*, Vol. 11, No.1, (Jan 1991) 25-29, ISSN 0272-1708

Maciejowski, J.M. (2002). *Predictive Control with Constraints*, Ed. Prentice Hall, ISBN 0-201-39823-0, Essex (England)

Nourbakhsh, I. R.; Andre, D.; Tomasi, C. & Genesereth, M. R. (1997). Mobile Robot Obstacle Avoidance Via Depth From Focus. *Proc. IEEE Robotics and Aut. Systems*, 151-158.

Murray R. M.; Aström K. J.; Boyd S. P.; Brockett R. W. & Stein G. (2003). Future Directions in Control in an Information-Rich World. *IEEE Control Systems Magazine*, Vol. 23, No. 2, (April 2003) 20-33, ISSN 0272-1708

Nesterov, I. E.; Nemirovskii, A. & Nesterov, Y. (1994). Interior_Point Polynomial Methods in Convex Programming. Siam Studies in Applied Mathematics, Vol 13, Publications, ISBN 0898713196

Norton, J. (1986). An Introduction to Identification. *Academic Press*, London and New York, 1986

Ögren, P. & Leonard, N. (2005). A convergent dynamic window approach to obstacle avoidance. *IEEE Transaction on Robotics*, Vol. 21, No. 2., (April 2005) 188-195

Ortega, J. M. & Rheinboldt, W.C. (2000). Iterative Solution of Nonlinear Equations in Several Variables. Society for Industrial and Applied Mathematics, ISBN-10: 0898714613

Pacheco, L. & Luo, N. (2007) Mobile Robot Local Predictive Control Using a Visual Perception Horizon, *Int. Journal of Factory Autom. Robotics and Soft Comp.*, Vol. 1, No. 2, (Dec 2007) 73-81

Pacheco, L.; Luo, N.; Ferrer, I. and Cufí, X. (2008). Control Education within a Multidisciplinary Summer Course on Applied Mobile Robotics. *Inter. 17th IFAC World Congress*, 11660-11665, Seoul (Korea), July 2008

Rimon, E. & Koditschek, D. (1992). Exact robot navigation using artificial potential functions. IEEE Transaction Robotics and Automation, Vol. 8, No. 5, (Oct. 1992) 501-518

Schilling, R.J. (1990). Fundamental of Robotics. *Prentice-Hall (Ed.)*, New Jersey (USA) 1990, ISBN 0-13-334376-6

Thrun, S. (2002). Robotic mapping: a survey. Exploring Artificial Intelligence in the New Millennium, *Exploring Artificial Intelligence in the New Millennium*, G. Lakemeyer and B. Nebel, editors. Morgan Kaufmann, San Mateo, CA

Wan, J. (2007) Computational reliable approaches of contractive MPC for discrete-time systems, *PhD Thesis, University of Girona*

# 16

# New Trends in Evaluation of the Sensors Output

Michal Pavlik, Jiri Haze and Radimir Vrba
*Brno University of Technology*
*Czech Republic*

## 1. Introduction

The area of physical quantity processing in recent sensoric applications demands for high accuracy, low distortion, fast response and low power consumption. There are many physical quantities, which could be processed, mainly pressure, temperature, humidity etc. The pressure processing is one of the most important issues for designers. There exist many problems, which have to be solved i.e. how to measure relative and absolute pressure, how to process signal from sensor with low amplitude, how to withdraw noise and parasitics and so on.
The pressure is usually measured using capacitive principle (Lee & Wise., 1982), (Chavan & Wise, 2001), (He et al., 2007). The capacitor is formed by measuring membrane and substrate, see Fig. 1. When the pressure is applied, the membrane is deformed and the distance between membrane and substrate changes. It causes the change of capacity, which is processed in next stages.

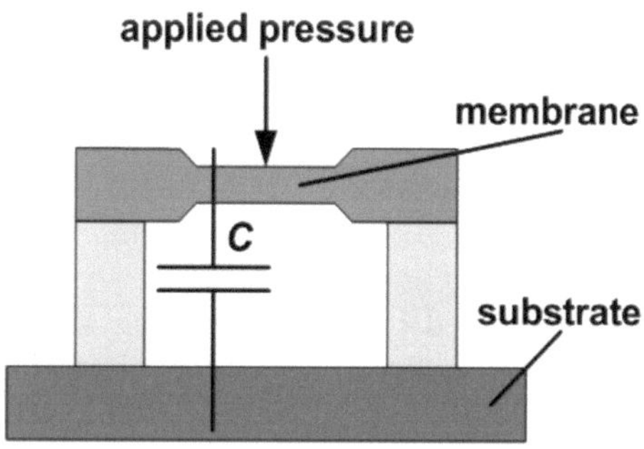


Fig. 1. Capacitive principle of pressure measurement

The capacity is defined as

$$C = \varepsilon_0 \varepsilon_r \frac{S}{d} \tag{1}$$

where $C$ is measured capacity, $\varepsilon_0$ is permittivity of vacuum, $\varepsilon_r$ is relative permittivity of the material (usually silicium), $S$ is working area of the capacitor (membrane) and $d$ is distance between membrane and substrate. This kind of structure could be modified for differential pressure measurement, where the measurement is accomplished by a pair of the sensing capacitors (Burczyk et al. 2002).

The standard differential capacity half bridge, which represents the pressure sensor, includes several parasitics (Fig. 2). There are mainly parasitic capacity $C_{DP}$, which is formed between top plate of $C_{D+}$ and bottom plate of $C_{D-}$, the inductance of the wires $L_{D+}$,$L_{D-}$ between $C_{D+}$,$C_{D-}$ and their pins and resistance $R_{D+}$,$R_{D-}$ between $C_{D+}$,$C_{D-}$ and ground. These parasitics have to be taken into account, when the processing method and circuits are designed.

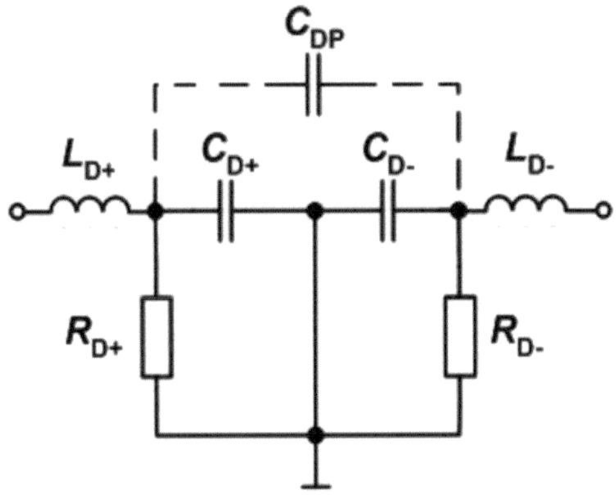


Fig. 2. Parasitics in differential pressure sensor

The ordinary value of the differential capacitors varies from units to hundreds of pF. The high measurement accuracy is important issue. Therefore the AD conversion with resolution of 16 bits or better is usually required. The measurement in order of fF is needed, because of the AD converter high accuracy. However these values are very low, so their measurement is very difficult.

Basically there are two well-known possibilities of pressure measurement. One is based on utilization of AD 7745 chip from Analog Devices Inc., which is capacitance-to-digital number converter (Whytock, 2005). The second is based on measurement using 555 timer/counter principle (Pavlik et al. 2007).

The block diagram of the principle, which uses AD 7745 chip is depicted in Fig. 3. The capacity $C_x$, which represents changes of pressure, is directly measured by the AD 7745. Since this chip includes sigma-delta converter, the accuracy of measurement is very high. The output digital signal is processed by microcontroller afterwards. Since this chip also includes temperature sensor, it is possible to realize temperature compensation of the output.

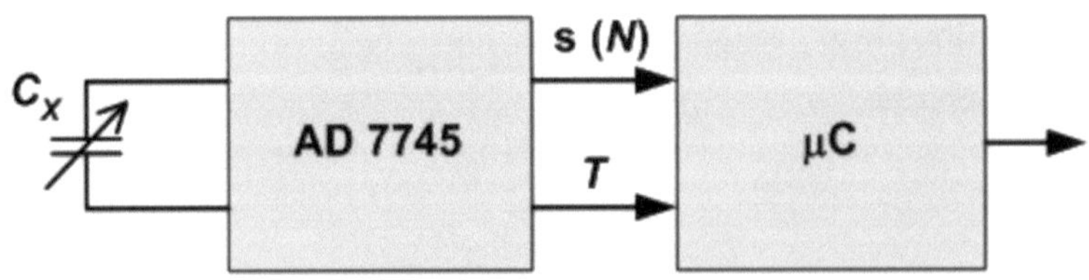


Fig. 3. The block diagram of the measurement principle using AD 7745

The method based on 555 timer/counter works as follows – Fig. 4. The signal processing of the differential pressure sensor is realized by a pair of oscillators. The oscillators are based on the two basic 555 circuits with several modifications. Their output frequencies reflect the value of the measured pressure. Consequently, galvanically separated stage, which includes microcontroller, converts the output frequency values of the oscillators onto the digital code values. The embedded microcontroller also calculates non-linear correction of the measured values and temperature calibration at the same time. The output quantity of this part of electronic circuitry is a digital calibrated value of pressure. The most important disadvantage of this method is influence of parasitic capacitance $C_{DP}$ between differential pair of measuring capacitors in sensor, which could make the 555 method unusable. It should have capacity of units of pF. Therefore the oscillators in the circuit do not measure exact value of the measuring capacity.

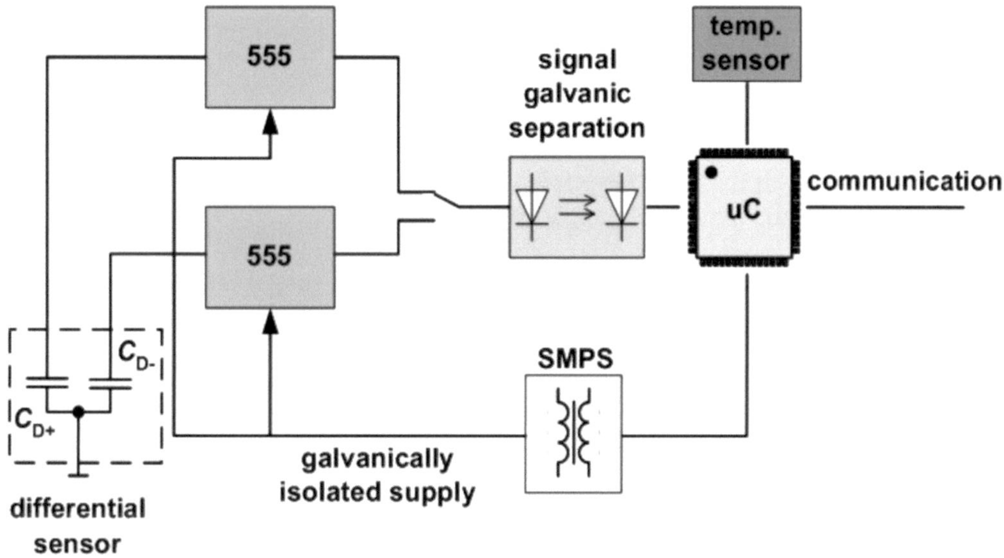


Fig. 4. The block diagram of the measurement method based on 555 timer/counter

Consequently, it is seen, that mentioned methods are relatively simple and effective, they occupy small area of the PCB with low costs. On the other hand they are intended for applications, where is no need for integration of all circuitries as System on Chip (SoC) and they suffer of parasitics influence.

There also exists one method, which is relatively new. This method uses bandpass sigma-delta modulation for pressure measurement (Ong & Wooley 1997), (Brychta, 2005), (Haze et al. 2007). The basic difference between this method and previous two techniques is integration of all functions as SoC. Fig. 5 shows the measurement chain containing a capacitive pressure sensor stage.

This block contains one sensing branch and one reference branch. The phase-locked loop circuit (PLL) provides coherent processing of the input signal in the bandpass sigma-delta modulator. The system works as follows. The pressure influences the $\Delta C_X$ capacity and humidity impacts the $\Delta G_X$ conductivity of the sensing branch. The $G_X + j\omega C_X$ complex conductance is converted by the bandpass sigma-delta modulator and two digital low pass filters into digital value of the real and the imaginary part. The digital resolution is proportional to the time window of the digital low pass filter.

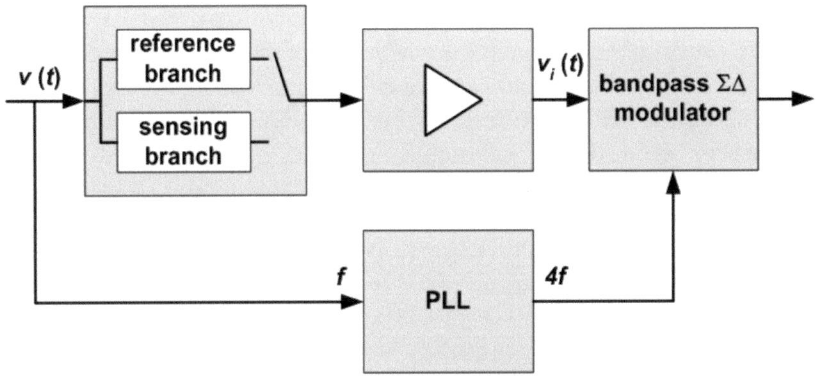


Fig. 5. Pressure sensing using bandpass sigma-delta modulation

The phase shift of the processing block is suppressed by the compensation principle. The difference of the digital values from the sensing branch of the low-pass filter normalized to the value measured in reference branch is expressed by the formula (2). Here measured changes of capacity and resistance changes are

$$\frac{v_1 - v_2}{v_3} = R(j\omega\Delta C_X + \Delta G_X) \qquad (2)$$

where $v_1$, $v_2$ and $v_3$ are voltages measured in each position of the switch, $R$ is resistance in amplifier stage. This principle is more sophisticated than previous methods. It is implemented as SoC with smaller die occupation, low power consumption and higher accuracy of the processed data. On the other hand, because this technique is quite new, some negative aspects could arise during its utilization. Probably it should be narrow area of application, because it is intended for customer defined signals and each change means a lot of changes in whole integrated circuit.
The authors would like to present another approach, which could help to improve the pressure measurement methodology.

## 2. Charge division approach

Since recent customer demands for low cost, accurate, low power and universal solution of pressure measurement and signal processing, the presented method is designed as SoC. The next reason, why the novel circuitry was designed is parasitic capacity $C_{DP}$ problem, which was discussed above.
The design and simulation of all circuitries were done using OrCAD PSPICE and CADENCE software tool with 035 µm AMIS technology.

### 2.1 The measurement system

The measurement system utilizes charge division technique on capacitors and subsequent measurement of steady voltage value on parallel combination of capacitors. The principle diagram of measuring circuitry is shown in Fig. 6. This approach requires lower power than other techniques. The main problem of presented technique is very low capacity of output capacitors $C_{D+}$ and $C_{D-}$. Small value of capacitors (typically 50 to 250 pF) leads to spontaneous looses of charge.

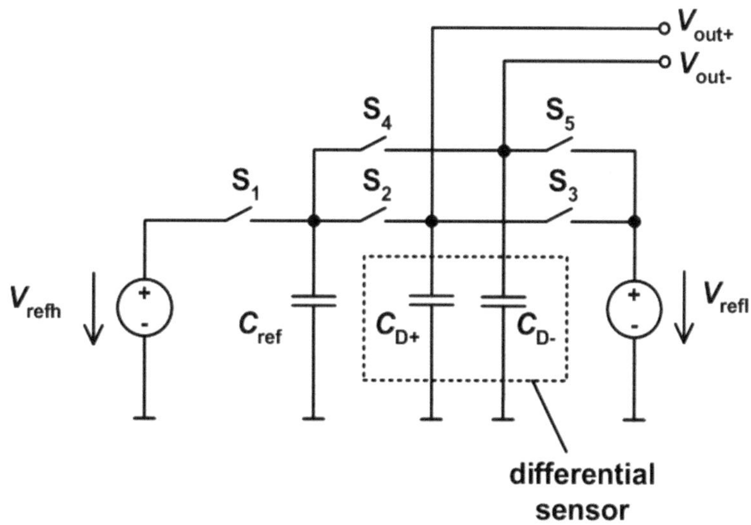


Fig. 6. The proposed charge division method

Many factors act very important role during measurement. The most crucial factors are parasitic leakage currents, charge injection through switches, noise (flicker, thermal) and temperature drift of values.

### 2.2 The description of measurement process

The whole measurement process is separated into the four phases. The first one is establishing of initial conditions, but only for $C_{D+}$ measurement. Switches $S_1$ and $S_3$ are turned on at time $t_1$. The referential capacitor $C_{ref}$ is connected to the reference voltage source $V_{refh}$ through the switch $S_1$ and charged on its voltage. The measured capacitor $C_{D+}$ is connected through switch $S_3$ to the reference voltage source $V_{refl}$ at the same time.

During the second phase the switches $S_1$ and $S_3$ are switched off and switch $S_2$ is switched on. Therefore the charge transport from the referential capacitor $C_{ref}$ on measured capacitor $C_{D+}$ is allowed. This procedure lasts until the voltages on both capacitors are equal. Now this voltage is sampled by sample and hold circuitry (S/H). The voltage is held for following processing by AD converter.

The third and the last phase are the same like mentioned procedures before with small modification. The capacity $C_{D-}$ is now measured and the working switches are $S_1$, $S_2$ and $S_3$. The output of each measurement (on $C_{D+}$ and $C_{D-}$) is then converted into digital number. These digital outputs are then subtracted, which leads to final digital value of the measured pressure.

### 2.3 The reference capacitor

The reference capacitor with high quality of dielectric has to be used. Moreover it is problem to implement capacitor with capacitance higher than tens of pF on chip. That is why it is used as external part of the chip. There should be the minimum parasitic leakage currents and the minimum properties changes depending on temperature. The very important parameter is the absolute value of the capacity of the reference capacitor.

The voltage for each differential capacitor is

$$V_{D+} = \frac{Q_{refh} + Q_{D+}}{C_{ref} + C_{D+}} = \frac{V_{refh}C_{ref} + V_{refl}C_{D+}}{C_{ref} + C_{D+}} \tag{3}$$

$$V_{D-} = \frac{Q_{refh} + Q_{D-}}{C_{ref} + C_{D-}} = \frac{V_{refh}C_{ref} + V_{refl}C_{D-}}{C_{ref} + C_{D-}} \tag{4}$$

After merging of both fractions

$$\Delta V = \frac{V_{refh}C_{ref} + V_{refl}C_{D-}}{C_{ref} + C_{D-}} - \frac{V_{refh}C_{ref} + V_{refl}C_{D+}}{C_{ref} + C_{D+}} \tag{5}$$

Fig. 7 shows dependence of voltage variation on reference capacitance and its derivative, which shows the optimal value of referential capacitor with maximum measurement voltage range. The derivative is

$$\begin{aligned} \frac{d\Delta V}{dC_{ref}} = & \frac{V_{refh}(C_{ref} + C_{D-}) + (V_{refh}C_{ref} + V_{refl}C_{D-})}{(C_{ref} + C_{D-})^2} + \\ & - \frac{V_{refh}(C_{ref} + C_{D+}) + (V_{refh}C_{ref} + V_{refl}C_{D+})}{(C_{ref} + C_{D+})^2} \end{aligned} \tag{6}$$

Finally after simplification

$$\begin{aligned} \frac{d\Delta V}{dC_{ref}} = & \frac{2V_{refh}C_{ref} + C_{D-}(V_{refh} + V_{refl})}{(C_{ref} + C_{D-})^2} + \\ & - \frac{2V_{refh}C_{ref} + C_{D+}(V_{refh} + V_{refl})}{(C_{ref} + C_{D+})^2} \end{aligned} \tag{7}$$

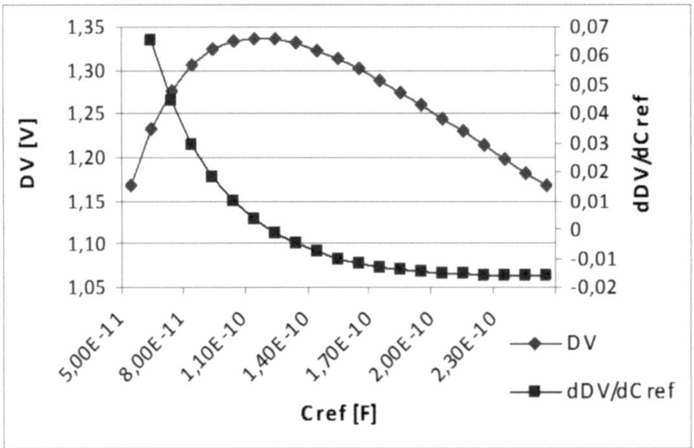


Fig. 7. Dependence of settled voltage after charge division on the value of the reference capacity and its derivative

Now it is simple to find the optimal value of $C_{ref}$ for defined voltage range. For range of measured capacities between 50 pF and 250 pF and reference voltage $U_{refh} = 3$ V the most wide range of final voltage is obtained when reference capacitor is $C_{ref} = 125$ pF. The final voltage range is from 1.33 V to 2.28 V.

The voltage step of one LSB is $V_{LSB} = 14.5$ µV for desired accuracy of 16 bits. Therefore the maximum uncorrectable error of measured voltage must not exceed 7.25 µV.

The confirmation of the formulas and the whole principle has been done in OrCAD PSPICE as shown in Fig. 8.

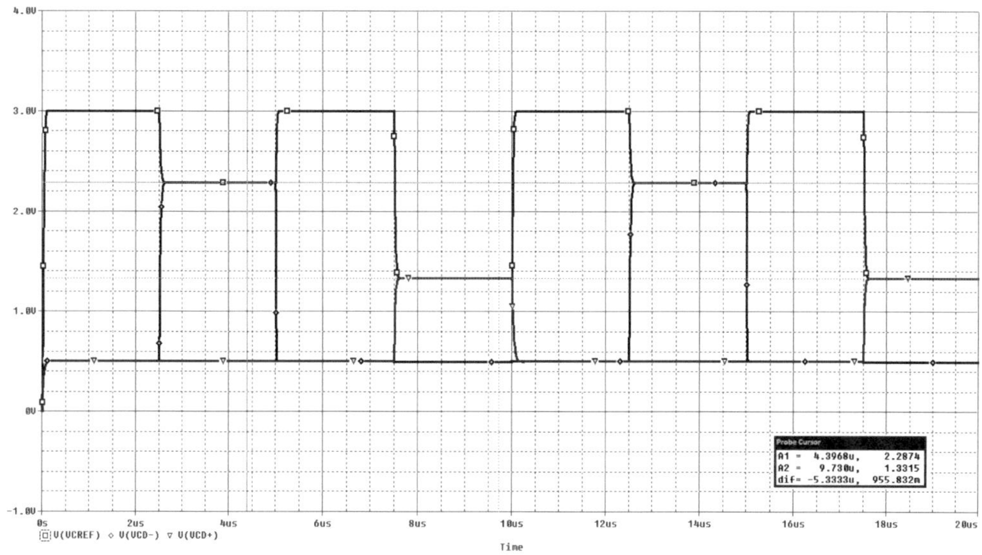


Fig. 8. The charge division method confirmation

The whole measurement process could be devided into two steps. The first includes settling of initial conditions for $C_{D+}$, measurement and charge division between reference capacitor and $C_{D+}$. Then this signal is sampled by S/H stage and processed by AD converter. Meanwhile, when the first step is in the middle of its processing period, the second step begins. The only change is that this procedure runs with $C_{D-}$.

### 2.4 The errors in proposed system

There arise various errors during measurement process. Some of them are systematic errors (i.e. offset, thermal drift of the switch parameters), which could be eliminated and corrected, for example using correction algorithm or utilizing of some improved circuit techniques (Unbehauen & Cichocki 1989), (Enz & Temes 1996). Fig. 9 shows the dependence of the measured voltage and charging time on the switch dimensions.

It is seen that switch dimensions have significant influence on conversion time and measurement voltage range. Therefore there exists the charge injection influence, too. These errors were corrected by means of dummy switch, utilizing of CMOS switch and fully differential stages (Unbehauen & Cichocki 1989), (Enz & Temes 1996).

There are also uncorrectable accidental errors. The sum of these uncorrectable errors must not exceed half of LSB. The most errors (uncertainty) are caused by the analog switches utilization in proposed measurement system shown in Fig. 6. There are mainly charge injection, parasitic capacitance and leakage current. The first one is charge injection from analog switch into the signal path. The charge is moved onto the switch during switch on phase, while during switch off phase the charge is removed from the switch.

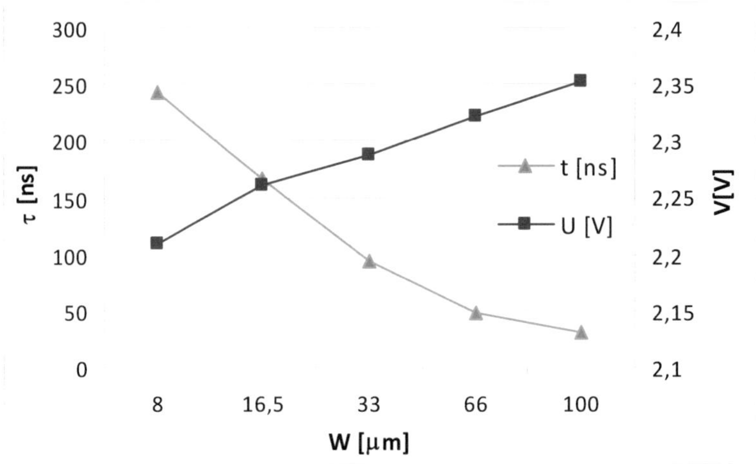


Fig. 9. The problem with switch dimensions

The next problem with real switches is the fact that the value of injected charge hardly depends on the value of the signal voltage. That is why the design shown in Fig. 5 considered this fact. The charge removing from the switches $S_1$ and $S_3$ ($S_5$) is moved on the switch $S_2$ ($S_4$). It leads to partial compensation of injected charge. It is clear, that the value of the injected charge depends mainly on dimensions of the analog switch. So it could be supposed, that the value of the injected charge is time independent or depends on time very marginally. Therefore the measurement error by the charge injection could be corrected effectively.

Another error, which is produced by the analog switch, is input parallel parasitic capacity. It is supposed that the error capacity is time independent. This capacity leads to certain capacity offset (about tens of fF) on measured capacitor.

One last important source of measurement error is leakage current of the analog switch between input and ground during switch on and switch off stage. It depends on switch dimensions and it is in order of hundreds of pA. This error is relatively well-known short time independent variable, but its impact leads to gradual discharging of parallel combination of measured and referential capacitor. The conversion time of high resolution and low power ADCs is usually from tens to hundreds of microseconds. The voltage drop on the parallel combination of the capacitors $C_{ref}$ and $C_D$ is high, which leads to worthless results of AD conversion. The S/H held the value of measured voltage during the whole measuring process. This voltage must not decrease under presented 0.5 LSB voltage $V_{HLSB}$ = 7.25 μV of ADC. The same or less drop must be ensured on parallel combination of measured and reference capacitors. There is only short time for acquisition needed, approximately $10^5$ shorter than for AD conversion.

### 2.5 Differential pressure sensor output

The output of differential pressure sensor consists of a pair of variable capacitors as has been presented. Dependence of output capacitors on differential pressure is shown in Fig. 9. The changes of applied pressure are represented by changes of membrane deflection. The basic measurement stage on Fig. 6 was extended by the second part, which serves for measuring of the second capacitor. This modification leads to self compensation of errors caused by parasitic capacities of switches and also partial self compensation of errors produced by leakage currents.

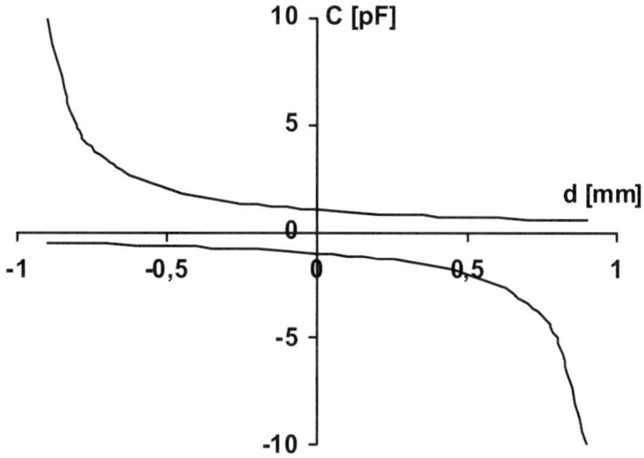


Fig. 9. The dependence of capacitors value on changes of applied pressure

### 2.6 The design of S/H stage

The input buffer was designed for S/H circuitry because of minimization of input current; otherwise the measured voltage would be influenced. The input of the buffer must have parasitic capacity as small as possible regarding the same problem. The simplified version of the sampling stage without input buffer is shown in Fig. 10. Since it utilizes switched-capacitors approach, the errors caused by this technique were considered and suppressed (Unbehauen & Cichocki 1989), (Enz & Temes 1996).

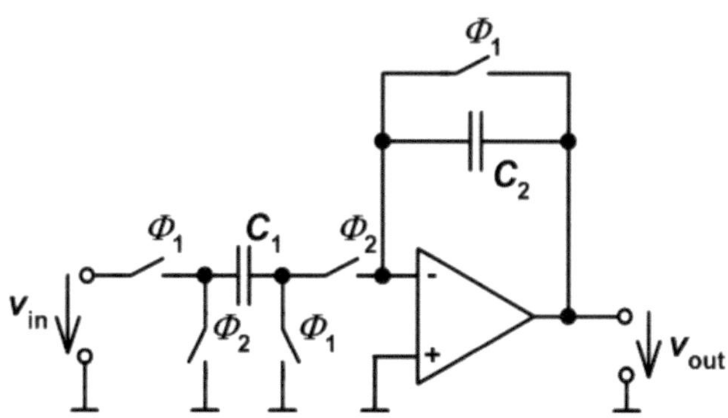


Fig. 10. The sampling stage

Since there was used only one S/H stage for both measured voltages, some additional errors (offset, temperature dependence) are partially suppressed after the processing.

### 2.7 The design digital part

The modern trend in sensors area tends to utilize a nonvolatile digital memory, which includes information about sensor and calibration values directly on chip. Thanks to this setup each used sensor is uniquely identifiable. Since the calibration values are stored also in the memory, it is possible to calibrate the measured data faster and directly utilizing calibration process. It is clear that due to flexibility requirement and accuracy of the designed device the best solution is placing the chip directly to the sensor.

Even the output of the proposed circuitry is digital output, there is another possibility of the signal processing, which is using of low power microprocessor. The microprocessor could calculate mathematical operations, which are needed for final calibration and its large memory EEPROM or FLASH would be used as storage of calibration data.

### 2.8 The calibration of the measured values

There are many ways how to calibrate the measured values. The easiest way is a correction of measured values by correction table (Okinori et al. 2003). It is relatively fast method for calibration of nonlinear course of the function, but it is not applicable for calibration, where the high resolution converters above 12-bits are used. It means that the table occupies large part of the memory.

Another fast calibration technique is linear regression (Cohen et al. 2002), but it could be used only for the linear course of the function. The modification based upon the linear regression is a piecewise linear conversion characteristic (Cohen et al. 2002). The accuracy of the calibration depends on number of linear parts. The number of linear parts depends only on character of conversion characteristic therefore individual parts would have not same length. It leads to decrease of memory seriousness and saved place, which would be used for more individual linearized parts. Therefore the most sophisticated solution is an approximation by *n*-order multinomial. It is possible to linearize the whole conversion characteristic with variable reliability till R = 99,995 %. On the other hand this feature is paid by very difficult computing of decimal numbers and power functions.

The polynomial of fifth to eighth order is used for calibration. The basic form of the polynomial is

$$y = a_0 + a_1 x + a_2 x^2 + .. + a_n x^n \tag{6}$$

The Lagrange's polynomial is used for calculation of the calibration constants. The Lagrange's polynomial is the lowest order polynomial which goes through specified values. The Lagrange's polynomial can be calculated by

$$\sum_{i=1}^{n} f(x_i)\lambda_i \tag{7}$$

where

$$\lambda_i = \frac{(x - x_1)(x - x_2)..(x - x_{i-1})(x - x_{i+1})..(x - x_n)}{(x_i - x_1)(x_i - x_2)..(x_i - x_{i-1})(x_i - x_{i+1})..(x_i - x_n)} \tag{8}$$

The calibration data are stored in on chip memory.

### 2.9 The temperature measurement and power supply

The most physical properties of materials depend on temperature variation. That is why the chip design considered this fact. The complete circuitry is proposed as temperature compensated within the range from - 40 to 70 °C.
The power supply of the whole SoC is 0 and 3.3 V and it consumes 37 mW.

## 3. Conclusion and future research

The pressure measurement nowadays tends to processing of the signal from sensor with high accuracy, low cost, low power and high effectivity. Usually the pressure sensor is formed as differential pair of measurement capacitors. There exist several methods of the signal processing, but they suffer from some unwanted effects like parasitic capacitances, inability of integration as SoC, special kind of application etc.
The chapter presents a novel method for pressure processing called charge division. This method is implemented as SoC, so it does not suffer of nonideal effects, which arise using traditional measurement approaches. The switched-capacitors errors are suppressed by means of combination of dummy switches, fully differential circuitries and CDS technique. The main advantages are its complexity, ability to be configured for customers demand, high accuracy and low power of the measurement. The disadvantage is mainly the usage of several external parts, which cannot be implemented on the chip.
The SoC is already finished and it will be fabricated. The testing and redesign will proceed in future towards higher integration a lower power consumption.

## 4. Acknowledgements

The research has been supported by the Czech Ministry of Education within the framework of the Research Program MSM0021630503 MIKROSYN and by the Czech Science Foundation under the 102/08/1116 Methods of Signal Digitizing for Advanced Sensors Project.

## 5. References

Brychta, M. Measure Capacitive Sensors with a Sigma-Delta Modulator, Electronic design online journal, April 2005, http://electronicdesign.com/Articles/Index.cfm?AD=1&AD=1&ArticleID=10185

Burczyk, D., Dannhauer, W., Nurnberger, R. Differential pressure sensor and method of differential pressure measurement. US Patent 6581469, 2002

Cohen, J., West, G., S., Cohen, P., Aiken, L. *Applied Multiple Regression/Correlation Analysis for the Behavioral Sciences* Lawrence Erlbaum Assoc Inc., 2002, ISBN 9780805822236

Enz, CH., C. & Temes, G., C. *Circuit Techniques for Reducing the Effects of Op-Amp Imperfections: Autozeroing, Correlated Double Sampling, and Chopper Stabilization. Proceedings of the IEEE,* Vol. 84, No. 11, (November 1996), 1584 - 1614, ISSN 0018-9219

Haze, J., Vrba, R., Fujcik, L., Forejtek, J., Zavoral, P., Pavlik, M. & Michaeli, L. *BandPass Sigma-Delta Modulator for Capacitive Pressure Sensor.* Proceedings of IMTC 2007 Conference. Warsaw, Poland, ISBN 1-4244-0589-0

He, F.; Huang, Q.A. & Qin, M. A silicon directly bonded capacitive absolute pressure sensor. *IEEE Sensors and Actuators A*. Vol. 135, Issue 2, (November 2007), 507 - 514, ISSN 0924-4247

Chavan, A. & Wise, K.D. A batch-processed vacuum-sealed capacitive pressure sensor, *IEEE Journal of Microelectromechanical Systems*, Vol. 10, No. 4, (Dec. 2001), 580 - 588, ISSN 1057-7157

Lee, Y. S. & Wise, K. D. A batch-fabricated silicon capacitive pressure transducer with low temperature sensitivity. *IEEE Transactions on Electron Devices*, Vol. 29, Issue 1, (January 1982), 42 - 48, ISSN 0018-9383

Ong, A., K. & Wooley, B., A. A Two-Path Bandpass SD Modulátor for Digital IF Extraction at 20 MHz, *IEEE Journal of Solid-State Circuits*, Vol. 32, No. 12, Dec., 1997, ISSN 0018-9200

Okinori, T., Shigeyasu, N., Akihiko, N. *Correction table generation method and method of controlling correction table generation apparatus*. US Patent 6945626, 2003

Pavlik, M., Haze, J., Vrba, R. & Sveda, M. *Smart Differential Pressure Sensor*. Proceedings of 4th International Conference on Informatics in Control, Automation and Robotics. Angers, France, 244 - 248, INSTICC, ISBN 978-972-8865-87-0

Unbehauen, R. & Cichocki, A. *MOS Switched-Capacitor and Continuous-Time Integrated Circuits and Systems*. Springer-Verlag, 1989, ISBN 3-540-50599-7, Berlin

Whytock, P. Capacitance and Impedance Converters - A High-Precision First. Electronic design online journal, April 2005, http://europe.elecdesign.com/Articles/ArticleID/10238/10238.html

Pavlik, M., Vrba, R., Kuban, M. Switched Current Flash Analog to DigitalConverter. In *Third International Conference on Systems*. 2008. p. 341 - 344. ISBN 978-0-7695-3105-2

# 17

# Modelling and Simultaneous Estimation of State and Parameters of Traffic System

Pavla Pecherková[1], Jindřich Duník[2] and Miroslav Flídr[2]
*[1]Research Centre Data-Algorithms-Decision Making, Dept. of Adaptive Systems, Institute of Information Theory and Automation, AV ČR, Pod vodárenskou věží 4, 182 08 Prague 8, Czech Republic*
*[2]Research Centre Data-Algorithms-Decision Making, Dept. of Cybernetics, Faculty of Applied Sciences, University of West Bohemia, Univerzitní 8, 306 14 Pilsen, Czech Republic*

## 1. Introduction

Majority of large cities suffers from traffic congestion, which results in both higher economic costs (caused by larger travel times, increased fuel consumption and more traffic accidents) and degradation of the environment (air pollution, noise, injuries). The traffic congestion forms because of insatisfactory capacity of traffic network or its insufficient exploitation.
The capacity of traffic network can be increased in several ways. These ways can be divided into three groups: (i) construction of new roads, (ii) additional traffic restrictions and (iii) improved traffic control. The first way, i.e. construction of new roads, lanes and intersections, is very complicated and expensive. The capacity is increased but at the price of vast investment costs. In historical centres of cities, constructions of new roads is even impossible because of protected historical areas. The second way is based on the decreasing number of vehicles in the traffic network by restrictions, e.g. decreasing number of entry and exit points of the traffic network. This measure is, however, highly unpopular among the drivers. The last way is deployment of a suitable traffic control system.
The main aim of traffic control is not to increase the capacity of whole traffic network but to increase troughput and thus to alleviatte the influence of bottlenecks. Bottleneck is a place (road, lane, intersection) in traffic network where the capacity is much smaller than capacity of the neighboring roads. In urban traffic networks the intersections, accidents, and closures are the main bottlenecks. The important intersections are equipped with traffic lights which can influence the behaviour of traffic flow. Thus, increasing traffic network capacity by traffic lights control is one possible way for cities with historical centres, such as Prague. The accidents and closures are time-limited restrictions and so their detection is a difficult task.
Strategies of the traffic control have been developed for more than 40 years. The first designed traffic controls were fixed control strategies (Papageorgiou et al., 2003). Fixed control strategies are based on a signal-timing plan defined by a typical daily traffic courses. Such control does not require any traffic detectors at the intersection. This type of control can be found mainly in smaller towns.

However, the pre-determined settings are not sufficient. It was found that the capacity of traffic network increases markedly when the traffic control via the signal lights is changing according to the current traffic situation. The newer approaches, so-called dynamic traffic control strategies, model the traffic situation from real time or near real time traffic information .

The first dynamic control strategies were designed to control of a single intersection only. Such controller does not co-operate with controller of the adjacent intersections. Uncoordinated traffic control can cause traffic congestion. New solution of the traffic control co-operates between adjacent intersections controllers. It means that e. g. the green wave is keeped. Around the world, several types of dynamic traffic control strategies are used. The most often used strategies are: TRANSYT (Courage, K. and Wallace, C.E., 1991; Robertson, 1969; ,Wong et al., 2000), SCOOT, SCATS (Hunt et al., 1982; Lowrie, P. R., 1982; Moor et al., 2005; Kosmatopoulos et al., 2006; Chilukuri et al., 2004) and MOTION (Pearson, 2000; Tichý, 2007). The strategy TRANSYT is based on principle of minimisation of a special criterion characterising the traffic flow. Then, the optimal signal plan is chosen from set of pre-defined signal plans according to the criterion value. The strategies SCOOT and SCATS are available for high-capacity traffic networks where each road has several lanes. The strategy MOTION is based on adaptive principle and it is designed for full co-operation between intersections. Also several other traffic control strategies have been developed, e.g. PRODYN, RHODES, TUC (Papageorgiou et al., 2003).

All types of dynamic traffic control strategies need information about the current traffic situation, i.e. about the actual traffic flow. The current traffic situation can be monitored by using several types of measurement devices, e.g. inductive detectors[1], radars and video detectors. Inductive detectors are placed under the road surface and they are activated/deactivated by passing vehicles. Radars and video detectors are usually used for detection and monitoring of vehicles at intersections. These types of traffic detectors require a long visual range and their acquisition costs are high. This limitations lead to the usage of the inductive detectors in the cities and aerial detectors are mostly used on motorways.

Most of the previously introduced dynamic control strategies are commercial tools with all associated advantages and disadvantages. The main advantage is in availability of full service. The main disadvantage is problematic adaptation of such product to the needs of a particular customer. These commercial strategies are usually geared to traffic control of big cities where the traffic network consists of the multilanes and the capacity of traffic network is huge. In the historical cities, such as Prague, these control strategies cannot be employed without problems (for example, application of the traffic control system MOTION in Prague shows its sensitivity on driver habits (Tichý, 2007)). For this reason, in Prague, a new control

---

[1] The road (or lane) can be equipped with one detector on the output and three detectors on the input: (i) detector on the stop line, (ii) outlying detector (30 m before the stop line), (iii) strategic detector - (100 m before the stop line). Ideally, each lane has all three types of the detectors but in real traffic system, the lane is usually equipped with one or two types of such detector, due to the constrained conditions e.g. in the old urban areas with short lanes. The strategic detectors, which are the most remote from a stop line, give the best information about the traffic flow at present. Unfortunately, the detector cannot provide any useful information about the real traffic flow if the queue length is greater then the distance between stop line and the detector placement.

strategy is being designed (Kratochvílová, 2004; Homolová, 2005; Pecherková et al., 2006). This traffic control strategy has been designed especially for historical urban areas, characteristic by a traffic network formed from many narrow one-way roads which are equipped mainly by the inductive detectors.

The designed traffic control strategy is based on the principle of sum of queue lengths minimization. The queue lengths are hardly measurable and predictable quantities. Currently, this quantity can be considered as unmeasured. For this reason, the queue lengths have to be estimated from available data by a suitable estimation method on the basis of proper traffic model. The main problem is thus to specify the exact model of the traffic flow behaviour on intersection or micro-region. The traffic model is considered as a non-linear state space model, where the quantity representing the queue length belongs to the state. Due to the nonlinearities in the state equation, the traffic system state is estimated by means of the nonlinear estimation methods on the basis of quantities such as intensity of the traffic flow or the occupancy measured by the detectors.

The main aim of this chapter is twofold. First, the focus will be directed to the design and in-depth description of the traffic system model. This model should be designed to properly describe the behaviour of a traffic flow on an arbitrarily complex micro-region. Second, the designed model will be validated and thus the methods for the model validation will be presented and applied.

The chapter is organised as follows. Section 2 provides a complete description of traffic model and its design. Section 3 deals with verification of a traffic model and there will be a short overview of suitable estimation methods employed for validation. In Section 4, the experiments presenting the properties and validation of the designed model using the real and the synthetic data will be shown. Finally, Section 5 comprises concluding remarks.

## 2. Traffic model design

This section will be devoted to the description of the traffic system model. First, basic quantities that describe the traffic system will be introduced. Second, the model of the traffic system based on the conservation principle will be presented. The description of the model will start with the case of a simple microregion and then it will be generalized.

### 2.1 Traffic quantities

There are many quantities that characterize the traffic system. These quantities can be divided into two basic groups:

i. Quantities determined by the intersection layout and configuration

- **Saturation flow** $S_k$ corresponds to the maximal number of vehicles flowing through the intersection arms per hour - given in $[uv/h]$, where $uv$ represents a unit vehicle. This quantity mainly depends on the road width, number of traffic lanes in one direction, and turning movements.
- **Turning movement** $\alpha_{i,j}$ is the ratio of vehicles going from the $i^{th}$ arm to the $j^{th}$ arm [%].
- **Cycle time** $t_c$ is a period of a phase change of the traffic light $[s]$.
- **Green time ratio** $z_k$ is the ratio of the effective green time to the measurement time period. Note that the green time ratio is usually defined as a ratio of the effective green time to the cycle time (Ackerman, 2000). However, such definition can lead to the possible discrepancy between the cycle time and measurement period which can lead to significant problems.

- **Offset** is the difference between the start (or end) of green at the two adjacent signalized intersection [$s$].

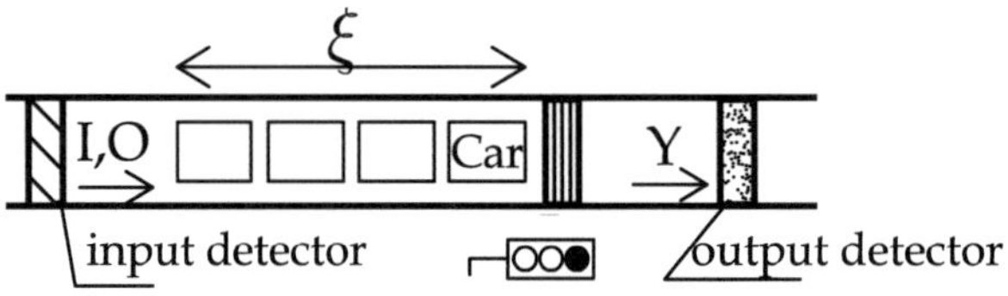


Fig. 1. Example of one way-road.

ii. Quantities describing the *traffic flow*

- **Input intensity** $\mathcal{I}_k$ or **output intensity** $\mathcal{Y}_k$ (at time instant $k$) is the amount of passing unit vehicles per hour [$uv/h$] measured by the detector placed in the input or output lane, respectively.
- **Occupancy** $O_k$ is the proportion of the period when the detector is occupied by vehicles [%].
- **Queue length** $\xi_k$ is a maximal number of vehicles waiting in one lane at period [$uv/period$][2].

Although, majority of quantities in the previous list is assumed as units per hour, in the designed traffic model all these quantities are recalculated with respect to the respective time period of the measurement. It means, that saturation flow $\mathcal{S}_k$ is given in unit vehicle per hour ([$uv/h$]), but in the model the corresponding recalculated quantity $S_k$ is given in unit vehicle per period ([$uv/period$]). The same goes for the input intensity $\mathcal{I}_k$ and the output intensity $\mathcal{Y}_k$, respectively. As the model is discrete then the dependence on period can be omitted, i.e. the quantities are given only in their respective units.

## 2.2 Traffic model

The fundamental idea of the described traffic model design technique is based on the traffic flow conservation principle. It means that the queue at time $k$+1 is equal to the sum of the previous queue at time $k$ and input intensity minus output intensity from the arm.

**Simple micro-region**

For the sake of simplicity the proposed technique for traffic system modelling will firstly be shown on the simple micro-region. This micro-region comprises a road with one input and one output detector and one traffic light as it is depicted in Figure 1. The traffic situation at time instant $k$ is completely described with the state $x_k$. The state $x_k$ is formed from the queue length $\xi_k$, the input intensity $I_k$ and the occupancy $O_k$. The measurement vector $y_k$ is in the considered micro-region formed from the input intensity $I_k$ and occupancy $O_k$ measured on the strategic detector, the output intensity $Y_k$ measured on output detector and the intensity $I_k^{SL}$ measured on the stop-line detector.

The state-space model is given as follows:

---

[2] The queue length can be also considered in meters. The length of unit vehicle is supposed to be 6 m.

$$x_{k+1} = \begin{bmatrix} x_{1,k+1} \\ x_{2,k+1} \\ x_{3,k+1} \end{bmatrix} = \begin{bmatrix} \xi_{k+1} \\ I_{k+1} \\ O_{k+1} \end{bmatrix} = \begin{bmatrix} \xi_k + I_k - I_k^{\pi}(\xi_k, I_k, V_k) + w_{1,k} \\ I_k + w_{2,k} \\ \kappa_k \xi_k + \beta_k O_k + \lambda_k + w_{3,k} \end{bmatrix} =$$

$$= \begin{bmatrix} x_{1,k} + x_{2,k} - I_k^{\pi}(x_{1,k}, x_{2,k}, V_k) \\ x_{2,k} \\ \kappa_k x_{1,k} + \beta_k x_{3,k} + \lambda_k \end{bmatrix} + \begin{bmatrix} w_{1,k} \\ w_{2,k} \\ w_{3,k} \end{bmatrix}, \tag{1}$$

$$y_k = \begin{bmatrix} y_{1,k} \\ y_{2,k} \\ y_{3,k} \\ y_{4,k} \end{bmatrix} = \begin{bmatrix} I_k \\ O_k \\ Y_k \\ I_k^{SL} \end{bmatrix} = \begin{bmatrix} x_{2,k} \\ x_{3,k} \\ I_k^{\pi}(x_{1,k}, x_{2,k}, V_k) \\ I_k^{\pi}(x_{1,k}, x_{2,k}, V_k) \end{bmatrix} + \begin{bmatrix} v_{1,k} \\ v_{2,k} \\ v_{3,k} \\ v_{4,k} \end{bmatrix}. \tag{2}$$

The queue length $\xi_{k+1}$ is given by the queue at previous time instant $\xi_k$, the input intensity $I_k$ representing the number of arrived cars and by the function $I_k^{\pi}(\xi_k, I_k, V_k)$ describing the number of passing vehicles. The occupancy $O_{k+1}$ depends on the occupancy and the queue length at previous time instant and on the parameters $\kappa_k$, $\beta_k$, and $\lambda_k$ which cannot be exactly determined from physical properties of a micro-region and they are generally unknown. The remaining state variable $I_k$ is modelled as a random walk. The probability density functions (pdfs) of the state noises $w_{i,k}$ and measurement noises $v_{i,k}$ are currently supposed to be zero mean with unknown covariance matrices. The state and measurement noises are supposed to be mutually independent and independent of the traffic system initial state. Note that in this simple example the measured intensity on the output detector $Y_k$ and on the stop-line detector $I_k^{SL}$ are the same. These intensities are different for micro-regions with more than one arm.

The function $I_k^{\pi}(\xi_k, I_k, V_k)$ represents the number of departed vehicles and depends on three quantities: (i) the queue length $\xi_k$, (ii) the input intensity $I_k$ and (iii) the maximal number of passing vehicles $V_k$ which can pass through the intersection in the measurement period. In short, the function represents a continuous approximation of the theoretical throughput of the intersection as it is shown in Figure 2. For details, see (Pecherková et al., 2007). The function $I_k^{\pi}(\xi_k, I_k, V_k)$ has following form

$$I_k^{\pi}(\xi_k, I_k, V_k) = V_k \cdot \left(1 - e^{-\frac{1}{V_k}(\xi_k + I_k)}\right) = V_k \cdot \left(1 - e^{-\frac{1}{V_k}(x_{1,k} + x_{2,k})}\right). \tag{3}$$

The maximal number of passing vehicles $V_k$ is given by the saturation flow $S_k$ and the green time ratio $z_k$. The quantity $V_k$ can be written as

$$V_k = S_k \cdot z_k. \tag{4}$$

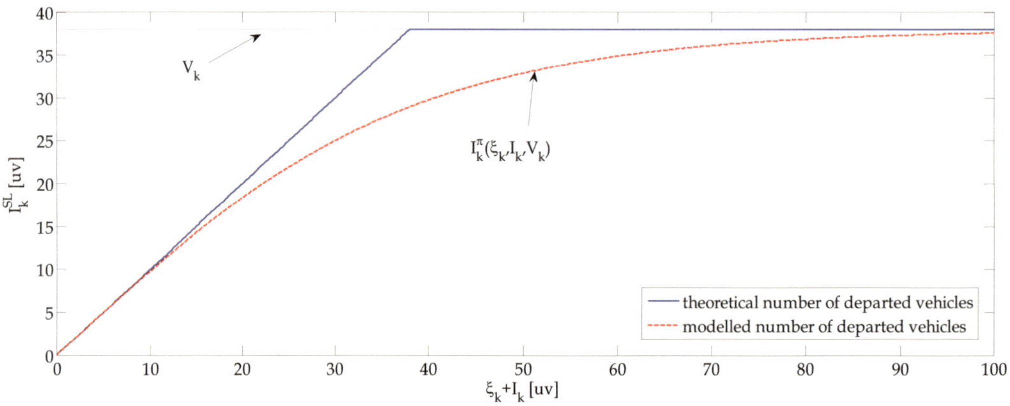


Fig. 2. Theoretical and modelled number of the departed vehicles.

In equation (4), it can be seen that saturation flow $S_k$ is supposed to be time-variant. In majority of the traffic control strategies, the saturation flow is assumed as time-invariant because working with time-variant one is more complicated. In this chapter, the saturation flow is assumed to be time-variant. Precondition of the time-invariant saturation flow is inaccurate because the actual saturation flow depends on the both invariant and time-variant quantities. The following relation shows how the saturation flow at time instant $k$ will be evaluated

$$S_k = \left(S_0 - \alpha^R c_{1,k} - \alpha^L \left(c_{2,k} + c_{3,k} I_k^P\right)\right) \cdot \gamma_k \cdot PHF. \tag{5}$$

The actual saturation flow $S_k$ depends on the theoretic saturation flow $S_0$, ratio of heavy vehicles $\gamma_k$ in the average traffic flow, right and left turning movements $\alpha^R$ and $\alpha^L$, respectively, the parameters of the traffic flow behaviour $c_{1,k}$, $c_{2,k}$, $c_{3,k}$, intensity of oncoming vehicles $I_k^P$ and peak hour factor $PHF$.

The theoretic saturation flow $S_0$ is time-invariant and depends on width of roads, speed limit and the shape of an intersection. The time-variant turning movements are not directly measurable. However, it is possible to find their typical daily values by means of analysis of measured and simulated data.

The last time-invariant traffic quantity is the peak hour factor *PHF*. This factor respects the fact that in oversaturation the vehicles have to start, stop and brake more often than usually. This behaviour reduces the speed of the traffic flow and the capacity of the road.

The ratio of heavy vehicles $\gamma_k$ and parameters of the traffic flow behaviour $c_{1,k}$, $c_{2,k}$, $c_{3,k}$ belong to the group of time-variant quantities and parameters. The parameter $\gamma_k$ describes the ratio between heavy and light vehicles (Kara, K. & Shabin, R., 2000). Heavy vehicles have different driving properties than light vehicles and so the traffic flow has different behaviour for the different ratios. The ratio of heavy vehicles $\gamma_k$ is an unmeasured quantity but unlike turning movement, the changes are not so significant and so it is possible to estimate this factor relatively well. The parameters of the traffic flow behaviour $c_{1,k}$, $c_{2,k}$

and $c_{3,k}$ can compensate faulty estimation of the actual turning movements. This fault is caused by precondition of time-invariant turning movement. These three parameters model the reduction of speed and the number of passing vehicles according to the turning of vehicles. The parameter $c_{1,k}$ models the right turning movement and the apriori setting of this parameter is given by typical turning movement and radius of right turn. The same function has parameter $c_{2,k}$ which models the left turning movement. The last parameter $c_{3,k}$ models intensity of oncoming vehicles with respect to the left turning movement. For example, strong left turning movement in combination with high oncoming intensity can cause saturation flow to be several times smaller than the theoretic saturation flow. In the micro-region described by relations (1) and (2), the actual saturation flow is identical to theoretical saturation flow since no turning movements and oncoming intensities.

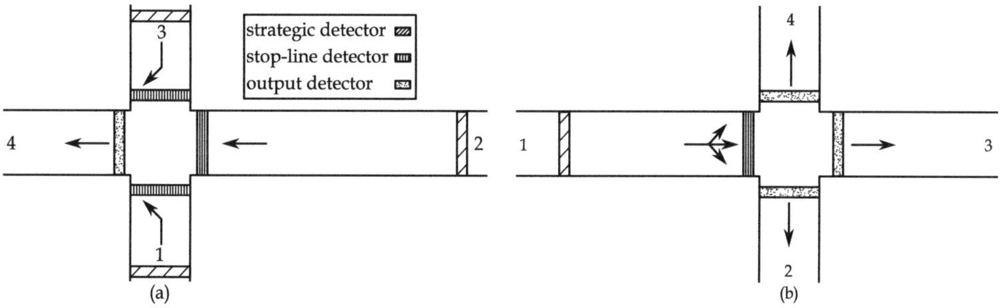


Fig. 3. Outline of more complex micro-regions: (a) four arms intersection with three input arms and one output arm, (b) four arms intersection with one input arm and three output arms.

**More complex micro-regions**

To illustrate the situation when either the actual and theoretical saturation flows are not the same or the intensities $I_k^{SL}$ and $Y_k$ are different, the model design for two other micro-regions will be shown. The first micro-region consists of one intersection with three input arms and one output arm, see Figure 3a. The second micro-region consists of one intersection with one input arm and three output arms, see Figure 3b. All arms are one-way with one lane only.

The first micro-region consists of one intersection with three input arms and one output arm. The arms number 1, 2 and 3 are the inputs arms and the arm number 4 is the output arm. This system is not uncommon in historical centres of citites. Such system can work with two- or three-phrase control. In this case, the two-phase control is used, in the first phase the arms number 1 and 3 have green simultaneously and in the second phase the arm 2 has green. The influence of the green time on the traffic model is not seen in the state or the measurement relations explicitly. The green time influences the nonlinear function $I_k^{\pi}(\xi_k, I_k, V_k)$ and is used for computation of the quantity $V_k$.

The construction of the model of such micro-region is based on the modelling of each arm separately and then on the modelling of particular relations among all arms. In this particular micro-region, the model has the dimension of the state $dim(x_k) = 9$ and dimension of measurements $dim(y_k) = 10$. The relation describing the state update is then given as

$$x_{k+1} = \begin{bmatrix} x_{1,k+1} \\ x_{2,k+1} \\ x_{3,k+1} \\ x_{4,k+1} \\ x_{5,k+1} \\ x_{6,k+1} \\ x_{7,k+1} \\ x_{8,k+1} \\ x_{9,k+1} \end{bmatrix} = \begin{bmatrix} \xi_{1,k+1} \\ \xi_{2,k+1} \\ \xi_{3,k+1} \\ I_{1,k+1} \\ I_{2,k+1} \\ I_{3,k+1} \\ O_{1,k+1} \\ O_{2,k+1} \\ O_{3,k+1} \end{bmatrix} = \begin{bmatrix} x_{1,k} + x_{4,k} - I^{\pi}_{1,k}(\cdot) + w_{1,k} \\ x_{2,k} + x_{5,k} - I^{\pi}_{2,k}(\cdot) + w_{2,k} \\ x_{3,k} + x_{6,k} - I^{\pi}_{3,k}(\cdot) + w_{3,k} \\ x_{4,k} + w_{4,k} \\ x_{5,k} + w_{5,k} \\ x_{6,k} + w_{6,k} \\ \kappa_{1,k} x_{1,k} + \beta_{1,k} x_{7,k} + \lambda_{1,k} + w_{7,k} \\ \kappa_{2,k} x_{2,k} + \beta_{2,k} x_{8,k} + \lambda_{2,k} + w_{8,k} \\ \kappa_{3,k} x_{3,k} + \beta_{3,k} x_{9,k} + \lambda_{3,k} + w_{9,k} \end{bmatrix} \quad (6)$$

and the measurement vector is given as

$$y_k = \begin{bmatrix} y_{1,k} \\ y_{2,k} \\ y_{3,k} \\ y_{4,k} \\ y_{5,k} \\ y_{6,k} \\ y_{7,k} \\ y_{8,k} \\ y_{9,k} \\ y_{10,k} \end{bmatrix} = \begin{bmatrix} I_{1,k} \\ I_{2,k} \\ I_{3,k} \\ O_{1,k} \\ O_{2,k} \\ O_{3,k} \\ Y_{4,k} \\ I^{SL}_{1,k} \\ I^{SL}_{2,k} \\ I^{SL}_{3,k} \end{bmatrix} = \begin{bmatrix} x_{4,k} + v_{1,k} \\ x_{5,k} + v_{2,k} \\ x_{6,k} + v_{3,k} \\ x_{1,k} + v_{4,k} \\ x_{7,k} + v_{5,k} \\ x_{8,k} + v_{6,k} \\ \alpha_{1,4} I^{\pi}_{1,k}(\cdot) + \alpha_{2,4} I^{\pi}_{2,k}(\cdot) + \alpha_{3,4} I^{\pi}_{3,k}(\cdot) + v_{7,k} \\ I^{\pi}_{1,k}(\cdot) + v_{8,k} \\ I^{\pi}_{2,k}(\cdot) + v_{9,k} \\ I^{\pi}_{3,k}(\cdot) + v_{10,k} \end{bmatrix}. \quad (7)$$

The intensities $I^{\pi}_{i,k}(\cdot)$ represent the number of passing vehicles from the $i^{th}$ arm where $i = 1,2,3$ and the turning movements $\alpha_{i,4}$ are equal to one because the traffic flow does not divide into several streams.

The second micro-region describes more usual situation where the traffic flow from one arm is divided into several streams. In this traffic system, the input intensity $I_k$ is divided into three output intensities $Y_{1,k}$, $Y_{2,k}$ and $Y_{3,k}$ according to particular turning movements $\alpha_{i,j}$. The subscript $i$ denotes the number of the input arm (in this case, $i$ =1) and $j$ the number of output arm.

In this case, the state update and measurament are described by the relations

$$x_{k+1} = \begin{bmatrix} x_{1,k+1} \\ x_{2,k+1} \\ x_{3,k+1} \end{bmatrix} = \begin{bmatrix} \xi_{1,k+1} \\ I_{1,k+1} \\ O_{1,k+1} \end{bmatrix} = \begin{bmatrix} x_{1,k} + x_{2,k} - I^{\pi}_{1,k}(\cdot) + w_{1,k} \\ x_{2,k} + w_{2,k} \\ \kappa_{1,k} x_{1,k} + \beta_{1,k} x_{3,k} + \lambda_{1,k} + w_{3,k} \end{bmatrix} \quad (8)$$

$$y_k = \begin{bmatrix} y_{1,k} \\ y_{2,k} \\ y_{3,k} \\ y_{4,k} \\ y_{5,k} \\ y_{6,k} \end{bmatrix} = \begin{bmatrix} I_{1,k} \\ O_{1,k} \\ Y_{2,k} \\ Y_{3,k} \\ Y_{4,k} \\ I_{1,k}^{SL} \end{bmatrix} = \begin{bmatrix} x_{2,k} + v_{1,k} \\ x_{3,k} + v_{2,k} \\ \alpha_{1,2} I_{1,k}^{\pi}(\cdot) + v_{3,k} \\ \alpha_{1,3} I_{1,k}^{\pi}(\cdot) + v_{4,k} \\ \alpha_{1,4} I_{1,k}^{\pi}(\cdot) + v_{5,k} \\ I_{1,k}^{\pi}(\cdot) + v_{6,k} \end{bmatrix} \tag{9}$$

The turning movements $\alpha_{i,j}$ influence not only the output intensity but also the saturation flow. The turning movements are assumed to follow the typical daily values and their sum have to be equal to one.
The previously described micro-regions can form together typical two-ways four-arm intersection with one lane in each direction. In such a case, the resulting model is obtained by simple coupling of the two previous models, i.e the three input arms with one output arm and one input arm with three output arms. Such traffic model has then the dimension of state $dim(x_k) = 12$ and dimension of the measurement $dim(y_k) = 16$. Usually the micro-region consists of 3 - 4 four-arm controlled intersections and equal number of uncontrolled intersections. Typically, uncontrolled intersections are not equipped with the detectors and so the intensities and occupancies are unmeasured quantities and they have to be estimated as well as queue lengths. Note that more complex micro-region will be discussed in Section 4.

## 3. Validation of the traffic model

In the previous section, a technique for the model design of an arbitrary micro-region was introduced. The aim of this section is to present procedures suitable for validation of the traffic system models.
Two criteria for validation of the models designed by means of the proposed technique are considered. The first one compares the "true" system state with its estimate. The second criterion compares the measured and predicted system output.

### 3.1 Validation via state estimation

The validation is based on the comparison of the true state of the traffic system with its estimate. The true state of the traffic system, particularly the unknown part of the true state representing the immeasurable queue lengths, can be determined by simulation software AIMSUN[3]. The estimates of the traffic system state can be found using various estimation techniques. However, the quality of the state estimates produced by the techniques strongly

[3]AIMSUN is a simulation software tool which is able to reproduce the traffic condition of any traffic network. It is mainly used for testing new traffic control system and management strategies, but it can be also used for traffic state prediction and other real time applications. The validation and calibration process of the simulator was made with respect to the particularities of the local traffic system. The validation of the queue length reconstruction was made on several types of micro-regions in Prague in accordance to the guidelines specified in AIMSUN (AIMSUN: Users manual, 2004).

depends on the accuracy of the traffic system model. The considered state estimation techniques are briefly described in the following text.
The aim of the state estimation is to find an estimate of the state $x_k$ conditioned by the measurements $y(k)=[y_0, y_1, \ldots, y_k]$ up to the time instant $k$. The state estimate is usually given by the conditional pdf $p(x_k \mid y(k))$ or by the conditional mean $\hat{x}_{k|k} = E[x_k \mid y(k)]$ and covariance matrix $P_{k|k} = cov[x_k \mid y(k)]$.
Utilisation of the state estimation methods is conditioned by the complete knowledge of the model. However, the nonlinear model presented in previous section contains several unknown parameters, in both the "deterministic" and the "stochastic" part, which cannot be determined from the physical properties of the traffic system, namely parameters $\beta_k, \kappa_k, \lambda_k$ and the statistics of the state and measurement noises. Therefore, the unknown parameters have to be identified somehow.
Generally, there are two possibilities how to estimate the state and the parameters in the deterministic part of the system. The first possibility is based on an off-line identification of the unknown parameters, e.g. by prediction error methods (Ljung, 1999), and subsequently on an on-line estimation of the state by the nonlinear state estimation techniques. However, off-line identified parameters represent average values rather than the actual (true) parameters and this approach is therefore suitable for traffic systems where intensity of the traffic flow is almost constant. The second possibility is based on the simultaneous on-line estimation of the state and the parameters by extension of the state with vector of the unknown parameters (Wan et al., 2000). This leads to the extended nonlinear model of the traffic.
There are two main groups of the nonlinear estimation methods, namely local and global methods. Although, the global methods are more sophisticated than local methods, they have significantly higher computation demands. Due to the computational efficiency, the stress will be mainly laid on the derivative-free local filter methods, namely the divided difference filter first order (Nørgaard et al., 2000). The comparison of various other local filtering methods (Julier & Uhlmann, 2004; Mihaylova et al., 2006; Hegyi at al., 2006) in traffic area can be found in e.g. (Pecherková et al., 2007).
The application of the local methods is also conditioned by the knowledge of the second-order statistics of the state noise $w_k$ and the measurement noise $v_k$. The state and measurement noise covariance matrices, can be hardly determined from the physical properties of the traffic system and they have to be estimated somehow. The noise covariance matrices can be generally estimated online or offline. Due to the extensive computational demands of online noise covariance matrices estimation methods (Mehra, 1972; Verdú & Poor, 1984), they were estimated offline by the estimation technique based on the multi-step prediction (Šimand & Duník, 2008) for both nonlinear models and extended nonlinear models for various intensities of the traffic flow.
Note that the state estimation techniques with state inequality constrains were used (Simon & Simon, 2003; Simon & Chia, 2002) in order to ensure that the state quantities are in an admissible region of the state space. The admissible region is defined on the basis of the physical consideration, for example the queue lengths cannot be negative.
The true and estimated queue lengths are compared via the root mean square error criterion

$$J^S = \sqrt{\frac{\sum_{k=0}^{K}\sum_{i=1}^{n_q}(x_{i,k} - \hat{x}_{i,k|k})}{(K+1)\times n_q}} \tag{10}$$

where $n_q$ is the number of estimated queue lengths, $x_{i,k}$ represents the $i^{th}$ component of the state, $\hat{x}_{i,k|k}$ its filtering estimate, and $K$ is the number of measured data. Thus, the $J^S$ represents an average error of the queue length estimate on one arm at one sample period. The value of the criterion depends not only on the applied estimation technique, but also on the quality of the proposed traffic model.

### 3.2 Validation via system output prediction

The state estimate $\hat{x}_{k-t|k-t}$ can be used to compute the $t$-step prediction of the output $\hat{y}_{k|k-t}$. The multi-step prediction can be obtained as multiple application of the one-step prediction which is an essential part of the state estimation algorithms (Šimand & Duník, 2006).

The multi-step prediction of the output depends not only on the measured output data up to the time instant $k-t$, i.e. $y(k-t)=[y_0, y_1, \ldots, y_{k-t}]$, and measured input data up to the time instant $k$, i.e. $u(k)$, but it also strongly depends on the quality of the model. Thus, the quality of the traffic system output can be validated also according to the following criterion

$$J^P_{y_i,t} = \frac{\sum_{k=t}^{K}(y_{i,k} - \hat{y}_{i,k|k-t})}{(K-t+1)\times n_y}, \tag{11}$$

where $y_{i,k}$ represents the $i^{th}$ component of the measurement and $\hat{y}_{i,k|k-t}$ is the $t$-step prediction of $y_{i,k}$. The criterion compares the measured data with predicted data, which are computed according to the model.

## 4. Experiments

In previous two sections, the traffic model design technique was introduced as well as the procedures for the model validation. The aim of this section is to demonstrate the behaviour and validity of the two designed models.

The real data, namely the input and output intensities, the occupancies and the cycle and green times, were collected on real places in the Prague traffic system. However, for the model validation the queue lengths are also needed. The queue lengths cannot be directly measured and thus they were syntetized using calibrated traffic simulator AIMSUN.

First, the validation of model of one way four-arm intersection where one arm is input arm and another arms are output arms will be presented. Second, typical two-way four-arm intersection will be shown. This type of intersection is very frequent intersection in the traffic networks all over the world.

### 4.1 One-way four-arm intersection

One-way four-arm intersection can be designed in different combinations of inputs and outputs. In the particlar case of this example, it is assumed that this intersection has one input and three ouput detectors, see Figure 3b. The basic model describing this intersection is given by the equations (8) and (9). For estimation and prediction the model is augmented with the unknown time-variant traffic parameters $\kappa_{1,k}$, $\beta_{1,k}$ and $\lambda_{1,k}$ and uknown traffic flow behaviour parameters $c_{1,k}$, $c_{2,k}$ and $c_{3,k}$.

This example demonstrates the quality of the model in case of the standard traffic flow on the arm 1, i.e without congestion. The model of this intersection will be validated using the criteria (10) and (11). The criterion $J^S$ evaluates only the root mean square error between the true and the estimated queue lengths on arm 1. The criterion $J^P$ is evaluated for the following quantities:

- The input intensity $I_1$ and its one-step prediction.
- The input occupancy $O_1$ and its one- and two-step prediction.
- The intensity on the stop-line $I_1^{SL}$ and its one-step prediction.
- The output intensity $Y_2$ and its one-step prediction.

The resulting criteria values are presented in Table 1. This table shows the actual values of the criterion accompanied with the maximal values. The maximal values are given in order to illustrate the scale of the errors with respect to the actual amplitudes of the corresponding quantities.

| | $J^P_{I_1=y_1,t=1}$ | $J^P_{O_1=y_2,t=1}$ | $J^P_{O_1=y_2,t=2}$ | $J^P_{I_1^{SL}=y_3,t=1}$ | $J^P_{Y_2=y_4,t=1}$ | $J^S$ |
|---|---|---|---|---|---|---|
| $J$ | 0.4331 | 0.0805 | 0.1027 | 0.9783 | 1.3279 | 3.5897 |
| Max. value of $y_i$ or $x_i$ | 19 | 18.08 | 18.08 | 21 | 6 | 40 |

Table 1. The values of criterion (10) for the input intensity $I_1$, output occupancy $O_1$, output intensity $Y_2$ and the intensity on the stop-line $I_1^{SL}$ and criterion (11) for the queue length. The last row contains maximal values of the corresponding quantities in the state or measurement.

Figure 4 depicts the typical daily courses of the queue length and occupancy on the arm 1. From the figure and from the ratio of the criterion and maximal value in Table 1 it can be seen that the quality of the estimates based on the designed model is adequate. So the model (8), (9) sufficiently accurately describes the considered micro-region.

### 4.2 Two four-arm intersections

In the second example, the micro-region comprises two interconnected four-arm intersection. Each of the intersection arms is two-way road. The micro-region is outlined in Figure 5. The model of each intersection in this micro-region is given by simple coupling of the two models described by relations (6), (7) and (8), (9). The output quantities on the arm 3 are equal to the input quantities on the arm 5 and vice versa. Again as in the previous example the model is augmented with the unknown time-variant traffic parameters.

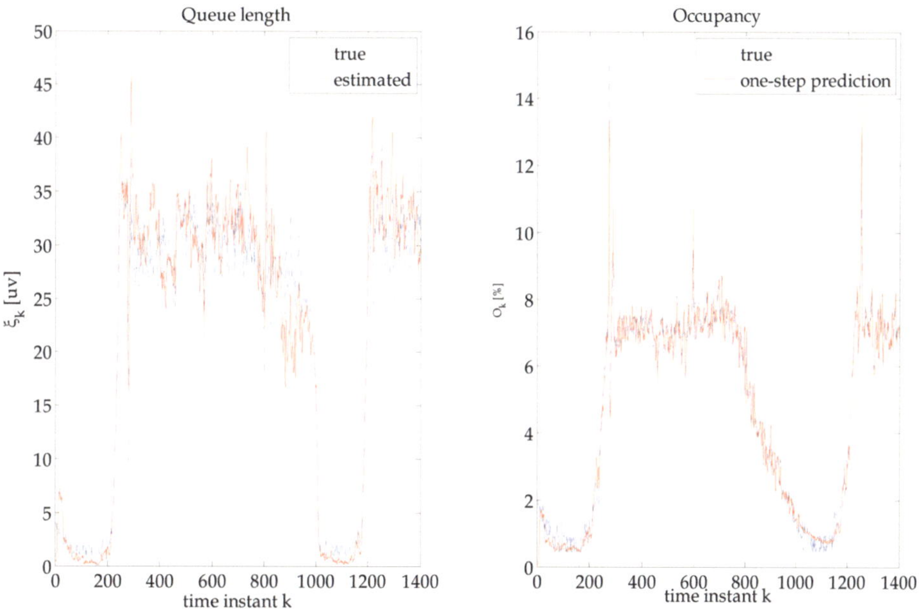


Fig. 4. An example of the true and estimated queue lengths and the true and predicted occupancy for the intersection arm 1.

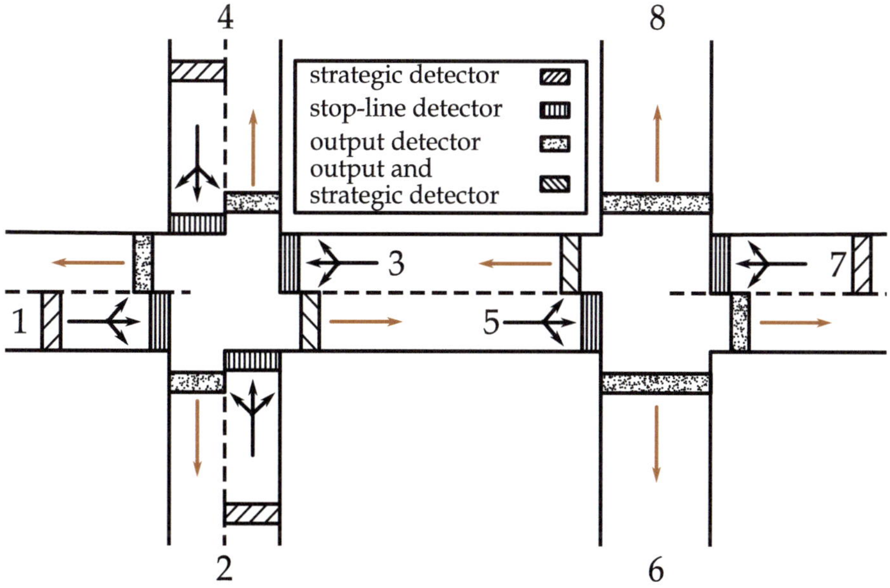


Fig. 5. Outline of two four-arm intersection.

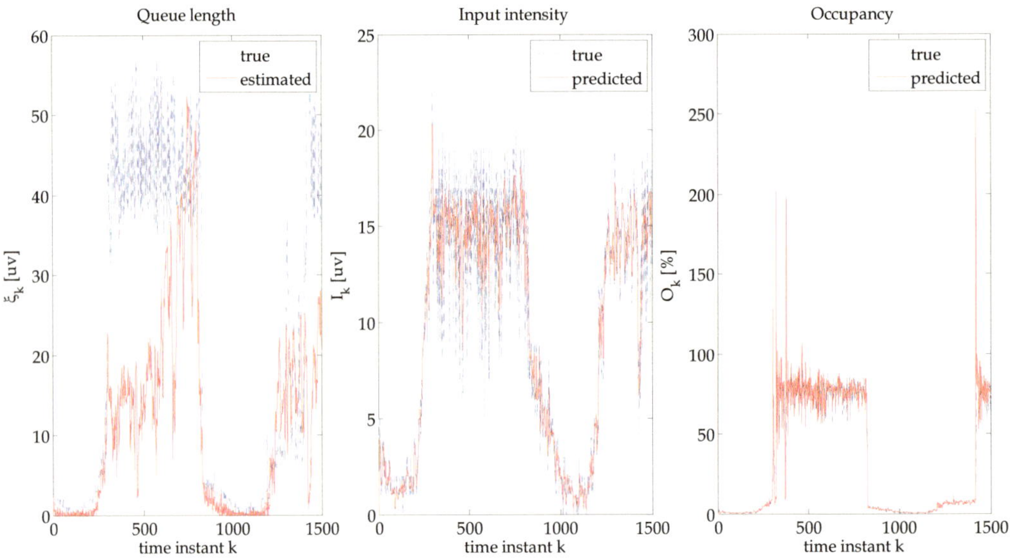


Fig. 6. The true and estimated queue length and the true and predicted input intensity and occupancy, respectively, on arm 1 (in the oversaturated situation the queue reaches over the strategic detector).

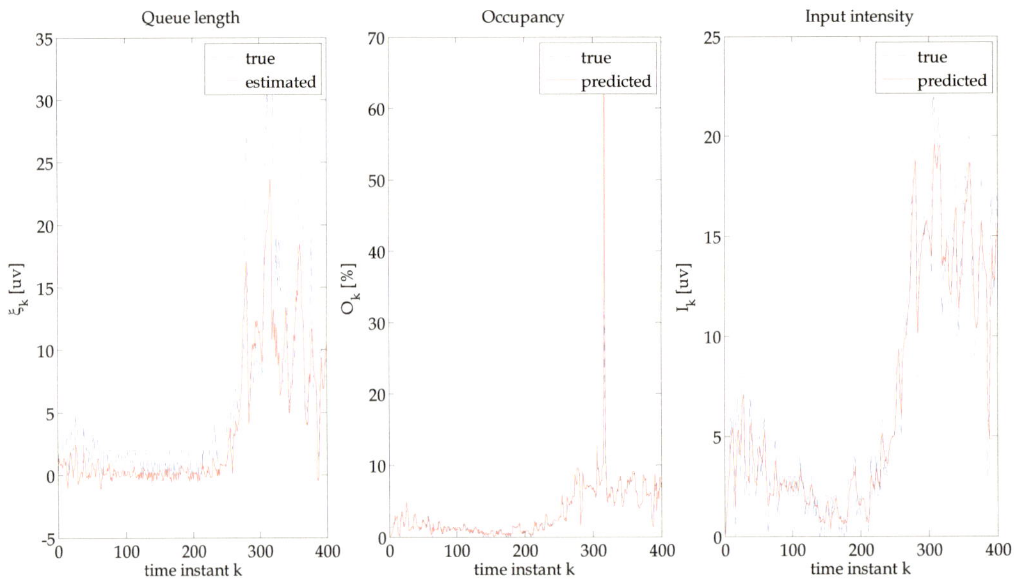


Fig. 7. The true and estimated queue length and the true and predicted input intensity and occupancy, respectively, on arm 3 (standard traffic situation, i.e. without oversaturation).

In this case some of the quantities do not always have the standard behaviour. On arms 1 and 7 the oversaturated traffic flow is occurring, i.e. the real queue often reaches over the strategic detector. This exceptional state can be verified on the basis of the occupancy. If the occupancy is higher than 70-80% then the traffic flow can be considered to be oversaturated (Gazis, D. C., 1964; Green, D. H., 1967). Then, the quantities measured by the traffic detectors do not contain any information about the real number of arrived vehicles. Therefore, the quality of the estimated and predicted quantities is significantly worsened. The situation with oversaturated traffic flow on arm 1 can be found in Figure 6. Note that in this example the state estimation techniques without state equality constrains were used. Therefore, the predicted occupancy could be greater than 100%.

| | $\frac{1}{n_q}\sum_i J^P_{I_i,t=1}$ | $\frac{1}{n_q}\sum_i J^P_{O_i,t=1}$ | $\frac{1}{n_q}\sum_i J^P_{Y_i,t=1}$ | $\frac{1}{n_q}\sum_i J^P_{I_i^{SL},t=1}$ | $J^S$ |
|---|---|---|---|---|---|
| $J$ (all arms) | 2.3468 | 33.4146 | 4.4227 | 2.7443 | 9.2226 |
| $J$ (arms 2, 3, 4, 5, 6 and 8 only) | 2.0566 | 11.7842 | - | 2.2878 | 4.6485 |
| Max. value of $y_i$ or $x_i$ | 26 | 92.6427 | 26 | 25 | 61 |

Table 2. The values of criterion (10) for the input intensities $I_1 - I_6$, output occupancies $O_1 - O_6$, output intensities $Y_1 - Y_8$ and the intensities on the stop-lines $I_1^{SL} - I_6^{SL}$ and criterion (11) for the queue lengths. The last row contains maximal values of the corresponding quantities in the states or measurements.

On the other hand, in the standard (not oversaturated) situation, the measured quantities by the detector represent the number of arrived vehicles quite well and thus the quality of the estimated queue length and the predicted intensity and occupancy is significantly better, see Figure 7, where the traffic flow on arm 3 is illustrated.

For completeness Table 2 presents the validation results for this micro-region. Contrary to the previous example the criteria are evaluated over quantities for all arms and for all arms with standard, i.e. not oversaturated, traffic flows. The last table row presents again the maximal values of the corresponding quantities.

## 5. Conclusion

This chapter deals with the model design of traffic systems. The main aims were to introduce the modelling technique and to present validation methods for the designed models.

The proposed modelling technique is based on the traffic flow conservation principle. The use of the technique was illustrated on three simple common micro-regions that can constitute base for larger and more complex micro-regions. For the purposes of the model validation two criteria based on estimated state and predicted system output were proposed. Two designed models were validated using these two criteria on base of true and synthetic data.

The illustrating examples showed that the proposed models represent a sufficiently accurate description of the traffic systems.

The future work will be geared toward simpler and more straightforward description of the occupancies and toward the use of the proposed modelling technique for purposes of traffic control system.

## 6. Acknowledgement

This work was supported by the Ministry of Education, Youth and Sports of the Czech Republic, project No. 1M0572.

## 7. References

Aboudolas, K.; Papageorgiou, M. & Kosmatopoulos, E.(2007). A quadratic programming approach for the signal control problem in large-scale congested road networks. *Proceedings of the Sixth Triennial Symposium on Transportation Analysis*, pp. 1-6, Phuket Island, Thailand, June 2007.

Ackerman, N. A. (2000). Highway capacity manual. Transportation research board, Washington. D.C., USA.

AIMSUN: Users manual, (2004). *Transport Simulation Systems*. http://www.aimsun.com/, 2004.

Chilukuri, B. R.; Perrin, J. & Martin, P. T. (2004). SCOOT and incidents - performance evaluation in simulated environment. *Freeway Operations and Traffic Signal Systems*, Vol. 1867, 2004, pp. 224-232, ISSN: 0361-1981.

Courage, K. and Wallace, C.E. (1991). TRANSYT-7F User's Guide, Federal Highway Administration, Washington, D.C.

Duník, J. & Šimandl, M (2008). Estimation of state and measurement noise covariance matrices by multi-step prediciton. *In preprints of 17th IFAC World Congress*, Seoul, Korea, July 2008.

Gazis, D. C. (1964). Optimum Control of a System of Oversaturated Source. *Operations Research*, Vol. 12, No. 6, Special Transportation Science Issue, pp. 815-831.

Green, D. H. (1967). Control of Oversaturated Intersections. *Operation Research*, Vol. 18, No. pp. 161-173.

Hegyi, A.; Girimonte, D.; Babuška, R. & De Schutter, B. (2006). A comparsion of filter configurations for freewaytraffic state estimation. In *Proceedings of the 2006 IEEE Intelligent Transportation Systems Conference* (ITSC 2006), pp. 1029-1034, Toronto, Canada, September 2006, ISBN: 1-4244-0094-5.

Homolová, J. & Nagy, I. (2005). Traffic model of a microregion. *Preprints of the 16th IFAC World Congress*, pp. 1-6, Prague, Czech Republic, July 2005, ISBN: 0-08-045108-X.

Hunt, P. B.; Robertson, D. I.; Bretherton, R. D. and Royle, M. C. (1982). The SCOOT on-line traffic signal optimization technique. *Traffic Eng. Control*. Vol.23, pp 190-192.

Julier, S. J. & Uhlmann, J. K. Unscented filtering and nonlinear estimation (2004). *Proceedings of* IEEE, Vol. 92, No. 3, Dec 2004, pp. 401-421, ISSN 0018-9219.

Kara, K. & Shabin, R. (2000). Effect of vehicle type on the capacity of signalized intersection: the case of light-duty trucks, *Journal of Transportation Engineering*, vol. 126, no. 6, pp. 506-512.

Kosmatopoulos, E. and et al. (2006) International comparative field evaluation of a traffic-responsive signal control strategy in three cities. Transportation research part A - policy and practice, Vol. 40, No. 5, pp. 399-413, 2006. ISSN 0965-8564

Kratochvílová, J. & Nagy, I. (2004). Traffic control of microregion. *CMP'04: Multiple Participant Decision Making, Theory, algorithms, software and applications*, ed.

Andrýsek, J. et al., pp. 161-171, Adelaide, May 2004. Advanced Knowledge International.

Ljung, L. (1999). *System identification: theory for the user*. UpperSaddle River, ISBN: 0-13-656695-2, NJ: Prentice-Hall.

Lowrie, P. R. (1982). SCATS: The Sydney co-ordinated adaptive traffic system - Principles, methodology, algorithms. In *Proceedings of IEE Int. Conf. Road Traffic Signalling*. London. England. Pp. 67-70.

Mehra, R.K. (1972). Approaches to adaptive filtering. *IEEE Transactions on Automatic Control*, Vol. 17, No. 10, pp. 693-698, ISSN: 0018-9286.

Mihaylova, L.; Boel, R. & Hegyi, A. (2006). Freeway traffic estimation within particle filltering framework. *Automatica*, Vol. 43, No. 2, pp. 290-300, ISSN: 0005-1098.

Moor, J.E.; Mattingly, S.P.; MacCArley, C.A. & McNally, M.G. (2005). Anaheim advanced traffc control system filed operation test: A technical evaluation of SCOOT. *Transportation Planning and Technology*, Vol. 28, No. 6, pp. 465-482, ISSN: 0308-1060.

Nørgaard, M.; Poulsen, N. K. & Ravn, O. (2000). New developments in state estimation for nonlinear systems. *Automatica*, vol. 36, No. 11, pp. 1627-1638, ISSN: 0005-1098.

Papageorgiou, M.; Diakaki, C.; Dinopoulou, V.; Kotsialos, A. & Wang, Y. (2003). Review of road traffic control strategies. *Proceedings of* IEEE, Vol. 91, No. 12, 2003, pp. 2043-2067, ISSN 0018-9219.

Pearson, R. (2000). *Traffic signal control*. ITS Decision Report, 2000.

Pecherková, P. & Duník, J. (2006). Using derivative-free methods for estimation of intersection queue lengths. In *Proceedings of the 11th International Conference of Hong Kong Society for Transportation Studies*, Sustainable Transportation, pp. 739, Hong Kong, December 2006. Hong Kong Society for Transportation Studies.

Pecherková, P.; Homolová, J. & Duník, J.(2007). Estimation of state and parameters of traffic system. *Proceedings of the Fourth International Conference on Informatics in Control, Automation and Robotics*, pp. 223-228, Angers, France, May 2007.

Robertson, D. I. (1969). *Transyt: a traffic network study tool*. Technical report, 1969.

Simon, D. & Chia, T. L. (2000). Kalman filtering state equality constrains. *IEEE Transactions on Aerospace and Electronic Systems*, vol. 38, No. 1, pp. 128-136, ISSN: 0018-9251.

Simon, D. & Simon, D. L. (2003). Aircraft turbofan engine health estimation using constrained Kalman filtering. *ASME Turbo Expo 2003*, pp. 1-8, Atlanta, GA, USA, June 2003.

Tichý, T. (2007). Evaluation methodology for traffic control system. *Preprints of the ITS Prague '07* (in Czech), Prague, Czech republic, May 2007, ISBN: 978-80-239-93111-0.

Verdú, S. & Poor, H.V. (1984). Minimax linear observers and regulators for stochastic systems with uncertain second-order statistics. *IEEE Transactions On Automatic Control*, vol. 29, No. 6, pp. 499-511, ISSN: 0018-9286.

Šimandl, M. & Duník, J. (2006). Design of derivative-free smoothers and predictors. *Preprints of the 14th IFAC Symposium on System Identification*, pp. 991-996, Newcastle, Australia, March 2006.

Wan, E. A.; van der Merwe, R. & Nelson, A. T. (2000). Dual estimation and the unscented transformation. *Advances in Neural Information Processing Systems* 12, ed. Solla, S. A. et al., pp. 666-672, MIT Press, ISBN: 0-262-19450-3.

Wong, S. C., Wong, W. T.; Xu, J. & Tong, C. O. A time-dependent TRANSYT traffic model for area traffic control. In *Traffic and Transportation Studies 2000*, Beijing, China.

# 18

# A Human Factors Approach to Supervisory Control Interface Improvement

Pere Ponsa[1], Ramon Vilanova[2], Marta Díaz[3] and Anton Gomà[4]
[1]*Automatic Control Department, Technical University of Catalonia*
[2]*Telecomunication and Systems Engineering Dept., Universitat Autònoma de Barcelona*
[3]*4all-L@b Usability Laboratory, Technical University of Catalonia*
[4]*S.A.F. Sports Service Area, Universitat Autònoma de Barcelona*
*Spain*

## 1. Introduction

In recent years, control systems and the role of control room human operators have changed dramatically. Human operator activity has evolved from manually performing the process, to control system supervision. Today, the human operator requires an in-depth knowledge of the process that he/she is overseeing and the ability to make effective decisions within demanding constraints.

The increased complexity of industrial process control calls for a new methodological approach (for research and design purposes), which reproduces the essential components of current control systems: the environment, the task at hand and human operator activity (Samad & Weyrauch, 2000).

The complexity of industrial process supervision makes it necessary to supplement the Control Engineering approach with a cross-disciplinary cooperation in order to integrate knowledge and methods from other fields, especially Human Factors, Human-Computer Interaction and Artificial Intelligence (Granollers et. al., 2005), (Holstom, 2000), (Petersen, 2000). Our view is that complete control systems engineering must encompass all these approaches.

Human Factors is concerned with the adaptation of technology to suit human operator need and ability so as to achieve effectiveness, efficiency and user/worker satisfaction and comfort (Wickens et. al., 1997). In the domain of human factors in control room design the aim is to help engineers to design better systems and processes, increase operational safety and efficiency (Noyes & Bransby, 2001). One of the main references in telerobotics, automation and human supervisory control is Thomas B. Sheridan, who had an influence on professionals of the areas of engineering of systems and human factors along two decades (Sheridan, 1992). One of this author's main ideas is that the progresses in robotics do not only depend on the changes in the technology, but also on the advances in the understanding of the relationship between humans and machines. One example is the development of human interaction for multiple robotic systems that incorporate intelligent decision support capabilities via distributed artificial intelligence (Adams, 2006).

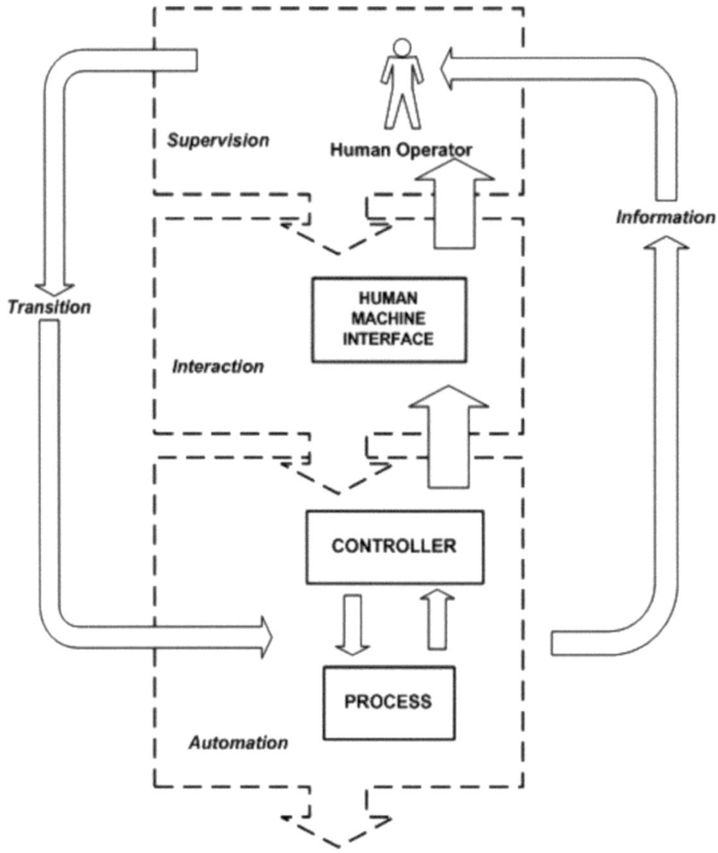


Fig. 1. A human supervisory control scheme in automation systems

Supervisory control is the set of activities and techniques developed over a set of controllers (programmable logic controllers and process controllers) which ensures the fulfilling of control goals. One of the main goals is to prevent possible plant malfunctions that can lead to economical lose and/or result in damage (Petersen & May, 2006). For this reason, other fields of knowledge concerned with manufacturing systems performance - such as maintenance and industrial security - are complementary in the study of supervision systems.

In Automation, Petri Nets (PN) have proved to be a successful approach on a broad range of applications (Kontogiannis, 2005). There is however one point that is not clears how to deal with within the PN formalism: the introduction of the operator (Lee & Hsu, 2006). Effectively, in every automation problem the fully automated part is just one part of the solution. It is customary that the operator can enter the loop in different ways, and such interaction needs to be considered as an integral part of the automation procedure as well as the communication of the automation device (usually a Programmable Logic Controller) with the operator. A human machine interface (computer display, industrial panel) is the connection between the human operator action and the input to algorithm control inside the controller (control based PC, control based PLC), (Ponsa et. al., 2007).

To these ideas it is necessary to add that the human and the interface cannot be considered in an isolated way, it is necessary to value that both, the human and the interface are in an

environment, for example the industrial environment of a petrochemical refinery. The human operator doesn't always understand or even feel the need to understand all the relationships behind the complex processes that they control via their interfaces. Furthermore, the human operators are not always aware of the constraints that affect the system that they work with, and discovering these constraints can take some extra effort (Burns & Hajdukiewicz, 2004). In this sense, the Ecological Interface Design EID incorporates this constraint based style in the design.
The cognitive ergonomics already includes this last idea inside the current conceptual framework named cognitive system engineering, in which it is described how the human centered design concept should be wrapped up from the perspective of the contextual design (Cañas, 2004), (Rasmussen et. al., 19994).
In this chapter a methodology for the creation of a human factor guideline for supervisory control interface design is proposed. In section 2 we present a checklist of indicators of the guideline called 'ergonomic guideline for supervisory control interface design' (GEDIS Guia ergonómica para el diseño de interfaz de supervision in Spanish version). The Sports Service Area project is described in section 3. The purpose is not to cover with detail the entire project but to give an idea of the different kind of topics that have been covered. In section 4, transition from the GEDIS model to Sports Service Area interface in control room is evaluated. In this section, a set of recommendations about graphical interface improvement are studied. Finally, conclusions and future research lines.

## 2. GEDIS guideline

Ergonomic criteria have been proved to increase the evaluation performance of experts (Reason, 1990). The previous research on human interface design guidelines includes for example the standard ISO 11064 that establishes ergonomic principles for the evaluation of control centers (ISO, 2004), the Human Factors Design Standards HFDS of the Federal Aviation Administration of the United States (Federal Aviation Administration, 1996), the Human Interface Design Review Guidelines NUREG 0700 in nuclear power plants (Nuclear Regulatory Commission, 2002), the I-002 Safety and Automation Systems NORSOK about Norwegian petroleum industry (Norsok, 2006) and the Man Systems Integration Standard NASA-STD-3000 about manned space programs (NASA, 1995).
An example of cognitive modelling in human computer interaction is the *Goals, Operators, Methods and Selection Rules* guideline, GOMS, in usability analysis (Card et. al., 1983). In combination with Keystroke-Level Model KLM an interface can be studied, also task execution time and human efficiency can be studied too.
In order to establish a relationship between ergonomics, usability and design process there is a variety of expert –review methods:

- Heuristic evaluation (the eight golden rules of interface design)
- Guidelines review (organising the display, getting the user's attention)
- Consistency inspection (the experts verify consistency of terminology, color, layout, etc.)
- Cognitive walkthrough (the experts simulate users walking through the interface to carry out typical tasks)
- Formal usability inspection (Palanque et. al., 2007)

A classical set of criteria for interface design to improve usability are the Ergonomic Criteria (Bastien & Scapin, 1995) and the Nielsen's Heuristic Guidelines (Nielsen, 1994).
The GEDIS guide is a method that seeks to cover all the aspects of the interface design (Ponsa & Díaz, 2007). From the initial point of view of strategies for effective human-computer interaction applied to supervision tasks in industrial control room (Nimmo, 2004), (Schneiderman, 1997).
The GEDIS guide offers design recommendations in the moment to create the interface. Also, already offers recommendations of improvement of interfaces created. The GEDIS guide is composed of two parts: description of ten indicators and measure of ten indicators. The indicators have been defined from extracted concepts of other generic human factors guidelines, and for aspects of human interface design in human computer interaction.
The method to continue for the use of the GEDIS guide is: analyze the indicator, measure the indicator, obtain the global evaluation index and finally offer recommendations of improvement.
For the correct use of the GEDIS guide it is necessary the collaboration between the control room technical team and the human factor technician, since in some cases to analyze the indicator is necessary the expert's opinion.
The subsection 2.1 shows a brief description of some relevant indicators, and the subsection 4.1 shows the measure of the ten indicators.

### 2.1 Indicators list

The GEDIS guide consists of ten indicators that seek to cover all the aspects of the interface design in the supervisory control domain. The indicators are: structure, distribution, navigation, color, text font, status of the devices, process values, graphs and tables, data-entry commands, and finally alarms. For example, the relationship between architecture and navigation indicators is illustrated in Fig. 1. The physical plant can separate in area, subarea, and local processes control[1]. In the same way, the interface presents four navigation levels. Fig. 2 shows a possible layout to locate all the connections between screens. The connection among screens is complex in a supervisory control interface. From the point of view of human computer interaction, is a typical example of cyclic network menu.
Distribution indicator of Fig. 3 shows a possible layout to locate all the objects inside the screen. The objects homogeneous distribution allows us to maintain the interface coherence when user changes the screen. The secondary objects are located in screen areas that don't require the user's attention (enterprise logo, and date/hour information). The user should recognize the screen title and the general navigation tool to move among screens. The main objects are located in visible screen areas (alarms, data-entry commands, subnavigation tool, and synoptic objects). The user can surveillance the process evolution without acting (human out of the loop), or he can decide to introduce changes in the set point or in the controller's parameters (human in the loop) inside a faceplate window in the data-entry command object. The user should have special attention to the alarm indicators, which should be located in a clear way in the screen so that the user can recognize the situation (situation awareness).

---

[1] Each process control has is own structure. In example a sugar mill has the following parts: diffusion, evaporation, purification, sugar room, boilers, dryer, liquor storage, crystallisation

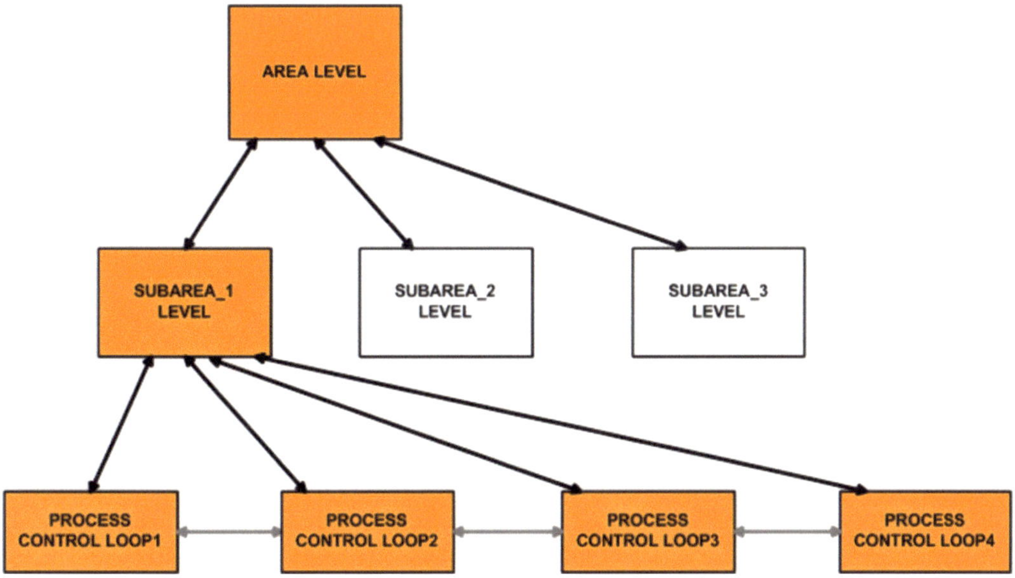


Fig. 2. A typical cyclic network menu in supervisory control interface associated to structure indicator. The arrow shows the navigation between screens.

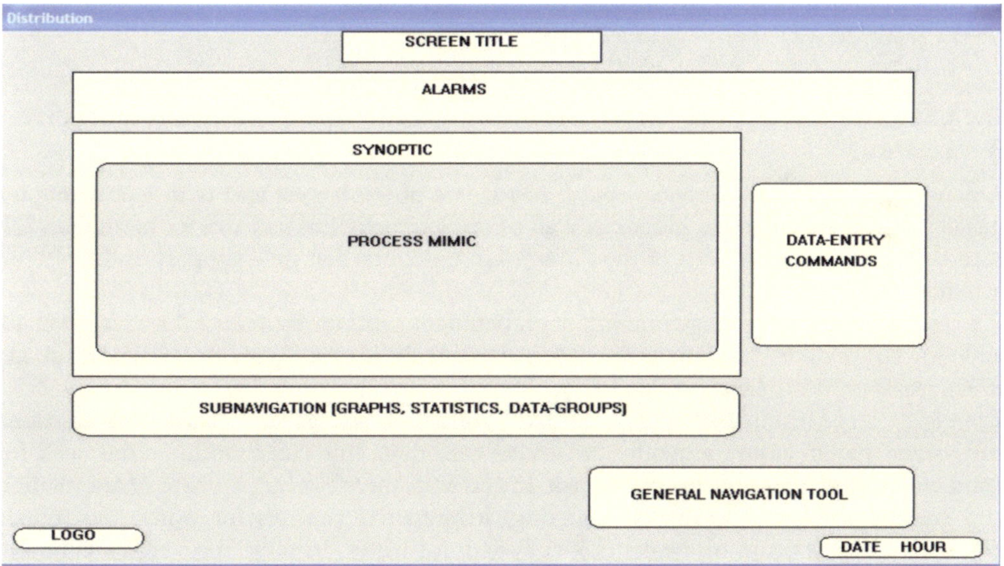


Fig. 3. An example of object's layout inside the screen for the distribution indicator

## 3. S.A.F. Project

This section presents the development of the supervisory control system, with special emphasis on the interface features, for the Sports Service Area (SAF in Catalan version) of the Universitat Autònoma of Barcelona (UAB). This supervisory control system has been developed by a team of Computer Science Engineers with common design guidelines. Even some basic principles on ergonomics and interface design were taken into account; the

GEDIS analysis will show existing weakness. An alternative presentation of the SAF project can be found in (Vilanova & Gomà, 2006).
First of all, it is worth to know that the UAB is a campus based university with more than 40.000 inhabitants (students, academics, staff, etc. In fact, this makes the University campus to behave like a city with some sort of facilities offered for their inhabitants (see Fig. 4).

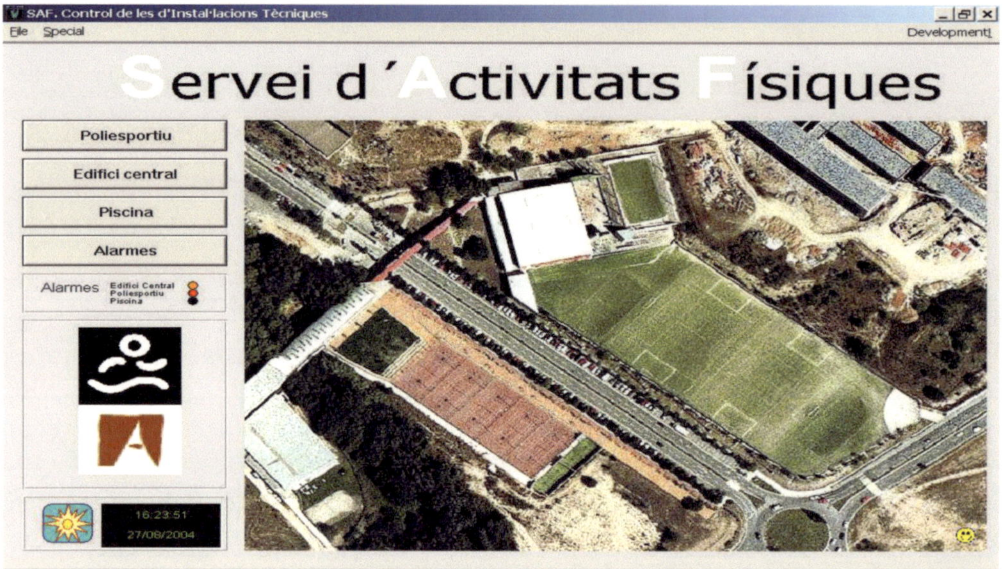


Fig. 4. Main window of the developed monitoring system with a global view of the Sports Service Area

Among them, the Sports Service Area (SAF) is one of the largest and with more complex installations. It encompasses indoor as well as outdoor activities that run for more than 12h each day. Just to give an idea of the different installations that give support to the offered activities.
We can find there: covered swimming pool, boulders, outdoor facilities for tennis, football, athletics, etc., indoor installations for fitness, basketball, aerobic, gym, etc. (Antsaklis et. al., 1999), (Astrom, 1999), (Kheir et. al., 1996).
Therefore, large complexes build up from different subsystems. Each one of these subsystems has to assure a quality of service each day. This fact introduces the need for good monitoring tools to help on this task. In addition, there is a hug number of automation and control problems (automated watering, temperature controls for water and indoor areas, lightning systems, ozone controlled system for water cleaning in covered swimming pools, etc.).
The SAF project has different automation levels: from field instrumentation and data collection, PLC programming and feedback loop configuration, to information integration on a SCADA system. The SAF project use PLC from different manufacturers (SIEMENS, GE-FANUC, Landys, and Mitsubishi).
All the data has been integrated through implementing the corresponding supervisory control interface with Wonderware suite called InTouch. The basic communications use specific drivers to connect PLC with PC based control; the advanced communications use the standard OPC protocol.

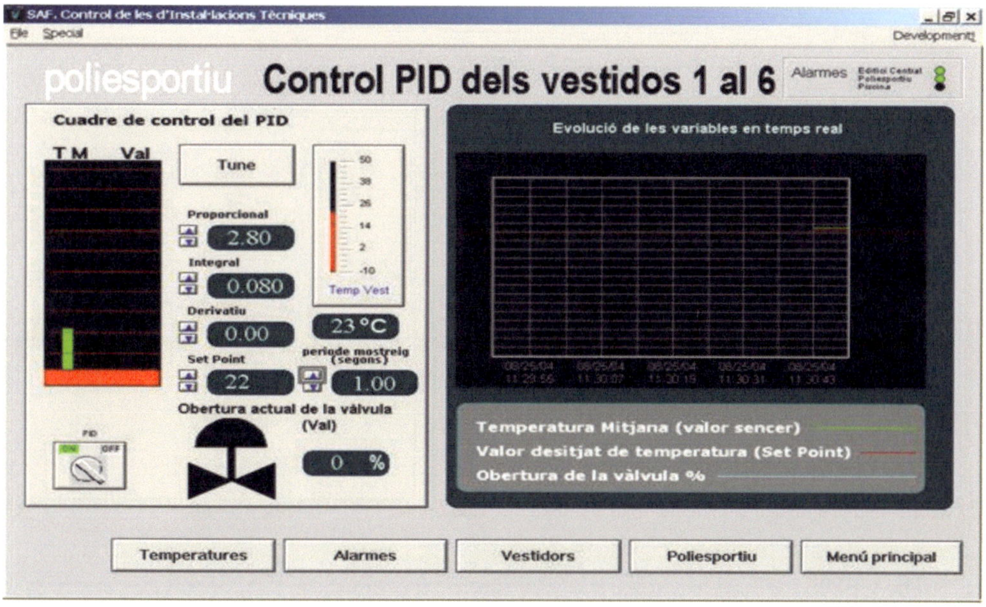


Fig. 5. An example of a Temperature feedback loop inside a SAF installation, and the ISA-PID control law used to close/open the regulation valve

An example: for indoor activities temperature control of both the SAF building and the water for the gym showers has been implemented. This means the students, in control room operator role, had to close some temperature feedback loops by using the ISA-PID control law present either in the PLC or in the software (see Fig. 5).

One important aspect of a monitoring system is how it deals with alarms. As this feature is a common feature, it should be incorporated in every part of the system according to the same rules. This way, in every SCADA window and alarm indicator has to be included that shows the human operator if an alarm is currently fired and can let you go directly to the main alarm window to process it.

Finally, design implementation and configuration of the InTouch based SCADA system has been done starting from zero. This allowed to think of a distributed application where from the different computers located either at the main SAF office or at the technical staff room the overall system can be accessed. In addition a special access, using terminal services, for the technical staff has been enabled so remote operation can also be done from outside the SAF installations.

## 4. S.A.F. evaluation

The connection between SAF designer and GEDIS guideline designer is necessary to define a global evaluation of the SAF interface and can give a set of recommendations about graphical screen improvement (see Fig. 6).

In general, heuristic evaluation is difficult for a single individual to do because one person will never be able to find all the usability problems in an interface (Nielsen, 1994). In this evaluation, the number of evaluators is seven groups of control engineering students. The heuristic evaluation session have been 6 hours. Some groups have been concentrating in a specific part of the SAF interface and a specific indicator (see Table 1). The advanced groups have been evaluating the SAF interface with the entire GEDIS guide.

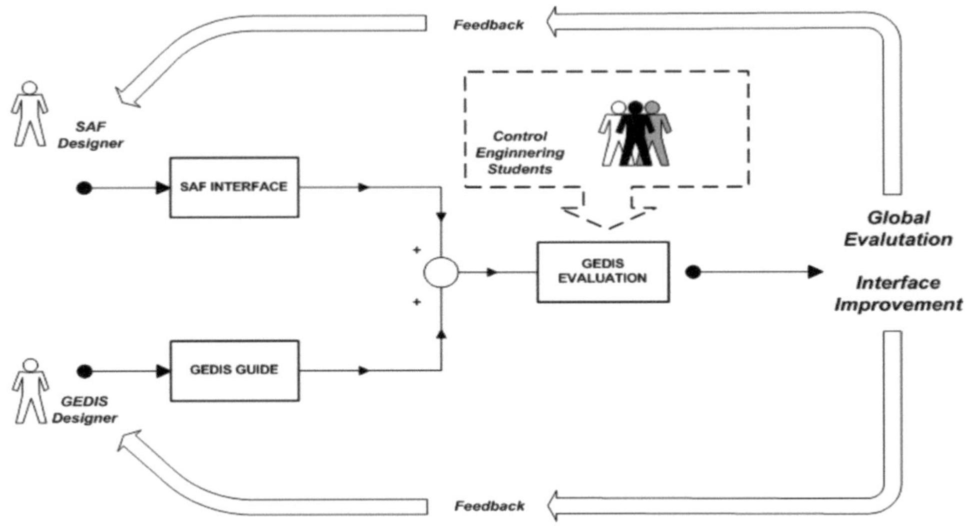


Fig. 6. The GEDIS method includes: the SAF interface evaluation, the control engineering students in the role of usability evaluators, the SAF technical staff and the GEDIS designers.

| Usability Evaluators (Control engineering students) | Evaluation of SAF Interface with the GEDIS guide |
|---|---|
| **Group (two students)** | |
| 1 | Alarms indicator 2,9 |
| 2 | GEDIS guide global evaluation 3,1 |
| 3 | Status of the devices indicator 3 |
| 4 | Structure indicator 3 |
| 5 | Navigation indicator 3 |
| 6 | GEDIS guide global evaluation 3,5 |
| 7 | Distribution indicator 3 |

Table 1. The number of usability evaluators is seven groups of control engineering students at the Technical University of Catalonia (EPSEVG school)

### 4.1 Evaluation

The evaluation expressed in quantitative numeric form or in qualitative format it seeks to promote the user's reflection that stuffs the GEDIS guide by way of questionnaire, so that it picks up the use experience that doesn't end up being verbalized in many occasions. Each one of the indicators of the Table 2 and Table 3 can substructure in diverse subindicators. For example, the indicator Color can be detailed in: absence of non appropriate combinations (5), number of colors (5), blink absence (no alarm situation) (5), contrast screen versus graphical objects (3), relationship with text (3).

| Indicator name and Subindicator name | Numeric/qualitative range and SAF numeric value |
|---|---|
| **Architecture** | **1,7** |
| Map existence | [YES, NO] [5,0] 0 |
| Number of levels le | [le<4, le>4] [5,0] 0 |
| Division: plant, area, subarea, team | [a, m. na] [5,3,0] 5 |
| **Distribution** | **3** |
| Model comparison | [a, m. na] [5,3,0] 3 |
| Flow process | [clear, medium, no clear] [5.3,0] 3 |
| Density | [a, m. na] [5,3,0] 3 |
| **Navigation** | **3** |
| Relationship with architecture | [a, m. na] [5,3,0] 3 |
| Navig. between screens | [a, m. na] [5,3,0] 3 |
| **Color** | **5** |
| Absence of non appropriate combinations | [YES, NO] [5,0] 5 |
| Color number c | [4<c<7, c>7] [5,0] 5 |
| Blink absence (no alarm situation) | [YES, NO] [5,0] 5 |
| Contrast screen versus graphical objects | [a, m. na] [5,3,0] 5 |
| Relationship with text | [a, m. na] [5,3,0] 5 |
| **Text font** | **3,2** |
| Font number f | [f<4, f>4] 5 |
| Absence of small font (smaller 8) | [YES, NO] [5,0] 0 |
| Absence of non appropriate combinations | [YES, NO] [5,0] 5 |
| Abbreviation use | [a, m. na] [5,3,0] 3 |

Table 2. GEDIS guide indicators (part one), where a = appropriate , m = medium and na = no appropriate

For each subindicator it is recommended it is punctuated numerically in a scale from 1 to 5. In this example the number of of subindicators of the indicator Color is J = 5 (see formula 1). The formula necessary to calculate the numeric value of each indicator is the formula 1.

$$Indicator = \frac{\sum_{j=1}^{J} w_j Subind_j}{\sum_{j=1}^{J} w_j} \tag{1}$$

where, *Subind*= subindicator and *w* = weight.

The mean value that one obtains by the formula 1 with these numeric values is 4,2 . If it is rounded, the value is 4, so that to the indicator Color it is assigned the value 4 in this example, considering that each one of the subindicators has the same weight (w1 = w2... =wj = 1).

Each one of the indicators of the Table 2 and Table 3 are measured in a scale from 1 to 5. The human expert operator prepares in this point of concrete information on the indicator, so that it can already value the necessities of improvement.

The values of the indicators can group so that the GEDIS guide offers the global evaluation of the interface and it can be compared with others interfaces.

| Indicator name and Subindicator name | Numeric/qualitative range and SAF numeric value |
|---|---|
| **Status of the devices** | **4** |
| Uniform icons and symbols | [a, m. na] [5,3,0] 3 |
| Status team representativeness | [YES, NO] [5,0] 5 |
| **Process values** | **3** |
| Visibility | [a, m. na] [5,3,0] 3 |
| Location | [a, m. na] [5,3,0] 3 |
| **Graphs and tables** | **4** |
| Format | [a, m. na] [5,3,0] 3 |
| Visibility | [a, m. na] [5,3,0] 3 |
| Location | [a, m. na] [5,3,0] 5 |
| Grouping | [a, m. na] [5,3,0] 5 |
| **Data-entry commands** | **3** |
| Visibility | [a, m. na] [5,3,0] 3 |
| Usability | [a, m. na] [5,3,0] 3 |
| Feedback | [a, m. na] [5,3,0] 3 |
| **Alarms** | **3,8** |
| Visibility of alarm window | [a, m. na] [5,3,0] 3 |
| Location | [a, m. na] [5,3,0] 3 |
| Situation awareness | [YES, NO] [5,0] 5 |
| Alarms grouping | [a, m. na] [5,3,0] 5 |
| Information to the operator | [a, m. na] [5,3,0] 3 |

Table 3. GEDIS guide indicators (part two), where a = appropriate , m = medium and na = no appropriate

The formula necessary to calculate the GEDIS guide global evaluation index is the formula 2.

$$Global\_evaluation = \frac{\sum_{i=1}^{10} p_i ind_i}{\sum_{i=1}^{10} p_i} \tag{2}$$

where, *ind*= indicator and *p* = weight.

In a first approach it has been considered the mean value among indicators expressed in the formula 2. That is to say, to each indicator it is assigned an identical weight (p1 = p2... =p10 = 1) although it will allow it in future studies to value the importance of some indicators above others. The global evaluation is expressed in a scale from 1 to 5. Assisting to the complexity of the systems of industrial supervision and the fact that an ineffective interface design can cause human error, the global evaluation of a supervision interface it should be located in an initial value of 3-4 and with the aid of GEDIS guide it is possible to propose measures of improvement to come closer at the 5.

### 4.2 Experimental study

The experimental study is the evaluation of SAF interface with the collaboration of control engineering students from Technical University of Catalonia. From Vilanova i la Geltrú city,

twenty four students monitoring SAF interface around three weeks. The students define a numeric value for each indicator and propose interface improvement (see Tables 2 and 3).
The SAF interface global evaluation is 3,4. The global evaluation is expressed in a scale from 1 to 5, so it is necessary to indicate SAF designer a set of important recommendations:

- Revise the relationship between structure, distribution and navigation indicators. There is a lack of coherence in some subareas
- Create a global map screen to support aid to human operator
- Improve the feedback between interface and human operator in data-entry commands indicator
- Improve the easy location of alarm indicator in each screen
- Improve the alarms classification in each subarea
- Build an alarm log with the abnormal situations and create variables as: the average alarm rate, the most frequent alarm, the longest-standing alarm and any regulary repeating alarms (Andow, 1998), (Bransby, 1998), (Bransby & Jenkinson, 1998)
- It is necessary to create a survey and evaluate the human operator performance and satisfaction with the use of the SAF interface
- It is necessary to create a training program to support the transition from novice to expert operator

With GEDIS guide is possible too to indicate SAF designer a set of important recommendations about graphical screen improvement. For example, the Piscina ACS screen can improve with a set of changes in color and text font indicators (see Fig. 7).
A second example, the Fronton and Rocodrom screen can improve with a set of changes in distribution indicator (see Fig 8).

## 5. Conclusions

In tasks of human supervision in industrial control room is habitual that an external engineer, - by means of the commercial programs Supervisory Control and Data Acquisition SCADA -, take charge of designing the supervision interfaces in function to the knowledge on the physical plant and the group of physical-chemical processes contributed by the process engineers.
Although standards exist about security in the human machine systems that impact in aspects of physical ergonomics, interface design by means of rules of style, it is remarkable the absence of the design of interactive systems centered in the user where the engineering usability and the cognitive ergonomics can contribute significant improvements (Nielsen, 1993).
The GEDIS guide is an approach that tries to fill a methodological hole that joins the efforts of the systems engineering and the human factors for the improvement of the effectiveness of the human-machine system in industrial control room.
The application of the GEDIS guide to the study of cases contributes among other details the measure in form of indicators of aspects of interface design, the recommendation of changes for the improvement of the interface, and a global evaluation index that allows to quantify the current state of the interface regarding the future state after applying the correct measures.
The studied case presented shows a Spanish academic application, Sports Service Area S.A.F., but with the same characteristics of an industrial project. With the GEDIS guide approach it's possible to perceive diverse anomalies and to propose improvements in the interface design. The technical staff of the S.A.F. project are working in introduce the important recommendations in order to improve the interface design and the usability.

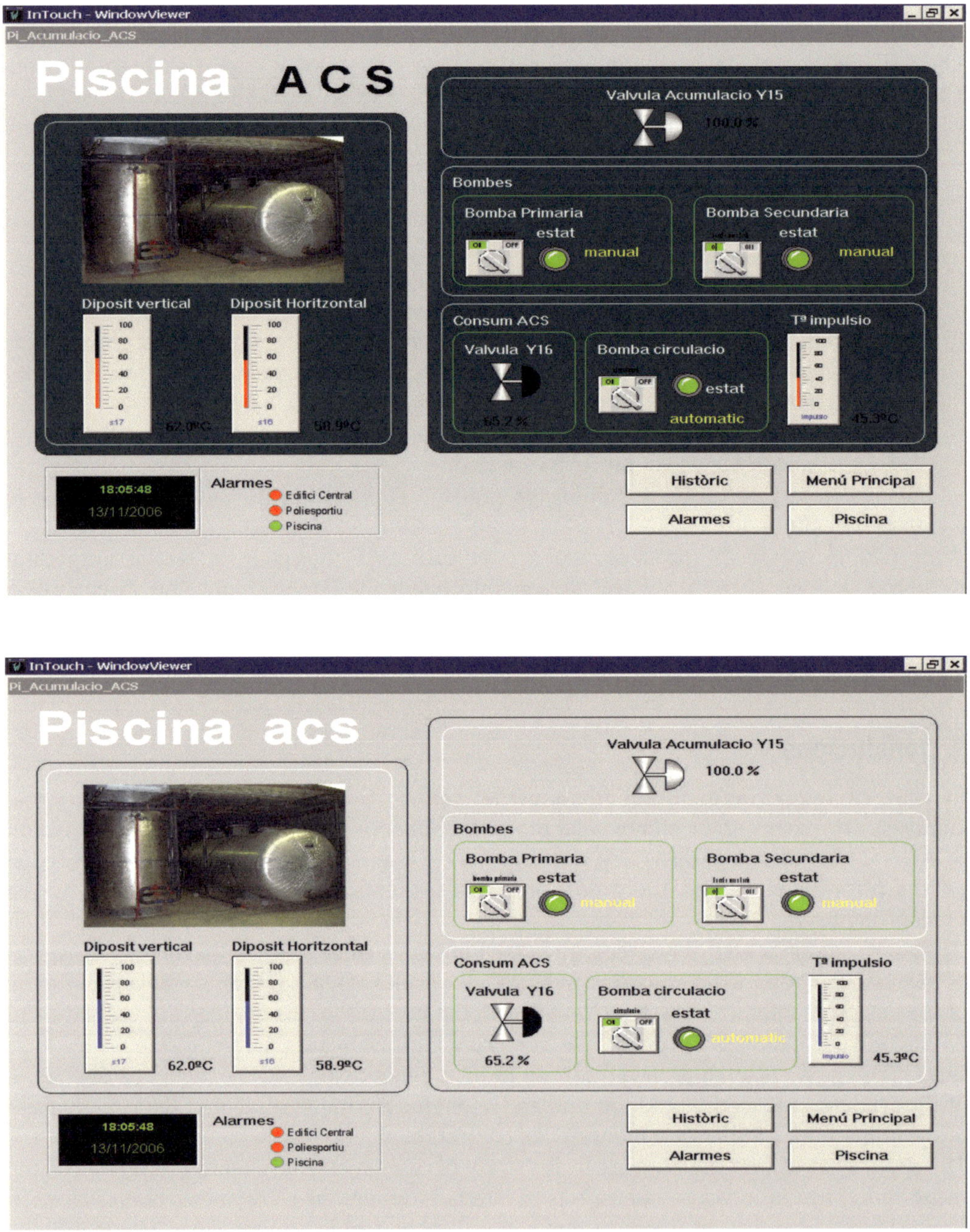


Fig. 7. Original Piscina ACS screen (top): There is confusion between the green screen color and the green color of the pumps display. The tank level display (left) shows a red bar in a situation without alarm. Piscina ACS revisited (bottom) with changes in color indicator. The screen color is now grey and the user can visualize better the states of the pumps display. The tank level display (left) is now a blue bar (the red color is only for alarm displays).

Fig. 8. Original Fronto and Rocodrom screen (top). There is a disorder between switches and photos in this screen. Fronto and Rocodrom revisited (bottom) with changes in distribution indicator. In this screen the user doesn't need to remember where each switch is; the distribution reduces the workload.

Another current study with the GEDIS guide is the analysis of a sugar mill interface. The Sugar Technology Center (CTA) in Spain has been developed a training simulator to modeling and simulating the production process and the human operators' supervisory tasks. The simulator developed in this center is an example of full scale simulator, a type of simulator that reproduces the whole operating environment (Merino et. al., 2005). This simulator emulates the control room of a sugar mill. A series of object oriented modelling library tools are used to create each part of the sugar mill: diffusion, evaporation, purification, sugar room, boilers, dryer, and liquor storage.
In these moments the 4all-L@b Usability Laboratory of the Technical University of Catalonia is analyzing the GEDIS guide to simplify the number of indicators of the guide, to improve the evaluation method, create new usability metrics, and to promote the use of the guide inside the cycle of life of the software engineering, in this case in the early phases of the supervisory control interface design.

## 8. References

Adams, J.A. (2006). Supporting human supervision of multiple robots. *The Journal of the Robotics Society of Japan*, Vol 24(5), pp. 17-19, ISSN: 0169-1864

Andow, P. (1998). Abnormal situation management: a major US programme to improve management of abnormal situations. Paper presented at IEE Colloquium on *Bet Practices in Alarm Management*, London. March.

Antsaklis, P.; Basar, T.; DeCarlo, R.; McClamroch,N.H.; Spong, M. & Yurkovich, S. (1999). Report on the nsf/css workshop on new directions in control engineering education. *IEEE Control Systems Magazine*, (10), pp 53-58, ISSN: 0272-1708

Astrom, K.J. (1999). Automatic control: the hidden technology. P.M. Frank ed. *Advances in Control: Highlights of ECC-99*, pp. 1-28, ISBN: 1852331224

Bastien, C. & Scapin, D. (1995). Evaluating a user interface with ergonomic criteria. *International Journal of Human-Computer Interaction*, Vol 7(2), pp-105-121, ISSN: 1044-7318

Bransby, M.L. (1998). Explosive lessons. *Computing and Control Engineering Journal*, 9, pp. 57-60.

Bransby, M.L. & Jenkinson, J. The management of alarm systems. HSE Research Report no CRR 166/1998.

Burns, C.M. & Hajdukiewicz, J.R. (2004). *Ecological Interface Design*. CRC Press, ISBN: 0-415-28374-4

Cañas, J.J. (2004). *Personas y máquinas*. Editorial Pirámide, Colección Psicología, ISBN: 84-368-1893-8

Card, S.; Moran, T.P. & Newell, A. (1983). *The psychology of human computer interaction*. Lawrence Erlbaum Associates, Inc., ISBN: 0898598591

Federal Aviation Administration (1996). *Human factors design guide for acquisition of commercial-off-the-shelf subsystems, non-developmental items, and developmental systems (DOT/FAA/CT-96/01)*. Atlantic City International Airport, DOT/FAA Technical Center

Granollers, T.; Lorés, J. & Cañas, J.J. (2005). *Diseño de sistemas interactivos centrados en el usuario*. Editorial UOC, Colección Informática, No 43, ISBN: 84-9788-320-9

Holstom, C. (2000). *Human factors and control centre issues. What lessons we have learned*. Institute for Energy Tecnology, OECD Halden Reactor Project

ISO International Organization for Standarization (2004). *Ergonomic design of control centres, parts I, II, III, IV.* In URL: http://www.iso.org

Kheir, N.A.; Astrom, K.J.; Auslander, D.; Cheok, K.C.; Franklin. G.F.; Masten, M. & Rabins, M., (1996). Control systems engineering education. *Automatica* 32 (2), pp. 147-166, ISSN: 0005-1098

Kontogiannis, T. (2005). Adaptable task modelling and its application to job design for safety and productivity in process control. In *ACM International Conference Proceedings*, Vol 132, pp. 27-34, ISBN: 9-60254-656-5

Lee, J.S. & Hsu, P.L. (2006). Applications of Petri Nets to human-in-the-loop control for discrete automation systems. In *Manufacturing the future: concepts, technologies, visions.* Edited by Kordic, V.; Lazinica, A & Merdan, M., Pro Literatur Verlag, pp. 167-194, ISBN: 3-86611-198-3

Merino, A.; Alves, R. & Acebes. L.F. (2005). A training Simulator fot the evaporation section of a beet sugar production. *European Simulation Multiconference, ESM-05*, Oporto, Portugal.

NASA (1995). *Man system integration standards.* NASA-STD-3000, In URL: http://msis.jsc.nasa.gov/

Nielsen, J. (1993). *Usability engineering.* Academic Press, Boston, ISBN: 0-12-518406-9

Nielsen, J. (1994). Heuristic Evaluation. In Nielsen, J. & Mack, R.L. (Eds), *Usability Inspection Methods.* John Wiley and Sons, New York, ISBN: 0-471-01877-5

Nimmo, I. (2004). Designing control rooms for humans. *Control Magazine*, pp. 47-53, ISSN: 1049-5541

Norsok Standard (2006). *I-002 Safety automation system.* Norwegian Technology Centre Oscarsgt. 20, Postbox 7072 Majorstua N-0306 Oslo. In URL: http://www.olf.no y http://trends.risoe.dk/detail-organisation.php?id=52#corpus

Noyes, J. & Bransby, M. (2001). *People in control: human factors in control room design.* IEE Control Engineeering Series 60, London-

Palanque, P.; Basnyat, S. & Navarre, D. (2007). Improving interactive systems usability using formal description techniques: application to HealthCare. *HCI and Usability for Medicine and Health Care, LNCS, November, pp. 21-40, ISSN:0302-9743*

Petersern, J. (2000). *Knowledge based support for situation assessment in human supervisory control.* Ph.D. Thesis. Department of Automation. Technical University of Denmark.

Petersen, J. & May, M. (2006). Scale transformations and information presentation in supervisory control. *International Journal of Human-Computer Studies*, Vol 64, 5, May, pp. 405-419, ISSN: 1071-5819

Ponsa, P., Vilanova, R. & Díaz, M. (2007). GEMMA Guide Approach for the Introduction of the Human Operator into the Automation Cycle. *Inf. tecnol.*, 2007, vol.18, no.5, p.21-30. ISSN 0718-0764.

Ponsa,P. & Díaz, M. (2007). Creation of an ergonomic guideline for supervisory control interface design. *12 th International Conference on Human-Computer Interaction.* July, Beijing, P.R. China. In URL: http://www.hcii2007.org , LNCS 4562, pp. 137-146, ISSN: 0302-9743

Rasmussen, J.; Pejtersen, A.M. & Goodstein, L.P. (1994). *Cognitive Systems Engineering.* New York, Wiley, ISBN: 0-471-01198-3

Reason, J.T. (1990). Human error. Cambridge University Press, Cambridge.

Samad, T. & Weyrauch, W. (2000). *Automation, control and complexity.* John Wiley and Sons, ISBN: 978-0-471-81654-6

Schneiderman, B. (1997). *Designing the user interface. Strategies for effective human-computer interaction.* Addison-Wesley, third edition, ISBN: 0-201-69497-2

Sheridan, T.B. (1992). *Telerobotics, automation and human supervisory control.* MIT Press, ISBN: 0-26219-316-0

U.S. Nuclear Regulatory Commission (2002). *NUREG-0700, Human-system interface design review guidelines.* Office of Nuclear Regulatory Research, Washington DC 20555-0001. In URL: http://www.nrc.gov/reading-rm/doc-collections/nuregs/staff/sr0700/nureg700.pdf

Vilanova, R.; Gomà, A. (2006). A collaborative experience to show how the University can play the industrial role. *7th IFAC Symposium on Advances in Control Education,* June, Madrid, Spain. URL: http:www.dia.uned/ace2006

Wickens, C.D.; Gordon, S.E. & Liu, Y. (1997). *An introduction to human factors engineering.* Longman, ISBN: 0-321-01229-1

# 19

# An Approach to Tune PID Fuzzy Logic Controllers Based on Reinforcement Learning

Hacene Rezine[1], Louali Rabah[1], Jèrome Faucher[2] and Pascal Maussion[2]
[1]*Unit of Control, Robotic and Productic Laboratory, Polytechnical Military School*
[2]*Ecole Nationale d'Electrotechnique, d'Electronique, d'Informatique et d'Hydraulique de Toulouse*
*Algeria*

## 1. Introduction

In the traditional control theory, an appropriate controller is designed based on a mathematical model of the plant under the assumption that the model provides a complete and accurate characterization of the plant. However, in some practical problems, the mathematical models of plants are difficult or time-consuming to be obtained because the plants are inherently nonlinear and/or exhibit uncertainty. Thus, new methods are proposed to process these caracteristics [1]. In recent years, increased efforts have been centered on developing intelligent control systems that can perform effectively in real-time. These include the development of non-analytical methods of Artificial Intelligence (AI) such as neural networks, fuzzy logic and genetic algorithms [1] but their combinations are also introduced such as Neuro-Fuzzy and Genetic-Fuzzy techniques [2], [3]. Fuzzy logic is a mathematical approach which has the ability to express the ambiguity of human thinking and translate expert knowledge into computable numerical data. It has been shown that fuzzy logic based modeling and control could serve as a powerful methodology for dealing with imprecision and non-linearity efficiently [4]. Also, for real-time applications, its relatively low computational complexity makes it a good candidate.Therefore, fuzzy logic control has emerged as one of the most successful nonlinear control techniques. Fuzzy Logic Controllers (FLC) are based on *if – then* rules integrating the valuable experiences of human. These rules use linguistic terms to describe systems. The mechanism of a FLC is that the uncertainty is represented by fuzzy sets and an action is generated co-operatively by several rules that are triggered to some degree, and produce smooth and robust control outputs. Recently, many authors proved that it is possible to reproduce the operation of any standard continuous controller using fuzzy controller [5] - [8].

Fuzzy logic controllers has shown good performances on the controlling of the complex, ill-defined and uncertain systems [9] and are being used siccessfully in many application areas such as mobile robots, subway system, nuclear reactor control and automobile transmission control, etc .

During the building of the FLC, the important tasks are the structure identification and parameters tuning [10]. The structure identification of the FLC includes the input-output variables of a controller, the rule base, the determination of the number of rules, the antecedent and consequent membership functions and their partition on their spaces respectively, the inference mechanism and the defuzzification method. The parameters tuning includes determing the optimal parameters of membership functions antecedent and consequent but also the scaling factors. [11].

The main problem arises from there not being a systematic approach to improve system performance. In conventional approach, the problem of generation of rules is solved by exploiting the knowledge of an expert or obtaining knowledge base (i.e, training data) by investigating relationship between an existing controller and the target system and forming the rule-base by a trial-and-error approach. An important number of choices is given a priori, these choices are carried with empirical methods, and then the design of the FLC can prove to be long and delicate towards the important number of parameters to determine, and can lead then to a solution with poor performance [12].

With this subjective approach, it is difficult for a designer to examine complex systems to find the necessary number of rules, and to determine appropriate parameters of the rules for implementing the fuzzy controller [13]. Also, it isn't easy to design an optimized fuzzy controller. Therefore, there has been a strong motivation to automate this process and consequently many researchers have been working to find learning algorithms for fuzzy system design.

Several approaches have been presented to learn and tune the fuzzy rules to achieve the desired performance. These automatic methods may be divised into two categories of supervised and unsupervised learning by whether the teaching signal is needed or not.

In the supervised learning approach, at each time step, if the input-output training data can be acquired, the FLC can be tuned based on the supervised learning methods. The artificial neural network (ANN)-based FLC can automatically determines or modifies the structure of the fuzzy rules and parameters of fuzzy membership functions with unsupervised or supervised learning by representing a FLC in a connectionist way such as ANFIS or other [14]- [17].

The other category contains genetic algorithm (GA) [18]-[23] and reinforcement learning (RL) systems [24]-[26] which are unsupervised leaming algorithms with the self-learning ability [9]. The GA-based and RL-based FLCs are two equivalent learning schemes which need a scalar response from the environment to provide the action performance [28], *that* value is easier to collect than the desired-output data pairs in the real application [11].

The difference between the GA-based and RL-based FLCs lies in the manner of state-action space searching. The GA-based FLC is a population based approach that encodes the structure and/or parameterof each FLC into chromosomes to form an individual, and evolves individuals across generations with genetic operators to find the best one. The RL-based FLC uses statistical techniques and dynamic programming methods to evaluate the value of FLC actions in the states of the world. However, the pure GA-based FLC can not proceed to the next generation until the arrival of the external reinforcement signal and dit is not easy pratical in real time applications. In contrast, the RL-based FLC can be employed to deal with the delayed reinforcement signal that appeares in many situations [11]. Recently, some researches on combining the advantages of GAs and RL have been proposed [28]-[30].

The basic idea of the reinforcement learning is to learn, through trial-and-error interaction with a dynamic environnement which returns a critic, called reinforcement, which can be thought of as a reward or a punishment, the control actions to determine desired changes in the control output that will increase the index of performance. Reinforcement learning techniques assume that, during the learning process, no supervisor is present to directly judge the quality of the selected control actions, and therefore, the final evaluation of process is only known after a long sequence of action. Also, the problem involves optimizing not only the direct reinforcement, but also the total amount of reinforcements the agent can receive in the future.This leads to the temporal credit assignment problem, i.e., how to distribute reward or punishment to each individual state-action pair to adjust the chosen action and improve its performance [31].

Supervised learning is more efficient than the reinforcement learning when the input-output training data are available [32], [33]

However, in most real-world application, precise training data is usually difficult and expensive to obtain or may not be available at all [12].

For the above reasons, reinforcement learning can be used to tune the fuzzy rules of fuzzy systems. Kaelbling, littman and Moore [34], and more recently Sutton and Barto [35], characterize two classes of methods for reinforcement learning: methods that search the space of value functions and methods that search the space of policies. The former class is exemplified by the temporal difference (TD) method and the latter by the genetic algorithm (GA) approach [36]. To solve reinforcemnt learning problem, the most approach is TD method [37]-[39]. Two TD based reinforcement learning approaches have been proposed the Adaptive Heuristic Critic (AHC) [40], [41] and Q-learning [42], [43]. The AHC consists of two separate networks: an action network actor) and an evaluation network (critic). Based on the AHC, many learning approaches have been proposed [20], [26], [40], [44]. One drawback of these actor-critic architectures is that they usually suffer from the local minimum problem in network learning due to the use of gradient descent learning method. Besides the aforementioned AHC algorithm based learning architecture, more and more advances are being dedicated to learning schemes based on Q-learning [45]. Some Q-learning based reinforcement learning structures have also been proposed [46] - [52]. Q-Learning is also modified to Dyna [53], TPQ-Learning [54], CQ-Learning [55], Q(λ)-Learning [56], and so on. Glorennec and Jouffe [51],[52],[57] extented the original Q-Learning method into a fuzzy environnment and introduced two fuzzy reinforcement learning methods, i.e., Fuzzy Actor-Critic Learning (FACL) and Fuzzy Q-Learning (FQL), to select the optimal conclusion for each fuzzy from an associated discrete action set. In these methods, the antecedent parameters are set using the *a priori* task knowledge of the user. From the point of view of reinforcement learning, a fuzzy inference system (FIS) is a means to introduce generalization in the state space and generate continuous actions in the reinforcement-learning problem whereas from the point of view of FISs, reinforcement learning is a learning method used to tune a fuzzy controller in a flexible way [58]. Fuzzy Q-learning collapses the two measures used by fuzzy actor/critic algorithms into one measure referred to as the Q-value []. It may be considered as a compact version of the FACL, also we adopt Fuzzy *Q*-learning in this work because it is conceptually simpler in implementation, and has been found empirically to converge faster in many cases [59], [60], for each fuzzy rule, a $q$ value is defined for each fuzzy consequence, which is the estimated cumulative reward for the fuzzy antecedents and fuzzy consequence pair of the rule. Q-learning is used to update these $q$ values. Optimal or sub-optimal FLC can be constructed by choosing the fuzzy consequence with the highest $q$ value for each rule. However the predefined value set needs to be set up by human experts and it is kept unchanged during learning, also if an improper value set is assigned, then those algorithms may not succeed at all [48], [50].

Horiuchi and *al* [49] consider a similar algorithm, termed fuzzy interpolation based Q-learning and further propose an extended roulette selection method so that continuous-valued actions can be selected stochastically based on the distribution of Q-values [61] proposes another version of Q-learning dealing with fuzzy constraints. In this case, we do not have fuzzy rules, but "fuzzy constraints" among the actions that can be done in a given state. These works, however, only adjust the parameters of FIS online. Structure identification, such as partitioning the input and output space and determination of the number of fuzzy rules are still carried out offline and it is time consuming. In [4] a novel online self-organizing learning algorithm is developed so that structure and parameters identification are accomplished automatically and simultaneously based only on Q-learning. In [45], [48], a dynamic fuzzy Q-learning is proposed for fuzzy inference system design. In this method, the consequent parts of fuzzy rules are randomly generated and the best rule set is selected based on its corresponding Q-value based genetic reinforcement learning. The problem in these approachs [4], [45], [50] is that if the optimal solution is not present in the randomly generated set, then the performance may be poor.
In order to solve these problems, this paper provides a systematic procedure for designing Fuzzy PID (FPID) controllers based on a reinforcement learning method. It is an automatic method capable of self-tuning parameters of a FLC based only on reinforcement signals. Continuous states are handled and continuous actions are generated by fuzzy reasoning. Prior knowledge can be embedded into the fuzzy rules, which can reduce the training time significantly. The proposed method is an efficient learning method whereby not only the conclusion part of a FLC can be tuning online, but also the parameters of antecedent part of a FLC can be tuning. We employ this approach for testing output voltage control of a DC/DC buck converter which is a traditional benchmark for testing nonlinear controllers, due to their inherent nonlinear characteristics [62], [63].
The best-known industrial process controller is the proportional-integral-derivative (PID) controller because of its simple structure, easy of design, inexpensive maintenance, low cost, and robust performance in a wide range of operations. However, it has been known that conventional PID controllers generally do not work well for nonlinear systems, higher order and time-delayed linear systems, and particularly complex and vague systems that have no precise mathematical models. To overcome these difficulties, the FPID controllers were developed and their improvement is still investigated [64]-[82]. This paper is devoted to this problem and describes some of the design aspects of the FPID.
The key concept of the proposed learning scheme is to evaluate all the principal parameters FPID in a procedure in three stages. The idea is to start with a basic FPID controller with its structure is chosen *a* priori and fixed during learning. In this work, we employ a Takagi-Sugeno of order zero as the controller of the system and the parameters tuning of fuzzy controler are the main issue researched. The membership functions or consequent parameters of each input/ouput variable are determinated with an equidistant partition. The neccessary scaling factors of the basic FPID are deduced from an initially one open-loop experimental response indicial as Ziegler-Nichols or Broida methods. This simple experimental on-site can be thought of as initial knowledge of the system and this basic FPID controller can yield an action that is feasible but far from optimal. In view of this, Reinforcement Learning is added to tune the fuzzy controller online. The predefined settings are used as starting points, also it is possible to determine the optimal parameters without too many iterations and the system can be operated safely even during learning.

Also in stage second FQL algorithm is used to select the optimal parameters of the FPID from an finite discrete set around the precedent predefined settings. This can be thought of as roughly tuning. The FQL algorithm proposed by Jouffe [52] is here extended for the antecedent parameters. Finally, in the third stage, a fine tuning procedure is follow to improve the FPID performance. This fine tuning is developped into an architecture composed of two integrated feedforward networks is proposed. One network (Q estimator **QE-FIS)** acts as a critic network to guide the learning of the other network (the action network). The action network is our FPID controller. Using the temporal difference (**TD**) prediction method, the critic network can predict the external reinforcement signal and provide a more informative internal reinforcement signal to the action network. The action network uses the gradient-descent algorithm to adapt itself in continuos according to the internal reinforcement signal. With the proposed architecture, the best parameters of the FPID in considering an IAE-criterion are determinated. This stage can be seen of as fine tuning and this way can solve the local minima problem. As a result, unlike many fuzzy Q-learning approaches that select the optimal action based on finite discrete actions [83]-[85], our proposed algorithm allows to obtain a continuos control output, the agent to learn more effectively, and helps reduce the time spent acting randomly. The salient features in our method are: (1) antecedents parameters of the fuzzy rules also can be updated, (2) not only discrete-valued antecedents/consequents parameters but also continuous-valued can be treated. (3) our technique can be used a precise simulator to speed up the learning process after final learning is achieved through the real system. Simulation and experiment results of a DC/DC Buck Converter indicate that efficiency and effectiveness of the proposed approach. Furthermore, the FPID controller learned by this approach has robust and adaptability, and can be applied to the different environments.

This paper is organized as follows. Section II briefly introduces reinforcement learning. The implementation and the limits of the Fuzzy-Q-Learning algorithm are introduced in section III. The architecture of the controller is described in section IV. The learning algorithms and parameter update laws are presented in section **V.** Section VI illustrates the performance of our proposed method through a static converter and compares experimental results with related works. Finally, conclusions and prospects are drawing in section VII.

## 2. Fuzzy PID system presentation

### 2.1 Control structure

The aim of this paper is the implementation of a FPID controller achieving the following properties:

1. Robustness around the operating point (e.g., in the case of a load change);
2. Good dynamic performance (i.e., rise time, overshoot, settling time, and limited output ripple) in the face of input voltage variations (and load changes);
3. Invariant dynamic performance in presence of varying output operating points.

We use a FPID based on the Takagi-Sugeno-type zero order method. In the FLC literature many forms of FPID structures [66], [71], [74] have been proposed. The controller in our work is a simple and classical FPID controller drawn on Fig.1. It is divided in two parts: a fuzzy part which performs the proportional-derivative action and a crisp integrator which is placed parallel of this fuzzy part so as to ensure a zero steady state error. Such a FPID

combines a high effciency with implementation easyness and is userfriendly because of its FPID-like action.
The two inputs of the controller are the error *e(k)* between the reference signal $y^*$ and the measured signal $y$ and the variation of this error *de(k)*. The output variable is the change in the control quantity $\Delta c_n(k)$. So as to ensure a good portability of the FPID controller, normalisation factors called *em, dem* and *gm* are used (Fig.1).

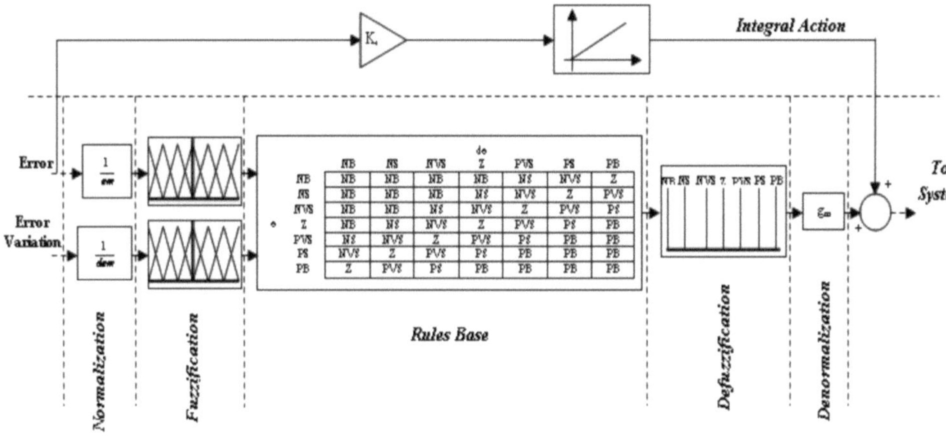


Fig 1. Structure of the Fuzzy PID Controller

The FPID considered here uses triangular membership functions with strong fuzzy partition because of its facilties and excellent approximation properties, and have been shown to be sufficient in a number of applications. We adopt for the two normalised input values $e_n(k)$ and $de_n(k)$ seven triangular membership functions on each input and seven singletons at the output. We use the Mac Vicar-Whelan 'base rules (1977) with 49 fuzzy rules. The membership functions are called PB, PS, PVS, Z, NVS, NS andNB (P: Positive, N: Negative, B: Big, S: Small, VS: Very Small). So as to warrant a similar response of the system for positive or negative sollicitations, a zero-symmetry can be imposed for both input membership functions and output singletons and a classical antidiagonal rules table is used. In addition, for a good portability of the FPID, the two inputs and the output are normalized on a [-1, +1] universe of discourse. Furthermore, the FPID is supposed to be well-normalized, it implies that the position of PB's and NB's apex are assigned to ±1 respectively. For the two inputs and the output of the FPID, the positions of the PS's and PVS's membership function's apex are mobile. Furthermore, the and-method is based on the product and the defuzzyfication method on the center of gravity is considered.

### 2.2 Factors definitions

In this part, we suppose that we don't know the dynamic of the system to control. Even for such a very simple fuzzy control structure, the tuning parameters of such a FPID are very numerous (positions of membership functions, normalisation gains, fuzzy rules, ...). In this paper and so as to limit the number of tuning parameters, we just hold 15 or 26 of these parameters back, which contributions to the optimization process according to the IAE criterion seam to be the greatests. These 15 or 26 parameters, that constitute the set of controlable factors, are the following:

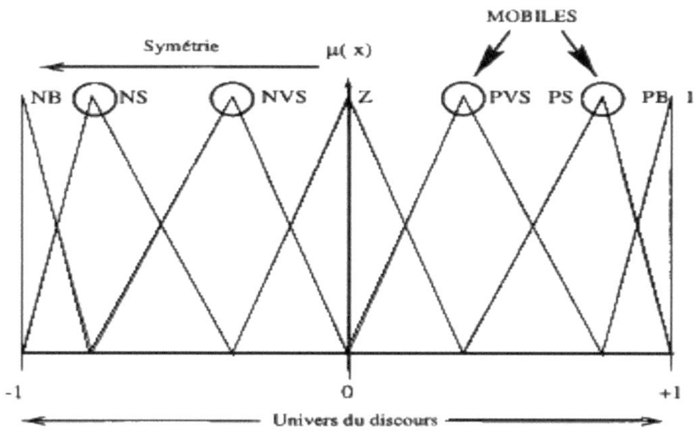


Fig 2. Membership functions for the two inputs e, de

- On the input *e* of the FPID: the position of the membership functions apex PS and PVS, the position of the membership functions apex NS and NVS is obtained by symmetry (fig.2.). The normalization factor em is supposed to be equal to the magnitude of the step sollicitation.
- On the input *de* of the FPID: the position of the membership functions apex PS and PVS and the position of the membership functions apex NS and NVS is obtained by symmetry.
- On the output $\Delta c_n$ of the FPID: the positions of the PS's and PVS's singletons. The principle of reinforcement learning allows considering for each fuzzy rule an individual discrete action set. Also at the end of the learning processus, the same linguistic label (Table I.4.) can have other significance in the base rule. Consequently, we obtain 11 or 22 tuning parameters in the case of a classical antidiagonal rules table or not respectively.

The normalization factor $de_m$, the denormalization gain $g_m$ and the integrator gain $K_i$ are fixed during the learning processus. They are determited by an open-loop identification test.

## 3. Q-learning algorithms

As previously mentioned, there are two ways to learn either you are told what to do in different situations or you get reward or punishment for doing good respectively bad actions. The former is called supervised learning and the latter is called learning with a critic, of which reinforcement learning (RL) is the most prominent representative with the self learning ability. It is shown that supervised learning is more efficient than reinforcement learning [32]. However, reinforcement learning only needs the critic information (evaluative signal) with respect to the different states of the controlled system [35]. This evaluative signal contains much less information than the reference signal used in supervised learning; also the reinforcement learning is appropriate for systems operating in a knowledge-poor environment [28].

The basic idea of reinforcement learning is that agents learn behaviour through trial-and-error interactions with the controlled system, and receive a critic, called reinforcement, which can be thought of as a reward or a punishment for behaving in such a way that a goal is fulfilled. This learning method is based on the common-sense idea that if an action is followed by a satisfactory state, or by an improvement, then the tendency to produce that action is strengthened, i.e., reinforced [58]. Reinforcement learning does not need teacher signal to guide action, since the learner is not told which action to take, it must discover the policy most effective, i.e. to know, in each possible situation, which action is achieved to maximize the expected cumulative reward in the long-term. In reinforcement learning, the final evaluation of process can be only known after a long sequence of actions. Thus, an internal evaluation function that is more informative than the evaluation function by the external critic is considered. This internal evaluation function takes the form of the expected sum of infinite horizon discounted payoffs, called the evaluation value of a policy:

$$R = \sum_{t=0}^{\infty} \gamma^k r_t \tag{1}$$

Where $\gamma$ is the discount factor ($0 \leq \gamma \leq 1$) used to determine the present value of future rewards.and $r_t$ is the external reinforcement signal received at time $t$.
The idea of Reinforcement Learning can be generalized into a model, in which there are two components: an agent that makes decisions and an environment in which the agent acts. For every time step $t$, the agent is in a state $s_t \in S$ where $S$ is the set of all possible states, and in that state the agent can take an action $u_t \in (U_t)$, where $(U_t)$ is the set of all possible actions in the state $s_t$. As the agent transits to a new state $s_{t+1}$ at time $t + 1$ it receives a numerical reward $r_{t+1}$,. It up to date then its estimate of the evaluation function of the action $Q(s_t, u_t)$ using the immediate reinforcement, $r_{t+1}$, and the estimated value of the following state, $Q_t^*(s_{t+1}, u_t')$, which is defined by:

$$Q_t^*(s_{t+1}, u_t') = \max_{u'_t \in U_{s_{t+1}}} Q(s_{t+1}, u') \tag{2}$$

The Q-value of each state/action pair is updated by

$$Q_{t+1}(s_t,u_t) = Q_t(s_t,u_t) + \beta\left\{r_{t+1} + \gamma Q_t^*(s_{t+1},u_t') - Q_t(s_t,u_t)\right\} \tag{3}$$

Where $r_{t+1} + \gamma Q^*_t(s_{t+1}, u_t') - Q_t(s_t, u_t)$ is the temporal difference (TD) error and $\beta$ is the learning rate. This algorithm is called Q-Learning. It shows several interesting characteristics. The estimates of the function Q, also called the Q-values, are independent of the policy pursued by the agent. To calculate the evaluation function of a state, it is not necessary to test all the possible actions in this state but only to take the maximum Q-value in the new state (eq.4). However, the too fast choice of the action having the greatest Q-value:

$$u'_t = \arg\max_{u_t \in U_{s_{t+1}}} Q_t(s_{t+1}, u_t) \tag{4}$$

Can lead to local minima. To obtain a useful estimate of Q, it is necessary to sweep and evaluate the whole of the possible actions for all the states: it is what one calls the phase of exploration [35].

In the previous equation, at state, only the Q-value of the activating action is updated, and the actions taken during the past time steps are not considered. But as is often the case with real, a reinforcement signal may not be available at a time long after the occurrence of a sequence of actions. This requires improving long-term consequences of an action or of a strategy for performing actions, in addition to short-term consequences. This problem is known as temporal credit assignment problem, i.e., how to distribute reward or punishment to each individual state/action pair to adjust the chosen action and improve its performance Also, to speed learning Sutton [35] extended the evaluation in all the states, according to their eligibility traces that memorise the previously visited state/action pairs weighted by their proximity to time step. They work like a short-term memory process activated by the occurrence of state/action pairs. The eligibility traces combined with Q-learning can be defined in several ways [48], [51]. Accumulating eligibility is defined by:

$$e_t(s) = \begin{cases} 1+\gamma\lambda e_{t-1}(s) & if \quad s = s_t \\ \gamma\lambda e_{t-1}(s) & \text{otherwise} \end{cases} \tag{5}$$

where λ is the eligibility rate used to weight time, since it accumulates whenever a state/action pair is selected and decays gradually when is not selected.
The algorithm Q (λ) is a generalization of Q-Learning (when λ=0) which uses the eligibilities traces. With eligibility trace, equation (3) is changed to:

$$Q_{t+1}(s_t,u_t) = Q_t(s_t,u_t) + \beta\left\{r_{t+1} + \gamma Q^*_t(s_{t+1},u_t') - Q_t(s_t,u_t)\right\} e_t(s) \tag{6}$$

## 4. Fuzzy Q-learning algorithms

The discrete Q-Learning such as we described it uses a discrete space of states and actions which must be have reasonable size to enable the algorithms to converge in an acceptable time in practice. In this case, a look-up table can be built up by listing the state/action pairs with their *Q* values. However in many applications, the number of state/action pairs is very large. Thus a method that is able to make Q-Learning applicable to the continuous problem domain is necessary. The Fuzzy Inference System (FIS) learner is one existing generalization methods which can be introduce generalization in the state space and generate continuous actions in the reinforcement-learning problem.

### A- FQL algorithm for consequents part (discrete parametrs tuning)

The principle of Fuzzy Q-Learning (FQL) proposed by Jouffe [51] is reinforcement learning method that tunes with only the consequent part in an incremental way based only on reinforcement signals. Each fuzzy rule *Ri* has possible *k* discrete candidate consequents (actions) $U_i = \left(u_1^i,\ u_2^i, \ldots\ldots,\ u_k^i\right)$ and it memorizes the parameter vector q associated with each of these actions. Local actions $(u_1, \ldots\ldots, u_k)$ selected from *U* compete with each other based on their q-values so as to maximize the discounted sum of rewards obtained while achieving the task. Each *Rule* $R_i$ of the FPID can be described as follow:

*If* e is $L_1^i$ and *de* is $L_2^i$ then $u^i$ is $u_1^i$ with $q(S,u_1^i)$

or $u^i$ is $u_2^i$ with $q(S,u_2^i)$ $L_m^i$ is linguistic label related to each

- - - - - - - - - -

or $u^i$ is $u_k^i$ with $q(S,u_k^i)$

input state, $u_j^i$ is a rule consequent (action) which corresponds to the consequent part of i-th fuzzy rule $R_i$ and has its own q-value $q_j^i$.

**1-*Generation of Continuous Actions***

The current state $S_t$ is perceived by the means of its activated degree of its firing rules. The winning local action cooperates to produce the global continuos output action which is defined by the weighted sum of the local actions elected in the fired rules that describe this state. It is given by

$$U_t(S_t) = \sum_{i=1}^{n} \pi_U(S_{t,}q_t)\alpha_t^i = \sum_{i=1}^{n} u_t^i \alpha_t^i \tag{7}$$

Where $u_t^i$ is the selected action of rule $R_i$, at time step *t* by a policy $\pi_U(S_{t,}q_t)$ and $\alpha_t^i$ is the rule's normalized firing strength.

After application of the global action, $U_t(S_t)$, the states change to $S_{t+1}$.

Optimal or sub-optimal FPID can be constructed by choosing the fuzzy consequence with the highest $q$ value for each rule. But, at the beginning stage for training, q-value can not correctly describe the valuation of action and taking the greedy actions or exploiting the previous experience too much during the learning would lead to local optima. The learnin algorithm would fail to find good actions On the other hand, taking random actions or exploring the spaces too much would affect both the learning convergence and the learning rate. Therefore in order to explore the set of possible actions and acquire experiences through the reinforcement signals, the actions are selected using an Exploration-Exploitation strategy. There are some random policies and the Boltamann probability distribution and -greedy method are the effective exploration/exploitation policy (EEP) to choose action. The algorithm FQL proposed in this paper uses an EEP [50], [51], combining an undirected exploration part $\rho^i(u)$ and directed exploration part $\eta^i(u)$ which are introduced by a random vector and a counter associated to actions. The proposed exploration-exploitation policy selects a local action from possible discrete actions vector, as follows:

$$\pi_U(S_t,q_t) = \arg\max_{u \in U}(q_t(S_t,u) + \eta(S_t,u) + \rho(S_t,u)) \tag{8}$$

The undirected term of exploration $\eta$ stems from a vector of random values $\Psi$, (exponential distribution) scaled up or down to take into account the range of $q$ values.

$$s_f = \begin{cases} 1 \quad \textit{if} \quad \max_u(q\,(S_t,u)) = \min_u(q(S_t,u)) \\ \dfrac{s_p(\max_u(q(S_t,u)) - \min_u(q(S_t,u\,)))}{\max(\psi)}, \quad \textit{otherwise,} \end{cases}$$

$$\eta(S_t, u) = s_f \psi \tag{9}$$

where $s_p$ is the noise size, with respect to the range of q qualities, and $s_f$ is the corresponding scaling factor. Decreasing the factor $s_p$ implies reducing the undirected exploration
The directed term $\rho$ gives a bonus to the actions that have been rarely elected

$$\rho(S_t, u) = \frac{\theta}{e^{n_t(S_t, u)}} \tag{10}$$

where $\theta$ represents a positive factor used to weight the directed exploration and $n_t(S_t,u)$ is the number of time steps in which action $u$ has been elected. This term is not directly available; it is approximated by the following fuzzy inference:

$$n_t(S_t, u) = \sum_{i=1}^{n} n_t(u^i)\alpha_t^i \tag{11}$$

where $n_t(u^i)$ is the number of applications, at time step $t$, of action $u$ in rule $R_i$. Let be $u_t^i$ the elected discrete action at time step $t$ in rule $R_i$. $n_t(u^i)$ is then defined according to

$$n_t(u^i) = n_{t-1}(u^i) + 1, \tag{12}$$

**2-.*Update of Q-Values***
We define also a function Q, which gives the action quality with respect to states. Q-values are also obtained by the FPID outputs, which are inferred from the quality of local discrete actions that constitute the global action.

$$Q_t(S_t, U_t) = \sum_{i=1}^{n} q_t(S_t, u_t^i)\alpha_t^i \tag{13}$$

where $U_t$ is the global continuos action, $u_t^i$ is the selected action of rule $R_i$ at time step t and $q_t$ is the q-value associated with fuzzy state, $S_i$ and action, $u_t^i$

Based on TD Learning, the Q-values corresponding to the rule optimal action are used to estimate the TD error, which is defined as follows :

$$Q_t^*(S_{t+1}) = \sum_{i=1}^{n} (\max_{u' \in U} q_t(S_{t+1}, u'_t))\alpha_t^i \tag{14}$$

and the TD error is defined only with quantities available at time step $t + 1$ as follow

$$\tilde{\varepsilon}_{t+1} = r_{t+1} + \gamma Q_t^*(S_{t+1}) - Q_t(S_t, u_t) \tag{15}$$

This TDerror can be used to evaluate the action just selected. If the TD error is positive, it suggests that the quality of this action should be strengthened for future use, whereas if the TD error is negative, it suggests that the quality should weakened [4]. The learning rule by taking the eligibity traces is given by

$$q_{t+1}(S_i,u_t^i) = q_t(S_i,u_t^i) + \beta_t^i \tilde{\varepsilon}_{t+1} \alpha_t^i e_t(S_i,u_j) \tag{16}$$

Where $\beta_t^i$ is the adaptative learning rate of rule $R_i$ at time step $t$ which is updating as follow:

$\delta_t^i = \tilde{\varepsilon}_{t+1}.\alpha_t^i$ Integrates eligibility trace

$$\beta_{t+1}^i = \begin{cases} \beta_t^i + k & if \;\; \bar{\delta}_{t-1}^i.\delta_t^i > 0, \\ \beta_t^i(1-\psi) & if \;\; \bar{\delta}_{t-1}^i.\delta_t^i < 0, \\ \beta_t^i & otherwise \end{cases} \tag{17}$$

$\bar{\delta}_t^i = (1-\varphi)\delta_t^i + \varphi\bar{\delta}_{t-1}^i$ Represents the geometric average

$\beta_t^i$ is used in order to increase the convergence rate during the FPID approximation and prevent oscillations and it is based on the Delta-Bar-Delta learning rule [51]. The rule (17) increments the critic learning rates linearly to prevent them from becoming too large too fast and decrements them exponentially to ensure that they always stay positive and to enable them to decrease rapidly

and $e_t(S_i,u_j)$ is the trace associated with discrete action $u_i$ of rule $R_i$ at time step $t$,

$$e_t(S_i,u_j) = \begin{cases} \gamma\lambda e_{t-1}(S_i,u_j) + \alpha_t^i & if \;\; u_i = u_t^i \\ \gamma\lambda e_{t-1}(S_i,u_j), & \text{otherwise} \end{cases} \tag{18}$$

The traces are updated between the action computation and its application.

**B- FQL algorithm for antecedents part ( discrete parametrs tuning)**

In the previous method, the antecedent parameters are set using the *a priori* task knowledge of the user. To restrict a FPID optimization to the only tuning of the parameters of the consequent part is often insufficient to reach high performances. Also in this paper, we bring a slight difference to the algorithm proceeding by introducing an extension to the Fuzzy Q-Learning algorithm to also allow the online tuning of the parameters of the antecedent part (positions of input fuzzy sets).

The principle of the FQL algorithm applied to the antecedent parameters will consist in selecting for each membership function a modal point from a possible discrete candidate modal points set basing on the evaluation function of the action $Q(s_t,u_t)$ because as the equations (7, 13) show it, the calculation of the action U and its quality Q(S,u) is closely linked to the antecedent parameters by the means of the degrees of activation of the fuzzy rules $\alpha_t^i$. The algorithm principle is depicted in Fig.3, 4.

Let us consider the FPID controller with the two input variables *e* or *de*. Each universe of discourse of *e* or *de* is partitioned in $Nmf_i$ membership functions

$F_i^j : \{F_i^1, F_i^2, \cdots, F_i^j, \cdots, F_i^{Nmfi}\}$ with each membership function. $F_i^j$ has an possible discrete modal points set $A(i, j) : \{a_i^{j,1}, a_i^{j,2} \cdots, a_i^{j,k}, \cdots, a_i^{j,Nniv_{i,j}}\}$, where $a_i^{j,k}$ is the $k^{eme}$ point modal possible of the $j^{eme}$ membership function of the input i $F_i^j$ and $Nniv_{i,j}$ is the cardinal of the set. Each *Rule* $R_i$ of the FPID can be described as follow:

*If* $e$ *is* $F_1^i(a_1^{i,1})$ *and* $de$ *is* $F_2^i(a_2^{i,1})$ *Then* $u$ *is* $u_t^i$
*or* $F_1^i(a_1^{i,2})$ *or* $F_2^i(a_2^{i,2})$
……… ………
*or* $F_1^i(a_1^{i,k})$ *or* $F_2^i(a_2^{i,k})$

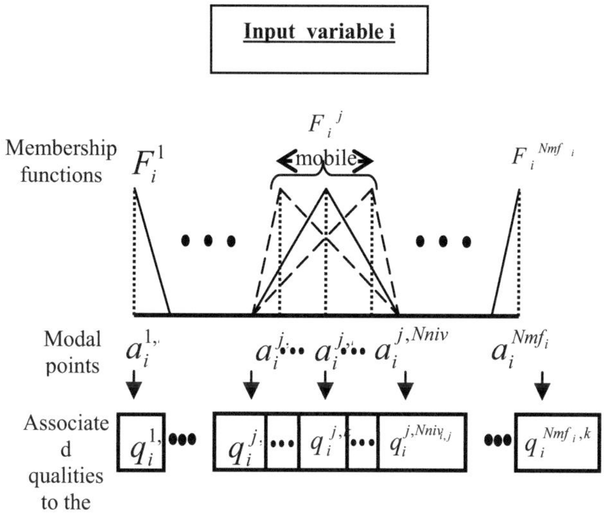


Fig. 3. FQL Structure for the antecedent part

As for the previous algorithme (section IV-A) local modal points $(a_1, \ldots\ldots, a_k)$ selected from $A$ compete with each other based on their q-values so as to maximize the discounted sum of rewards obtained while achieving the task. Global quality $Q$ for the state $S_t$ is then defined by the inference of these qualities locally elected

$$Q_t(S_t) = \sum_{i=1}^{2} \sum_{j=1}^{Nmf_i} q_t(a_i^j) \cdot \mu_t^{F_i^j}(x_i) \tag{19}$$

where $a_i^j$ is the elected modal point at the time step t

for the membership function $F_i^j$ by an exploration-exploitation policy and $\mu^{F_i^j}(x_i)$ is the membership degree of the input variable $x_i$ ($x_1=e$ , $x_2=de$) to $F_i^j$ and it measures the contribution of the modal points to the generation of the globale action .

***Execution Procedure***

In order to describe the working principle of the FQL algorithm applied to the antecedent parameters, a one-time-step global execution procedure is presented. Let $t+1$ be the current time step; the learner has performed the action $U_t$ and has received a reinforcement signal $r_{t+1}$according to the state transition $S_t$ to $S_{t+1}$. After the fuzzification step and rule strength computations (describing the new state ) by using the modal points ($a_t$) elected with the step of previous time $t$,, the eight stages are as follows:

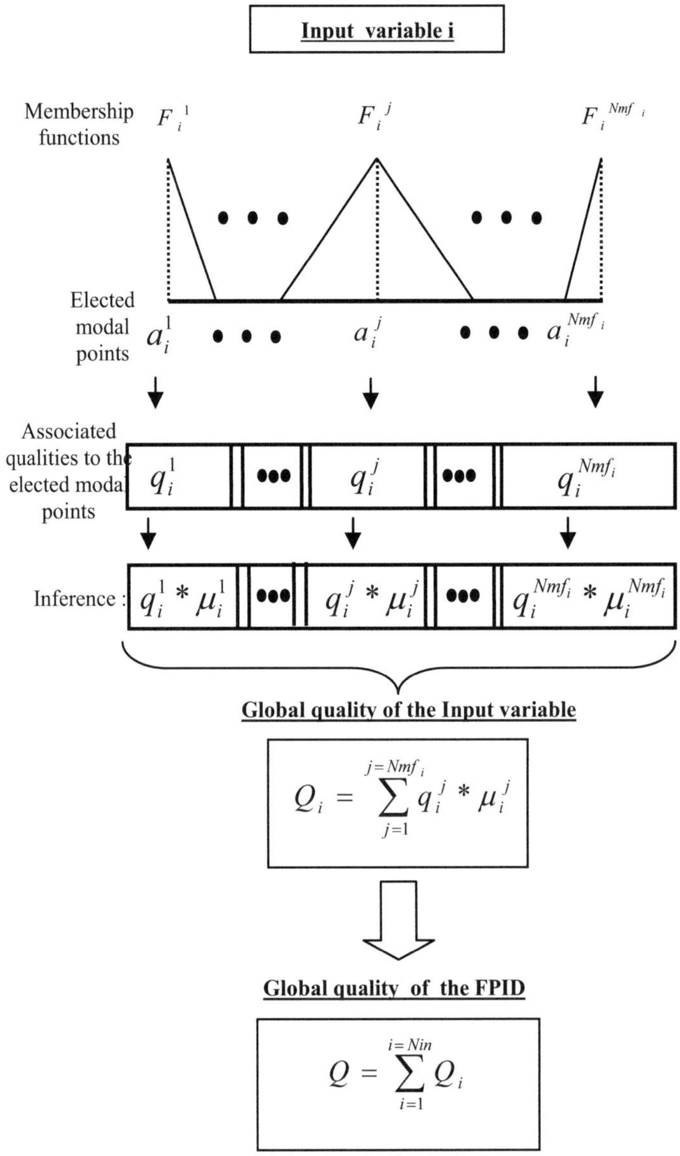


Fig. 4. FQL Algorithm Principle for the antecedent part

1. Estimation of the t-optimal evaluation function corresponding to the current state from the best modal points of each membership function, as follows. :

$$Q_t^*(S_{t+1}) = \sum_{i=1}^{2} \sum_{j=1}^{Nmf_{i,j}} \left( \underset{k=1,\text{Nniv}_{i,j}}{Max} \left( q_t \left( a_i^{j,k} \right) \right) \right) \cdot \mu_t^{F_i^j} (x_i) \tag{20}$$

2. Compute the TD error ;

$$\tilde{\varepsilon}_{t+1} = r_{t+1} + \gamma Q_t^*(S_{t+1}) - Q_t(S_t) \tag{21}$$

3. Update the adaptative learning rate $\vartheta_i^{j,k}(t+1)$ with the Delta-Bar rule by using the equation defined in (17) ;
4. Tune parametr vector $q$ based on on current eligibility trace ;

$$q_i^{j,k}(t+1) = q_i^{j,k}(t) + \vartheta_i^{j,k}(t+1) \cdot \tilde{\varepsilon}_{t+1} \cdot e_i^{j,k}(t) \tag{22}$$

where: $i = 1,\ldots,n$ , $j = 1,\ldots,Nmf_i$ , $k = 1,\ldots,Nniv_{i,j}$

5. Elect the modal points $a_i^j$ for each membership function $F_i^j$ by using an exploration-exploitation strategy identical to that used in algorithm FQL for the consequents part and defined as (8) ;
6. Estimation of the new evaluation functions for the current state *t+1* with the new modal points vector. We recompute the membership degrees of the inputs to the new membership functions ($\mu_{t+1}^{F_i^j}$).
7. Update the eligibility trace which will be used in parameter updating at the next time step:

$$e_i^{j,k}(t+1) = \begin{cases} \gamma\lambda e_i^{j,k}(t) + \mu_{t+1}^{F_i^j}(x_i), & if \quad a_i^{j,k} = a_i^j \\ \gamma\lambda e_i^{j,k}(t), & otherwise \end{cases} \tag{23}$$

Eligibility trace values need to be reset to zeros at the end of each episode ;

8. New estimation of the evaluation function corresponding to the elected modal points at the current state for the current state with the new updating parametrs:

$$Q_{t+1}(S_{t+1}) = \sum_{i=1}^{2} \sum_{j=1}^{Nmf_i} q_{t+1}(a_i^j) \cdot \mu_{t+1}^{F_i^j}(x_i) \tag{24}$$

$Q_{t+1}(S_{t+1})$ will be used in the error computation at the next time step.

## 5. Continuos fuzzy Q-learning algorithm (continuous parametrs tuning)

Until now, the definition of controlable factors for the FPID in the reinforcemnt learning algorithms has just been led on discrete levels [83]- [85] which are kept unchanged during learning. This is the basic idea of most existing algorithms which adopt Q-learning. In general, however, this is not case and then the algorithm may fail. Indeed, for complex systems like static converter, *priori* knowledge are not available, then it becomes difficult to determine a set of parameters in which figure the optimal controlable factors for each fuzzy rule and thus the FPID controller can't accomplish the given task through Fuzzy Q-Learning.

To ensure a fine optimization of these parameters in the vicinity of those obtained in the last section, a continue reinforcement learning algorithm [58] has been proposed. Compared to the idea in this article, we employ an FIS, in the place of a neural network, to estimated Q(S, u) which is only used to tune the parameters of the FPID. As a result, our proposed algorithm is more applicable to real systems because we combine reinforcement learning with an Fuzzy Logic Controller (FLC) and it learns more effectively, and helps reduce the time spent acting randomly.
The overall architecture of this FLC, inspired of [58] is shown in Fig. 5. It is similar to that of the FACL [51]. The proposed controller is constructed of two parts: a critic, a Q(S, u) estimator FIS (QE_FIS ) and an actor, the FPID which have two main responsabilities respectively : the critic is to estimate the optimal-action value function $Q^*(S,u)$, and the actor is to get the control output based on the estimated-action value function $Q(S,u)$. There is an indirect coupling between the two. Indeed, when action U acts in the system, the state $S_t$ transites towards the new state $S_{t+1}$. According to this transition, the system emits a reinforcement $r_{t+1}$which is used in the error TD $\tilde{\varepsilon}_{t+1}$. According to the latter, we adjust the parameters of the evaluation function $Q(S,u)$ which is then used to tuning in a continuous way the controlable parameters of the FPID in such manner to maximize quality $Q(S,u)$by a rise of gradient. The architecture of this FLC is explained in detail in the following parts of this section.

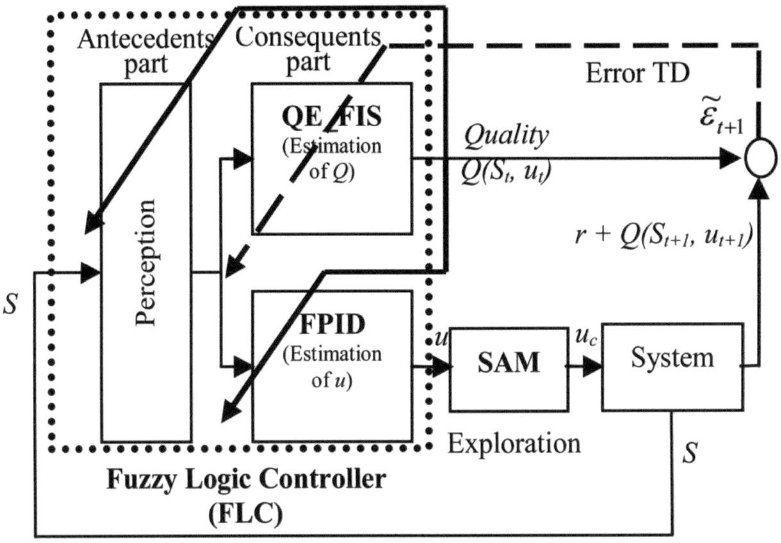


Fig. 5. Architecture proposed for the continuous tuning of the parameters FPID

### A. FIS System for Estimating the Optimal-Action Value Function

As the approximator of the value function for reinforcement learning, different types of approximators have been proposed for different tasks and different objectives. Interested readers can see references indicated in [58] for more information about approximator methods. In this paper, we use a Fuzzy Inference System (FIS) to approximate and to predict the optimal-action value function $Q^*(S,u)$ due its good approximation property. The $Q^*(S,u)$estimator fuzzy (QE-FIS) is associated with the same inputs $S$ of the FPID also they can be to amalgamate as an only one fuzzy controller which allows a considerable profit in the computing time.The QE-FIS is based on the prediction TD error $\tilde{\varepsilon}_{t+1}$ (Fig. 5) for the optimization of its parameters. It is used to guide the FPID controller to tune parameters so that the fuzzy controller will achieve better performance.

### B. FPID for producing the control output

The FPID generates the control output $u$ that maximizes the action value function $Q$ $(S,u)$ produced by the QE-FIS incrementally. The final output provided by FPID maximizes $Q$ $(S,u)$ with respect to all possible $u$, instead of the finite candidate discrete action set $U$ in the last section. By its robustness, the FPID can achieve acceptable performance even in the early stage of learning, in which the approximation error of $Q$ $(S,u)$is large [58] and the information, obtained in the section preceding, about the candidate discrete factors set $U$ and $A$ allow it to learn more effectively by narrowing the research space and to speed up the learning processus.

Consider a multiple-input single-output (MISO) Fuzzy Logic Controller (FLC) that performs a mapping from a state vector $\underline{x} = (x_1, x_2, \cdots, x_n)^T \in \Re^n$ to a control input $u \in \Re$. Using the Takagi–Sugeno model, the If–Then rules of the FLC may be expressed as

$$R^l : If \quad x_1 \quad is \quad F_1^l, \cdots and \quad x_n \quad is \quad F_n^l \quad Then \begin{cases} u^l = K_0^l + K_1^l x_1 + \cdots + K_n^l x_n & \rightarrow FPID \\ q = q^l & \rightarrow QE\_FIS \end{cases}$$

where $F_i^l$ is the label of the fuzzy set in $x_i$ for $l$= 1, ...$M$ ($M$ is rules number), $K_0^l, K_1^l, \cdots and \quad K_n^l$ are the constant coefficient of the consequent part of the fuzzy rule and $q^l$ is local action quality $u^l$. We consider in a way general the first input $x_1$ has $Nmf_1$ membership functions and the second input $x_2$ has $Nmf_2$ membership functions. In our application $n = 2, (x_1 = e \quad and\, x_2 = de)$ , $Nmf_1$=7, $Nmf_2$=7, $K_0^l \neq 0, K_1^l = 0, \cdots and \quad K_n^l = 0$ have different consequents and the local quality $q^l$ entirely qualifies the $R^l$ rule, from where the idea to use it to tune the controlable parameters of the FPID. We use product inference for the fuzzy implication and $t$ norm, singleton fuzzifier, and center-average defuzzifier; consequently, the final output value of the FPID and the final quality value of the QE_FIS are respectively

$$u(x) = \frac{\sum_{k=1}^{Nmf_1} \sum_{t=1}^{Nmf_2} \left\{ \alpha^{k,t} \cdot \left( \sum_{i=0}^{n} K_i^{k,t} x_i \right) \right\}}{\sum_{k=1}^{Nmf_1} \sum_{t=1}^{Nmf_2} \alpha^{k,t}} \tag{25}$$

$$Q(x,u) = \frac{\sum_{k=1}^{Nmf_1} \sum_{t=1}^{Nmf_2} \alpha^{k,t} \cdot q^{k,t}}{\sum_{k=1}^{Nmf_1} \sum_{t=1}^{Nmf_2} \alpha^{k,t}} \tag{26}$$

where $\alpha^l$ is the firing strength of the fuzzy rule $R^l$ calculated as follow $\alpha^l = \prod_{i=1}^{n} \mu^{F_i^l}(x_i)$, $\mu F_i^l$ is the membership degree of the fuzzy set $F_i^l$, $\left(\sum_{i=0}^{n} K_i^l x_i\right)$ is the conclusion part of the fuzzy rule $R_l$, ( $x_0$ = 1) and $q^l$ is the local action quality of the fuzzy rule $R_l$ . With the choice of strong fuzzy partitions $\sum_{k=1}^{Nmf_1} \sum_{t=1}^{Nmf_2} \alpha^{k,t} = 1$.

**C. Stochastic Action Modifier (SAM)**

In order to guarantee that Q(x, a) converges to Q*(x, a) with a probability equal to 1, we implement an exploration strategy for the control output $u$ provided by the FPID which deals with continuos actions. For that, we add a stochastic action modifier (SAM) after the FPID and before the system input [58]. The SAM generates the control command $u_c$, which is a gaussian random variable with mean $u$ recommended by the FPID and standard deviation $\sigma_u$, and $\sigma_u$ satisfies the condition that it will converge to zero gradually, *i.e.*

$$u_c = u + \sigma_u(t) \cdot n(t), and \qquad \lim_{t \to \infty} \sigma_u(t) = 0 \tag{27}$$

where $n(t)$ represents a gaussian random variable normalised and centred.
The adopted standard deviation is as follow:

$$\sigma_u(t) = \exp(-\alpha t), avec \qquad \alpha \succ 0. \tag{28}$$

**D. Learning algorithms for FLC**

In this section, we develop the learning algorithms for the QE-FIS and the FPID. The learning mechanism (estimation of local qualities $q^l$) of the QE-FIS is the combination of the TD methods and the gradient descent algorithm. The learning mechanism of the FPID is based on gradient rise algorithm. Indeed, the global quality $Q$ ($x$, $u$) (see equation 26) qualifies the action $u$ but also the FPID [45]. Thus more its value is high, plus the performances of FPID are better. We adjust consequently the antecedents parameters (modal points of the membership functions) and the consequents parameters (the $K_i^l$ coefficients parameters) of the FPID in such manner to maximize global quality $Q$ ($x$, $u$).
While starting of a SIF optimized in the phase of discrete reinforcement learning, the proposed algorithm uses two stages to avoid all risk of instability. These two stages constitute the continuous reinforcement learning algorithm: - the first stage is the simultaneous optimization of the antecedent and consequent parameters of the QE_FIS, while fixing the consequent parameters of FPID ([values found in the phase of the discrete reinforcement learning) - the second stage is the simultaneous optimization of the parameters of the FPID and the QE_FIS, while fixing the antecedent parameters to the values found in the stage one.

**1. *Learning algorithm for consequent (QE-FIS)***

From the previous section on reinforcement learning, we know TD methods learn their estimates, in part, on the basis of other estimates. We can also tune the consequent

parameters of our proposed QE-FIS based on this method. We can achieve the task of approximating the optimal-action value function with the QE-FIS by reducing the TD error $\tilde{\varepsilon}_{t+1}$, continuously with a descent gradient algorithm.
The quality-update rule of the QE_FIS is then given by:

$$q_{t+1}^{l} = q_{t}^{l} + \eta_{t+1}\tilde{\varepsilon}_{t+1} \cdot e_{t}^{l} \tag{29}$$

The adaptative learning rate $\eta_{t+1}$ and the eligibility traces for the QE-FIS are determided on the rules by using respectively the equations (17) and (18) ;
The TD error is calculated as follows:

$$\tilde{\varepsilon}_{t+1} = r_{t+1} + \gamma \max_{u'} Q(x_{t+1}, u') - Q(x_t, u_t) \tag{30}$$

In this expression, we consider that $\max_{u'} Q(x_{t+1}, u') \approx Q(x_{t+1}, u_{t+1})$ because $u_{t+1}$ is the ouput FPID tuned continuously to maximize $Q\ (x, u)$, therefore $\tilde{\varepsilon}_{t+1}$ is written as follows:

$$\tilde{\varepsilon}_{t+1} = r_{t+1} + \gamma Q(x_{t+1}, u_{t+1}) - Q(x_t, u_t) \tag{31}$$

**2.** ***Learning algorithm for FPID and antecedent parameters QE-FIS***
Now, we consider how to improve the control policy using the associated value function. In other words, we consider how to tune the parameters of the FPID controller based on the approximated $Q\ (x, u)$obtained from the previous section.
In order to optimize the output of the FPID, we can update the parameters of the FPID to maximize the action value function $Q(x, u)$ with respect to the control output $u$ for the current state. As a result, the learning algorithms of the FPID can be derived using gradient rules

**2.1** ***Learning algorithm for antecedent parameters***
The parameters of the modal point vector ($a_i^j \quad i = 1,....,n \ ; \ j = 1,..., Nmf_i$ ) are updated by a rise gradient algorithm :

$$a_i^j(t+1) = a_i^j(t) + \rho \frac{\partial Q(x_{t+1}, u_{t+1})}{\partial a_i^j} \tag{32}$$

Let us consider the case of a modal point of the first input. The calculation of the $\dfrac{\partial Q(x,u)}{\partial a_1^j}$ revealed the three expressions $\mu^{F_1^{j-1}}(x_1), \mu^{F_1^{j}}(x_1)\ et\ \mu^{F_1^{j+1}}(x_1)$ where the modal point $a_1^j$ can intervene in two expressions in maximum among these three. The calculation of the membership degrees of different fuzzy set $F_i^{j-1},\ F_i^{j}\ et\ F_i^{j+1}$) shows that it exists four possible cases :
The table 1 recapitulates every case to be considerate.

| | $\frac{\partial Q(x,u)}{\partial a_i^j} =$ | Explanation of the notations | |
|---|---|---|---|
| | | *First input e* ($i=1$) | *Second input de* ($i=2$) |
| *1.3 case :* *If* $x_i \le a_i^{j-1}$ *or* $x_i \ge a_i^{j+1}$ *or* $x_i = a_i^j$ | $0$ | $q_1^{j-1} = \sum_{l=1}^{Nmf_2} \mu^{F_2^l}(x_2) \cdot q^{j-1,l}$ | $q_2^{j-1} = \sum_{k=1}^{Nmf_1} \mu^{F_1^k}(x_1) \cdot q^{k,j-1}$ |
| *2 case :* *If* $a_i^{j-1} < x_i < a_i^j$ | $\frac{(q_i^j - q_i^{j-1}) \cdot (a_i^{j-1} - x_i)}{(a_i^j - a_i^{j-1})^2}$ | $q_1^{j} = \sum_{l=1}^{Nmf_2} \mu^{F_2^l}(x_2) \cdot q^{j,l}$ | $q_2^{j} = \sum_{k=1}^{Nmf_1} \mu^{F_1^k}(x_1) \cdot q^{k,j}$ |
| *4 case :* *If* $a_i^j < x_i < a_i^{j+1}$ | $\frac{(q_i^j - q_i^{j+1}) \cdot (a_i^{j+1} - x_i)}{(a_i^j - a_i^{j+1})^2}$ | $q_1^{j+1} = \sum_{l=1}^{Nmf_2} \mu^{F_2^l}(x_2) \cdot q^{j+1,l}$ | $q_2^{j+1} = \sum_{k=1}^{Nmf_1} \mu^{F_1^k}(x_1) \cdot q^{k,j+1}$ |

TABLE I Summary of the calculation of $\frac{\partial Q(x,u)}{\partial a_i^j}$ Some remarks are to be noted.

1. There is a discontinuity to the point $a_i^j$ (superior limit of $F_i^{j-1}$ , modal point of $F_i^j$ and inferior limit of $F_i^{j+1}$ ), indeed :

$$\lim_{x_i \to a_i^{j-}} \frac{\partial Q(x,u)}{\partial a_i^j} \neq \lim_{x_i \to a_i^{j+}} \frac{\partial Q(x,u)}{\partial a_i^j} \neq \left. \frac{\partial Q(x,u)}{\partial a_i^j} \right|_{x=a_i^j}$$

Superior limit of $F_i^{j-1}$ : $\lim_{x_i \to a_i^{j-}} \frac{\partial Q(x,u)}{\partial a_i^j} = \lim_{x_i \to a_i^{j-}} \left\{ \frac{(q_i^j - q_i^{j-1}) \cdot (a_i^{j-1} - x_i)}{(a_i^j - a_i^{j-1})^2} \right\} = \frac{(q_i^j - q_i^{j-1})}{(a_i^{j-1} - a_i^j)}$

Inferior limit of $F_i^{j+1}$: $\lim_{x_i \to a_i^{j+}} \frac{\partial Q(x,u)}{\partial a_i^j} = \lim_{x_i \to a_i^{j+}} \left\{ \frac{(q_i^j - q_i^{j+1}) \cdot (a_i^{j+1} - x_i)}{(a_i^j - a_i^{j+1})^2} \right\} = \frac{(q_i^j - q_i^{j+1})}{(a_i^{j+1} - a_i^j)}$

Modal point of $F_i^j$ : $\left. \frac{\partial Q(x,u)}{\partial a_i^j} \right|_{x=a_i^j} = 0$

In order to solve this problem of discontinuity, we impose the following constraint:

$$\lim_{x_i \to a_i^j} \frac{\partial Q(x,u)}{\partial a_i^j} = 0$$

In this case the modal point is not adjusted and it keeps its old value $a_i^j(t+1) = a_i^j(t)$. This constraint doesn't really affect the performances of the tuning algorithm because the parameters of the FPID are sufficiently close to the optimum.

2. The calculation of the $\frac{\partial Q(x,u)}{\partial a_i^j}$ becomes indefinite when $a_i^j = a_i^{j-1}$ and $a_i^j = a_i^{j+1}$.

   We can ovoid this case by imposing constraints of non overlappings of the membership functions to keep the legibility of the fuzzy rules

3. In order to insure the legibility of the fuzzy rules ovoiding labels inversions of the inputs membership functions, we imposed constraints during the adjustment process. Therefore we have to fix a superior limit and an inferior limit for each mobile modal points as can be shown in Fig. 6.

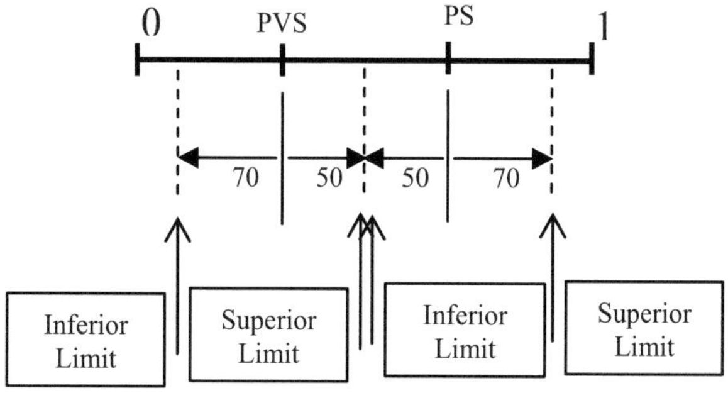


Fig. 6. Fixation of the variation limits modal points

The modal points of PS and PVS are mobile, so we have fixed their limits in order to have a maximal movement field which explains the choice of percentage (70%) to the right and to the left of respectively PS and of PVS .The field between the two modal points is equally shared (50%). The modal points of the negative part (NS et NVS) are obtained by symetry.

**2-2** ***Learning algorithm for consequents parameters***

The constant coefficient of the consequent part of the fuzzy rule is also updated by a rise gradient algorithm:

$$K_i^l(t+1) = K_i^l(t) + \beta \frac{\partial Q(x_{t+1}, u_{t+1})}{\partial K_i^l} \tag{33}$$

With

$$\frac{\partial Q(x_{t+1}, u_{t+1})}{\partial K_i^l} = \frac{\partial Q(x_{t+1}, u_{t+1})}{\partial u} \cdot \frac{\partial u}{\partial K_i^l} \tag{34}$$

Not having a mathematical model of the process, we approximate the Jacobian $\frac{\partial Q(x,u)}{\partial u}$ as follow:

$$\frac{\partial Q(x_{t+1}, u_{t+1})}{\partial u} = \frac{Q(x_{t+1}, u_{t+1}) - Q(x_t, u_t)}{u_{t+1} - u_t} \tag{35}$$

The previous expression becomes very large if $u_t \approx u_{t-1}$ which can cause an inflation of the FPID parameters. To remedy it, we adopt an important and frequently used simplification which consists in replacing the Jacobian by its sign [86]. Finaly, we obtain the following equation:

$$\frac{\partial Q(x_{t+1}, u_{t+1})}{\partial K_i^l} = sign\left(\frac{\partial Q(x_{t+1}, u_{t+1})}{\partial u}\right) \cdot \alpha^l x_i \quad \text{with } x_0 = 1; \tag{36}$$

**2.3.** ***Execution Procedure***

We present the detailed of an one-time-step global execution procedure as follows. Let $t+1$ be the current time step; the FPID obtain the new control output $u_{t+1}$ based on (25) by using perception $\alpha_t(x_t)$. The action $u_{t+1}$ after its transformation by the SAM becomes the $u_{c,t+1}$ (27). The action $u_{c,t+1}$ is applied on the system and makes it transits from the state $x_t$ towards $x_{t+1}$. The performance of the FPID controller is evalueted by the signal of reinforcement $r_{t+1}$. The perception organ in common to the FPID and **QE_FIS, perceives** then this state $x_{t+1}$ by the means of the firing strength of the rules $\alpha_t^l(x_{t+1})$. Let us note that $\alpha_t^l(x_{t+1})$ is subscripted by $t$, for it is the perception of $x_{t+1}$ but using the perception parameters (modal points) which we adjusted to the time step $t$.

**A.** The one stage of the learning algorithm consists in the execution procedure of the continuous tuning of the antecedent parameters of the FPID/QE-FIS. The eight steps are as follows:

1. Estimation of the function of evaluation t-optimal of the current state that is provided by the QE_FIS;

$$Q_t^*(\underline{x}_{t+1}, u_{t+1}) = \sum_{i=1}^{Nmf_1}\sum_{j=1}^{Nmf_2} \alpha_t^{i,j}(x_{t+1}) \cdot q_t^{i,j} \tag{37}$$

   The indice $t$ indicates that we have used the parameters adjusted to the previous time step.

2. Compute the TD error $\tilde{\varepsilon}_{t+1}$ using (21);
3. Tune parameter vector $q$ of the QE_FIS according to (22) and (17);
4. Tune the antecedent parameters (modal points) of the FPID as defined in (32**)**. The term of adjustment $\dfrac{\partial Q(x,u)}{\partial a_i^j}$ is given by table 1 and the learning rate $\rho$ is also adaptive according to the rule Delta Bar Delta seen in equation (17);
5. Recompute the perception since we adjusted the position of the modal points which makes change perception ;

$$\alpha_{t+1}^{i,j}(x_{t+1}) = \mu^{F_1^i}(x_1(t+1)) \cdot \mu^{F_2^j}(x_2(t+1)) \tag{38}$$

$$i = 1,\ldots, Nmf_1; \quad j = 1,\ldots, Nmf_2$$

6. Update the eligibility trace used for the parameters tuning of the QE_FIS according to (23);
7. New calculation of the evaluation function corresponding to the action ($u_{t+1}$) and the current state ($x_{t+1}$) with the new parameters ;

$$Q_{t+1}(\underline{x}_{t+1}, u_t) = \sum_{i=1}^{Nmf_1}\sum_{j=1}^{Nmf_2} \alpha_{t+1}^{i,j}(x_{t+1}) \cdot q_{t+1}^{i,j} \tag{39}$$

   This value will be used for the calculation of the TD error at the next step of time.

8. Calculation of the control ouput $u_{t+1}$, by using new perception;

$$u_{t+2} = \sum_{i=1}^{Nmf_1} \sum_{j=1}^{Nmf_2} \alpha_{t+1}^{i,j}\left(x_{t+1}\right) \cdot u^{i,j} \tag{40}$$

**B.** The stage seconde of the learning algorithm consists in the execution procedure of the continuous tuning of the consequent parameters of the FPID. The antecedent parameters are fixed to the values found in the stage one The sevent steps are as follows:

1. Estimation of the function of evaluation t-optimal of the current state $Q_t^*(\underline{x}_{t+1}, u_{t+1})$ as defined in (38) , that is provided by the QE_FIS ;
2. Compute the TD error $\tilde{\varepsilon}_{t+1}$ using (21 );
3. Tune parameter vector $q$ of the QE_FIS according to (22) and (17);
4. New calculation of the t-optimal evaluation function of the current state by using the new parameters of the QE_FIS:

$$Q_{t+1}(\underline{x}_{t+1}, u_{t+1}) = \sum_{i=1}^{Nmf_1} \sum_{j=1}^{Nmf_2} \alpha^{i,j}\left(x_{t+1}\right) \cdot q_{t+1}^{i,j} \tag{41}$$

5. Update the eligibility trace used for the parameters tuning of the QE_FIS according to (23) ; This stage completes the learning part of the QE_FIS
6. Tune the parameters of the FPID according to (33) and (36) ;

We use perception $\alpha^l\left(x_t\right)$and $x_j(t)$ for the tuning parameters of FPID, for it is them which are responsible for the issue of the control ouput $u_{t+1}$

7. Calculation of the control ouput $u_{t+2}$, by using the new parameters of FPID:

$$u_{t+2} = \sum_{i=1}^{Nmf_1} \sum_{j=1}^{Nmf_2} \alpha^{i,j}\left(x_{t+1}\right) \cdot u_{t+1}^{i,j} \tag{42}$$

## 6. Simulations-experimentations

The proposed fuzzy reinforcement-learning controller has been simulated for the DC/DC buck converter control problem and some simulation and experimentation results are presented.

### 1. DC/DC Buck Converter

The system is a 1kW DC/DC Buck Converter with the control of its output voltage in current mode which consists in imposing the current peak which crosses the smoothing inductance L therefore also the output current in the variable resistive load. It has an impulse response under deadened. We will carry out a voltage fuzzy control of the cascade type. To restrict the influence of measure noise, the output voltage is filtered using a 50 Hz low-pass first-order filter. The general diagram of such a chopper is given in fig.7. The values of the different components of the system are given by table 2.

The inner structure of the FPID part and its Fifteen (or twenty six) resulting tuning factors are defined on the normalised universe of discourse by the user. Thus values for the scaling factors *dem* and *gm* and the integrator gain called *Ki* have to be defined. The scaling factor *em* is supposed to be defined considering the benchmark step magnitude. Due to their global effect on the control performance and robustness, optimal input and output scaling factors play critical role in the FPID controller and they have the highest priority in terms of tuning and optimisation [68],[79]. To tune theme a lot of strategies have been proposed. Only these last three parameters pre-established values that we will provide in this part will depend on the open-loop identification of the process and on the sample period. In this way, this methodology can be seen as a Ziegler-Nichols -like tuning strategy for fuzzy controllers. In order to evaluate the performances of the control strategy, this pre- established set will be provided for a very common industrial benchmark: no-load step response, nominal load regulation, no-load regulation. It will warrant a high performances process response on this benchmark combined with a high robustness. The non-linear charge will be connected or disconnected according to the following functioning cycle fixed for the whole of our tests at 0.5 second: the no-load starting ($R_0$ = 20Ω) is carried out at t=0s; to sudden maximum load connection ($R$ = 150Ω) at t=0.166s (at the third of the horizon); and sudden disconnection ($R_0$ = 20Ω) intervenes at t=0.333s (at the two thirds of the horizon).

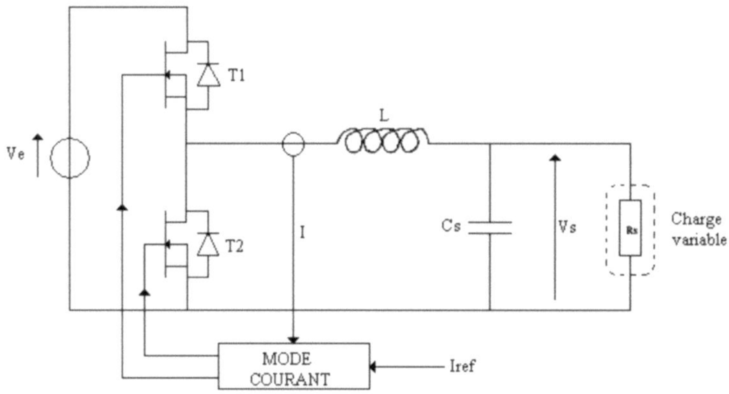


Fig. 7. General diagram of the Buck chopper

| | | | |
|---|---|---|---|
| $V_e$ | 60V | $\alpha_{min}$ | 0.05 |
| L | 1.3 mH | $R_{load}$ | 20 Ω |
| $C_s$ | 165 $\mu$F | $R_0$ | 150 Ω |
| I | ≤10A | $V_s$ | 0 to 10 V |

TABLE II. System Parameters Values

## 2.Open-loop identification test

In this paper, in the first part we want to propose pre-established settings for this FPID that are based on just one open-loop step identification test of the process.These predefined settings already exist in case of use of the FLC with an open-loop stable or evolutive process [62-63], that are widely found in electrotechnical applications. The control quality for this system will be evaluated considering the Integrative Absolute Error (IAE) between the reference and the measured signal on the specified benchmark. The FPID will then be all the more robust since these IAE-criterion remains insensitive to uncontrolable factors (white noise, process misindentification ...).

The open-loop identification of this stable process like DC/DC Buck converter will provide three parameters called *K*, *T* and τ for an open-loop transfer function of the process that can be expressed as Eq. 4. Figure 8 presents the open-loop response of the considered system.

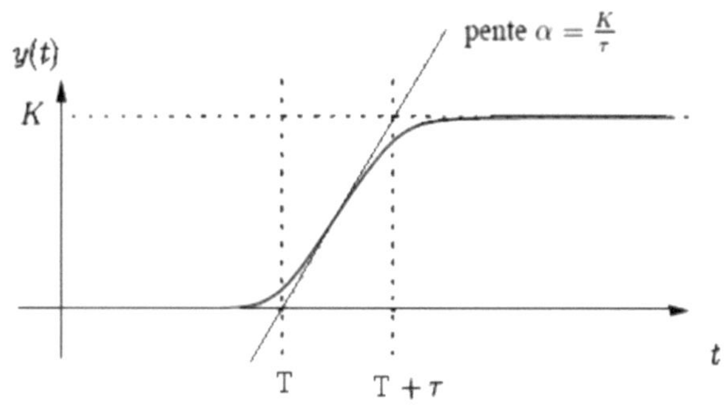


Fig. 8. Open-loop response of the considered system

$$FT(p) == K.\frac{e^{-Tp}}{1+\tau p} = FT^{H}_{50Hz} = 14.7 \cdot \frac{e^{-0.0028p}}{1+0.0174p} \tag{43}$$

with: K=14.7, T=0.0028s, τ=0.002.8s

Pre -established settings for the FPID-like FLC is provided in Table 3. The choice is inspired from the Broïda predefined settings [62] for conventional PID controllers by imposing an equal distribution of the membership functions of the inputs/outputs variables and in the case of a value step magnitude c with Tech the sampling period equal to 0.00015 s.

| Factors | Level |
|---|---|
| $PS_e$ | 0.67 |
| $PVS_e$ | 0.33 |
| $PS_{de}$ | 0.67 |
| $PVS_{de}$ | 0.33 |
| $PS_s$ | 0.67 |
| $PVS_s$ | 0.33 |
| $e_m$ | $c = 60V$ |
| $K_i$ | $K_i = \frac{0.8}{K \cdot T} = 19.4363$ |
| $de_m$ | $de_m = \frac{2.5}{\tau \cdot T} \cdot (\tau + 0.4 \cdot T) \cdot T_{ech} \cdot c = 8.5530$ dV/dt |
| $g_m$ | $g_m = \frac{0.8}{K \cdot T} \cdot (\tau + 0.4 \cdot T) \cdot c = 21.5977$ |

TABLE III. Pre-Defined Factors Levels

Fig. 9 and Fig. 10 give the simulation results for the DC/DC converter controlled either by a conventional PID controller whose parameters have been set according to the Broida settings or by our FPID with pre-established settings (table 2) i.e. equally -distributed for antecedent memberships function and consequent values. With the FPID, all the simulations lead to an outstanding improvement of the behavior with respect to the standard tunings of the Broida PID controller. The measured IAE-criterion is 0.9729 V.s in the FPID case and 1.134 V.s in the classical PID case. Beyond this 17%-criterion improvement, we can underline the similar behavior of the two controllers.

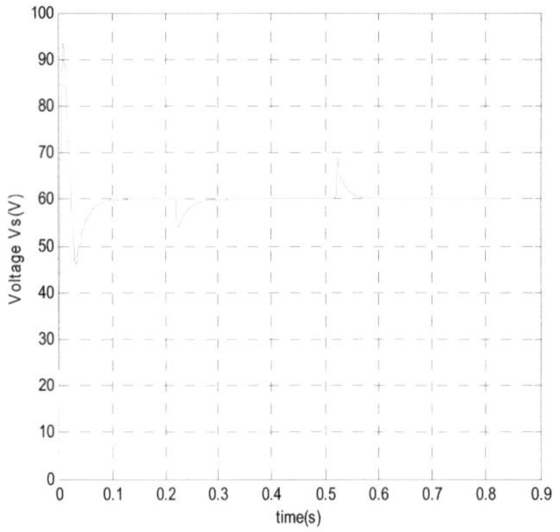


Output voltage control

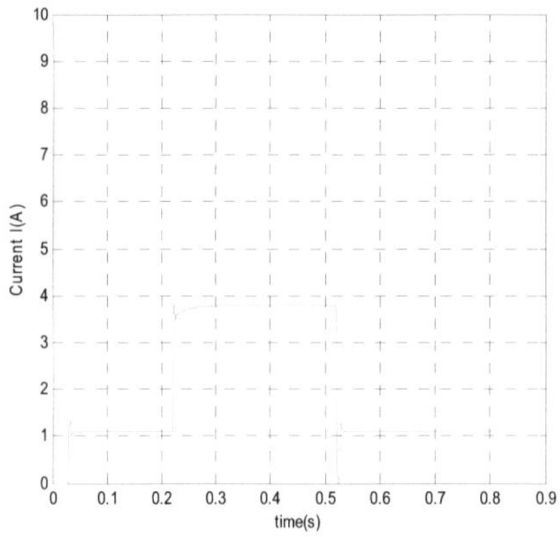


Imposed reference current

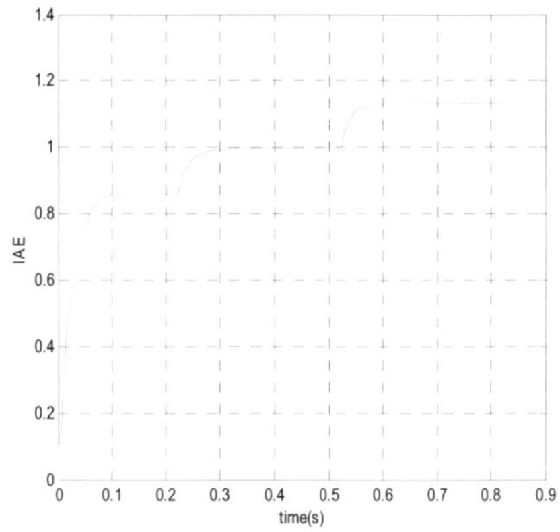


Fig. 9. Broida PID controller

### 3. Discret parameter tuning

Now, we apply the discret FQL algorithms to tuning the antecedent parameters then the consequent parameters of the FPID controller determined previously. The values for the scaling factors *dem, gm, Ki* and *em* are fixed during the learning processus and therefore we have eight parameters tuning (table 3). This doesn't constitute a difficulty because the reinforcement learning method doesn't depend on the parameter number to tune , which constitute a considerable advantage as compared to the other tuning methods such as the experimental product-plan method [62],[63].

First of all, it's necessary to define the levels (two in our case) for each of the considered controlable factors. Referring to Figure 2 and seeing the 50%-overlapping rate as a mechanical constraint, the positions of the PVS's and PS's membership functions apex aren't totally independent. When the position of PS is fixed at $x$ (x between 0 and +1), the position of PVS can only be between 0 and $x$. Thus, it's more advisable to define the two levels of these input factors relatively [62]. Under his assumption, the two chosen levels for each of the eight tuning factors are provided in Table 4.

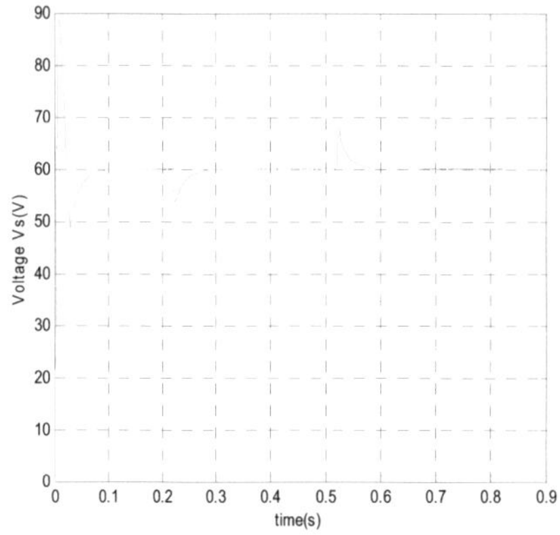


Output voltage control

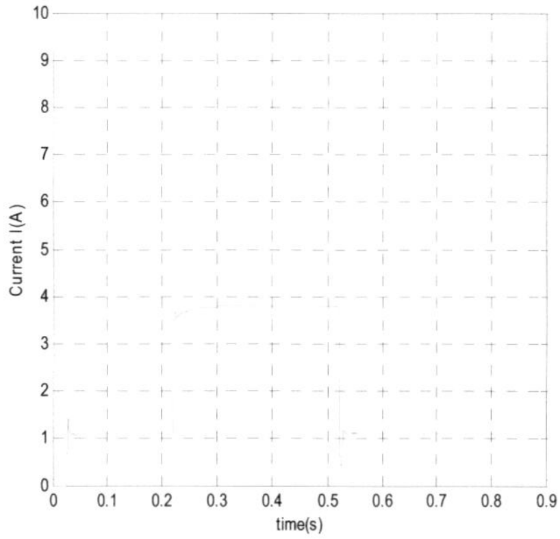


Imposed reference current

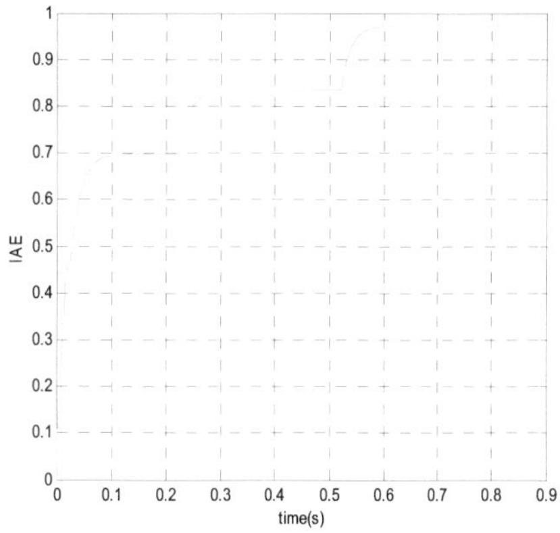


Fig. 10. Broida FPID controller

| Factors | Level 1 | Level 2 | Level 3 | Level 4 |
|---|---|---|---|---|
| $PS_e$ | 0.30 | 0.40 | 0.60 | 0.75 |
| $PVS_e$ | 0.12 | 0.17 | 0.23 | 0.28 |
| $PS_{de}$ | 0.30 | 0.40 | 0.60 | 0.75 |
| $PVS_{de}$ | 0.12 | 0.17 | 0.23 | 0.28 |
| $PS_s$ | 0.20 | 0.40 | 0.60 | 0.80 |
| $PVS_s$ | 0.20 | 0.40 | 0.60 | 0.80 |
| $NS_s$ | -0.20 | -0.40 | -0.60 | -0.80 |
| $NVS_s$ | -0.20 | -0.40 | -0.60 | -0.80 |

TABLE IV. Controlable Factors Levels

The choice of these levels has been achieved so a to sweep the area of researches.

### 3.1 Reinforcement function

The next step is to define the reinforcement signal *r*. The process performances are evaluated by the maximum absolute value of the error deviation *e*, because the objective of the FPID controller is to minimise this error as small as possible in tracking and in regulation. The reinforcement signal *r* may be defined as follows:

$$r(t) = \begin{cases} -\dfrac{|e|}{e_{\max}} & \textit{If} \quad |e| \leq e_{\max}, \text{success} \\ -1 & \textit{If} \quad |e| \succ e_{\max}, \text{failure} \end{cases} \tag{46}$$

We adopt a reinforcement function with a punitive policy in order to obtain optimal parameters in a short time period. In our simulations, we assume the maximum allowed error as 0.05. The parameters used in the simulations and experimentations are summarized in Table 5. The second value of parameter indicated in this table relates to the learning algorithm for consequents parameters.

| Parametres | Symbol | Value |
|---|---|---|
| Discount factors | $\gamma$ | 0.9/0.95 |
| Proximity factors of the omission function (eligibility traces ) | $\lambda$ | 0.9 |
| Initial learning rate | $\rho$ | 0.1 |
| Interferer in the adaptative learning rate | $K_\rho$ | 0.0001/ 0.001 |
| Interferer in the adaptative learning rate | $\psi_\rho$ | 0.5/ 0.001 |
| Interferer in the adaptative learning rate | $\varphi_\rho$ | 0.5 |
| Tolerance Noise/Amplitude of the quality | $S_p$ | 0.1 |
| Proportioning of the exploration | $\theta$ | 1 |

TABLE V. Parameters of the Discrete FQL algorithm

At first we tried to put into functions the two algorithms (FQL applied to the premises and FQL applied to the conclusions) similarly. But different attempts have shown that a simultaneous tuning of antecedent and consequent parameters lead to instability, therefore we opted for a separated tuning procedure beginning with the FQL for antecedent part followed by those of the consequent part.

The final tunings are provided in Table 6. Figures 11, 12 and 13 present the simulations results on the DC/DC converter. The "before optimisation" results have been obtained when the membership functions and the singletons are equidistributed on the universe of discourse. This kind of "medium"setting can be seen as a preliminary setting but constitutes unfortunately often the final setting of the FLC too.

### 3.2 Simulations and experimental results

**Stage one: antecedent parameters tuning**

We use the FQL algorithm applied to the antecedents with equally -distributed consequent values. We obtain the following results.

| Factors | $PS_e$ | $PVS_e$ | $PS_{de}$ | $PVS_{de}$ |
|---|---|---|---|---|
| Level selected | 0.75 | 0.12 | 0.3 | 0.17 |

TABLE VI. Modal points determined by discrete tuning

The results of simulations (fig.11) during the learning stage give an IAE criterion equals to 0.8544. We can notice then an improvement due to a larger space of solutions.

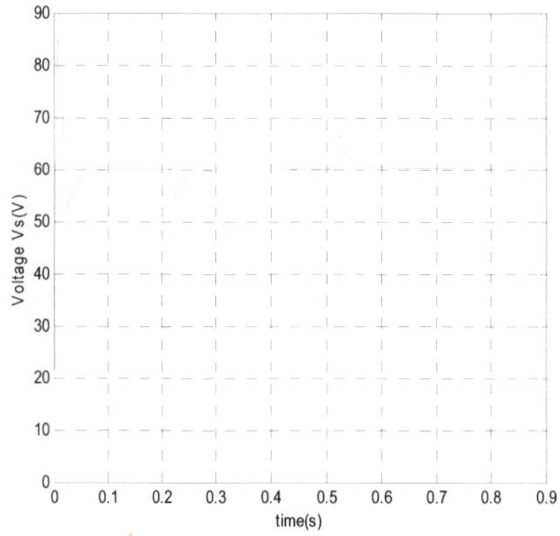


Output voltage control

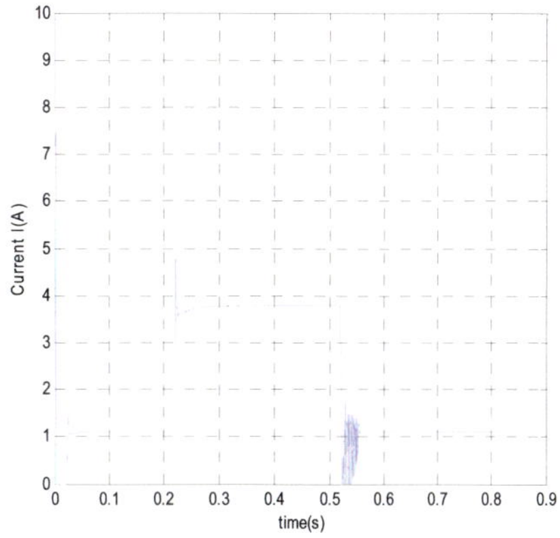


Imposed Reference Current

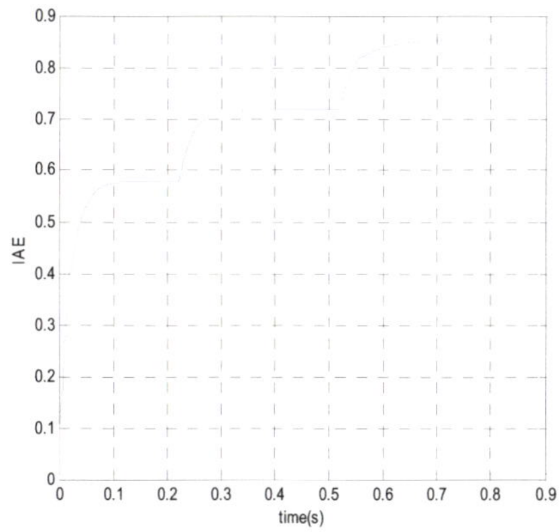


Fig.11. tecedent tuning of the FPID controller

**Stage second: consequent parameters tuning**

Initially, to show the importance to tune the antecedents parameters, we execute classic algorithm FQL with equally -distributed for antecedent memberships function. The results of simulations obtained during the learning stage give an IAE criterion equals 0.938, what corresponds to a deterioration of 10% of the optimization criterion as compared to the value obtained with classic FQL algorithm. Then we use the consequent FQL algorithm with the modal point values of the input membership functions determined in stage one.We get then the rule base showned in the table 7. The results of the simulations obtained during the learning (fig 12 and 13) show an improvement of the IAE criterion, equals now to 0.5318 compared to respectevely 0.8544 for the during the learning and validation phase. Therefore we have reached an improvement of the criterion of 12% as compared to the stage one. These simulations results show an important decreasing of the optimisation criterion (about 45%) as compared to the value obtained before the tuning by FQL algorithms. More over the behaviour of tracking is improved because the response time is shorter.

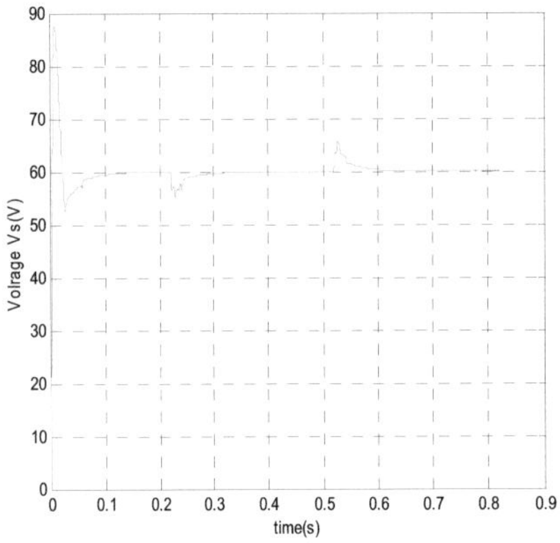


Output voltage control

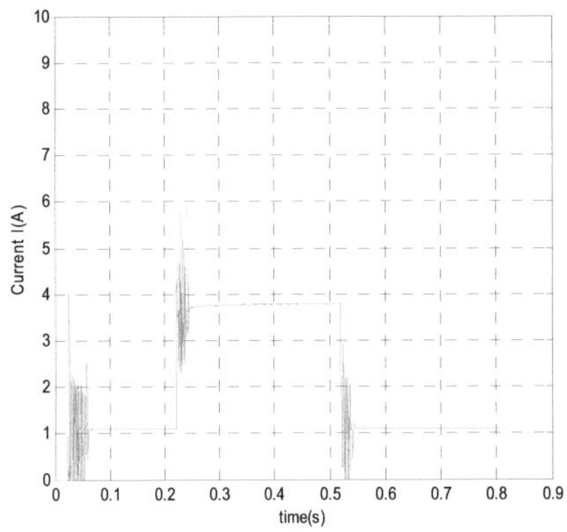


Imposed rference current

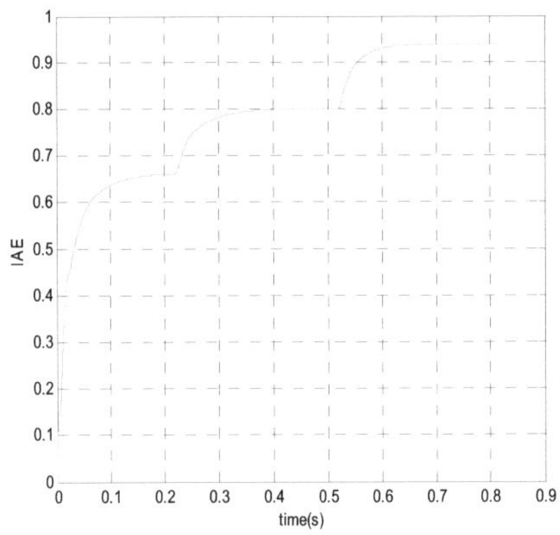


Fig. 12. Consequent tuning of the FPID controller with classic FQL algorithm

| | | *de* | | | | | | |
|---|---|---|---|---|---|---|---|---|
| | | NB | NS | NVS | Z | PVS | PS | PB |
| *e* | NB | -1 | -1 | -1 | -1 | -0.2 | -0.8 | 0 |
| | NS | -1 | -1 | -1 | -0.4 | -0.4 | 0 | 0.8 |
| | NVS | -1 | -1 | -0.8 | -0.6 | 0 | 0.4 | 0.2 |
| | Z | -1 | -02 | -0.8 | 0 | 0.6 | 0.4 | 1 |
| | PVS | -0.8 | -0.8 | 0 | 0.8 | 0.8 | 1 | 1 |
| | PS | -0.4 | 0 | 0.8 | 0.2 | 1 | 1 | 1 |
| | PB | 0 | 0.4 | 0.8 | 1 | 1 | 1 | 1 |

TABLE VII Antidiagonal rule base

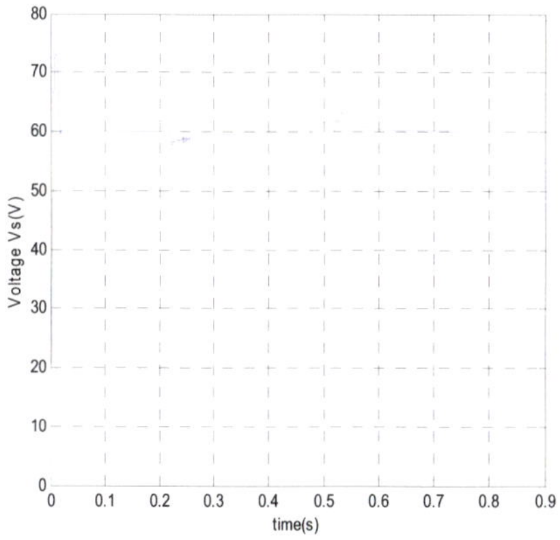


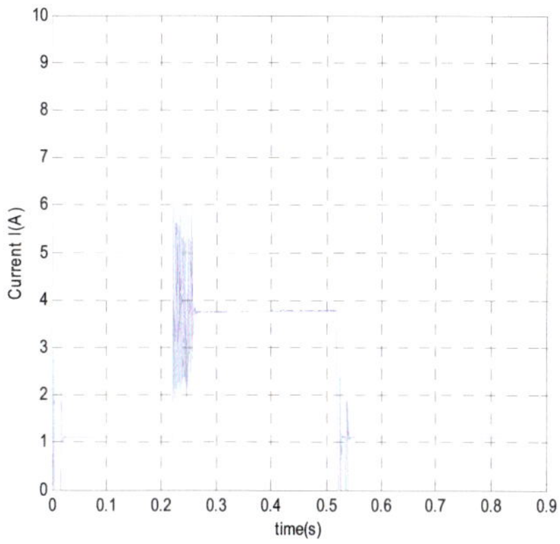


Output voltage control

Imposed rference current

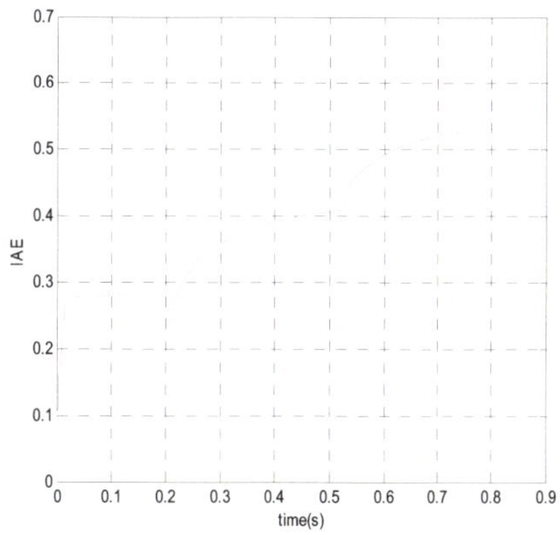


Fig.13. Consequent tuning of the FPID controller (table 6 with antidiagonal rules table 7)

### 4. Continuous parameter tuning

The previous obtained results (tables 6 and 7) constitute values close to the optimum which just need to be refined by means of the proposed continuous tuning algorithm. The parameters of the continuous FQL algorithm used in the simulations/experimentations are summarized in Table 8.

The simulation results obtained with non antidiagonal table 10 shown on system responses of the fig.14 , an IAE criterion of 0.4789 during the learning phase, which constitute a decreasing of 10% in comparison to the discreet tuning and of 58% in comparison to the classic Broïda PID. Let's remind also that the method of experimentation plans measure an IAE criterion of 1.05 (standard tuning) and 0.56 (robust tuning) with 16 experimental tests, therefore we have reached an improvement of 54% (standard tuning) and 15% (robust tuning) in relation to this method.

| Parametres | Symbol | Value |
|---|---|---|
| Discounted factor | $\gamma$ | 0.9 |
| Proximity factor of the omission function (eligibility traces) | $\lambda$ | 0.9 |
| Learning rate QE_FIS | $\rho$ | 0.0001 |
| Interfere in the adaptative learning rate | $K_\eta$ | 0.00001 |
| Interfere in the adaptative learning rate | $\psi_\eta$ | 0.5 |
| Interfere in the adaptative learning rate | $\varphi_\eta$ | 0.5 |
| Learning rate FPID control | $\beta$ | 0.00001 |
| Learning rate of the antecedents | $\rho$ | 0.00001 |

TABLE VIII Parameters of the Continuous FQL algorithm

We obtain the following results with the continuous FQL algorithm.

| Factors | $PS_e$ | $PVS_e$ | $PS_{de}$ | $PVS_{de}$ |
|---|---|---|---|---|
| Level selected | 0.75 | 0.07 | 0.3 | 0.171 |

TABLE IX Modal points determined by continuous tuning

| | | *de* | | | | | | |
|---|---|---|---|---|---|---|---|---|
| | | NB | NS | NVS | Z | PVS | PS | PB |
| *e* | NB | -1 | -1 | -1 | -1 | -0.2 | -0.8 | 0 |
| | NS | -1 | -1 | -1 | -0.4001 | -0.4 | 0 | 0.6 |
| | NVS | -1 | -1 | -0.8 | -0.6006 | 0 | 0.4 | 0.4 |
| | Z | -1 | -0.2 | -0.8 | 0 | 0.6 | 0.4 | 1 |
| | PVS | -0.8 | -0.8 | 0 | 0.8005 | 0.8 | 1 | 1 |
| | PS | -0.4 | 0 | 0.8 | 0.2 | 1 | 1 | 1 |
| | PB | 0 | 0.4 | 0.8 | 1 | 1 | 1 | 1 |

TABLE X Non Antidiagonal rules base

The best values controllable factors determined by simulations are then used to carry out the experimental tests.The experimental results (fig 16) are very similar to the simulation results, which confirms the performances of experimental methodology proposed. Moreover, it has better results on the the system responses and IAE criterion in comparison with standard and robust methods experimental plans as can be seen in Fig.15. The Table 11 relating to the IAE criterion confirmed the quality of the obtained responses by our approach. The different results show that it carries out one compromised for the two types of operation in tracking and in regulation. Also, the performances of the system can be greatly improved by executing a specific FPID controller for each operation type in comparison to one FPID controller functioning on all the operating range.The figure 17 shows that the time of response and the time of reject of the perturbations have considerably decreased. These simulations results show an important decreasing of the optimisation criterion (about 2%) as compared to the value obtained with the only using of one FPID controller tuning by FQL algorithms.

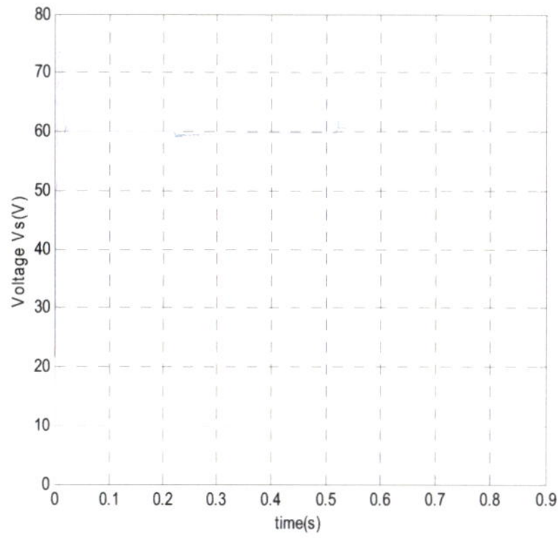


Output voltage control

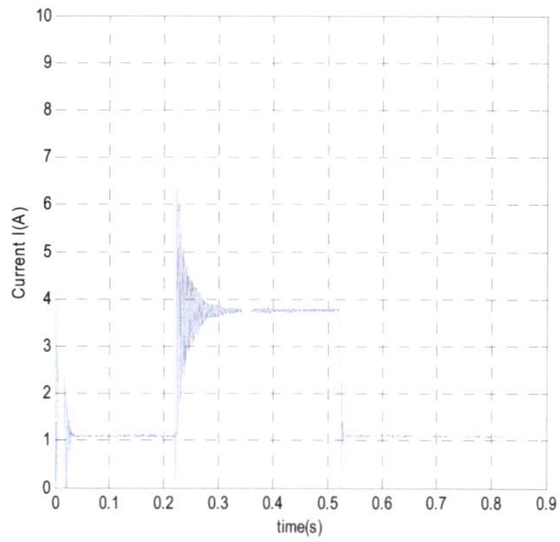


Imposed reference current

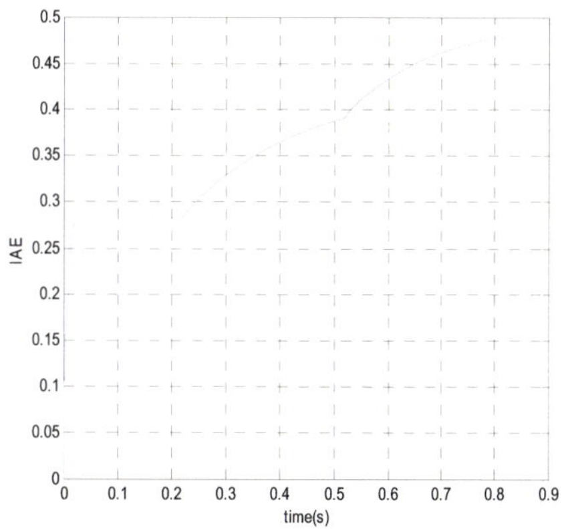


Fig. 14. Continuous tuning of the FPID controller

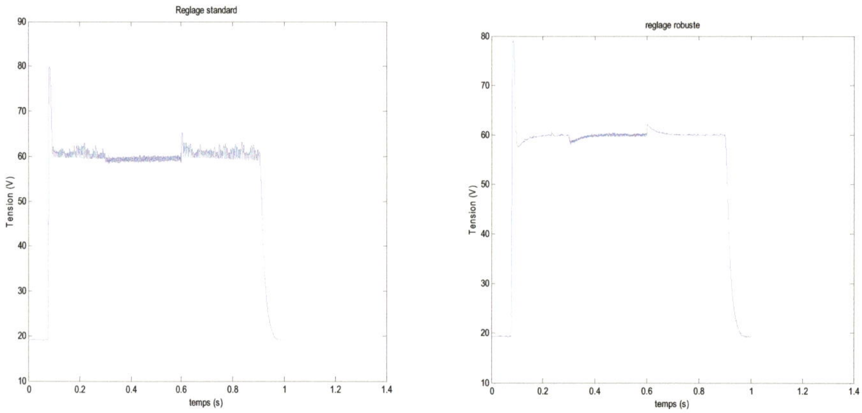


Output voltage control

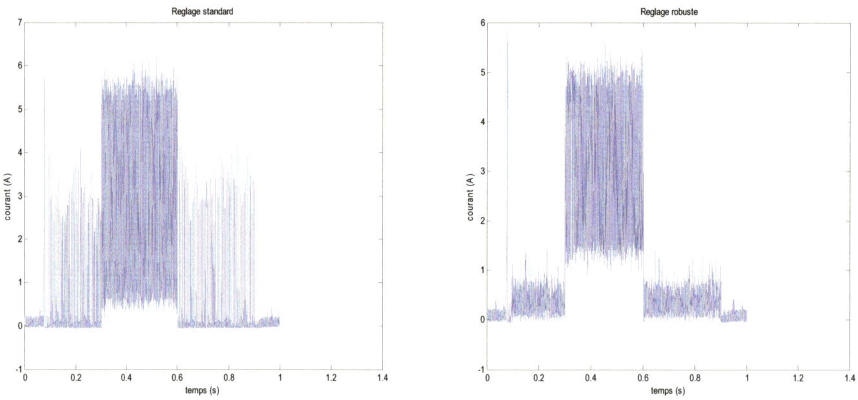


Imposed reference current

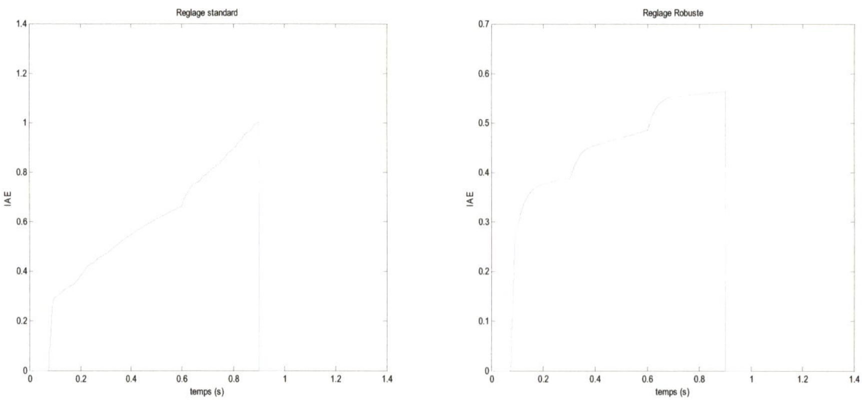


IAE criterion

Fig.15. Experimental results of FPID controller (standard and robust methods of experimental plans)

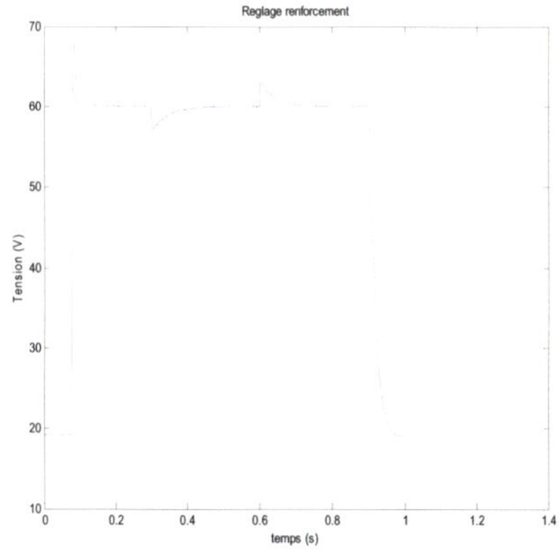


Output voltage control

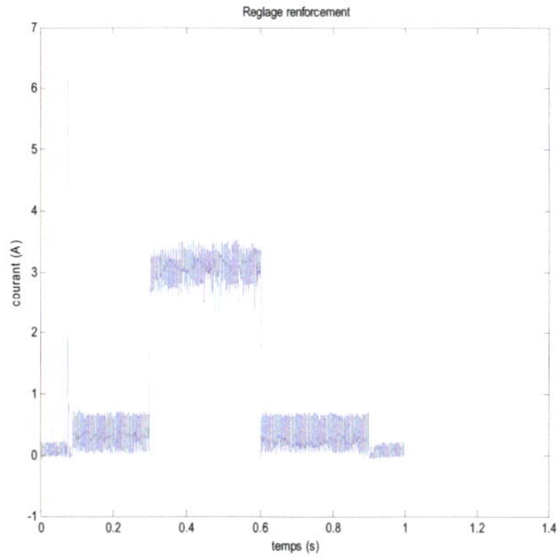


Imposed reference current

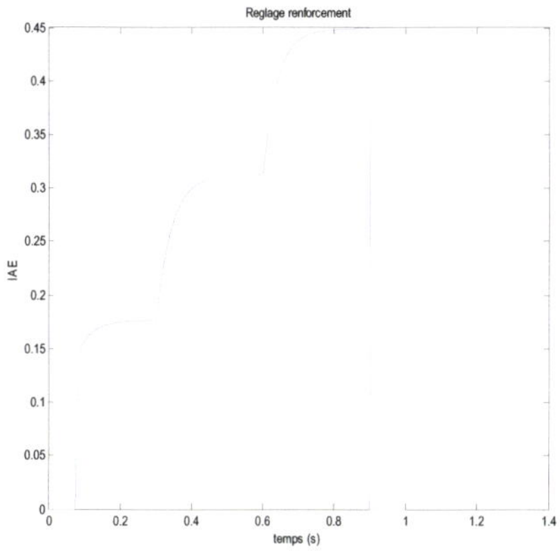


IAE criterion

Fig.16. Experimental results of the FPID controller continuous tuning

| Methods | IAE |
|---|---|
| PID Broida | 1.134 |
| FPID Broida | 0.9729 |
| Classic FQL | 0.9 |
| Discret FQL | 0.5454 |
| Continous FQL (one FPID), | 0.5396 |
| Continous FQL (two FPID) | 0.4693 |
| Expermental plans (standard) | 1.05 |
| Expermental plans (robust) | 0.56 |

TABLE XI Comparative table of the various methods

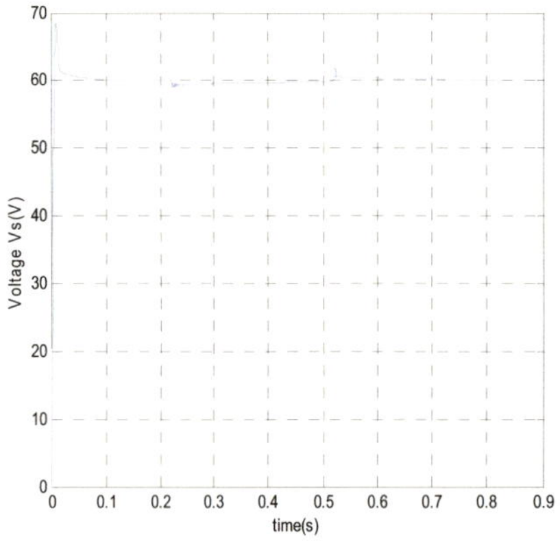


Output voltage control

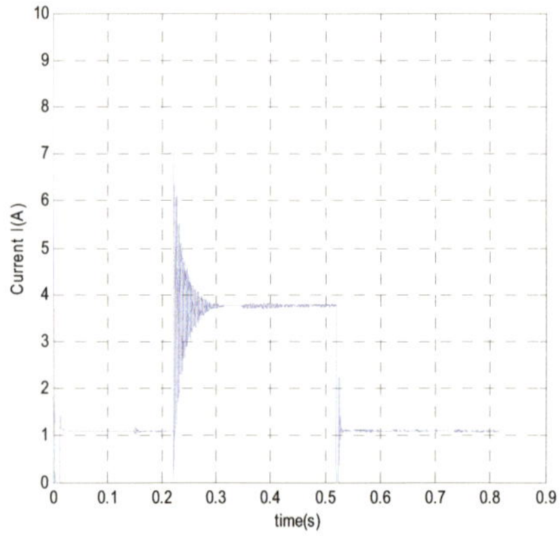


Imposed reference current

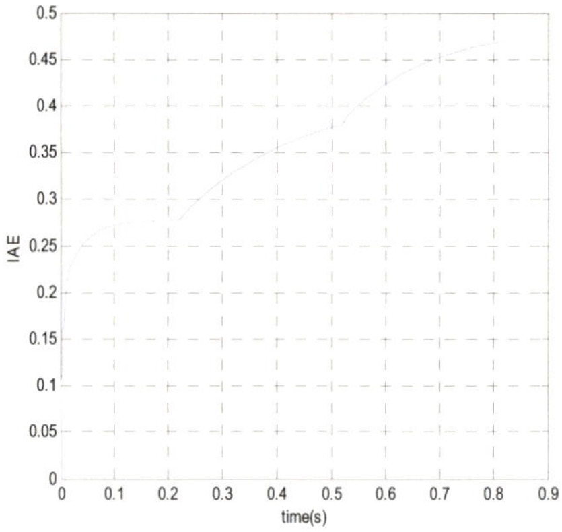


IAE criterion

Fig.17. Simulation results ofcommutation between two FPID controller obtained by method continuous tuning

## 7. Conclusion

On the whole, this paper provides some new and original headlines for the on-site tuning PID like fuzzy logic controllers.With the Broida methodology for conventional PID controllers; it is possible to obtain satisfying basic FPID. These settings depend only on two or three parameters (K, T and eventually $\tau$) extracting from an open-loop identification test of the process.

A second set of pre-defined settings for the controllable factors FPID is proposed and selected from the reinforcement learning design procedure. Two discrete tuning algorithms are developed based FQL for antecedent parameters and consequent parameters tuning respectively. The process performances are always evaluated considering the IAE-criterion between the reference and the measured signal. The FPID will then be all the more robust since this IAE-criterion remains insensitive to uncontrollable factors (here white noise, process misidentification or high order process). This second stage tuning could be seen as an on-site roughly FPID tuning strategy.

Finally the paper presents the controller architecture composed of the QE-FIS and the FPID for solving continuous space reinforcement learning problems and fine tuning controllable factors. The QE-FIS is used to estimate the optimal action-value function, and the FPID is used to get the control output based on the estimated action value function provided by the QE-FIS. With the proposed architecture, the parameters learning algorithms for the QE-FIS and the FPID are developed based on techniques of temporal difference and rise gradient algorithm. The two stages (discrete tuning and continuous tuning) complete one another because the first stage does a preliminary tuning which reduces the field of researches and provides an initialization to the second stage.This initialization of the parameter vector close to the optimum enables to ovoid the local optimums of the gradient methods. Finally, the simulations and the experimentation of the DC/DC converter demonstrate the validity and performance of the proposed learning algorithms and give better results as compared to the method of experimentation plans.

## 8. Acknowledgment

**Hacène Rezine**: was born in 1955 in El-Kala, Algeria. He received an engineering degree from the Ecole Nationale d'Ingénieurs et de Techniciens, Alger in 1979, the Engineer Degree from the Ecole Nationale Supérieur d'Electrotechnique, d'Electronique, d'Informatique, et d'Hydraulique, Toulouse, France in 1980 and the Doctor-Engineer Degree from the Institut Nationale Polytechnique of Toulouse in 1983. He is currently the Head of Unit of teaching and control research of the Algerian Polytechnic Military School. His current research areas include Electrical drives, Mobile robotic, Fuzzy control and Artificial Intelligent Control.
**Rabah Louali**: was born in October, 15th, 1982 in Blida (Algeria). He received the Engineering degree in 2006, from the Military Polytechnic School (Algeria). Currently, he is Magister student at this school. His current research areas include Mobile Robotic, Control of System and Fuzzy control.

## 9. References

[1] M. Fei, SLHo, "Progress in On-line Adaptive, Learning and Evolutionary Strategies for Fuzzy Logic Control", IEEE 1999 International Conference on Power Electronics and Drive Systems, Hong Kong. , pp 1108-1113. July 1999]

[2] Chin-Teng Lin, " A neural fuzzy control system with structure and parameter learning", Fuzzy Ssets and Systems, (70), pp 183-212, 1995

[3] F. Herrera, M. Lozano & J.L. Verdegay, "Tuning fuzzy logic controllers by genetic algorithms," Int. J. Approximate reasoning, v. 12, pp 299-315, 1995.

[4] Meng Joo Er, Member, IEEE, and Chang Deng, "Online Tuning of Fuzzy Inference Systems Using Dynamic Fuzzy Q-Learning ", IEEE Transactions on Systems, Man, and Cybernetics—PartT B: Cybernetics, Vol. 34, No. 3, June 2004

[5] L. Jouffe: "Actor-Critic Learning Based on Fuzzy Inference System", Proc of the IEEE Conference on Systems1996. Man and Cybernetics. Beijing, China, pp.339-344, 1996

[6] C. Watkins, P. Dayan: "Q-Learning Machine Learning", pp.279-292, 1992.

[7] P. Y. Glorennec and L. Jouffe, "Fuzzy Q-learning," Proc. OfIEEE Int. Con$ On Fuzzy Systems, pp. 659-662, 1997

[8] Dongbing Gu- Huosheng Hu- Libor Spacek, ": Learning Fuzzy Logic ontroller for Reactive Robot Behaviours", Proceedings of the 2003 IEEEASME International Conference on Advanced Intelligent Mechatronics, pp 46-51, 2003.
[9] C.C. Lee, "Fuzzy logic in control systems: fuzzy logic controller- part I&II," IEEE T-SMC, v.20, n.2, pp404-435, 1992.
[10] Mitra, S., Hayashi, Y. (2000). Neuro-Fuzzy Rule Generation: Survey in Soft Computing Framework, IEEE Trans. On Neural Networks, Vol. 11, No. 3...
[11] Dongbing Gu and Huosheng Hu, « Reinforcement Learning of Fuzzy Logic Controllers for Quadruped Walking Robots", Copyright 2002 IFAC 15th Triennial World Congress, Barcelona, Spain.
[12] Chia-Feng Juang, « Construction of Dynamic Fuzzy If-Then Rules through Genetic einforcement Learning for Temporal Problems Solving », IEEE, pp 2341-2346, 2001.
[13] Meng Joo Er, Member, IEEE, and Chang Deng, "Online Tuning of Fuzzy Inference Systems Using Dynamic Fuzzy Q-Learning ", IEEE Transactions On Systems, Man, And Cybernetics—PartT B: Cybernetics, Vol. 34, No. 3, June 2004.
[14] Jang J. C, ''. ANFIS: Adaptive-Neural-Based Fuzzy Inference System'', IEEE Trans. on SMC, Vol. 23, pages 665- 685, Aug. 1993.
[15] J. S. R. Jang, C. T. Sun, and E. Mizutani, Neuro-Fuzzy and Soft Computing.Englewood Cliffs, NJ: Prentice-Hall, 1997.
[16] S. Wu, M. J. Er, and Y. Gao, "A fast approach for automatic generation of fuzzy rules by generalized dynamic fuzzy neural networks," IEEE Trans. Fuzzy Syst., vol. 9, pp. 578–594, Aug. 2001.].
[17] C. W. Anderson, "Strategy learning with multilayer connectionist representations," in Proc. 4th Int. Workshop on Mach. Learn., Irvine, CA, June 1987, pp. 103–114.
[18] Bonarini, A. Evolutionary Learning of Fuzzy rules:competition and cooperation, In Pedrycz, W. (Ed.), FuzzyModelling: Paradigms and Practice, Kluwer Academic Press,Norwell, MA, pp265 - 284,1997.
[19] Juang, C. F., Lin, J.Y., and Lin, C. T. (2000). Genetic Reinforcement learning through Symbiotic Evolution for Fuzzy Controller Design, IEEE Transactions on SMC-Part B, Vol. 30, No. 2, pages 290-301.
[20] C. Karr, "Genetic algorithms for fuzzy controllers," AI Expert, vol. 2, pp. 27–33, 1991.
[21] M. Lee and H. Takagi, "Integrating design stages of fuzzy systems using genetic algorithm," in Proc. 2nd IEEE Int. Conf. Fuzzy Systems, San Francisco, CA, 1993, pp. 612–617.
[22] K. Kropp, "Optimization of fuzzy logic controller inference rules using a genetic algorithm," in Proc. EUFIT'93, Aachen, Germany, 1993, pp.1090–1096.
[23] T. Kawabe, T. Tagami, and T. Katayama, "A genetic algorithm based minimax optimal design of robust I-PD controller," in Proc. IEE Int.Conf. Control, London, U.K., 1996, pp. 436–441.
[24] Berenji, H. R., Khedkar, P. (1992). Learning and Tuning Fuzzy Logic Controller through Reinforcements, IEEE Trans. OnNeural Networks, Vol. 3, No. 5, pages 724-740. September, 1992.
[25] Chin-Teng Ling, C. S. George Lee, "Reinforcement Structure/Parameter Learning for Neural-Network- Based Fuzzy Logic Control Systems", IEEE Trans. on Fuzzy Systems, vol. 2, no. 1, Feb. , 1994.

[26] C. T. Lin and C. S. G. Lee, "Reinforcement structure/parameter learning for an integrated fuzzy neural network," IEEE Trans. Fuzzy Syst., vol. 2, pp. 46–63, Feb. 1994

[27] Kaelbling, L. P., Moor, A. W. (1996). Reinforcement Learning: A Survey, Journal of Artificial Intelligence Research, Vol. 4, pages 237-285

[28] C. K. Chiang, H. Y. Chung, and J. J. Lin, "A self-learning fuzzy logic controller using genetic algorithms with reinforcements," IEEE Trans. Fuzzy Syst., vol. 5, no. 3, pp. 460–467, Jun. 1997.

[29] C. T. Lin and C. P. Jou, "Controlling chaos by GA-based reinforcement learning neural network," IEEE Trans. Neural Networks, vol. 10, no. 4,pp. 846–859, Jul. 1999.

[30] Lin, C. T., C. P. Jou,. " GA-Based Fuzzy Reinforcement Learning for Control of a Magnetic Bearing System", IEEETrans. On SMC-Part B, Vol. 30, No. 2, pages 276-289, 2000.

[31] Hyo-Byung Jun and Kwee-Bo Sim, "Behavior Learning and Evolution of Collective Autonomous Mobile Robotsbased on Reinforcement Learning and Distributed Genetic Algorithms",IEEE International Workshop on Robot and Human Communication 1997 IEEE,pp 248,253

[32] A. G. Barto and M. I. Jordan, "Gradient following without backpropagation in layered networks," in Proc. of IEEE I" Annual Con$ Neural Networks, vol. 2, San Diego, CA, 1987, pp. 629-636

[33] C. W. Anderson, ''Learning and Problem Solving With Multilayer Connectionist Systems '' , Ph.D. disseration, Univ. Massachusetts, Amherst, 1986.

[34] Kaelbling, Littman, and Moore (1996)[ Kaelbling, L. P., Littman, M. L., & Moore, A. W. (1996). Reinforcement learning: A survey.Journal of Arti_cial Intelligence Research, 4, 237{285.

[35] Sutton, R. S., & Barto, A. (1998). Reinforcement Learning: An Introduction. MIT Press, Cambridge, MA

[36][David E Moriarty- Alan C.Schultz-John J.Grefenstette, "Evolutionary Algorithms for Reinforcement Learning", Journal of Arti_cial Intelligence Research 11 (1999) 241-276 .

[37] R. S. Sutton, "Learning to predict by the methods of temporal differences," Mach. Learn., vol. 3, no. 1, pp. 9–44, 1988.

[38] R. S. Sutton and A. G. Barto, Reinforcement Learning. Cambridge, MA: MIT Press, 1998.

[39] J. N. Tsitsiklis and B. V. Roy, "An analysis of temporal-difference learning with function approximation," IEEE Trans. Autom. Control, vol. 42, no. 5, pp. 674–690, May 1997

[40] A. G. Barto, R. S. Sutton, and C. W. Anderson, "Neuron like adaptive elements that can solve difficult learning control problems," IEEE Trans. Systems, Man, Cybern., vol. SMC-13, no. 5, pp. 834–846, Sep. 1983.

[41] C.W. Anderson, "Learning to control an inverted pendulum using neural networks," IEEE Control Syst. Mag., vol. 9, pp. 31–37, 1989.

[42] C. J. Wakins, "Learning from delayed rewards," Ph.D. dissertation, King's College, Cambridge, U.K., 1989.

[43] C. J.Wakins and P. Dayan, "Technical note: Q-learning," Mach. Learn., vol. 8, pp. 279–292, 1992.

[44] C.W. Anderson, "Learning to control an inverted pendulum using neural networks," IEEE Control Syst. Mag., vol. 9, pp. 31–37, 1989

[45] Chia-Feng Juang, « Combination of Online Clustering and Q-Value Based GA for Reinforcement Fuzzy System Design", IEEE Transactions On Fuzzy Systems, Vol. 13, No. 3, June 2005, pp 289-302.
[46] H. R. Berenji, "Fuzzy Q-learning for generalization of reinforcement,"in Proc. IEEE Int. Conf. Fuzzy Systems, 1996, pp. 2208–2214.
[47] Rummery, G. A., On-Line Q-Learning Using Connectionist Systems, Tech. Rep., CUED/F-INFENG/TR 166, CambridgeUniversity, 1994.
[48] P. Y. Glorennec, "Fuzzy Q-learning and dynamic fuzzy Q-learning," in Proc. IEEE Int. Conf. Fuzzy Systems, vol. 1, Orlando, FL, 1994, pp. 474–479.
[49] T. Horiuchi, A. Fujino, O. Katai, and T. Sawaragi, "Fuzzy interpolationbased Q-learning with continuous states and actions," in Proc. IEEE Int. Conf. Fuzzy Systems, 1996, pp. 594–600.
[50] P. Y. Glorennec and L. Jouffe, "Fuzzy Q-learning," in Proc. IEEE Int. Conf. Fuzzy Systems, 1997, pp. 659–662.
[51] L. Jouffe, "Fuzzy inference system learning by reinforcement methods," IEEE Trans. Syst., Man, Cybern. C, Appl. Rev., vol. 28, no. 3, pp. 338–355, Aug. 1998.
[52] M. S. Kim and J. J. Lee, "Constructing a fuzzy logic controller using evolutionary Q-learning," in Proc. 26th Annu. Conf. Industrial Electronics Society, 2000, pp. 1785–1790
[53] R.Ssutton, ''Integrated Architecture for Learning, Planing, and Reacting Based on Approximating Dynamic programming'', Proc.of 7th international Conference on machine Learning, pp 216-224, 1990
[54] R.A Mccallum, ''Using Transition Proximity for Faster Reinforcement Learning, Proc.of 9th international Conference on machine Learning, pp 316-321, 1992
[55] S.P.Singh, ''Transfert of learning by composing solutions of elemental sequential tasks, Proc.of 7th international Conference on machine Learning, pp 323-339, 1992
[56] J.Peng, "Efficient memory-based Dynamic programming, Proc.of 12th international Conference on machine Learning, pp 438-446, 1995
[57] P. Y. Glorennec and L. Jouffe, "A reinforcement learning method for an autonomous robot," in Proc. EUFIT'96, 4th Eur. Congr. Intell. Tech. Soft Comput. Aachen, Germany, pp. 1100–1104
[58] Xiaohui Dai, Chi-Kwong Li, « An Approach to Tune Fuzzy Controllers Based on Reinforcement Learning for Autonomous Vehicle Control", IEEE Transactions On Intelligent Transportation Systems, Vol. 6, No. 3, September 2005, pp 285-293
[59] R. S. Sutton, "Integrated architectures for learning, planning, and reacting based on approximating dynamic programming," in Proc. 7th Int. Workshop Machine Learning, 1990, pp. 216–224.
[60] L. J. Lin, "Self-improving reactive agents: Case studies of reinforcement learning frameworks," in Proc. 1st Int. Conf. Simulation of Adaptive Behavior: From Animals to Animals, 1991, pp. 297–305
[61] H. R. Berenji, "Fuzzy Q-Learning: A new approach for fuzzy dynamic programming," in Proc. IEEE Int. Conf. Fuzzy Systems, 1994, pp. 486–491
[62] D. Hissel, P. Maussion, G. Gateau, J. Faucher,"Fuzzy Logic Control Optimization of Electrical Systems using Experimental Designs", in Proceedings of the EPE'97 Conference, Trondheim, Norv_ege, pp. 1.090-1.095, 1997.

[63] P. Maussion, D. Hissel, "Optimized Fuzzy Logic Controller Parameters for Open-loop Stable or Evolutive Electromechanical Systems", in Proceedings of the IECON'98 Conference, Aachen, 1998.
[64] S. Joe Qin, '' Auto-Tuned Fuzzy Logic Control'', Proc.of the America Control Conference Baltimore, Maryland, June 1994, pp 2465-2469
[65] Yang Xinxin" Yang Lei, He Kezhong, Huang Shengle, Guo Muhe, Zhang Bo, "A Double-Level Fuzzy Controller with an Intelligently Adjusting Strategy of Quantization and Scale Factors,1996 IEEE, pp 280-285.
[66] George K. I. Mann, Bao-Gang Hu, and Raymond G. Gosine, "Analysis of Direct Action ,Fuzzy PID Controller Structures", IEEE Transactions On Systems, Man, and Cybernetics—Part b: Cybernetics, Vol. 29, No. 3, June 1999, pp 371-388.
[67] Baogang Hu,- George K. I. Mann, and Raymond G. Gosine, "New Methodology for Analytical and Optimal Design of Fuzzy PID Controllers », IEEE Transactions On Fuzzy Systems, VOL. 7, No. 5, October 1999, pp 521-539
[68] Zhi-Wei Woo, Hung-Yuan Chung, jin-Jye Lin, '' A PID type Fuzzy controller with self-tuning scaling factors'', Fuzzy sets and Systems , 2000, pp 321-326
[69] P.T.Chan, W.F.Xie, A.B.Rad, '' Tuning of Fuzzy controller for an open-loop unstable system: a genetic approach '', Fuzzy sets and Systems , 2000, pp 137-152
[70] K. S. Tang,- Kim Fung Man,- Guanrong Chen, Sam Kwong, "An Optimal Fuzzy PID Controller", IEEE Transactions On Industrial Electronics, Vol. 48, No. 4, August 2001, pp 757-765.
[71] Bao-Gang Hu,- George K. I. Mann, and Raymond G. Gosine, "A Systematic Study of Fuzzy PID Controllers—Function-Based Evaluation Approach »,IEEE Transactions On Fuzzy Systems, vol. 9, No. 5, October 2001, , pp 669-712
[72] A. Balestrino, A. Landi, and L. Sani, "CUK Converter Global Control Via Fuzzy Logic and Scaling Factors » IEEE Transactions On Industry Applications, Vol. 38, No. 2, March/April 2002, pp 406-413.
[73] Michail Petrov, Ivan Genchev, and Albina Taneva, " Fuzzy PID control of nonlinear plants", 2002 first international IEEE Symposium "Intelligent systems", September 2002, pp 30-35.
[74] P.G.Escamilla-ambrosio, N. Mort, " A novel design and tuning procedure for PID type fuzzy logic controllers", 2002 First International IEEE Symposium "Intelligent Systems", September 2002, pp 36-41.
[75] Ning wang, " A Fuzzy PID Controller For Multi-Model Plants", proceedings of the first international conference on machine learning and cybernetics, bijing, 4-5 November 2002, , pp 1401-1404
[76] I.S. Akkizidisa,*, G.N. Robertsa, P. Ridaob, J. Batlleb , "Designing a Fuzzy-like PD controller for an underwater robot", Control Engineering Practice , pp 417-480, 2003.
[77] Yu Yongquan- Huang Ying- Zeng Bi, "The Dynamic Fuzzy Method to Tune the Weight Factors of Neural Fuzzy PID Controller", IEEE 2004, pp 2397-2402.
[78] Jingwei Xu- Xin Feng, « Design of Adaptive Fuzzy PID Tuner Using Optimization Method", Proceedings of the 5" World Congress on Intelligent Control and Automation, June 15-19, 2004. Hangzhou, PR. China, IEEE, pp 2454-2458.
[79] Huiwen Deng, Yi Wang, « An Adaptive Fuzzy Logic Controller with Self-tuning Scaling Factors Based on Neural Networks*, 2005 IEEE, pp 392-396.

[80] Guihua Han, Lihua Chen, Junpeng Shao, Zhibin Sun, "Study of Fuzzy PID Controller for Industrial Steam Turbine Governing System", Proceedings of ISCIT2005, 2005 IEEE, pp 1228-1232.

[81] Jian Pei[1], Li-Ming Zhao[1], De-Jun Wang[1], Liang Chu, " Fuzzy PID Control Of Traction System For Vehicles", Proceedings of the Fourth International Conference on Machine Learning and Cybernetics, Guangzhou, 18-21 August 2005, IEEE , pp 773-777.

[82] Han-Xiong Li, , Lei Zhang, Kai-Yuan Cai, Guanrong Chen," An Improved Robust Fuzzy-PID Controller With Optimal Fuzzy Reasoning", IEEE Transactions on Systems, Man, and Cybernetics—Part b: Cybernetics, Vol. 35, No. 6, December 2005, pp 1283,1294

[83] Y. Koike and K. Doya, "Multiple state estimation reinforcement learning for driving model: Driver model of automobile," in Proc. IEEE Int. Conf. Systems, Man, and Cybernetics (SMC), Tokyo, Japan, 1999,vol. 5, pp. 504–509,

[84] I. H. Suh, J. H. Kim, and F. C.-H. Rhee, "Fuzzy Q-learning for autonomous robot systems," in Proc. Int. Conf. Neural Networks, Houston, TX, 1997, vol. 3, pp. 1738–1743

[85] I. H. Suh, J. H. Kim, and S. R. Oh, "Region-based Q-learning for intelligent robot systems," in IEEE Int. Symp. Computational Intelligence Robotics and Automation (CIRA), Monterey, CA, 1997, pp. 172–178.

[86] Renders JM, "Metaphores biologiques appliquées à la commande de processus, Thèse de doctorat, Université libre de Bruxelles, 1994

# 20

# Autonomous Robot Navigation using Flatness-based Control and Multi-Sensor Fusion

Gerasimos G. Rigatos
*Unit of Industrial Automation,*
*Industrial Systems Institute,*
*26504, Rion Patras,*
*Greece*

## 1. Introduction

This research work contains results on flatness-based control and sensor fusion for mobile robot systems. Flatness-based control is applicable to *differentially flat systems* i.e. to systems the behavior of which is determined by the trajectory of a finite collection of quantities, consisting of the *flat output* and its derivatives [Mounier, H. & Roudolf, J. (2001)],[Rouchon, P. (2005)]. Flatness-based control is equivalent to the feedback linearization method where the error dynamics of the closed-loop system can be described by a linear ODE after state feedback and subsequently can be stabilized using methods from linear control theory [Fliess, M. & Mounier, H. (1999)], [Roudolf, H. (2003)]. An advantage of flatness based control is that it simplifies trajectory planning and enables open-loop controller design. For linear finite dimensional systems flatness coincides with controllability, while this property can be generalized in the case of infinite dimensional systems [Laroche, B.; Martin, P. & Petit, N. (2007)], [Martin, P. & Rouchon, P. (1999)], [Meurer, T. & Zeitz, M. (2004) ,[Lévine, J. & Nguyen, D.V. (2003)].
Motion planning and control of autonomous vehicles is an important research topic in robotics (results in [Rigatos, G.G. (2003)], [Rigatos, G.G.et al., (2001)], [Rigatos, G.G. (2008)]) and flatness based control has been proposed as a suitable methodology for this class of problems [Martin, P. & Rouchon, P, (1999)]. Using the concept of flatness-based control, motion control algorithms have been developed that permit steering of the robotic vehicle along any desirable path in the 2D plane [Oriolo, G.et al., (2002)]. It will be shown that the kinematic model of the robotic vehicle is a flat system and thus can be expressed using a flat output and its derivatives. Moreover, the case in which the mobile robot's state vector is estimated through fusion of measurements from distributed sensors will be examined [Caron, F. et al., (2007)], [Jetto, L. et al., (1999)], [Yang, N. et al., (2005)]. To this end, the state vector of the robotic vehicle will be reconstructed with the use of Gaussian or non-parametric state estimators (such as Extended Kalman Filtering or Particle Filtering) [Rigatos, G.G. (2007)], [Rigatos, G.G. & Tzafestas, S.G. (2007a)], [Rigatos, G.G. (2007b)]. Simulation experiments in the case of the completely measurable state vector can show that flatness-based control enables the mobile robot to follow any reference path. Additionally, simulation experiments can show that using the state vector which is estimated from sensor

fusion, flatness-based control is also efficient in making the mobile robot track any desirable trajectory.
The structure of the paper is as follows: In Section 2 principles of flatness-based control are analyzed and examples are given for finite dimensional systems. In Section 3, the applicability of flatness-based control to mobile robot systems is examined. In Section 4 the problem of sensor fusion for mobile robot navigation is studied. Sensor fusion of measurements coming from odometer and sonar sensors is performed for the estimation of the mobile robot's state vector. The reconstructed state vector is then used in the flatness-based control algorithm to make the robotic vehicle track the desirable trajectory. In Section 5, simulation experiments are carried out. First, it is shown how flatness-based control succeeds tracking of the desirable trajectory for the mobile robot when the complete state vector is measurable. Next, flatness-based control generates the control signal using the state vector which is reconstructed after the fusion of measurements from distributed sensors. Finally, in Section 6, concluding remarks are stated.

## 2. Flatness-based control

### 2.1 Differential flatness for finite dimensional systems

A finite dimensional system is considered. This can be written in the general form of an ODE, i.e.

$$S_i(w, \dot{w}, \ddot{w}, \ldots, w^i), \quad i = 1, 2, \ldots, q \tag{1}$$

The quantity $w$ denotes the system variable, while $w^i$, $i = 1, \ldots, q$ are its derivatives (these and can be for instance the elements of the system's state vector). The system of Eq. (1) is said to be *differentially flat* if there exists a collection of $m$ functions $y = (y_1, \ldots, y_m)$ of the system variables $w_i$, $i = 1, \ldots, s$ and of their time-derivatives, i.e.

$$y_i = \varphi(w, \dot{w}, \ddot{w}, \ldots, w^{\alpha_i}), \quad i = 1, \ldots, m \tag{2}$$

such that the following two conditions are satisfied [Roudolf, J. (2003)]:

1. There does not exist any differential relation of the form

$$R(y, \dot{y}, \ldots, y^{(\beta)}) = 0 \tag{3}$$

which implies that the derivatives of the flat output are not coupled in the sense of an ODE, or equivalently it can be said that the flat output is differentially independent.

2. All system variables, i.e. the components of $w$ (elements of the system's state vectors) can be expressed using only the flat output $y$ and its time derivatives

$$w_i = \psi_i(y, \dot{y}, \ldots, y^{\gamma_i}), \qquad i = i, \ldots, s \tag{4}$$

An equivalent definition of differentially flat systems is as follows:

Definition: The system $\dot{x} = f(x, u)$, $x \in R^n$, $u \in R^m$ is differentially flat if there exist relations $h$: $R^n \times (R^m)^{r+1} \rightarrow R^m$, $\varphi$: $(R^m)^r \rightarrow R^n$ and $\psi$: $(R^m)^{r+1} \rightarrow R^m$, such that

$y=h(x,u,\dot{u},...,u^{(r)})$, $x=\varphi(y,\dot{y},...,y^{(r-1)})$ and $u=\psi(y,\dot{y},...,y^{(r-1)},y^{(r)})$. This means that all system dynamics can be expressed as a function of the flat output and its derivatives, therefore the state vector $x$ and the control input $u$ can be written as $x(t)=\varphi(y(t),\dot{y}(t),...,y^{(r)}(t))$ and $u(t)=\psi(y(t),\dot{y}(t),...,y^{(r+1)}(t))$.

It is noted that for linear systems the property of differential flatness is equivalent to that of controllability. Next, two examples is given to clarify the design of a differentially flat controller for finite dimensional system.

*Example 1*: Flatness-based control of a nonlinear system [Laroche, B. et al., (2007)]

$$\begin{aligned}\dot{x}_1 &= x_3 - x_2u \\ \dot{x}_2 &= -x_2 + u \\ \dot{x}_3 &= x_2 - x_1 + 2x_2(u - x_2)\end{aligned} \tag{5}$$

The candidate flat differential output is $y_1 = x_1 + \frac{x_2^2}{2}$. Thus one gets:

$$\begin{aligned}y_1 &= x_1 + \frac{x_2^2}{2} \\ y_2 = \dot{y}_1 &= (x_3 - x_2u) + x_2(u - x_2) - x_2^2 \\ y_3 = \dot{y}_2 = \ddot{y}_1 &= x_2 - x_1 + 2x_2(u - x_2) - 2x_2(u - x_2) = -x_1 + x_2 \\ \upsilon = \dot{y}_3 = y_1^{(3)} &= -x_3 + x_2u - x_2 + u = -x_2 - x_3 + u(1 + x_2)\end{aligned} \tag{6}$$

It can be verified that property (1) holds, i.e. there does not exist any differential relation of the form $R(y,\dot{y},...,y^{(\beta)})=0$, and this implies that the derivatives of the flat output are not coupled. Moreover, it can be shown that property (2) also holds i.e. the components $w$ of the system (elements of the system's state vector and control input) can be expressed using only the flat output $y$ and its time derivatives $w_i=\psi_i(y,\dot{y},...,y^{\gamma_i})$, $i=1,...,s$. For instance to calculate $x_1$ with respect to $y_1$, $\dot{y}_1$, $\ddot{y}_1$ and $y_1^{(3)}$ the relation of $\ddot{y}_1$ is used, i.e.:

$$x_1^2 + 2x_1(1 + \ddot{y}_1) + \ddot{y}_1^2 - 2y_1 = 0 \tag{7}$$

from which two possible solutions are derived, i.e.: $x_1 = -((1+\ddot{y}_1) - \sqrt{1+2(y_1+\ddot{y}_1)})$, $x_1 = -((1+\ddot{y}_1) + \sqrt{1+2(y_1+\ddot{y}_1)})$. Keeping the largest of these two solutions one obtains:

$$x_1 = -(1+\ddot{y}_1+\sqrt{1+2(y_1+\ddot{y}_1)})$$
$$x_2 = \ddot{y}_1+x_1$$
$$x_3 = \dot{y}_1+\ddot{y}_1^2+2x_1\ddot{y}_1+x_1^2 \tag{8}$$
$$u = \frac{y_1^3+\ddot{y}_1^2+\ddot{y}_1+\dot{y}_1+x_1+2x_1\ddot{y}_1+x_1^2}{1+x_1+\ddot{y}_1}$$

Using the flat output and its derivatives, the system of Eq.(5) can be written in Brunovsky (canonical) form:

$$\frac{d}{dt}\begin{pmatrix} y_1 \\ y_2 \\ y_3 \end{pmatrix} = \begin{pmatrix} 0 & 1 & 0 \\ 0 & 0 & 1 \\ 0 & 0 & 0 \end{pmatrix}\begin{pmatrix} y_1 \\ y_2 \\ y_3 \end{pmatrix} + \begin{pmatrix} 0 \\ 0 \\ 1 \end{pmatrix}\upsilon \tag{9}$$

Therefore, a transformation of the system into a linear equivalent is obtained and then it is straightforward to design a a controller based on linear control theory. Thus given the reference trajectory $[x_1^r \;\; x_2^r \;\; x_3^r]^T$ one can find the transformed reference trajectory $[y_1^r \;\; \dot{y}_1^r \;\; \ddot{y}_1^r]^T$ and select the appropriate control input $v$ that succeeds tracking. Knowing $\upsilon$ the control $u$ of the initial system can be found.

*Example 2*: Flatness based Control of $N$ linear coupled oscillators [Rouchon, P. (2005)]

The generalized coordinates $z_i$ are considered and $n$ oscillators are taken. The oscillators can be coupled through an interaction term $f_i(z_1,z_2,...,z_N)=0$ and through the common control input $u$. This means that the general oscillator model can be written as

$$\frac{d^2}{dt^2}z_i = -(\omega_i)^2 z_i + f_i(z_1,z_2,...,z_N)+b_i u, \;\; i=1,...,N \tag{10}$$

or, for $f_i(z_1,z_2,..,z_N)=0$ one obtains

$$\frac{d^2}{dt^2}z_i = -(\omega_i)^2 z_i + b_i u, \;\; i=1,..,N \tag{11}$$

The terms $\omega_i > 0$ and $b_i \neq 0$ are constant parameters, while $T>0$ and $D \neq 0$ are also defined. The objective is to find open-loop control $[0,T]$ with $t \rightarrow u(t)$ steering the system from an initial to a final state. In [Rouchon, P. (2005)] it has been shown that such control can be obtained explicitly, according to the following procedure: the Laplace transform of Eq. (11) gives

$$(s^2+(\omega_i)^2)z_i = b_i u, \quad i=1,..,N \tag{12}$$

Then the system can be written in the form [Lévine, J. & Nguyen, D.V. (2003) ]:

$$z_i = Q_i(s)y, \quad u = Q(s)y, \quad \text{wirh } y = \sum_{k=1}^{N} c_k z_k$$
$$Q_i(s) = \frac{b_i}{(\omega_i)^2} \prod_{k=1}^{N} (1+(\frac{s}{\omega_k})^2) \text{ for } k \neq i, \quad Q(s) = \prod_{k=1}^{N}(1+(\frac{s}{\omega_k})^2), \quad c_k = \frac{1}{Q_k(j\omega_k)} \in R \tag{13}$$

The real coefficients $q_k^i$ and $q_k$ are defined as follows [Rouchon, P. (2005)]: $Q_i(s) = \sum_{k=0}^{N-1} q_k^i s^{2k}$ and $Q(s) = \sum_{k=0}^{N} q_k s^{2k}$. This enables to express both the system parameters $x(t)$ and the control input $u(t)$ as functions of the flat output $y(t)$, i.e.

$$x_i(t) = \sum_{k=0}^{n-1} q_k^i y^{(2k)}(t), \quad \upsilon(t) = \sum_{k=0}^{n} q_k y^{(2k)}(t) \tag{14}$$

which satisfy Eq. (11), with $z_i = x_i$ and $u = v$. Moreover it holds that

$$y(t) = \sum_k c_k x_k(t) \tag{15}$$

There is a one to one linear correspondence between the trajectories of Eq. (11) and the arbitrary smooth time function $y$. More precisely, any piecewise continuous open-loop control $u(t),\ t \in [0,T]$ steering from the steady-state $z_i(0) = 0$ to the steady-state $z_i(T) = \frac{b_i}{(\omega_i)^2} D$ can be written as

$$u(t) = \sum_{k=0}^{N} q_k y^{(2k)}(t) \tag{16}$$

for all functions $y(t)$ such that $y(0) = 0$, $y(T) = D\ \forall\ i\ \{1,2,..,2n-1\}$, $y^{(i)}(0) = y^{(i)}(T) = 0$.

The results can be extended to the case of a harmonic oscillator with damping. In that case Eq. (11) is replaced by [Rouchon, P. (2005)]

$$\frac{d^2 z_i}{dt^2} = -\omega_i^2 z_i - 2\xi_i \omega_i \dot{z}_i + b_i u, \quad i = 1,\dots,n \tag{17}$$

where the damping coefficient is $\xi_i \geq 0$. In that case one obtains

$$z_i = Q_i(s)y, \quad u = Q(s)y \tag{18}$$

with

$$Q_i(s) = \frac{b_i}{(\omega_i)^2} \prod_{k=1}^{n} (1+2\xi_k(\frac{s}{\omega_k})+(\frac{s}{\omega_k})^2) \ k \neq i, \quad Q(s) = \prod_{k=1}^{n}(1+2\xi_k(\frac{s}{\omega_k})+(\frac{s}{\omega_k})^2) \tag{19}$$

which proves again that the system's parameters (state variables) and the control input can be written as functions of the flat output *y* and its derivatives. In that case the flat output is of the form:

$$y = \sum_{k=1}^{n} c_k z_k + d_k s z_k \tag{20}$$

where $s = d/dt$ and the coefficients $c_k$ and $d_k$ can be computed explicitly. According to [Lévine, J. & Nguyen, D.V. (2003)] explicit descriptions of the system parameters via an arbitrary function $y$ (flat output) and its derivatives are possible for any controllable linear system of finite dimension (controllability is equivalent to flatness).

### 2.2 Flatness-based control for the unicycle robot

Several mobile robots with non-holonomic motion constraints, are differentially flat systems. The conditions for motion without sliding are given in the sequel:

$$\begin{aligned} \dot{x} &= v\cos\theta \\ \dot{y} &= v\sin\theta \\ \dot{\theta} &= \frac{v}{l}\tan\varphi \end{aligned} \tag{21}$$

where $v$ is the velocity of the vehicle, $l$ is the vehicle's length, $\theta$ is the angle between the transversal axis of the vehicle and axis $OX$, and $\phi$ is the angle of the steering wheels with respect to the transversal axis of the vehicle. The flat output is the cartesian position of the vehicle's center of gravity, denoted as $\eta = (x, y)$, while the other model parameters can be written as:

$$v = \pm \| \dot{\eta} \|, \quad \begin{bmatrix} \cos(\theta) \\ \sin(\theta) \end{bmatrix} = \frac{\dot{\eta}}{v}, \quad \tan(\varphi) = l\det(\dot{\eta}\ddot{\eta})/v^3 \tag{22}$$

These formulas show simply that is the tangent angle of the curve traced by $P$ and $\tan(\phi)$ is the associated curvature.
With reference to a generic driftless nonlinear system

$$\dot{q} = G(q)w, \quad q \in R^n, \quad w \in R^m \tag{23}$$

the dynamic feedback linearization consists in finding a feedback compensator of the form

$$\begin{aligned} \dot{\xi} &= \alpha(q,\xi) + b(q,\xi)u \\ w &= c(q,\xi) + d(q,\xi)u \end{aligned} \tag{24}$$

with state $\xi \in R^v$ and input $u \in R^m$, such that the closed-loop system of Eq. (23) and Eq. (24) is equivalent under a state transformation $z = T(q,\xi)$ to a linear system. The starting point is the definition of an $m$-dimensional output $\eta = h(q)$ to which a desired behavior can be assigned. One then proceeds by successively differentiating the output until the input appears in a non-singular way. If the sum of the output differentiation orders equals the dimension $n+v$ of the extended state space, full input-state-output linearization is obtained (In this case $\eta$ is also called a flat output). The closed-loop system is then equivalent to a set of decoupled input-output chains of integrators from $u_i$ to $\eta_i$ $(i = 1, 2, .., m)$. The exact linearization procedure is illustrated for the unicycle model of Eq. (21). As flat output the coordinates of the center of gravity of the vehicle is considered $\eta = (x, y)$. Differentiation with respect to time then yields [Oriolo, G. et al.,(2002)]

$$\dot{\eta} = \begin{bmatrix} \dot{x} \\ \dot{y} \end{bmatrix} = \begin{pmatrix} \cos(\theta) & 0 \\ \sin(\theta) & 0 \end{pmatrix} \cdot \begin{bmatrix} v \\ \omega \end{bmatrix} \quad (25)$$

showing that only $v$ affects $\dot{\eta}$, while the angular velocity $\omega$ cannot be recovered from this first-order differential information. To proceed, one needs to add an integrator (whose state is denoted by $\xi$) on the linear velocity input

$$v = \xi, \quad \dot{\xi} = \alpha \Rightarrow \dot{\eta} = \xi \begin{pmatrix} \cos(\theta) \\ \sin(\theta) \end{pmatrix} \quad (26)$$

where $\alpha$ denotes the linear acceleration of the robotic vehicle. Differentiating further one obtains

$$\ddot{\eta} = \dot{\xi} \begin{pmatrix} \cos(\theta) \\ \sin(\theta) \end{pmatrix} + \xi \dot{\theta} \begin{pmatrix} \sin(\theta) \\ \cos(\theta) \end{pmatrix} = \begin{pmatrix} \cos(\theta) & -\xi \sin(\theta) \\ \sin(\theta) & \xi \cos(\theta) \end{pmatrix} \begin{pmatrix} \alpha \\ \omega \end{pmatrix} \quad (27)$$

and the matrix multiplying the modified input $(a, \omega)$ is nonsingular if $\xi \neq 0$. Under this assumption one defines

$$\begin{pmatrix} \alpha \\ \omega \end{pmatrix} = \begin{pmatrix} \cos(\theta) & -\xi \sin(\theta) \\ \sin(\theta) & \xi \cos(\theta) \end{pmatrix}^{-1} \begin{pmatrix} u_1 \\ u_2 \end{pmatrix} \quad (28)$$

$\ddot{\eta}$ is denoted as

$$\ddot{\eta} = \begin{pmatrix} \ddot{\eta}_1 \\ \ddot{\eta}_2 \end{pmatrix} = \begin{pmatrix} u_1 \\ u_2 \end{pmatrix} = u \quad (29)$$

which means that the desirable linear acceleration and the desirable angular velocity can be expressed using the transformed control inputs $u_1$ and $u_2$. Then, the resulting dynamic compensator is (return to the initial control inputs $v$ and $\omega$)

$$\dot{\xi} = u_1 \cos(\theta) + u_2 \sin(\theta)$$
$$v = \xi \tag{30}$$
$$\omega = \frac{u_2 \cos(\theta) - u_1 \sin(\theta)}{\xi}$$

Being $\xi \in R$, it is $n + v = 3 + 1 = 4$, equal to the output differentiation order in Eq. (29). In the new coordinates

$$\begin{aligned} z_1 &= x \\ z_2 &= y \\ z_3 &= \dot{x} = \xi \cos(\theta) \\ z_4 &= \dot{y} = \xi \sin(\theta) \end{aligned} \tag{31}$$

The extended system is thus fully linearized and described by the chains of integrators, in Eq. (29), and can be rewritten as

$$\begin{aligned} \ddot{z}_1 &= u_1 \\ \ddot{z}_2 &= u_2 \end{aligned} \tag{32}$$

The dynamic compensator of Eq. (30) has a potential singularity at $\xi = v = 0$, i.e. when the unicycle is not rolling. The occurrence of such singularity is structural for non-holonomic systems. This difficulty must be obviously taken into account when designing control laws on the equivalent linear model.
A nonlinear controller for output trajectory tracking, based on dynamic feedback linearization, is easily derived. Assume that the robot must follow a smooth trajectory $(x_d(t), y_d(t))$ which is persistent, i.e. for which the nominal velocity $v_d = (\dot{x}_d^2 + \dot{y}_d^2)^{1/2}$ along the trajectory never goes to zeros (and thus singularities are avoided). On the equivalent and decoupled system of Eq. (32), one can easily design an exponentially stabilizing feedback for the desired trajectory, which has the form

$$\begin{aligned} u_1 &= \ddot{x}_d + k_{p_1}(x_d - x) + k_{d_1}(\dot{x}_d - \dot{x}) \\ u_2 &= \ddot{y}_d + k_{p_1}(y_d - y) + k_{d_1}(\dot{y}_d - \dot{y}) \end{aligned} \tag{33}$$

which results in the following error dynamics for the closed-loop system

$$\begin{aligned} \ddot{e}_x + k_{d_1}\dot{e}_x + k_{p_1}e_x &= 0 \\ \ddot{e}_y + k_{d_2}\dot{e}_y + k_{p_2}e_y &= 0 \end{aligned} \tag{34}$$

where $e_x = x - x_d$ and $e_y = y - y_d$. The proportional-derivative (PD) gains are chosen as $k_{p_i} > 0$ and $k_{d_i} > 0$ for $i = 1,2$. Knowing the control inputs $u_1$, $u_2$ for the linearized system

one can calculate the control inputs $v$ and $\omega$ applied to the robotic vehicle, using Eq. (24). The above result is valid, provided that the dynamic feedback compensator does not need the singularity $v = \xi_i = 0$. The following theorem assures the avoidance of singularities in the proposed control law [Oriolo, G. et al., (2002)]:
*Theorem*: Let $\lambda_{11}$, $\lambda_{12}$ and $\lambda_{21}$, $\lambda_{22}$, be respectively the eigenvalues of two equations of the error dynamics, given in Eq. (34). Assume that, for $i = 1,2$ it is $\lambda_{i1} < \lambda_{i2} < 0$ (negative real eigenvalues), and that $\lambda_{i2}$ is sufficiently small. If

$$\min_{t \geq 0} \left\| \begin{bmatrix} \dot{x}_d(t) \\ \dot{y}_d(t) \end{bmatrix} \right\| > \left\| \begin{bmatrix} \dot{\varepsilon}_x^0 \\ \dot{\varepsilon}_y^0 \end{bmatrix} \right\| \tag{35}$$

with $\dot{\varepsilon}_x^0 = \dot{\varepsilon}_x(0) \neq 0$ and $\dot{\varepsilon}_y^0 = \dot{\varepsilon}_y(0) \neq 0$, then the singularity $\xi = 0$ is never met.

## 3. Fusion of distributed measurements using the Extended Kalman Filter

The control law for the unicycle robot described in subsection 2.2. was based on the assumption that the vehicle's state vector $[x(k), y(k), \theta(k)]$ was measurable at every time instant. Here, the case in which the vehicle's state vector is reconstructed through the fusion of measurements received from distributed sensors (such as odometer or sonar sensors) will be examined.
The first approach to be analyzed is that of fusion of measurements coming from distributed sensors, with the use of nonlinear filtering methods such as Extended Kalman Filtering (EKF). The fused data are used to reconstruct the state vector of a mobile robot, and the estimated state vector is in turn used in a control-loop.
Extended Kalman Filtering for the nonlinear state-measurement model is revised. The following nonlinear time-invariant state model is now considered [Rigatos, G.G. & Tzafestas, S.G. (2007)]:

$$\begin{aligned} x(k+1) &= \phi(x(k)) + L(k)u(k) + w(k) \\ z(k) &= \gamma(x(k)) + v(k) \end{aligned} \tag{36}$$

where $w(k)$ and $v(k)$ are uncorrelated, Gaussian zero-mean noise processes with covariance matrices $Q(k)$ and $R(k)$ respectively. The operators $\phi(x)$ and $\gamma(x)$ are given by, $\phi(x) = [\phi_1(x), \phi_2(x), ..., \phi_m(x),]^T$, and $\gamma(x) = [\gamma_1(x), \gamma_2(x), ..., \gamma_p(x)]^T$, respectively. It is assumed that $\phi$ and $\gamma$ are sufficiently smooth in $x$ so that each one has a valid series Taylor expansion. Following a linearization procedure, $\phi$ is expanded into Taylor series about $\hat{x}$:

$$\phi(x(k)) = \phi(\hat{x}(k)) + J_\phi(\hat{x}(k))[x(k) - \hat{x}(k)] + ... \tag{37}$$

where $J_{\phi}(x)$ is the Jacobian of $\phi$ calculated at $\hat{x}(k)$:

$$J_{\phi}(x)=\frac{\partial\phi}{\partial x}|_{x=\hat{x}(k)}=\begin{pmatrix}\frac{\partial\phi_1}{\partial x_1} & \frac{\partial\phi_1}{\partial x_2} & \cdots & \frac{\partial\phi_1}{\partial x_N}\\ \frac{\partial\phi_2}{\partial x_1} & \frac{\partial\phi_2}{\partial x_2} & \cdots & \frac{\partial\phi_2}{\partial x_N}\\ \cdots & \cdots & \cdots & \cdots\\ \frac{\partial\phi_m}{\partial x_1} & \frac{\partial\phi_m}{\partial x_2} & \cdots & \frac{\partial\phi_m}{\partial x_N}\end{pmatrix} \tag{38}$$

Likewise, $\gamma$ is expanded about $\hat{x}^{-}(k)$

$$\gamma(x(k))=\gamma(\hat{x}(k))+J_{\lambda}(\hat{x}(k))[x(k)-\hat{x}(k)]+\ldots \tag{39}$$

where $\hat{x}^{-}(k)$ is the prior to time instant $k$ estimation of the state vector $x(k)$, and $\hat{x}(k)$ is the estimation of $x(k)$ at time instant $k$. The Jacobian $J_{\gamma}(x)$ is

$$J_{\gamma}(x)=\frac{\partial\gamma}{\vartheta x}|_{x=\hat{x}^{-}(k)}=\begin{pmatrix}\frac{\partial\gamma_1}{\partial x_1} & \frac{\partial\gamma_1}{\partial x_2} & \cdots & \frac{\partial\gamma_1}{\partial x_N}\\ \frac{\partial\gamma_2}{\partial x_1} & \frac{\partial\gamma_2}{\partial x_2} & \cdots & \frac{\partial\gamma_2}{\partial x_N}\\ \cdots & \cdots & \cdots & \cdots\\ \frac{\partial\gamma_p}{\partial x_1} & \frac{\partial\gamma_p}{\partial x_2} & \cdots & \frac{\partial\gamma_p}{\partial x_N}\end{pmatrix} \tag{40}$$

The resulting expressions create first order approximations of $\phi$ and $\gamma$. Thus the linearized version of the plant is obtained:

$$x(k+1)=\varphi(\hat{x}(k))+J_{\varphi}(\hat{x}(k))[x(k)-\hat{x}(k)]+w(k)$$
$$z(k)=\gamma(\hat{x}^{-}(k))+J_{\gamma}(\hat{x}^{-}(k))[x(k)-\hat{x}^{-}(k)]+v(k)$$

Now, the EKF recursion is as follows: First the time update is considered: by $\hat{x}(k)$ the estimation of the state vector at instant $k$ is denoted. Given initial conditions $\hat{x}^{-}(0)$ and $P^{-}(0)$ the recursion proceeds as:

- *Measurement update.* Acquire $z(k)$ and compute:

$$K(k)=P^{-}(k)J_{\gamma}^{T}(\hat{x}^{-}(k))[J_{\gamma}(\hat{x}^{-}(k))P^{-}(k)J_{\gamma}^{T}(\hat{x}^{-}(k))+R(k)]^{-1}$$
$$\hat{x}(k)=\hat{x}^{-}(k)+K(k)[z(k)-\gamma(\hat{x}^{-}(k))] \tag{41}$$
$$P(k)=P^{-}(k)-K(k)J_{\gamma}(\hat{x}^{-}(k))P^{-}(k)$$

- *Time update.* Compute:

$$P^{-}(k+1) = J_{\phi}(\hat{x}(k))P(k)J_{\phi}^{T}(\hat{x}(k)) + Q(k)$$
$$\hat{x}^{-}(k+1) = \phi(\hat{x}(k)) + L(k)u(k) \tag{42}$$

The schematic diagram of the EKF loop is given in Fig. 1(a).

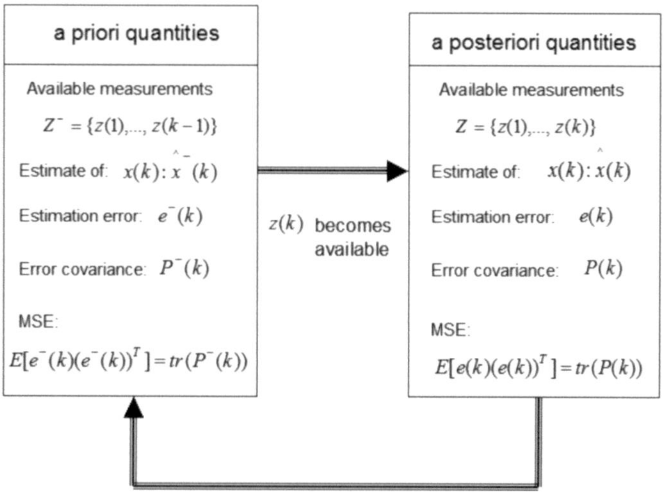


Fig. 1. (a) Schematic diagram of the Extended Kalman Filter Loop

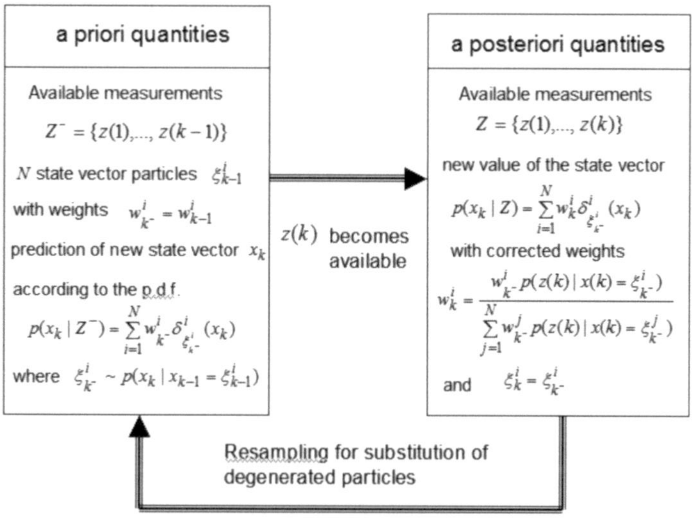


Fig. 1. (b) Schematic diagram of the Particle Filter loop

## 4. Particle Filtering for the nonlinear state-measurement model

### 4.1 Particle Filtering with sequential importance resampling

Next, the problem of fusing measurements coming from distributed sensors will be solved using the Particle Filtering method. The fused data are used again to estimate the state vector of a mobile robot and the reconstructed state vector will be used in closed-loop control.

In the general case the equations of the optimal filter used for the calculation of the state-vector of a nonlinear dynamical system do not have an explicit solution. This happens for instance when the process noise and the noise of the output measurement do not follow a Gaussian distribution. In that case, approximation through Monte-Carlo methods can used. As in the case of the Kalman Filter or the Extended Kalman Filter the particles filter consists of the measurement update (correction stage) and the time update (prediction stage) [Thrun, S. et al., (2005)].

a. *The prediction stage*:

The prediction stage calculates $p(x(k)\,|\,Z^-)$ where $Z^- = \{z(1),..,z(n-1)\}$, using:

$$p(x(k-1)\,|\,Z^-) = \sum_{i=1}^{N} w_{k-1}^i \delta_{\xi_{k-1}^i}(x(k-1)) \tag{43}$$

while from Bayes formula it holds $p(x(k-1)\,|\,Z^-) = \int p(x(k)\,|\,x(k-1))p(x(k-1)\,|\,Z^-)dx$. This finally gives

$$p(x(k)\,|\,Z^-) = \sum_{i=1}^{N} w_{k-1}^i \delta_{\xi_{k^-}^i}(x(k))$$
$$\text{with } \xi_{k^-}^i \sim p(x(k)\,|\,x(k-1)) = \xi_{k-1}^i \tag{44}$$

The meaning of Eq. (44) is as follows: the state equation of the nonlinear system of Eq. (36) is executed $N$ times, starting from the $N$ previous values of the state vectors $x(k-1) = \xi_{k-1}^i$ with the use of Eq. (36). This means that the value of the state vector which is calculated in the prediction stage is the result of the weighted averaging of the state vectors which were calculated after running the state equation, starting from the $N$ previous values of the state vectors $\xi_{k-1}^i$.

b. *The correction stage*

The a-posteriori probability density was performed using Eq. (44). Now a new position measurement $z(k)$ is obtained and the objective is to calculate the corrected probability density $p(x(k)\,|\,Z)$, where $Z = \{z(1), z(2),.., z(k)\}$. From Bayes law it holds that $p(x(k)\,|\,Z) = \frac{p(Z\,|\,x(k))p(x(k))}{p(Z)}$, which finally results into

$$p(x(k)\,|\,Z^-) = \sum_{i=1}^{N} w_{k-1}^i \delta_{\xi_{k^-}^i}(x(k))$$

$$\text{where } w_k^i = \frac{w_{k^-}^i \, p(z(k) \mid x(k) = \xi_{k^-}^i)}{\sum_{j=1}^{N} w_{k^-}^j \, p(z(k) \mid x(k) = \xi_{k^-}^j)} \tag{45}$$

Eq. (45) denotes the corrected value for the state vector. The recursion of the Particle Filter proceeds in a way similar to the update of the Kalman Filter or the Extended Kalman Filter, i.e.:

*Measurement update*: Acquire $z(k)$ and compute the new value of the state vector

$$p(x(k) \mid Z) = \sum_{i=1}^{N} w_k^i \delta_{k^-}^i (x(k))$$

$$\text{with corrected weights } w_k^i = \frac{w_{k^-}^i \, p(z(k) \mid x(k) = \xi_{k^-}^i)}{\sum_{j=1}^{N} w_{k^-}^i \, p(z(k) \mid x(k) = \xi_{k^-}^i)^i} \quad \text{and } \xi_k^i = \xi_{k^-}^i \tag{46}$$

Resample for substitution of the degenerated particles.

*Time update*: compute state vector $x(k+1)$ according to

$$p(x(k+1) \mid Z) = \sum_{i=1}^{N} w_k^i \delta_{\xi_k^i} (x(k))$$

$$\text{where } \xi_k^i \sim p(x(k+1) \mid x(k) = \xi_k^i) \tag{47}$$

The stages of state vector estimation with the use of the particle filtering algorithm are depicted in Fig. 1(b).

### 4.2 Resampling issues in particle filtering

*a. Degeneration of particles*

The algorithm of particle filtering which is described through Eq. (44) and Eq. (45) has a significant drawback: after a certain number of iterations $k$, almost all the weights $w_k^i$ become $0$. In the ideal case all the weights should converge to the value $1/N$, i.e. the particles should have the same significance. The criterion used to define a sufficient number of particles is $N_k^{eff} = 1/\sum_{i=1}^{N} w_k^{i\,2} \in [1, N]$. When $N_k^{eff}$ is close to value $N$ then all particles have almost the same significance. However using the algorithm of Eq. (44) and Eq. (45) results in $N_k^{eff} \to 1$, which means that the particles are degenerated, i.e. they lose their effectiveness. Therefore, it is necessary to modify the algorithm so as to assure that degeneration of the particles will not take place [Crisan, D. & Doucet, A. (2002)].

When $N_k^{eff}$ is small then most of the particles have weights close to $0$ and consequently they have a negligible contribution to the estimation of the state vector. The concept proposed to overcome this drawback of the algorithm is to weaken this particle in favor of particles that have a non-negligible contribution. Therefore, the particles of low weight factors are

removed and their place is occupied by duplicates of the particles with high weight factors. The total number of particles remains unchanged (equal to $N$) and therefore this procedure can be viewed as a "resampling" or "redistribution" of the particles set.
The particles resampling mentioned above maybe slow if not appropriately tuned. There are improved versions of it which substitute the particles of low importance with those of higher importance [Arulampalam, S. et al., (2002)], [Kitagawa, (1996)]. A first choice would be to perform a multinomial resampling. $N$ particles are chosen between $\{\xi_k^1,...,\xi_k^N\}$ and the corresponding weights are $\{w_k^1,...,w_k^N\}$. The number of times each particle is selected is given by $[j_1,..,j_N]$. Thus a set of $N$ particles is again created, the elements of which are chosen after sampling with the discrete distribution $\sum_{i=1}^{N} w_k^i \delta_{\xi_k^i}(x)$. The particles $\{\xi_k^1,...,\xi_k^N\}$ are chosen according to the probabilities $\{w_k^1,..,w_k^N\}$. The selected particles are assigned with equal weights $1/N$.

b. *Other approaches to the implementation of resampling*

Although sorting of the particles' weights is not necessary for the convergence of the particle filter algorithm, there are variants of the resampling procedure of $(\xi_k^i, w_k^i,\ i=1,..,N)$ which are based on previous sorting in decreasing order of the particles' weights. It is noted that efficient sorting approaches make the complexity of the particle filtering to be $O(N\log(N))$, while the avoidance of resampling could result in a faster algorithm of complexity $O(N)$. Sorting of particles' weights gives $w^{s[1]} > w^{s[2]} > ... > w^{s[N]}$. A random numbers generator is evoked and the resulting numbers $u^{i:N} \sim U[0,1]$ fall in the partitions of the interval $[0,1]$. The width of these partitions is $w^i$ and thus a redistribution of the particles is generated. For instance, in a wide partition of width $w^j$ will be assigned more particles than to a narrow partition of width $w^m$.

Two other methods that have been proposed for the implementation of resampling in Particle Filtering are explained in the sequel. These are Kitagawa's approach and the residuals resampling approach [Kitagawa, G. (1996)]. In Kitagawa's resampling the speed of the resampling procedure is increased by using less the random numbers generator. The weights are sorted again in decreasing order $w^{s[j]}$ so as to cover the region that corresponds to the interval [0,1]. Then the random numbers generator is used to produce the variable $u_i$ according to according to $u_1 \sim U[0,1/N]$, and $u^i = u^1 + \frac{i}{N}$, $i=2,..,N$. The rest of the variables $u^i$ are produced in a deterministic way [Campillo, F. (2006)].

In the residuals resampling approach, the redistribution of the residuals is performed as follows: at a first stage particle $\xi^i$ is chosen in a deterministic way $[w^i/N]$ times (with rounding). The residual weights are $\tilde{w}^i = w^i - N[w^i/N]$ and are normalized. Thus, a probability distribution is generated. The rest of the particles are selected according to multinomial resampling. The method can be applied if the number $\tilde{N}$ which remains at the second stage is small, i.e. when $N^{eff} = 1/\sum_{i=1}^{N} w_i^2$ is small.

## 5. Simulation tests

### 5.1 Flatness-based control using a state vector estimated by EKF

The application of EKF to the fusion of data that come from different sensors is examined first [Rigatos, G.G. & Tzafestas, S.G. (2007)]. A unicycle robot is considered. Its continuous-time kinematic equation is:

$$\dot{x}(t) = v(t)\cos(\theta(t)), \quad \dot{y}(t) = v(t)\sin(\theta(t)), \quad \dot{\theta}(\mathrm{t}) = \omega(\mathrm{t}) \tag{48}$$

Encoders are placed on the driving wheels and provide a measure of the incremental angles over a sampling period $T$. These odometric sensors are used to obtain an estimation of the displacement and the angular velocity of the vehicle $v(t)$ and $\omega(t)$, respectively. These encoders introduce incremental errors, which result in an erroneous estimation of the orientation $\theta$. To improve the accuracy of the vehicle's localization, measurements from sonars can be used. The distance measure of sonar $i$ from a neighboring surface $P_j$ is thus taken into account (see Fig. 2(a) and Fig. 2(b)). Sonar measurements may be affected by white Gaussian noise and also by crosstalk interferences and multiples echoes.

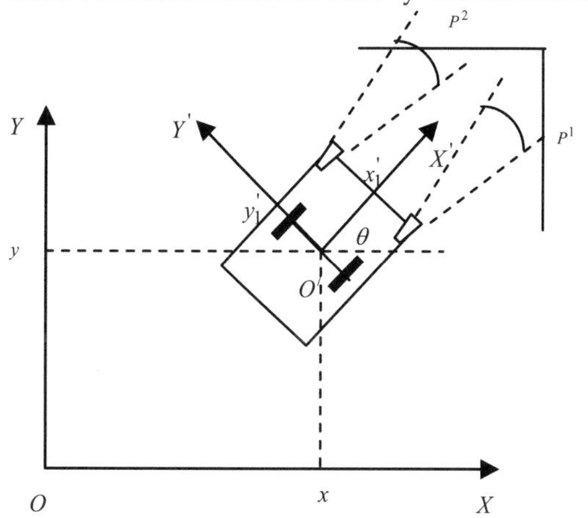


Fig. 2. (a). Mobile robot with odometric and sonar sensors

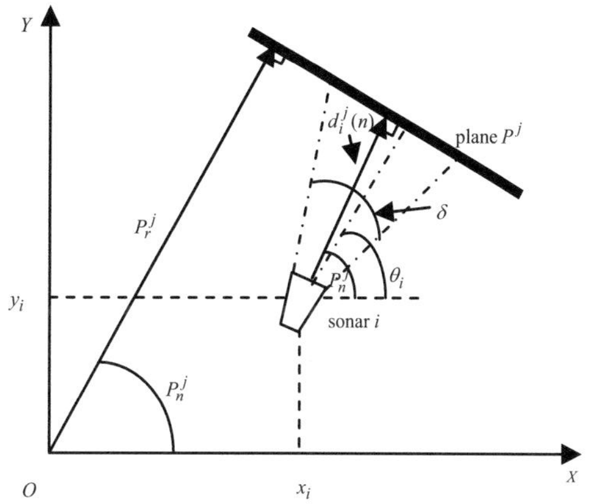


Fig. 2. (b). Orientation of the sonar $i$

The inertial coordinates system $OXY$ is defined. Furthermore the coordinates system $O'X'Y'$ is considered (Fig. 2(a)). $O'X'Y'$ results from $OXY$ if it is rotated by an angle $\theta$. The coordinates of the center of the wheels axis with respect to $OXY$ are $(x, y)$, while the coordinates of the sonar $i$ that is mounted on the vehicle, with respect to $O'X'Y'$ are $x_i', y_i'$. The orientation of the sonar with respect to $O'X'Y'$ is $\theta_i'$. Thus the coordinates of the sonar with respect to $OXY$ are $(x_i, y_i)$ and its orientation is $\theta_i$, and are given by:

$$\begin{aligned} x_i(k) &= x(k) + x_i' \sin(\theta(k)) + y_i' \sin(\theta(k)) \\ x_i(k) &= x(k) + x_i' \sin(\theta(k)) + y_i' \sin(\theta(k)) \\ \theta_i(k) &= \theta(k) + \theta_i \end{aligned} \tag{49}$$

Each plane $P^j$ in the robot's environment can be represented by $P_r^j$ and $P_n^j$ (Fig. 2(b)), where (i) $P_r^j$ is the normal distance of the plane from the origin O, (ii) $P_n^j$ is the angle between the normal line to the plane and the x-direction.
The sonar $i$ is at position $(x_i(k), y_i(k))$ with respect to the inertial coordinates system $OXY$ and its orientation is $\theta_i(k)$. Using the above notation, the distance of the sonar $i$, from the plane $P^j$ is represented by $P_r^j, P_n^j$ (see Fig. 2(b)):

$$d_i^k(k) = P_r^j - x_i(k)\cos(P_n^j) - y_i(k)\sin(P_n^j) \tag{50}$$

where $P_n^j \in [\theta_i(n) - \delta/2, \theta_i(n) + \delta/2]$, and $\delta$ is the width of the sonar beam. Assuming a constant sampling period $\Delta t_k = T$ the measurement equation is $z(k+1) = \gamma(x(k)) + v(k)$, where $z(k)$ is the vector containing sonar and odometer measures and $v(k)$ is a white noise sequence $\sim N(0, R(kT))$. The dimension $p_k$ of $z(k)$ depends on the number of sonar sensors. The measure vector $z(k)$ can be decomposed in two sub-vectors

$$\begin{aligned} z_1(k+1) &= [x(k) + v_1(k), y(k) + v_2(k), \theta(k) + v_3(k)] \\ z_2(k+1) &= [d_1^j(k) + v_4(k), ..., d_{n_s}^j(k) + v_{3+n_2}(k)] \end{aligned} \tag{51}$$

with $i = 1, 2, ..., n_s$, where $n_s$ is the number of sonars, $d_i^j(k)$ is the distance measure with respect to the plane $P^j$ provided by the $i$-th sonar and $j = 1,..,n_p$ where $n_p$ is the number of surfaces. By definition of the measurement vector one has that the output function $\gamma(x(k))$ is given by $\gamma(x(k)) = [x(k), y(k), \theta(k), d_1^1(k), d_2^2(k), ..., d_{n_s}^{n_p}]^T$. The robot state is $[x(k), y(k), \theta(k)]^T$ and the control input is denoted by $U(k) = [u(k), \omega(k)]^T$.
In the simulation tests, the number of sonar is taken to be $n_s = 1$, and the number of planes $n_p = 1$, thus the measurement vector becomes s. To obtain the Extended Kalman Filter (EKF), the kinematic model of the vehicle is linearized about the estimates $\hat{x}(k)$ and $\hat{x}^-(k)$, the control input $U(k-1)$ is applied.
The *measurement update* of the EKF is

$$K(k) = P^-(k) J_\gamma^T(\hat{x}^-(k))[J_\gamma(\hat{x}^-(k)) P^-(k) J_\gamma^T(\hat{x}^-(k)) + R(k)]^{-1}$$

$$\hat{x}(k) = \hat{x}^-(k) + K(k)[z(k) - \gamma(\hat{x}^-(k))]$$

$$P(k) = P^-(k) - K(k) J_\gamma(\hat{x}^-(k)) P^-(k)$$

The *time update* of the EKF is

$$P^-(k+1) = J_\phi(\hat{x}(k)) P(k) J_\phi^T(\hat{x}(k)) + Q(k)$$

$$\hat{x}^-(k+1) = \phi(\hat{x}(k)) + L(k) u(k)$$

where $L(n)=\begin{pmatrix} T\cos(\theta(k)) & 0 \\ T\sin(\theta(k)) & 0 \\ 0 & T \end{pmatrix}$ and $J_{\varphi}(\hat{x}(k))=\begin{pmatrix} 1 & 0 & -v(k)\sin(\theta)T \\ 0 & 1 & -v(k)\cos(\theta)T \\ 0 & 0 & 1 \end{pmatrix}$

while $Q(k)=diag[\sigma^2(k),\sigma^2(k),\sigma^2(k)]$, with $\sigma^2(k)$ chosen to be $10^{-3}$, and $\phi(\hat{x}(k))=[\hat{x}(k),\hat{y}(k),\hat{\theta}(k)]^T$, $\gamma(\hat{x}(k))=[\hat{x}(k),\hat{y}(k),\hat{\theta}(k),d(k)]^T$, i.e.

$$\gamma(\hat{x}(k))=\begin{pmatrix} \hat{x}(k) \\ \hat{y}(k) \\ \hat{\theta}(k) \\ P_r^j - x_i(k)\cos(P_n^j) - y_i(k)\sin(P_n^j) \end{pmatrix} \tag{52}$$

Assuming one sonar $n_s=1$, and one plane $P^1$, $n_p=1$ in the mobile robot's neighborhood one gets

$$J_{\gamma}^T(\hat{x}^-(k))=\begin{pmatrix} 1 & 0 & 0 \\ 0 & 1 & 0 \\ 0 & 0 & 1 \\ -\cos(P_n^j) & -\sin(P_n^j) & \{x_i^{'}\cos(\theta-P_n^j)-y_i^{'}\sin(\theta-P_n^j)\} \end{pmatrix} \tag{53}$$

The vehicle is steered by a dynamic feedback linearization control algorithm which is based on flatness-based control [Oriolo, G. et al., (2002)]:

$$\begin{aligned} u_1 &= \ddot{x}_d + K_{p_1}(x_d - x) + K_{d_1}(\dot{x}_d - \dot{x}) \\ u_2 &= \ddot{y}_d + K_{p_2}(y_d - y) + K_{d_2}(\dot{y}_d - \dot{y}) \\ \dot{\xi} &= u_1\cos(\theta) + u_2\sin(\theta) \\ v = \xi, \quad \omega &= \frac{u_2\cos(\theta) - u_1\sin(\theta)}{\xi} \end{aligned} \tag{54}$$

The following initialization is assumed (see Fig. 3(a)): (i) vehicle's initial position in $OXY$: $x(0)=0\,m$, $y(0)=0\,m$, $\theta(0)=45.0^o$, (ii) position of the sonar in $O'X'Y'$: $x_1^{'}=0.5\,m$, $y_1^{'}=0.5\,m$, $\theta_1^{'}=0^o$, (iii) position of the plane $P^1$: $P_r^1=15.5\,m$, $P_n^1=45^o$, (iv) state noise $w(k)=0$, $\hat{P}(0)=diag[0.1,0.1,0.1]$ and $R=diag[10^{-3},10^{-3},10^{-3}]$, (v) Kalman Gain $K(k)\in R^{3\times 4}$

The use of EKF for fusing the data that come from odometric and sonar sensors provides an estimation of the state vector $[x(t),y(t),\theta(t)]$ and enables the successful application of nonlinear steering control of Eq. (54). For the case of motion along a straight line on the 2D-plane, the obtained results are depicted in Fig. 3(a). Moreover, results on the tracking of a circular reference path are given in Fig. 4(a), while the case of tracking of an eight-shaped reference path is depicted in Fig. 5(a). Tracking experiments for EKF-based state estimation were completed in the case of a curved path as the one shown in Fig. 6(a).

### 5.2 Flatness-based control using a state vector estimated by Particle Filtering

The particle filter can also provide solution to the sensor fusion problem. The mobile robot model described in Eq. (48), and the control law given in Eq. (54) are used again. The number of particles was set to $N = 1000$.

The *measurement update* of the PF is $p(x(k) \mid Z) = \sum_{i=1}^{N} w_k^i \delta_{\xi_{k^-}^i}(x(k))$ with $w_k^i = \dfrac{w_{k^-}^i p(z(k) \mid x(k) = \xi_{k^-}^i)}{\sum_{j=1}^{N} w_k^j p(z(k) \mid x(k) = \xi_{k^-}^j)}$ where the measurement equation is given by $\hat{z}(k) = z(k) + v(k)$ with $z(k) = [x(k), y(k), \theta(k), d(k)]^T$, and $v(k) =$ measurement noise.

The *time update* of the PF is $p(x(k+1) \mid Z) = \sum_{i=1}^{N} w_k^i \delta_{\xi_k^i}(x(k))$ where $\xi_k^i \sim p(x(k+1) \mid x(k) = \xi_k^i)$ and the state equation is $\hat{x}^- = \varphi(x(k)) + L(k)U(k)$, where $\phi(x(k))$, $L(k)$ and $U(k)$ are defined in subsection 5.1. At each run of the time update of the PF, the state vector estimation $\hat{x}^-(k+1)$ is calculated $N$ times, starting each time from a different value of the state vector $\xi_k^i$. The measurement noise distribution was assumed to be Gaussian. As the number of particles increases, the performance of the particle filter-based tracking algorithm also improves, but this happens at higher demand for computational resources. Control of the diversity of the particles through the tuning of the resampling procedure may also affect the performance of the algorithm. The obtained results are given in Fig. 3(b) for the case of motion along a straight line on the 2D plane. Additionally, results on the tracking of a circular reference path are given in Fig. 4(b), while the case of tracking of an eight-shaped reference path is depicted in Fig. 5(b). Tracking experiments for PF-based state estimation were completed in the case of a curved path as the one shown in Fig. 6(b).

From the depicted simulation experiments it can be deduced that the particle filter for a sufficiently large number of particles can have good performance, in the problem of estimation of the state vector of the mobile robot, without being subject to the constraint of Gaussian distribution for the obtained measurements. The number of particles influences the performance of the particle filter algorithm. The accuracy of the estimation succeeded by the PF algorithm improves as the number of particles increases. The initialization of the particles, (state vector estimates) may also affect the convergence of the PF towards the real value of the state vector of the monitored system. It should be also noted that the calculation time is a critical parameter for the suitability of the PF algorithm for real-time applications.

When it is necessary to use more particles, improved hardware and parallel processing available to embedded systems, enable the PF to be implemented in real-time systems [Yang, N. et al., (2005)].

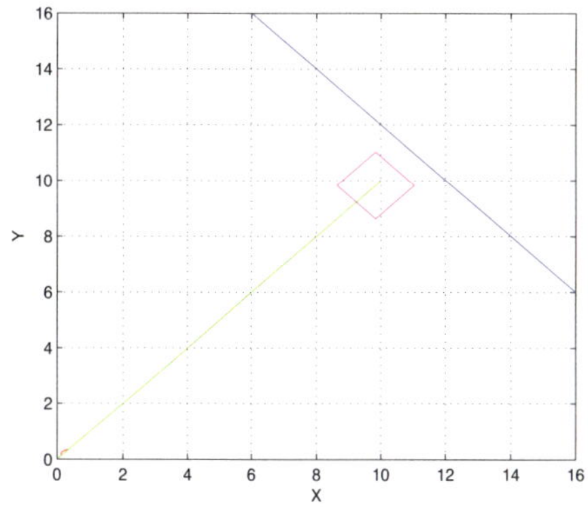


Fig. 5. (a) Desirable trajectory (continuous line) and trajectory using EKF fusion based on odometric and sonar measurements, when tracking a straight line.

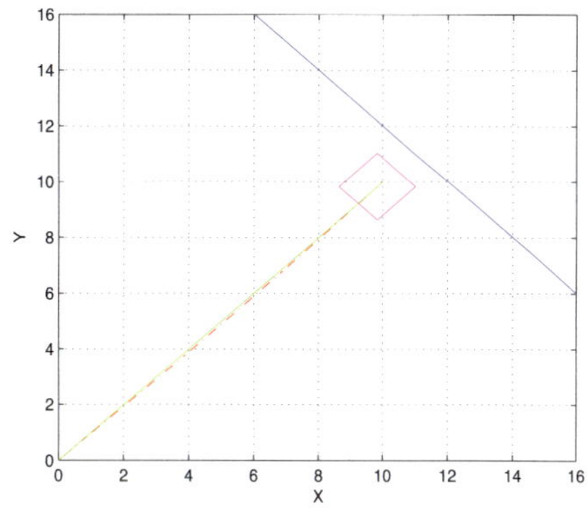


Fig. 5. (b) Desirable trajectory (continuous line) and trajectory using PF fusion based on odometric and sonar measurements, when tracking a straight line.

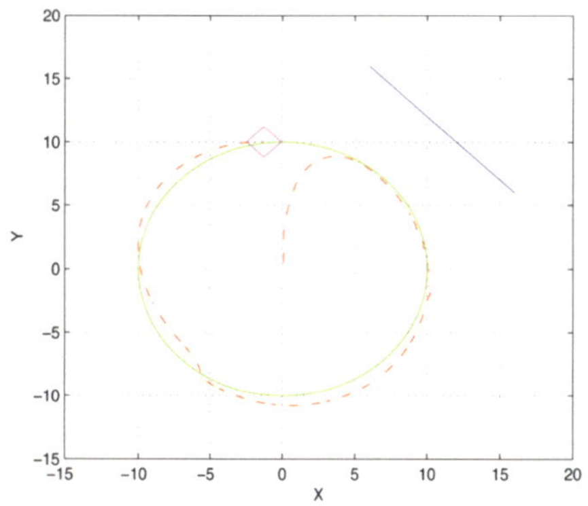


Fig. 4. (a) The trajectory of the mobile robot (dashed line) tracks the reference circular path (continuous line) when the robot's state vector is estimated with the use of Extended Kalman Filtering.

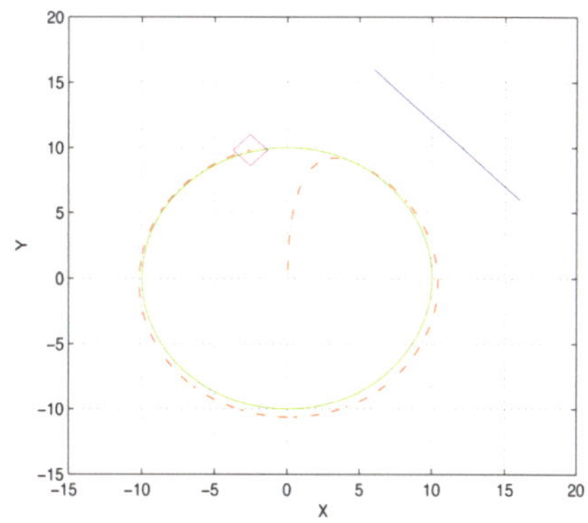


Fig. 4. (b) The trajectory of the mobile robot (dashed line) tracks the reference circular path (continuous line) when the robot's state vector is estimated with the use of particle filtering.

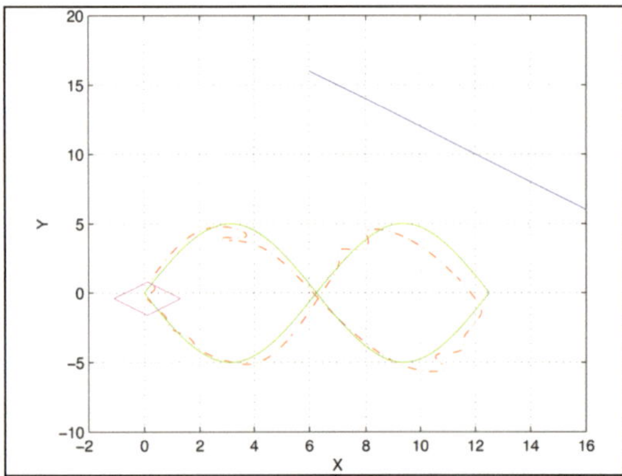


Fig. 5. (a) The trajectory of the mobile robot (dashed line) tracks the reference eight-shaped path (continuous line) when the robot's state vector is estimated with the use of Extended Kalman Filtering.

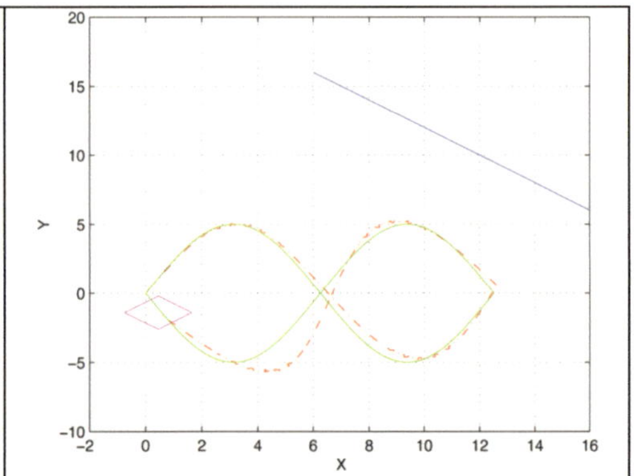


Fig. 5. (b) The trajectory of the mobile robot (dashed line) tracks the reference eight-shaped path (continuous line) when the robot's state vector is estimated with the use of Particle Filtering.

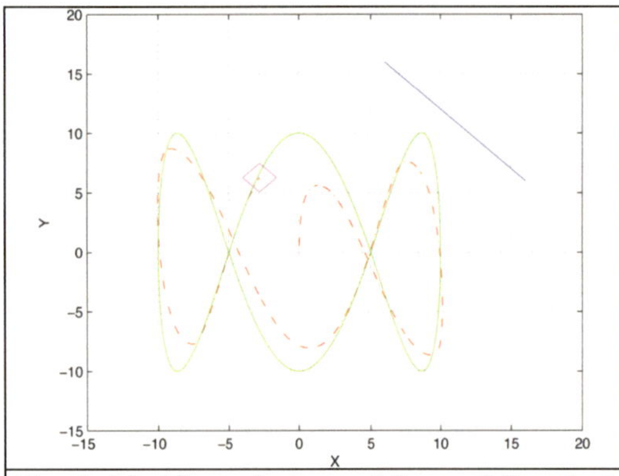


Fig. 6. (a) The trajectory of the mobile robot (dashed line) tracks the reference curve-shaped path (continuous line) when the robot's state vector is estimated with the use of Extended Kalman Filtering.

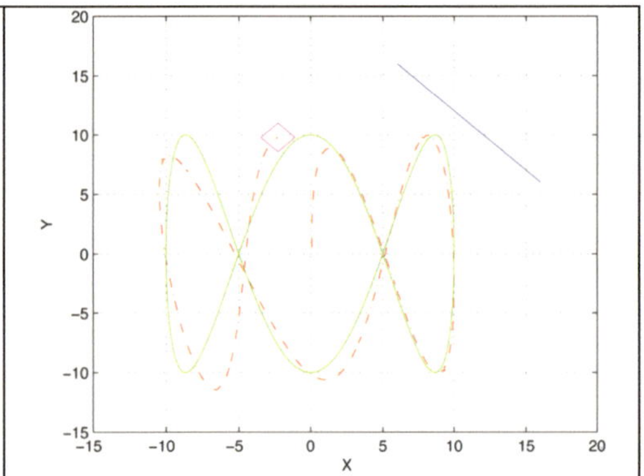


Fig. 6. (b) The trajectory of the mobile robot (dashed line) tracks the reference curve-shaped path (continuous line) when the robot's state vector is estimated with the use of Particle Filtering.

## 6. Conclusions

The paper has studied flatness-based control and sensor fusion for motion control of autonomous mobile robots. Flatness-based control stems from the concept of differential flatness, i.e. of the ability to express the system parameters (such as the elements of the state vector) and the control input as relations of a function $y$ called *flat output* and of its higher order derivatives. Flatness-based control affects the dynamics of the system in a way similar to control through feedback-linearization. This means that writing the system variables and the control input as functions of the flat output enables transformation of the system dynamics into a linear ODE and subsequently permits trajectory tracking using linear control methods. For linear systems differential-flatness coincides with the property of controllability. Flatness-based control is applicable to finite dimensional systems (linear or nonlinear) as well as to infinite dimensional systems, such as the ones usually described by PDE.

The problem of motion control of the mobile robots becomes more complicated when the robot's state vector is not directly measurable but has to be reconstructed with the use of measurements coming from distributed sensors. Consequently, the control input generated by the flatness-based control algorithm has to use the estimated state vector of the robotic vehicle instead of the real one. Extended Kalman and Particle filtering have been tested in the problem of estimation of the state vector of a mobile robot through the fusion of position measurements coming from odometric and sonar sensors. The paper has summarized the basics of the Extended Kalman Filter, which is the most popular approach to implement sensor fusion in nonlinear systems. The EKF is a linearization technique, based of a first-order Taylor expansion of the nonlinear state functions and the nonlinear measurement functions of the state model. In the EKF, the state distribution is approximated by a Gaussian random variable. Although the EKF is a fast algorithm, the underlying series approximations can lead to poor representations of the nonlinear functions and the associated probability distributions. As a result, the EKF can sometimes be divergent.

To overcome these weekness of the EKF as well as the constraint of the Gaussian state distribution, particle filtering has been introduced. Whereas the EKF makes a Gaussian assumption to simplify the optimal recursive state estimation, the particle filter makes no assumptions on the forms of the state vector and measurement probability densities. In the particle filter, a set of weighted particles (state vector estimates evolving in parallel) is used to approximate the posterior distribution of the state vector. An iteration of the particle filter includes particle update and weights update. To succeed the convergence of the algorithm at each iteration resampling takes place through which particles with low weights are substituted by particles of high weights.

Simulation tests have been carried out to evaluate the performance of flatness-based control for the autonomous mobile robot, when using the EKF and the particle filter for the localization of the robotic vehicle (through the fusion of measurements coming from distributed sensors). It has been shown, that comparing to EKF, the PF algorithm results in better estimates of the mobile robot's state vector as the number of particles increases, but on the expense of higher computational effort. Consequently, the flatness-based controller which used the robot's state vector coming from the particle filter, had better tracking performance than the flatness-based controller which used the robot's state vector that was estimated by the EKF. It has been also observed that the accuracy in the localization of the mobile robot, succeeded by the particle filter algorithm depends on the number of particles and their initialization.

## 7. References

Arulampalam, S.; Maskell, S.R., Gordon, N.J. & Clapp, T. (2002). A tutorial on particle filters for on-line nonlinear/non-Gaussian Bayesian tracking. *IEEE Transactions on Signal Processing*, Vol. 50, pp. 174-188.

Campillo, F. (2006). *Particulaire & Modèles de Markov Cachés, Master Course Notes Filtrage et traitment des donées*, Université de Sud-Toulon Var, France.

Caron, F.; Davy,M., Duflos, E. & Vanheeghe, P. (2007). Particle Filtering for Multi-Sensor Data Fusion with Switching Observation Models: Applications to Land Vehicle Positioning. *IEEE Transactions on Signal Processing*, Vol. 55, No. 6, pp., 2703-2719.

Crisan, D. & Doucet, A. (2002). A Survey of Convergence Results on Particle Filtering Methods for Practitioners. *IEEE Transactions on Signal Processing*, Vol. 50, No. 3, pp.736-746.

Fliess, M. & Mounier, H. (1999). Tracking control and $\pi$ -freeness of infinite dimensional linear systems, *Dynamical Systems, Control, Coding and Computer Vision*, G. Picci and D.S. Gilliam (Eds.), Vol. 258, pp. 41-68, Birkhaüser.

Jetto, L.; Longhi, S. & Venturini, G. (1999). Development and Experimental Validation of an Adaptive Extended Kalman Filter for the Localization of Mobile Robots. *IEEE Transactions on Robotics and Automation*, Vol. 15, No.2.

Kitagawa, G. (1996). Monte-Carlo filter and smoother for non-Gaussian non-linear state-space models. *J. Computat. Graph. Statist.*, Vol.5, No.1, pp. 1-25.

Laroche,B. ; Martin, P. & Petit,N. (2007). *Commande par platitude: Equations différentielles ordinaires et aux derivées partielles*, Ecole Nationale Supérieure des Techniques Avancées, Paris.

Lévine, J.; Nguyen, D.V. (2003). Flat output characterization for linear systems using polynomial matrices. *Systems & Control Letters*, Elsevier, Vol. 48, pp. 69-75.

Martin, P. & Rouchon,P. (1999). Systèmes plats: planification et suivi des trajectoires, *Journées X-UPS,Ecole des Mines de Paris*, Centre Automatique et Systèmes, Mai 1999.

Meurer T.; & Zeitz, M. (2004). A modal approach to flatness-based control of flexible structures. *PAMM Proceedings on Applied Mathematics and Mechanics*, Vol. 4, pp. 133-134.

Mounier, H. & Rudolph, J. (2001). Trajectory tracking for $\pi$ -flat nonlinear delay systems with a motor example, *Nonlinear control in the year 2000* (A. Isidori, F. Lamnabhi-Lagarrigue and W. Respondek, editors), Vol.1, Lecture Notes in Control and Inform. Sci.,vol. 258, pp. 339-352, Springer.

Oriolo,G.; De Luca, A. & Vendittelli, M. (2002). WMR Control Via Dynamic Feedback Linearization: Design, Implementation and Experimental Validation. *IEEE Transactions on Control Systems Technology*, Vol. 10, No.6 , pp. 835-852.

Rigatos, G.G. (2003). Fuzzy Stochastic Automata for Intelligent Vehicle Control. *IEEE Transactions on Industrial Electronics*, Vol. 50, No. 1, pp. 76-79.

Rigatos,G.G.; Tzafestas, S.G. & Evangelidis, G.A. (2001). Reactive parking control of a nonholonomic vehicle via a Fuzzy Learning Automaton. *IEE Proceedings : Control Theory and Applications*, Vol. 148, pp. 169-179.

Rigatos,G.G. (2008). Coordinated motion of autonomous vehicles with the use of a distributed gradient algorithm, *Journal of Applied Mathematics and Computation, Elsevier*, Vol 199, No. 2, pp 494-503.

Rigatos, G.G. (2007a). Particle Filtering for state estimation in nonlinear industrial systems, *2nd IC-EpsMsO 2007, 2nd International Conference on Experiments / Process / System Modelling / Simulation & Optimization*, Athens, Greece, July 2007 .

Rigatos, G.G. & Tzafestas,S.G. (2007). Extended Kalman Filtering for Fuzzy Modeling and Mutli-Sensor Fusion. *Mathematical and Computer Modeling of Dynamical Systems*, Vol. 13, No. 3, Taylor and Francis.

Rigatos,G.G. (2007b). Extended Kalman and particle filtering for sensor fusion in mobile robot localization, *PhysCon 2007, International Conference on Physics and Control*, Potsdam, Germany, Sep. 2007.

Rouchon, P. (2005). Flatness-based control of oscillators. *ZAMM Zeitschrift fur Angewandte Mathematik und Mechanik*, Vol. 85, No.6, pp. 411-421.

Rudolph,J. (2003). *Flatness Based Control of Distributed Parameter Systems*, Steuerungs und Regelungstechnik, Shaker Verlag, Aachen.

Yang, N.; Tian, W.F., Jin, Z.H. & Zhang, C.B. (2005). Particle Filter for sensor fusion in a land vehicle navigation system. *Measurement Science and Technology*, Institute of Physics Publishing, Vol. 16, pp. 677-681.

Thrun, S.; Burgard, W. & Fox, D. (2005). *Probabilistic Robotics*, MIT Press.

21

# An Improved Real-Time Particle Filter for Robot Localization

Dario Lodi Rizzini and Stefano Caselli
*Dipartimento di Ingegneria dell'Informazione, Università degli Studi di Parma, Parma, Italy*

## 1. Introduction

Robot localization is the problem of estimating robot coordinates with respect to an external reference frame. In the common formulation of the localization problem, the robot is given a map of its environment, and to localize itself relative to this map it needs to consult its sensor data. The effectiveness of a solution to the localization problem in an unstructured environment strongly depends on how it copes with the uncertainty affecting robot perception.

The *probabilistic robotics* paradigm provides statistical techniques for representing information and making decision, along with a unifying mathematical framework for probabilistic algorithms based on Bayes rule (Thrun et al., 2005). For this reason, bayesian filtering has become the prevailing approach in recent works on localization (Elinas & Little, 2005; Sridharan et al., 2005; Hester & Stone, 2008).

Bayesian filtering is a general probabilistic paradigm to arrange motion and sensor data in order to achieve a solution in the form of distribution of state random variables. Bayesian filters differ in the representation of the probability density function (PDF) of state. For example, the resulting estimation of Gaussian filters (Kalman Filter, Extended Kalman Filter) (Leonard & Durrant-Whyte, 1991; Arras et al., 2002) is expressed in the form of a continuous parametric function, while the state posterior is decomposed in discrete elements for nonparametric filters.

The main nonparametric algorithm is called *Particle Filter* (PF) (Fox et al., 1999) and relies on importance sampling (Doucet et al., 2001). With importance sampling, the probability density of the robot pose is approximated by a set of samples drawn from a proposal distribution, and an importance weight measures the distance of each sample from the correct estimation.

The nonparametric approach has the advantage of providing a better approximation of the posterior when a parametric model does not exist or changes during iteration, e.g. in initialization or when environment symmetries determine a multi-modal PDF. Even if techniques like Multi-Hypothesis Tracking (Arras et al., 2002) attempt to manage multi-modal distributions, particle filters are more efficient and can represent all kinds of PDFs, including uniform distributions. Moreover, particle filters limit errors due to the linearization of model equations that can lead to poor performance and divergence of the filter for highly nonlinear problems. Unfortunately, particle filters suffer from computational complexity due to the large number of discrete samples of the posterior: for each sample a pose update, a correction and a resample step are performed. Since localization can be performed slowly with respect to the usual movement and tasks of the robot, it would be conceivable to perform localization over a large time interval. Therefore, there have been attempts to adapt the number of samples (Fox, 2003). However, during an excessive time interval uncertainty increases and many useful observations are dropped; a proper interval to complete a particle filter iteration should be approximately equal to the rate of incoming data. A trade-off must therefore be reached between time constraints imposed by the need of collecting sensor data incoming with a given rate and the number of samples determining the accuracy of the representation of localization hypotheses. Performance depends both on the number of integrated observations and on the number of samples.

The *Real-Time Particle Filter* (RTPF) (Kwok et al., 2004) is a variant of a standard particle filter addressing this problem. Samples are partitioned into subsets among observations over an estimation window. The size of each partitioned subset is chosen so that a particle filter iteration can be performed before a new observation is acquired. The difference with standard PF with smaller sample set lies in the representation of the posterior as a mixture of samples: at the end of an estimation window the distribution consists of the samples from each subset of the window. Mixture weights determine how each partition set contributes to the posterior and are computed in order to minimize the approximation error of the mixture distribution.

While RTPF represents a remarkable step toward a viable particle filter-based localizer, there are a few issues to be addressed in developing an effective implementation. RTPF convergence is prone to bias problem and to some numerical instability in the computation of the mixture weights arising from the need to perform a numerical gradient descent. Furthermore, even adopting RTPF as the basic architecture, the design of a flexible and customizable particle filter remains a challenging task. For example, life cycle of samples extends beyond a single iteration and covers an estimation window in which mixture posterior computation is completed. This extended life cycle of samples impacts over software design. Moreover, RTPF addresses observations management and derived constraints. A good implementation should be adaptable to a variety of sensors.

In this chapter, we describe the application of RTPF to robot localization and provide three additional contributions: a formal analysis for the evolution of mixture of posterior in RTPF, a novel solution for the computation of mixture weights yielding improved stability and convergence, and a discussion of the design issues arising in developing a RTPF-based robot localization system.

The algorithm described in (Kwok et al., 2004) computes mixture weights by minimizing the Kullback-Leibler (KL) distance between the mixture distribution and the theoretically-correct one. Unfortunately, this criterion tends to promote partition sets of the estimation window that provide a poor representation of the distribution of robot state. In particular, we show that KL criterion favours sets with low *effective sample size* (Liu, 1996) and leads to a bias in the estimation. As an alternative way to compute mixture weights, we define a *weights matrix*, whose elements are related to the effective sample size. The mixture weight vector is then computed as an eigenvector of this matrix. This solution is more robust and less prone to numerical instability. Finally, we propose the design of a library that takes care of efficient life cycle of samples and control data, which is different between RTPF and standard particle filter, and supports multiple motion and sensor models. This flexibility is achieved by applying generic programming techniques and a policy pattern. Moreover, differing from other particle filter implementations (e.g., CARMEN (Montemerlo et al., 2003)), the library is independent from specific control frameworks and toolkits.

The remaining of the chapter is organized as follows. Section 2 contains an overview of RTPF with the original algorithm to compute mixture weights. Section 3 provides a formal description of the bias problem and a novel approach in the computation of mixture weights based on the effective number of samples. This approach simplifies RTPF and tries to avoid spurious numeric convergence of gradient descent methods. Section 4 illustrates design issues connected to RTPF and describes a localization library implementing a highly configurable particle filter localizer. Section 5 presents simulation and experimental results which are reported and compared with the original RTPF performance. Finally, section 6 gives conclusion remarks.

## 2. Real-time particle filters

In particle filters, updating the particles used to represent the probability density function (potentially a large number) usually requires a time which is a multiple of the cycle of sensor information arrival. Naive approaches, yet often adopted, include discarding observations arriving during the update of the sample set, aggregating multiple observations into a single one, and halting the generation of new samples upon a new observation arrival (Kwok et al., 2004). These approaches can affect filter convergence, as either they loose valuable sensor information, or they result in inefficient choices in algorithm parameters.

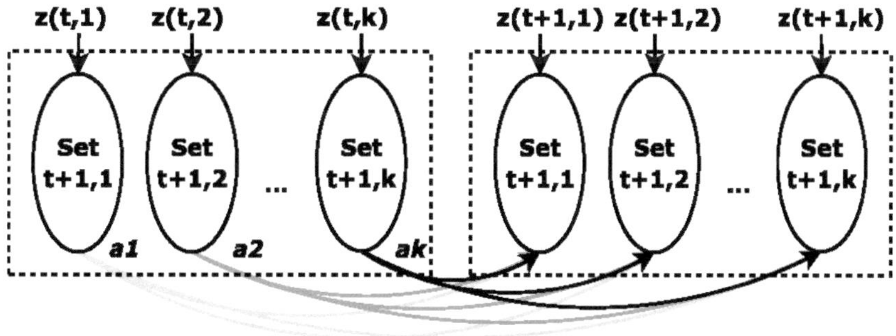


Fig. 1 RTPF operation: samples are distributed in sets, associated with the observations. The distribution is a mixture of the sample sets based on weights $\alpha_i$ (shown as *ai* in figure).

An advanced approach dealing with such situations is the Real-Time Particle Filters (RTPF) (Kwok et al., 2003; Kwok et al., 2004) which is briefly described in the following. Consider $k$ observations. The key idea of the Real-Time Particle Filter is to distribute the samples in sets, each one associated with one of the $k$ observations. The distribution representing the system state within an estimation window will be defined as a mixture of the $k$ sample sets as shown in Fig. 1. At the end of each estimation window, the weights of the mixture belief are determined by RTPF based on the associated observations in order to minimize the approximation error relative to the optimal filter process. The optimal belief could be obtained with enough computational resources by computing the whole set of samples for each observation. Formally:

$$Bel_{opt}(x_{tk}) \propto \int \prod_{i=1}^{k} p(z_{ti}|x_{ti}) p(x_{ti}|x_{ti-1},u_{ti}) Bel(x_{t0}) dx_{t0} \dots dx_{tk-1} \tag{1}$$

where $Bel(x_{t0})$ is the belief generated in the previous estimation window, and $z_{ti}$, $u_{ti}$, $x_{ti}$ are, respectively, the observation, the control information, and the state for the $i-th$ interval.
Within the RTPF framework, the belief for the $i-th$ set can be expressed, similarly, as:

$$Bel_i(x_{tk}) \propto \int p(z_{ti}|x_{ti}) \prod_{j=1}^{k} p(x_{tj}|x_{tj-1},u_{tj}) Bel(x_{t0}) dx_{t0} \dots dx_{tk-1} \tag{2}$$

containing only observation-free trajectories, since the only feedback is based on the observation $z_{ti}$, sensor data available at time $t_i$. The weighted sum of the $k$ believes belonging to an estimation window results in an approximation of the optimal belief:

$$Bel_{mix}(x_{tk} \mid \alpha) \propto \sum_{i=1}^{k} \alpha_i Bel_i(x_{tk}) \tag{3}$$

An open problem is how to define the optimal mixture weights minimizing the difference between the $Bel_{opt}(x_{tk})$ and $Bel_{mix}(x_{tk} \mid \alpha)$. In (Kwok et al., 2004), the authors propose to minimize their Kullback-Leibler distance (KLD). This measure of the difference between probability distributions is largely used in information theory (Cover & Thomas, 1991) and can be expressed as:

$$J(\alpha) \propto \int Bel_{mix}(x_{tk} \mid \alpha) \log \frac{Bel_{mix}(x_{tk} \mid \alpha)}{Bel_{opt}(x_{tk})} dx_{t_{k-1}} \tag{4}$$

To optimize the weights of mixture approximation, a gradient descent method is proposed in (Kwok et al., 2004). Since gradient computation is not possible without knowing the optimal belief, which requires the integration of all observations, the gradient is obtained by Monte Carlo approximation: believes $Bel_i$ share the same trajectories over the estimation windows, so we can use the weights to evaluate both $Bel_i$ (each weight corresponds to an observation) and $Bel_{opt}$ (the weight of a trajectory is the product of the weights associated to this trajectory in each partition). Hence, the gradient is given by the following formula:

$$\frac{\partial J}{\partial \alpha_i} \cong 1 + \sum_{s=1}^{Np} w_{ti}(x_{ti}^{(s)}) \frac{\sum_{j=1}^{k} \alpha_j w_{tj}(x_{tj}^{(s)})}{\prod_{j=1}^{k} w_{tj}(x_{tj}^{(s)})} \tag{5}$$

where $Bel_i$ is substituted by the sum of the weights of partition set $i-th$ and $Bel_{opt}$ by the sum of the weights of each trajectory. Unfortunately, (5) suffers from a bias problem, which (Kwok et al., 2004) solve by clustering samples and computing separately the contribution of each cluster to the gradient (5). In the next section, an alternative solution is proposed.

## 3. An enhanced RTPF

In this section we provide a formal investigation on the motivation of bias in RTPF estimation in (Kwok et al., 2004) and we propose a new solution for mixture weights computation.

### 3.1 A bias in RTPF

In RTPF, samples belonging to different partition sets are drawn from the same proposal, but their importance weights depend on different observation likelihood functions $p(z_{ti} \mid x_{ti})$, which are computed in different time instants $t_i$. Hence, the first source of disparity among partition sets is the degree of proposal dispersion during the correction step. A suitable measure of proposal dispersion at iteration $t_i$ is provided by the radius of the ball set $B(\eta_{xt_i}, r) \in \Re^d$, which is centered on expected value $x_{ti}$ and includes a consistent portion of the distribution of $x_{ti}$. The probability that a sample falls in $B(\eta_{xt_i}, r)$ can be bound by $r$ and the trace of the covariance matrix $\Sigma_{x_{ti}}$, since the following Chebychev-like inequality holds:

$$P(x_{t_i} \in B(\eta_{xt_i}, r)) > 1 - \frac{\mathrm{tr}\left(\Sigma_{x_{ti}}\right)}{r^2} \tag{6}$$

In the following, the probability of event given by $B(\eta_{xt_i}, r)$ will refer to a proposal density function arrested in $t_i$:

$$\pi(x_{ti}) = \int_{\Re^{d \times i}} \prod_{j=1}^{i} p(x_{tj} \mid x_{tj-1}, u_{tj}) dx_{t0} \ldots dx_{tk-1} \tag{7}$$

Then, given $0 < \varepsilon < 1$, a sample falls in a ball with at least probability $\varepsilon$ when its radius is larger than the *dispersion radius*:

$$r_{t_i,\varepsilon} = \sqrt{\frac{\mathrm{tr}(\Sigma_{x_{ti}})}{1-\varepsilon}} \tag{8}$$

Parameter $r_{t_i,\varepsilon}$ provides a rough estimation for dispersion because only for unimodal PDF the ball $B(\eta_{xt_i}, r_{t_i,\varepsilon})$ (briefly $B$ hereafter) limits a region around a local maximum. Furthermore, it is often the case that $x_{ti}$ is a vector of heterogeneous random variables (e.g. cartesian coordinates and angular values), whose variances are mixed in the trace, with the result that bound (8) largely overestimates the region. However, the dispersion radius is a synthetic value and can be adapted to multimodal distributions after decomposition into a sum of unimodal hypotheses. Empirically, this decomposition is achieved by clustering on samples. By applying command control and updating robot position, the dispersion radius increases together with the trace of the covariance matrix. If $G_{ti}$ is the Jacobian of motion model computed in $(\eta_{xti}, u_{ti})$, with $G_{ti} G_{ti}^T \geq I$ (hypotheses verified by a standard model like (Thrun et al., 2005)), and $\Sigma_{wti}$ is the covariance matrix of additive noise, then

$$\mathrm{tr}\left(\Sigma_{xt_{i+1}}\right) \approx \mathrm{tr}\left(G_{t_i}\Sigma_{xt_i}G_{t_i}^T\right) + \mathrm{tr}\left(\Sigma_{wt_i}\right) \tag{9}$$

Thus, we conclude that $\mathrm{tr}(\Sigma_{xt_i}) \le \mathrm{tr}(\Sigma_{xt_{i+1}})$ and that the dispersion radius increases over the estimation window. A more accurate estimation of how it increases could be obtained with further hypotheses on the motion model, e.g. Lipschitz continuity.
Since the proposal is more and more spread in the estimation window and correction is performed at different times for each partition, we want to investigate how the dispersion affects importance weights. Observation likelihood $w_{ti}(x) = p(z_{ti} \mid x)$ is usually more concentrated than the proposal, sometimes peaked as shown in (Grisetti et al., 2007). We assume that, given a proper $\delta > 0$, region

$$L = \left\{x \in B \mid w_{ti}(x) > \delta\right\} \tag{10}$$

covers a consistent portion of $w_{ti}(x)$. Thus, observation likelihood is bound in $L$ by $M = \sup_{x \in L} w_{ti}(x) < \infty$ (envelope condition) and in $B \setminus L$ by $\delta$. Hence, $w_{ti}(x) < \lambda(x)$

$$\lambda(x) = \begin{cases} M & x \in B \\ \delta & \text{else} \end{cases} \tag{11}$$

The bounding function $\lambda(x)$ and set $L$ are defined on ball $B$, and in the following we will restrict the sampling domain to $B$ using $\pi(x_{ti} \mid x_{ti} \in B)$ as proposal. This assumption allows us to consider the dispersion radius in the following discussion. Moreover, this approximation is not so rough when $\varepsilon$ is close to 1. The effective sample size (Liu, 1996) is a measure of the efficiency of a set of samples in the representation of a target posterior:

$$n_{eff_{ti}} = \frac{1}{\sum_{s=1}^{N} \tilde{\mathrm{w}}_{tj}^2(\mathrm{x}_{tj}^{(s)})} \tag{12}$$

$$= \frac{\left(\sum_{s=1}^{N} \mathrm{w}_{tj}(\mathrm{x}_{tj}^{(s)})\right)^2}{\sum_{s=1}^{N} \mathrm{w}_{tj}^2(\mathrm{x}_{tj}^{(s)})} \tag{13}$$

The above expression is achieved by substituting normalized weights $\widetilde{w}_{ti}(x)$ with their expression. Maximizing the effective sample size is equivalent to minimizing the variance of the weights: it is easy to show with Jensen inequality that $n_{eff}$ is bounded by the number of samples $N$, which is obtained when each weight is equal to 1 and the variance is small. Bounds on observation likelihood allow an approximation of expected values of weight and square weight:

$$E_{\pi}\left[w_{ti}(x) \mid x_{ti} \in B\right] \le M \cdot H_L + \delta \cdot H_{B\setminus L} \tag{14}$$

$$E_{\pi}\left[w_{ti}^{2}(x)\mid x_{ti}\in B\right]\leq M^{2}\cdot H_{L}+\delta^{2}\cdot H_{B\backslash L} \tag{15}$$

where $H_L = E_{\pi}[I_L(x)]$ and $H_{B\backslash L} = E_{\pi}[I_{B\backslash L}(x)]$ are the visit histograms of bins $L$ and $B\backslash L$ respectively; in our notation $I_D(x)$ is the indicator variable with value 1 when $x$ falls in $D$, zero otherwise. Equations (14) and (15) can be used to approximate numerator and denominator of (13):

$$n_{eff_{ti}}\cong N\frac{(M\cdot H_{L}+\delta\cdot H_{B\backslash L})^{2}}{M^{2}\cdot H_{L}+\delta^{2}\cdot H_{B\backslash L}}\cong \mathrm{N}\left(H_{B\backslash L}+2\frac{M}{\delta}H_{L}+\frac{M^{2}H_{L}^{2}}{\delta^{2}H_{B\backslash L}}\right) \tag{16}$$

The approximation given by (16) follows from the assumption that $H_L / H_{B\backslash L} << (\delta / M)^2$. When dispersion is large, the proposal can be considered almost constant on region $L$ and its visit histogram $H_L$ decreases proportionally with the ratio of hypervolumes of $L$ and $B\backslash L$: $H_L \propto 1/r_{ti,\varepsilon}^{d}$ in $d$-dimensional space. Thus, the last partition sets in the estimation window, i.e. those approximating better the distribution at the end of the estimation window, have a spread proposal and are represented by few effective samples, as shown by the trend of (16). From difference between effective sample size and KLD reduction, the bias in estimation follows.

The solution proposed in (Kwok et al., 2004) mitigates the effects of bias by considering the multimodal structure of samples distribution in KL-distance gradient estimation. The estimation of gradient given by (5) ignores samples dispersion in different bins. Formally, gradient (5) is the result of underestimation of KL-divergence: call $Bel_{mix}(C_j)$ and $Bel_{opt}(C_j)$ the mixture and optimal histograms for cluster $C_j$ respectively; from the convexity of KLD (Cover & Thomas, 1991), Jensen inequality holds

$$KL\Big(\sum_{j=1}^{M}Bel_{mix}(C_{j}),\sum_{j=1}^{M}Bel_{opt}(C_{j})\Big)\leq\sum_{j=1}^{M}KL(Bel_{mix}(C_{j}),Bel_{opt}(C_{j})) \tag{17}$$

Gradient estimation based on the second term of inequality (17) is better than the previous one based on the first term which underestimates the distance, but no optimality can be claimed since bin subdivision is empirical and gradient descent approaches easily incur in local minima problems. Furthermore, even if cluster detection is usually performed in PF to group localization hypotheses and no additional computational load is required, sample management is not at all straightforward.

### 3.2 Alternative computation of mixture weights

This section proposes an alternative criterion to compute the values of the weights for the mixture belief. Instead of trying to reduce the Kullback-Leibler divergence, our approach considers mixture weights as the assigned measure of relative importance of partitions that is transformed by processing at the end of estimation window. RTPF prior distribution is the result of two main steps: resampling of samples and propagation of trajectories along the previous estimation window. The effect of resampling is the concentration of previous estimation window samples in a unique distribution carrying information from each

observation. Conversely, the trajectories update given by odometry and observation spreads the particles on partition sets.
Our attempt is to build a linear map modeling the change of relative importance, i.e. mixture weights , due to resampling and propagation of samples. This map should depend on sample weights. Let $wi_j$ be the weight of the $i-th$ sample (or trajectory) of the $j-th$ partition set. Then, the weight partition matrix is given by

$$W = \begin{bmatrix} w_{11} & \cdots & w_{1k} \\ \vdots & & \vdots \\ w_{N_p 1} & \cdots & w_{N_p k} \end{bmatrix} \tag{18}$$

The weights on a row of this matrix trace the history of a trajectory on the estimation window; a group of values along a column depicts a partition handling sensor data in a given time. Resampling and trajectory propagation steps can be shaped using matrix $W$ and mixture weights.

- *Resampling*. The effect of resampling is the concentration of each trajectory in a unique sample whose weight is the weighted mean of the weights of the trajectory. In formula, the vector of trajectory weights is given by $t = W\alpha$ .
- *Propagation*. Projecting a sample along a trajectory is equivalent to the computation of the weight of the sample (i.e., the posterior) for each set, given the proper sensor information. Again, matrix $W$ gives an estimation of the weight. Trajectories projection can thus be done with a simple matrix product

$$\hat{\alpha} = W^T t = W^T W \alpha \tag{19}$$

Vector $\hat{\alpha}$ is a measure of the relative amount of importance of each partition set after resampling and propagation depending on the choice of coefficient $\alpha$ . Hence, $\hat{\alpha}$ is the new coefficient vector for the new mixture of believes. Some remarks can be made about the matrix $V = W^T W$ in (19). First, since we assume $w_{ij} > 0$ , $V$ is a symmetric and positive semi-definite (SPSD) matrix. Moreover, each element $j$ on the main diagonal is the inverse of the effective sample size of set $j$ . The effective sample size is a measure of the efficiency of importance sampling on each of the partition sets. Therefore, the off-diagonal elements of $V$ correspond to a sort of importance covariances among two partition sets. Thus we will refer to this matrix as *weights matrix*.
Hence, a criterion to compute the mixture weights consists of choosing the vector that is left unchanged by map (19) except for scale. Since (19) depends on square of sample weights, the resulting mixture weights reflects the importance of each partition set according to the effective sample size. The vector is thus obtained by searching for an eigenvector of matrix V. To achieve better stability we choose the eigenvector corresponding to the largest eigenvalue. The eigenvector can be computed using the power method or the inverse power method. This criterion can be interpreted as an effort to balance the effective number of samples keeping the proportion among different partition sets.
Fig. 2 illustrates the differences in mixture weights computed according to the original algorithm (RTPF-Grad) and the proposed variant (RTPF-Eig) with an example. When RTPT-

Eig is used to compute mixture weights, the weights of the last partition sets in the estimation window decrease with the effective sample size of the sets, while they increase with RTPF-Grad. Thus, the proposed criterion takes into account the effectiveness of representation provided by partition sets.

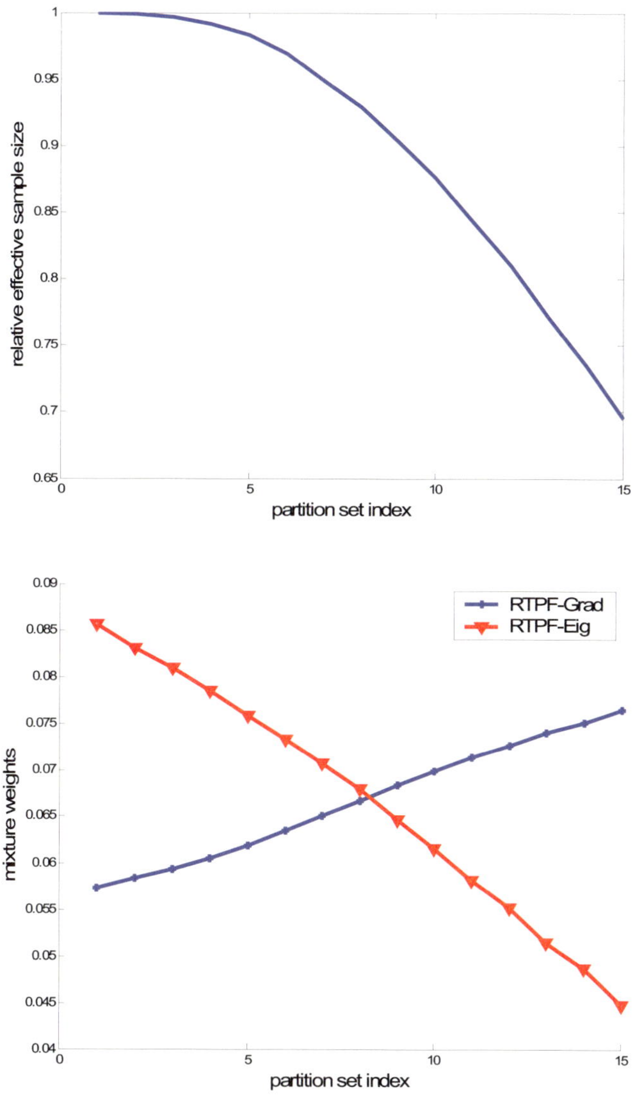


Fig. 2 Effective sample size (top) and mixture weights computed according to the original algorithm and to the proposed variant (bottom) in an estimation window of 15 partitions.

## 4. Complexity of RTPF implementation

As pointed out in the previous section, updating the whole set of samples can be quite demanding from a computational perspective. Together with advanced algorithms, able to

maximize the use of sensor data, the developer should pay great attention to the implementation issues. Inefficiencies due to poor analysis in object creation/destruction cycles and abstraction levels supporting the polymorphic behaviour can introduce drain of computing time preventing the successful application of the conceived algorithm.
This section describes a library designed to efficiently support the implementation of particle filter localization algorithms. The library aims at providing an efficient yet open infrastructure allowing the users to exploit the provided genericity to integrate their own algorithms. The library has been studied to be easily included in the majority of control systems for autonomous mobile robots. In a functional layer, or controller, with the basic computational threads for robot action and perception, the localization task can be simply configured as a computationally demanding, low priority thread.

### 4.1 Design of the library

Based on the functional analysis of the localization problem, four main components have been identified: the localizer, the dynamic model of the system, the sensor data model, and the maps. The interaction among these components enables the implementation of the prediction, matching, and resampling phases of particle filter algorithms. Three classes storing basic data information are required: system state, control command, and sensor data. The main component is the localizer implemented by the *Localizer* class, managing the localization cycle through the coordination of the classes composing the library. Listing 1 shows a simplified interface of the *Localizer* class, including just two *update()* methods.
In the prediction phase, *update()* uses the control command executed by the robot (*Control* class) to perform the computation on the *SystemModel* class, representing the dynamic model of the system. In the correction phase, the second *update()* method uses the perception of the environment (*SensorData* class) to update the density function through the evaluation of the weight. In the following, we describe only implementation strategies to face the main sources of computational overhead arising in the implementation of abstraction layers and object creation/destruction cycles.
The main goal of the library is to provide an open infrastructure aimed at integrating developer choices without their hard-wiring inside the code. Often this goal is achieved with the strategy pattern (Gamma et al., 1996). With this pattern the algorithms are implemented separately as subclasses of an abstract strategy class. While this implementation avoids the hard-wiring of user choices inside the library code, it causes inefficiency due to the abstraction levels introduced to support the polymorphic behaviour. This remark, together with the observation that the choices are immutable at runtime, suggested the use of static polymorphism in the implementation of the current version of the library. Static polymorphism is a more effective flavor of polymorphism based on the use of templates. Templates were originally conceived to support generic programming, as they are functions or classes that are written for one or more types not yet specified. Each template parameter models one degree of variability of the problem domain. This parameter must be fixed at compile time allowing the compiler to generate the proper code. This static polymorphism guarantees type checking and improves code optimization. Exploitation of templates to perform code generation is also known as generic programming (Alexandrescu, 2001).

```
template< State,
          SensorData,
          Control,
          SampleCounter,
          SampleManager,
          Fusion >
class Localizer : public SampleManager<State>
{
   Fusion<SensorData> fusion_;

  public:
   template< ... >      // StateConverter
   Localizer(Pdf &pdf, StateConverter<...> &converter, tods::TimeInterval &period);

   ~Localizer();

   template< ... >      // SystemModel Parameters
   void update(Control &u, SystemModel<...> &sys);

   template< ... >      // SensorModel Parameters
   void update(SensorData &z, SensorModel<...> &sen);
};
```

Listing 1 The Localizer class.

To reduce the second source of computational overhead, we focused our attention on the creation/destruction of the objects within the localization cycle to reduce their dynamic allocation. Fig. 3 presents the life cycle of the objects created and destroyed during a single localization cycle of the RTPF algorithm. During each step, new samples and new objects for representation of odometric and sensor data are created to be immediately destroyed. Note that the samples and controls are created in different iterations of the localizer in the same estimation window and survive after the end of this window: that is a rather different way of handling objects from a standard particle filter implementation. Thus a management policy and proper data structures for storage are required in a flexible implementation. Therefore observations, even if they are created and used in an iteration and their handling is quite straightforward, exhibit a variety of sensor models given by the range of sensors. To support such a variety, the library provides generic sensor data.

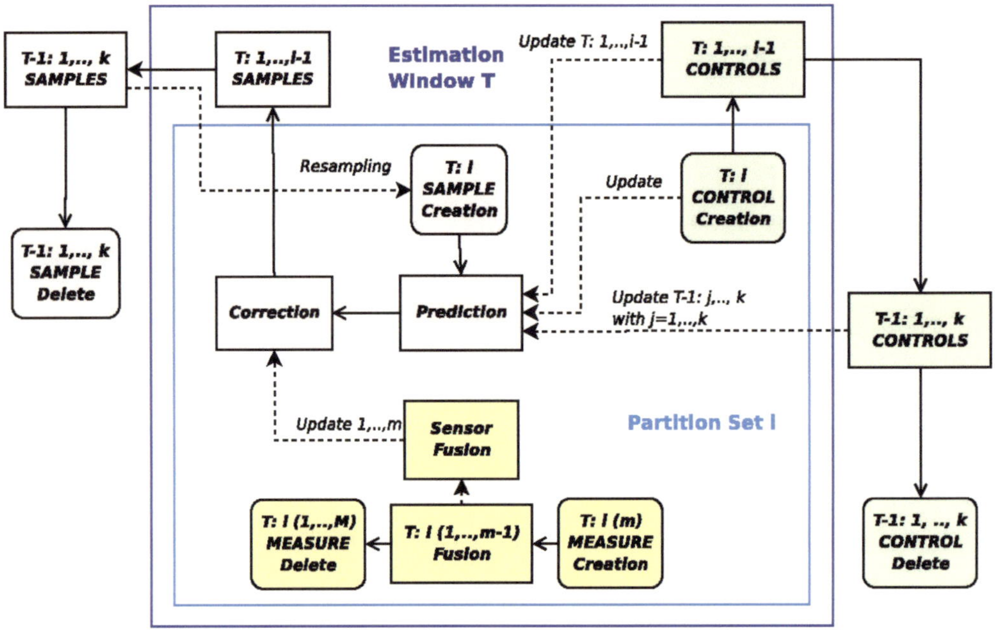


Fig. 3 Life cycle for particle, measurement, and control objects within a single step in a real-time particle filter.

## 5. Results

In this section performance of RTPF is evaluated both in simulated environment and using experimental data collected by robot sensor. An important purpose of this section is the comparison of the two RTPF versions differing in the method for computing mixture weights.

### 5.1 Simulation

Several tests were performed in the environments shown in Fig. 4 and Fig. 5. They correspond to the main ground floor hallway in the Computer Engineering Department of the University of Parma (Fig. 4) and to the hallway of the Department of Computer Science and Engineering of the University of Washington (Fig. 5, map adapted from (Kwok et al., 2004)). These environments allow verification of RTPF correctness while coping with several symmetric features, which may cause ambiguities in the choice of correct localization hypotheses. The environment of Fig. 5 had been exploited in (Kwok et al., 2004) to verify RTPF correctness and has therefore been considered as a reference.

In simulation, the map is stored as a grid with a given resolution (0.20 m) and is used both to create simulated observations and to compute importance weights in correction steps. Data provided to the localizer consist of a sequence of laser scans and measurements:

scanned ranges are obtained by ray tracing a beam on the discretized map. The measurement model is also based on ray tracing according to standard beam models for laser scanner (Thrun et al., 2005). In our tests we have used only three laser beams measuring distances to left, right and frontal obstacles; such poor sensor data stress the role of algorithm instead of sensor data quality. A gaussian additive noise was added to both range beams and robot movements representing environment inputs and robot state in simulation. Thus simulation tests are performed in an environment known in detail and are best suited for comparing performance between algorithms. The task of the robot is to achieve localization while moving in the environments of Fig. 4 and Fig. 5 along assigned trajectories. Simulated trajectories, labeled as Path 1 and Path 2 in Fig. 4 and Fig. 5, correspond to lengths of approximately 5 to 8 m.

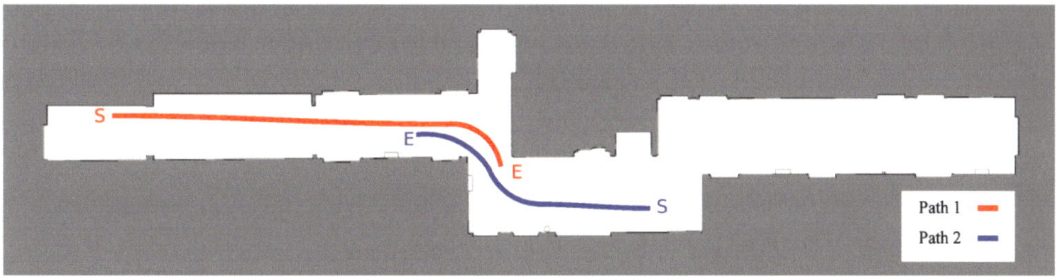


Fig. 4 Map 1 - Hallway and simulated paths in the Computer Engineering Department, University of Parma.

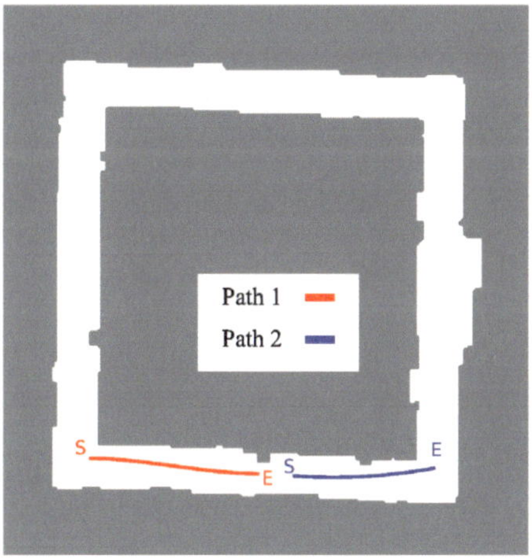


Fig. 5 Map 2 - Hallway and simulated paths in the Department of Computer Science and Engineering, University of Washington.

Localization algorithms investigated are RTPFs in the two versions: the original steepest descent-based one (RTPF-Grad) and the proposed one based on the effective number of samples (RTPF-Eig). During these tests the partition set size was 1000 samples.

A summary of simulation results is reported in Fig. 6 and Fig. 7, where curves show the localization error for the two algorithms at each iteration by considering convergence to the maximal hypothesis. For both curves, each value is obtained by averaging the distances of the estimated pose from the real pose over 10 trials where localization eventually converged to the correct hypothesis within the maximum number of iterations (set to 40). For both algorithms there were also a few instances where localization did not converge to the correct hypothesis within the length of the path, although the correct hypothesis was the second best. These unsuccessful experiments were approximately 10% of all simulated localization trials. We did not verify whether the robot would eventually recover its correct pose in the environment with further navigation.

On the average, the two versions of the RTPF-based localizer converge to some few hypotheses after three iterations, and the common samples distribution is multi-modal. Hence, cluster search leads to few hypotheses with different weight. In our tests a hypothesis close to the correct robot pose always exists, and when this hypothesis prevails there is a sudden change in localization error, as shown in Fig. 6 and Fig. 7.

Convergence is helped by recognizable features, e.g. the shape of scans, but when the environment is symmetric it can be difficult to reach, especially with limited or noisy sensoriality. Of course, the mean error trend in Fig. 6 and Fig. 7 does not correspond to any of the simulated trials; rather, it is the result of averaging trials with quick convergence and trials where the correct hypothesis could only be recovered after several more iterations.

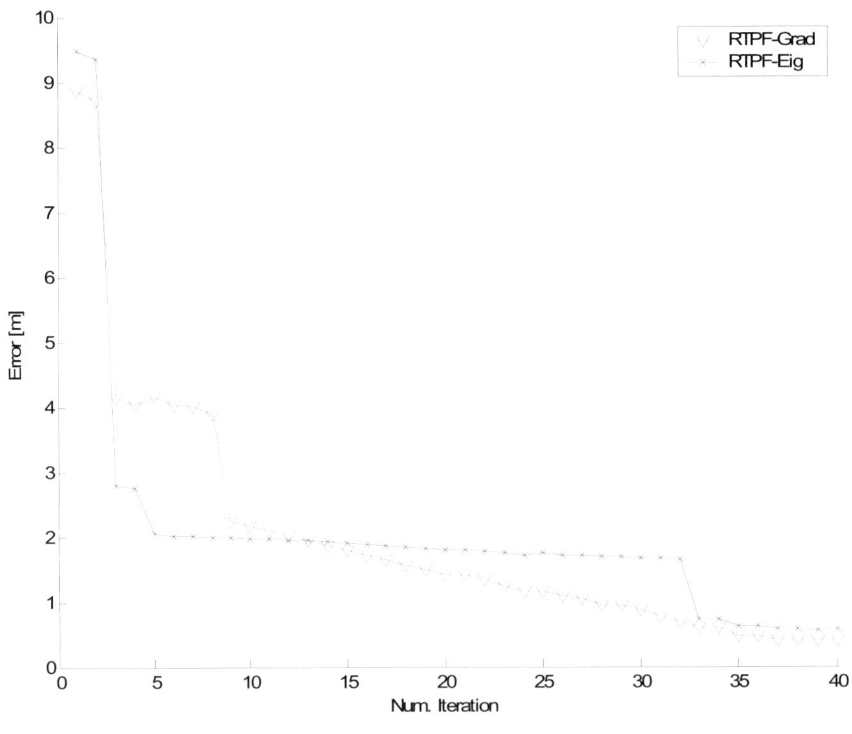


Fig. 6 Performance of the two RTPF versions in the simulated environment of Map 1. The x-axis represents the iterations of the algorithm. The y-axis shows the average error distance of the estimated pose from the actual robot pose.

Fig. 8 provides an alternative view of the same data, as curves show the percentage of simulation trials converging to the correct hypothesis (i.e. with localization error less than 1.5 m) at each iteration. For both environments, convergence is reached with only few

iterations in some simulation runs. In other simulations, the correct robot pose is recovered only after about 20 or 30 iterations, i.e. after sensing map features that increase the weight of the correct samples.
Empirically, for the examined environments RTPF-Eig seems to exhibit a slightly faster convergence, on the average, to the correct localization hypothesis, even though its average error at the last recorded iteration appears somewhat larger.

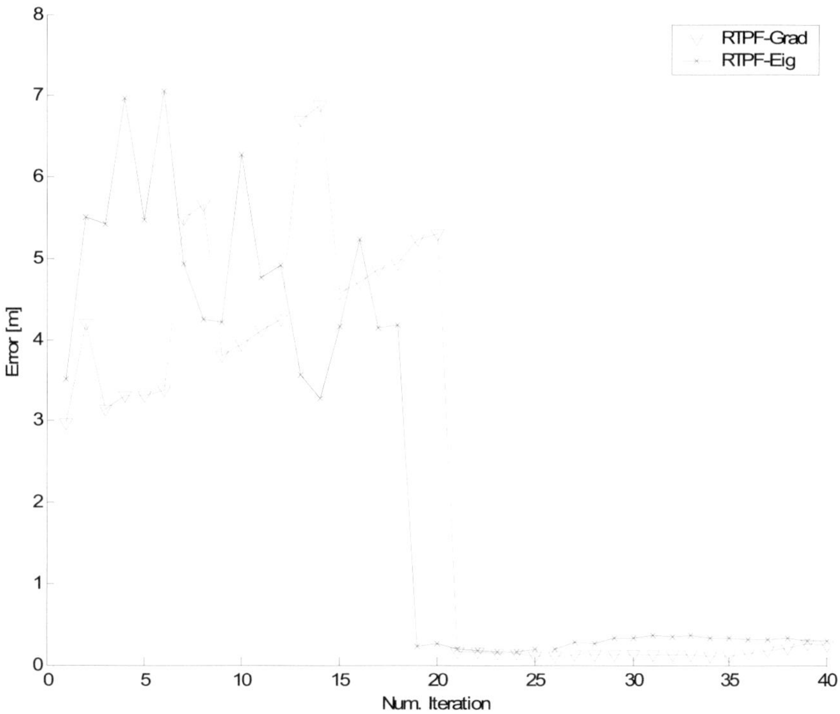


Fig. 7 Performance of the two RTPF versions in the simulated environmentof Map 2. The x-axis represents the iterations of the algorithm. The y-axis shows the average error distance of the estimated pose from the actual robot pose.

### 5.2 Experiments

Real experiments took place in the environment of Fig. 4 collecting data with a Nomad 200 mobile robot equipped with a Sick LMS 200 laser scanner. The robot moved along Path 1 for about 5 m, from the left end of the hallway in steps of about 15−20 cm and reading three laser beams from each scan in the same way of the simulation tests. In the real environment localization was always successful, i.e. it always converged to the hypothesis closer to the actual pose in less than 10 iterations (remarkably faster than in simulation). Localization error after convergence was measured below 50 cm, comparable or better than in simulation.
To assess the consistency of the localizer's output on a larger set of experiments, we compared the robot pose computed by the localizer (using the RTPF-Eig algorithm) with the one provided by an independent localization methodology. To this purpose, some visual

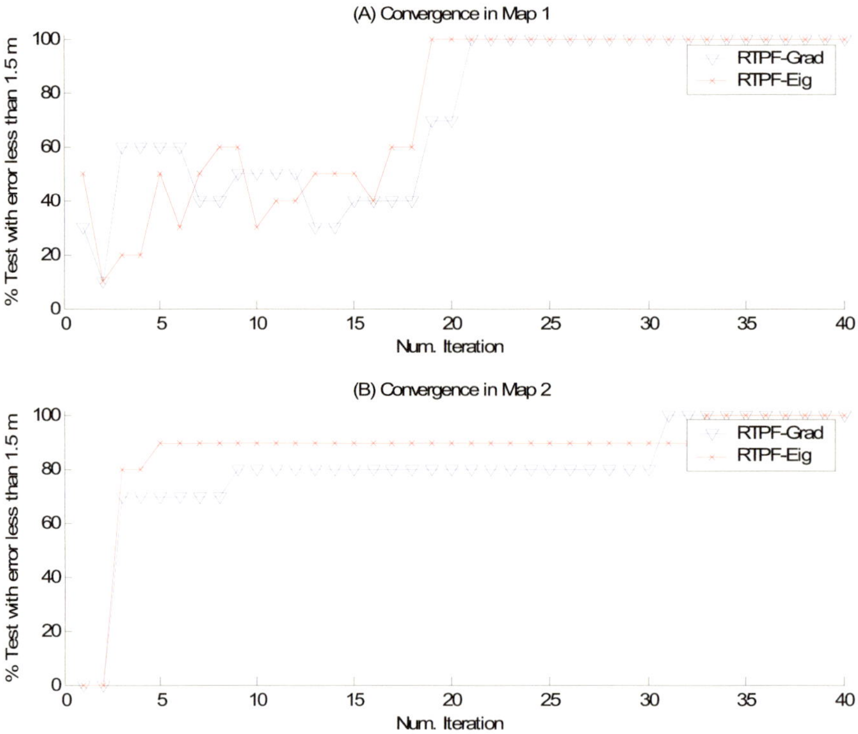


Fig. 8 Percentage of simulation trials converged to the correct hypothesis, i.e. with localization error less than 1.5 m, during iterations for Map 1 (a) and Map 2 (b).

landmarks were placed in the environment and on the mobile robot, and a vision system exploiting the ARToolKit framework (Kato & Billinghurst, 1999) was exploited to triangualate the robot position based on these landmarks. The vision system provided an independent, coarse estimate of the robot pose at any step, and hence allowed to establish convergence of the RTPF-based localizer. The two localization estimates were computed concurrently at each location and stored by the robot.
Fig. 9 shows the results of 10 tests of RTPF-Eig over about 20 iterations. These results confirm that RTPF-Eig achieves localization to the correct hypothesis very fast in most experiments. After convergence, the maximum distance between RTPF-based and vision based estimates is about 70 cm due to the compound error of the two systems.

## 6. Conclusion

In this chapter, we have described an enhanced Real-Time Particle Filter for mobile robot localization incorporating several formal and practical improvements. We have presented a formal discussion of computation of mixture weights in RTPFs, along with a new approach overcoming potential problems associated with the existing technique. The method proposed in this chapter computes mixture weights as the eigenvector of a matrix and thus avoids gradient descent, possibly prone to numerical instability. The method provides a balance of the effective sample size of partition sets on an estimation window. We have also

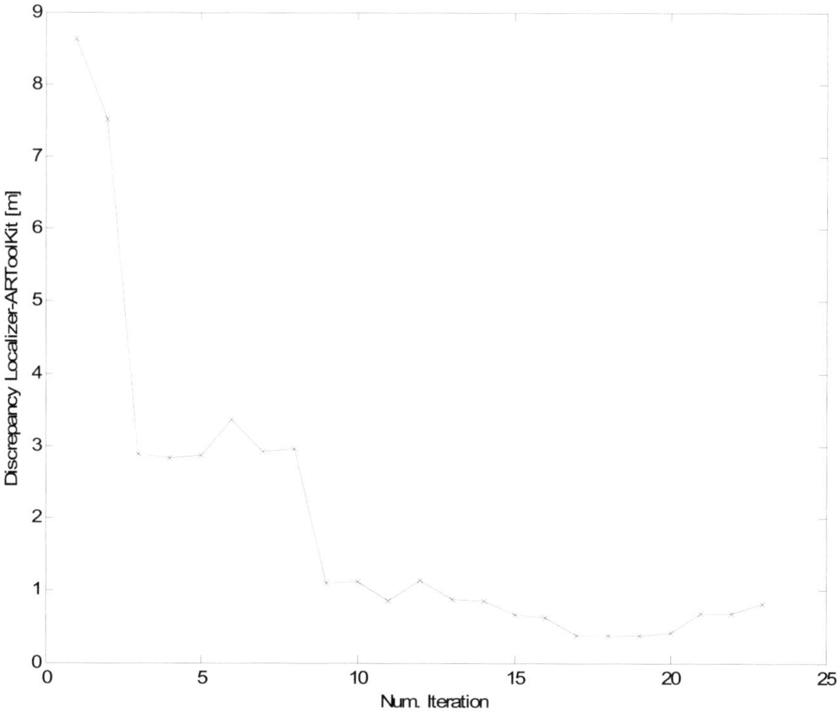


Fig. 9 Performance of RTPF-Eig using real data collected in the hallway of Map 1.

described a library efficiently supporting implementation of particle filter algorithms and independent from any specific robot control architecture. The library takes advantage from generic programming and from a carefully designed object lifecycle model to minimize overhead while providing flexibility.

The proposed approach has been implemented in a RTPF for localization with a Nomad 200 mobile robot equipped with a laser range scanner, and evaluated in both simulation tests and real experiments. In two simulation environments, the new approach has achieved a localization performance similar to the original KLD-based algorithm, while avoiding the potential problems associated with gradient search methods. In real experiments with the mobile robot, the modified RTPF-based localization system has proven very effective, yielding correct localization within a small number of filter iterations.

Of course, further experimental work is required to assess the relative merit of the improved RTPF over the original approach. Nonetheless, the research described in this paper shows how a through theoretical understanding of the problem and of the algorithmic solution should be combined with a careful software implementation to attain the potential of probabilistic localization methods.

## 7. Acknowledgement

This research has been partially supported by Laboratory LARER of Regione Emilia-Romagna, Italy.

## 8. References


Alexandrescu, A. (2001). Modern C++ Design: Generic Programming and Design Pattern Applied. *Addison-Wesley*.

Arras, K. O., Castellanos, H. F., and Siegwart, R.(2002). Feature-based multi-hypothesis localization and tracking for mobile robots using geometric constraints. *IEEE Int. Conf. on Robotics and Automation*, 2:1371–1377.

Cover, T. M. & Thomas, J. A. (1991). Elements of Information Theory. *Wiley*.

Doucet, A., de Freitas, J., & Gordon, N. (2001). Sequential Monte Carlo Methods in Practice. Springer.

Elinas, P. and Little, J. (2005). sMCL: Monte-carlo localization for mobile robots with stereo vision. *Proc. of Robotics: Science and Systems*.

Fox, D. (2003). Adapting the sample size in particle filters through KLD-sampling. *Int. J. of Robotics Research*, 22(12):985–1003.

Fox, D., Burgard, W., & Thrun, S. (1999). Monte Carlo Localization: Efficient position estimation for mobile robots. *Proc. of the National Conference on Artificial Intelligence*.

Gamma, E., Helm, R., Johnson, R., and Vlissides, J. (1995). Design Patterns: elements of reusable object-oriented software. *Addison-Wesley*.

Grisetti, G., Stachniss, C., & Burgard, W. (2007). Improved techniques for grid mapping with Rao-Blackwellized particle filters. *IEEE Trans. on Robotics*, 23(1):34–46, February 2007.

Hester, T., & Stone, P. (2008). Negative Information and Line Observations for Monte Carlo Localization. *IEEE Int. Conf. on Robotics and Automation*, pages 2764–2769.

Kato, H. and Billinghurst, M. (1999). Marker tracking and HMD calibration for a video-based augmented reality conferencing system. *Proc. of the Int. Workshop on Augmented Reality*.

Kwok, C., Fox, D., & Meilă, M. (2003). Adaptive real-time particle filters for robot localization. *IEEE Int. Conf. on Robotics and Automation*, 2:2836–2841, 2003.

Kwok, C., Fox, D., & Meilă, M. (2004). Real-time particle filters. *Proc. of the IEEE*, 92(3):469–484.

Leonard, J. J. & Durrant-Whyte, H. F. (1991). Mobile Robot Localization by Traking Geometric Beacons. *IEEE Int. Conf. on Robotics and Automation*.

Liu, J. (1996). Metropolized independent sampling with comparisons to rejection sampling and importance sampling. *Statistics and Computing*, 6(2):113–119.

Montemerlo, M., Roy, N., & Thrun, S. (2003). Perspectives on standardization in mobile robot programming: The Carnegie Mellon navigation (CARMEN) toolkit. *IEEE/RSJ Int. Conf. on Intelligent Robots and Systems*.

Se, S., Lowe, D., & Little, J. (2002). Mobile robot localization and mapping with uncertainty using scale invariant visual landmark. *Int. J. of Robotics Research*, 21(8):735–758.

Sridharan, M., Kuhlmann, G., & Stone, P. (2005). Practical vision-based Monte Carlo localization on a legged robot. *IEEE Int. Conf. on Robotics and Automation*, pages 3366–3371.

Thrun, S., Burgard, W., & Fox, D. (2005). Probabibilistic Robotics. *MIT Press*, Cambridge, MA.

# 22

# Dependability of Autonomous Mobile Systems

Jan Rüdiger, AchimWagner and Essam Badreddin
*Automation Laboratory, Dept. Mathematics & Computer Science,*
*University of Heidelberg Mannheim,*
*Germany*

## 1. Introduction

Computer systems pervade more and more our everyday life. They are found in workstations, mobile devices and in nearly every device - from consumer products such as coffee-machines to safety critical automotive systems such as cars, industrial production-lines etc. Due to increasing complexity, the controllability of such systems is a serious problem,while impacts and consequences on our daily life are continuously increasing. Therefore, non-functional system properties like dependability became a crucial factor within the computerized product design process. Although common unformal ideas in terms of dependability exist, a formal definition for dependability is still missing. One reason may be the historical growth of the definition of the term dependability, which has been added incrementally by a number of attributes. In the 40ies of the last century, the first computer based on vacuum tube technology with huge failure probabilities (in the ENIAC computer tubes failed every 7 minutes) were constructed, reliability became an issue. Due to increased interaction with the systems later on in the 60ies availability became even more important. Related to the requirements of controlling safety critical plants like nuclear power stations or space crafts the safety property of computer systems where getting into the focus during the 70ies. Internet connectivity, data bases and mobile services were the reason why security, integrity and maintainability have been added to the dependability concept. The state-of-the-art assessment of dependability is based on a binary fault model, which describes components on an operable - not operable basis and a logical error propagation using fault trees, event tress or binary block diagrams (Vesely et al., 1981). Modern approaches like Markov-Chain models (Flammini, 2006) or Stochastic Petri Nets (Filippini & Bondavalli, 2004) capture the time dependent probability of a combination of error states within a system. However, they are not able to detect the origin of an error resulting from the system dynamics. For instance, a light bulb mostly crashes during the switching on phase and not during stationary operation. This error scenario cannot be reflected by pure probabilistic modelling.

A further disadvantage of pure probabilistic models is that they are more or less decoupled from the original behaviour of the system. Thus, finding a valid fault model which starts from the functional model of the system is up to the design engineer. However, even a fault probability equal to zero does not guarantee, that systems operate according to what users expect because the requirements on the dynamic system behaviour are not modelled. Dependability is more than a collection of attributes related to a probabilistic - but static - error description.

This is particularly important when dealing with autonomous or semi-autonomous systems. With an increasing degree of autonomy and safety requirements, requirements for dependability increase. Hence, being able to measure and compare the dependability of a system becomes more and more vital. Since autonomous mobile systems are often described by their behaviour, it is straightforward to also define the dependability of such systems in a behavioural context. To show the link between conventional approaches and to allow the usage existing tools, the proposed dependability model is supplemented by the majority of the dependability attributes described above. The proposed approach of a behaviour based definition for dependability described in this chapter is focused on autonomous mobile systems. Nevertheless, the ideas behind it are more general.

## 2. Basics of dependable systems

According to (Candea, 2003) a general notion of what dependability is usually understood is summarizes as follows:

> turn your homework in on time, if you say you'll do something then do it, don't endanger other etc.

Computer controlled systems, in our case autonomousmobile systems, do, however, not understand these vague concepts. When it comes to machines we need amore precise understanding and definition of dependability. If system dependability must be expressed and measured in numbers, a formal definition is even more important. This definition must be mostly system-independent in order to have the opportunity to compare the dependability between different systems. Finally, this definition must agree with the general notion of dependability.

This chapter gives a broad overview of what is usually understood under the term *dependability* and discusses the sometimes different definitions of dependability used throughout literature. Based on the aforementioned and in combination with a behavioural system description, a dependability definition for autonomous mobile systems is proposed.

Non-functional properties reflect the overall quality of a system. Beside performance, robustness, usability etc., dependability is getting amore important non-functional system requirement. The general, qualitative, definitions for *dependability* used so far in literature are presented and discussed in the following. The most frequently cited definitions of dependability are the ones introduced by Carter and Laprie which are presented here together with others in chronological order.

**Carter** (Carter, 1982): A system is dependable if it is trustworthy enough that reliance can be placed on the service it delivers.

**Laprie** (Laprie, 1992): Dependability is that property of a computing system which allows reliance to be justifiably placed on the service it delivers.

**Badreddin** (Badreddin, 1999): Dependability in general is the capability of a system to successfully and safely fulfill its mission.

**Dubrova** (Dubrova, 2006): Dependability is the ability of a system to deliver its intended level of service to its users.

All four definitions have in common that they define dependability on the service a system delivers and the reliance that can be placed on that service. The service a system delivers is the behaviour perceived by the user, which can also be called the *mission of the system*.

Only few further definitions, following the idea of the definitions presented above exist. Among them, the dependability definition of the International Electrotechnical Commission (IEC) ("IEC 60050, IEV 191-02-03: Dependability and quality of service - Part 1: Common terms" see IEC, 1990)[1], the definition from the IFIP 10.4 Working Group on Dependable Computing and Fault Tolerance (see International Federation for Information Processing, ) and the definition from the Department of Defence (see Department of Defence, 1970).
These classical definitions of dependability do, however, offer two major drawbacks:
1. They do not define a directly applicable and repeatable way to compute the dependability of a system.[2]
2. The dynamics of the system are not directly taken into account when investigating the dependability of the system. Using the information gained from the mathematical system model, for example, seems obvious when investing the dependability of that system. Neglecting the dynamics of a system means ignoring all information coming from that model. Furthermore the dynamics of a system is crucial for fault scenarios not only in the case of autonomous mobile systems.

In this chapter a theoretical framework is proposed, that does not only describe a dependability definition and dependability means in relation to the dynamics of a system but also presents a repeatable and system-independent way of how to measure dependability.
Dependability is mainly understood as an integrated concept that additionally consists of different attributes (see also Figure 1). According to (Avizienis et al., 2004b; Avizienis et al., 2004a; Randell, 2000) dependability consists of the following attributes:
- **Availability** readiness for correct service,
- **Reliability** continuity of correct service,
- **Safety** absence of catastrophic consequences for both user(s) and environment,
- **Confidentiality** absence of unauthorized disclosure of information,
- **Integrity** absence of improper system state alteration and
- **Maintanability** ability to undergo modifications and repairs.

For further information as to the impact of these attributes on the dependability of the complete system, please refer to (Avizienis et al., 2004a) and (Dubrova, 2006).
Further definitions with slightly different attributes exist in the literature (see i.e. Candea, 2003; Dewsbury et al., 2003). The main idea, i.e. dependability consists of different attributes, is still of value and will be part of the definition proposed below.
The authors would like to express, however, that more than a static aggregation of attributes is needed for the description of the dependability of an autonomous mobile system. These attributes are always related to a static model, even if error propagation is dynamic. The proposal for dependability measurement also includes the attribute approach as a special case for being compatible with the qualitative dependability definitions.
Unfortunately a few of the attributes have different, not necessarily similar, definitions. Still missing is a formal and comprehensive definition for a few attributes.
As for autonomous mobile systems, not all attributes are of equal importance. For a discussion about the set of attributes important for autonomous mobile systems the reader is referred to (Rüdiger et al., 2007b).

---

[1] This definition of dependability is often referred to as ISO 1992 or IEC 50(191)
[2] According to IEC norm the definition is „used only for general descriptions in non-quantitative terms"

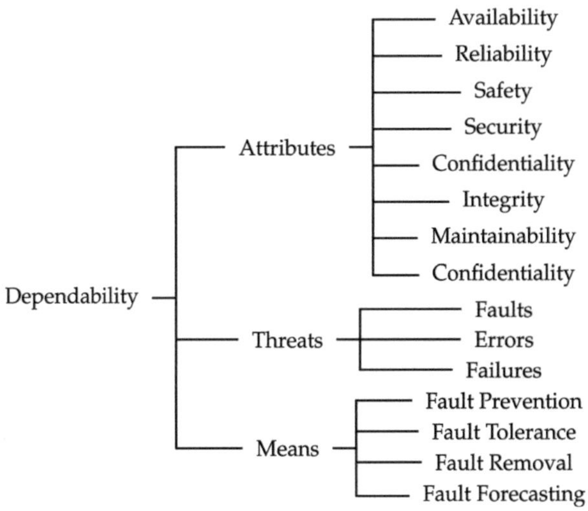


Figure 1. The Dependability Tree

## 3. Framework for a theory of dynamical systems

In order to develop a formal definition for the dependability of autonomous mobile systems, first of all the following terms

- service,
- system,
- and reliance

mentioned in the non-formal definitions need a mathematical description. Therefore, a formal, mathematical, theory that is capable of describing these terms is needed.

There are different techniques to describe a system mathematicely, that is modeling a system. Among them, the state-space approach can be mentioned, where a system is modelled by a set of input, output and state variables related by first order equations (time domain), and, for example, the frequency domain approach. Another approach is the *Behavioural Modelling* approach (see Willems, 1991) where a system is modelled only by describing its behaviour.

Since the service a system delivers is the behaviour as it is perceived by the user, the later modelling technique was used and is shortly introduced in the following. Additionally, the behaviour based approach is quite common in the field of autonomous mobile robots. This goes back to 1980 where Rodney Brook introduced his subsumption architecture (see Brooks, 1986). Finally, the behavioural modeling approach offers the opportunity to modell the complete system including both environment and user.

Willems (Willems, 1991) defines a system in a universum $\mathbb{U}$. The elements of $\mathbb{U}$ are called outcomes of the system. A mathematical model of a system from a behavioural or black-box point of view claims that certain outcomes are possible, while others are not. The model thus defines a specific subset $\mathfrak{B} \subset \mathbb{U}$. This subset is called the *behaviour* of the system. Subsequently, a (deterministic) mathematical model of a system is defined as:

**Definition 3.1** A mathematical model *is a pair* $(\mathbb{U}, \mathfrak{B})$ *with the universum* $\mathbb{U}$ *- its elements are called* outcomes *- and* $\mathfrak{B}$ *the behaviour.*

Given the above definition of a mathematical model, a dynamical system is a set of trajectories describing the system behaviour during the time instants of interest in its signal space $\mathbb{W}$.

In contrast to the state space representation, like $\dot{x} = f \circ x$, a dynamical system is defined as:
**Definition 3.2** A dynamical system $\Sigma$ *is a triple* $\Sigma = (\mathbb{T}, \mathbb{W}, \mathfrak{B})$ *with* $\mathbb{T} \subseteq \mathbb{R}$ *the time axis,* $\mathbb{W}$ *the signal space, and* $\mathfrak{B} \subseteq \mathbb{W}^{\mathbb{T}}$ *the behaviour.*
In the above definition a dynamical system is described by a period of time $\mathbb{T}$, the signal space $\mathbb{W}$ and a set of time trajectories $\mathfrak{B}$. The behaviour of a dynamic system is described by a subset $\mathfrak{B}$ of all possible time trajectories $\mathbb{T} \Rightarrow \mathbb{W}$. For a trajectory (an event) $w : \mathbb{T} \rightarrow \mathbb{W}$ the following applies:

- $w \in \mathfrak{B}$ the model allows the trajectory $w$
- $w \notin \mathfrak{B}$ the model forbids the trajectory $w$

The reader is referred to (Willems, 1991) for examples of this modeling technique.

## 4. Behaviour-based dependability of autonomous mobile systems

As stated earlier, the service a system delivers is the behaviour as it is perceived by the user. A framework for a mathematical system description according to its behaviour was introduced in the last section.
So far the system and its service, from the dependability definition, can be mathematical described. What remains is a fundamental definition of the *service a system should deliver* and an additional description of the *reliance*, in terms of the service offered, in the given framework. Finally the attributes of dependability as discussed in section 2 need to be defined in the framework (see also Rüdiger et al., 2007a).

### 4.1 Behaviour and mission of autonomous mobile systems

As afore mentioned, the behaviour of a system is the set of all possible time trajectories of its external states (see section 3). Due to limitations, either deliberately, caused by a fault in the system, or changes in the environment, this set $\mathfrak{B}$ could be slightly changed or reduced to a set of behaviours actually available to the system. Although it will probably be only a subset of the behaviours of the original system. This set, which may vary over time, will be defined as follows:
**Definition 4.1** *Let* $\Sigma = (\mathbb{T}, \mathbb{W}, \mathfrak{B})$ *be a dynamical system then* $\mathbb{B} \subseteq \mathbb{W}^{\mathbb{T}}$ *is called the set of* available behaviours $w_i(t) : \mathbb{T} \rightarrow \mathbb{W}$, $i = 1 \ldots n$ *to the system at time t.*
Again, the set $\mathbb{B}$ is a set of time trajectories within the original signal space $\mathbb{W}$. For the set $\mathbb{B}$ it must not necessarily apply $\mathbb{B} \subset \mathfrak{B}$ since it can also contain trajectories as a result of a fault in the system or a change in the environment not previously modelled. The set $\mathbb{B}$ may also vary over time.
Additionally, the set $\mathbb{B}$ must not cover all trajectories from the set $\mathfrak{B}$ since due to implementation reasons not all possible trajectories must be available to the system.
Autonomous mobile systems are usually build as general purpose machines that handle different tasks. In the following, these tasks will also be called missions. For estimating the dependability of an autonomous mobile system it is, however, not important to handle all kinds of missions dependable, but only the mission in terms of the *service the system should deliver* or what the system is *intended* to do. Such missions are be defined as:
**Definition 4.2 (Mission)** Let $\Sigma = (\mathbb{T}, \mathbb{W}, \mathfrak{B})$ be a time-invariant dynamical system. We say the mission $w_m$ of this system is the map $w_m : \mathbb{T} \rightarrow \mathbb{W}$ with $w_m \in \mathfrak{B}$.

Note that even if an autonomous mobile system is usually build for the handling of different tasks, the actual mission the system has to accomplish is defined as only one special behaviour from the set $\mathfrak{B}$.

Thus, the mission, here defined, is just a special trajectory or, more precisely, a special behaviour in $\mathfrak{B}$. *Weak controllability* (see Hermann, 1977) is assumed since it does not make any sense to define a mission to a system which by definition is not able to accomplish it.

If the system is capable of fullfilling its mission $w_m$ is defined as follows:

**Definition 4.3** *A mission* $w_m \in \mathfrak{B}$ *for a given dynamical system* $\Sigma = (\mathbb{T}, \mathbb{W}, \mathfrak{B})$ *with the behaviours* $\mathbb{B}$ *is said to be* accomplishable *by this system if for all* $w_1 \in \mathfrak{B}$ *there exists a* $t \in \mathbb{T}$, $t \geq 0$, *a behaviour* $w \in \mathbb{B}$, $w : \mathbb{T} \cap [0, t] \rightarrow \mathbb{W}$ *and a behaviour* $w_2 \in \mathbb{B}$ *such that* $w' \in \mathfrak{B}$, *with* $w' : \mathbb{T} \rightarrow \mathbb{W}$ *defined by:*

$$w'_{(t')} = \begin{cases} w_{1(t')} & \text{for } t' < 0 \\ w_{(t')} & \text{for } 0 \leq t' \leq t \\ w_{2(t'-t)} & \text{for } t' > t \end{cases}$$

*and*

$$w'_{(t')} = w_m \quad \text{for } t' > t$$

Based on the definition of controllability according (Willems, 1991) a mission $w_m$ is accomplishable if the system can be steered from any past trajectory (light green trajectory in Fig. 2) to the beginning of the mission trajectory $w_m$ (black trajectory in Fig. 2) and can then be steered along the mission trajectory (red and blue trajectories in Fig. 2).

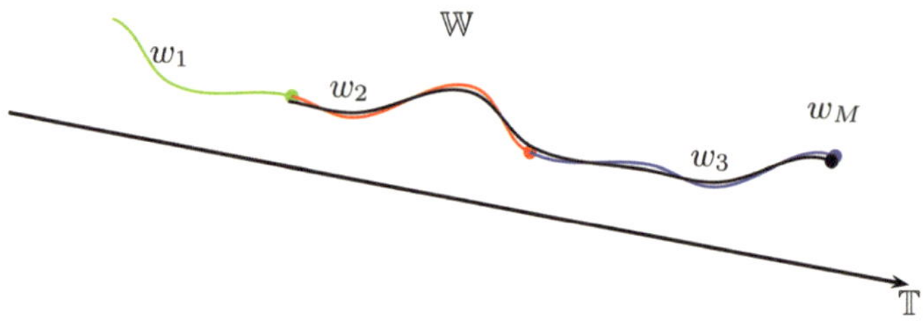


Figure 2. A mission (black line) is accomplished by steering the system to the mission trajectory by behaviour $w_1$ and subsequently by steering along the mission trajectory with behaviours $w_2$ and $w_3$

### 4.2 Behaviour based attributes of dependability

Before continuing with the definition of dependability for autonomous mobile systems, the basic attributes of dependability must be defined in a behavioural context. Only the main important attributes for dependability of autonomous mobile systems are introduced. Please refer to (Rüdiger et al., 2007a) for an advanced description and (Rüdiger et al., 2007b) for the subset of attributes needed for measuring the dependability of autonomous mobile systems.

#### 4.2.1 Reliability

A common (see e.g. Dubrova, 2006) unformal definition for reliability is:

> Reliability $R|_t$ is the probability that the system will operate correctly in a specified operating environment in the interval [0, *t*], given that it worked at time 0.

An autonomous system is, thus, said to be reliable if the system state does not leave the set of admissible trajectories $\mathfrak{B}$. The reliability of a system can be defined as:

**Definition 4.4** *Let* $\Sigma = (\mathbb{T}, \mathbb{W}, \mathfrak{B})$, $\mathbb{T} = \mathbb{Z}$ *or* $\mathbb{R}$, *be a time-invariant dynamical system. The system is said to be* reliable *in the period* $[0, t]$ *if for all* $0 \leq t_1 \leq t$ *the system state is* $w(t_1) \in \mathfrak{B}$. *Correspondingly, the* reliability *of the system is the probability that the system is reliable.*

### 4.2.2 Availability

Availability is typically important for real-time systems where a short interruption can be tolerated if the deadline is not missed.

> Availability $A|_t$ is the probability that a system is operational at the instant of time t.

In contrast to reliability the availability is defined at a time instant t while the reliability is defined in a time interval.

**Definition 4.5** *Let* $\Sigma = (\mathbb{T}, \mathbb{W}, \mathfrak{B})$, $\mathbb{T} = \mathbb{Z}$ *or* $\mathbb{R}$, *be a time-invariant dynamical system. The system is said to be available at time t if* $w(t) \in \mathfrak{B}$. *Correspondingly, the availability of the system is the probability that the system is available.*

### 4.2.3 Safety

From the reliability point of view, all failures are equal. In case of safety, those failures are further divided into *fail-safe* and *fail-unsafe* ones. Safety is reliability with respect to failures that may cause catastrophic consequences. Therefore safety is unformaly defined as (see e.g. Dubrova, 2006):

> Safety $S(t)$ of a system is the probability that the system will either perform its function correctly or will discontinue its operation in a fail-safe manner.

For the formal definition of safety an area $\mathfrak{S}$ is introduced, as in (Badreddin & Abdel-Geliel, 2004), which leads to catastrophic consequences when left. In the latter case it is, however, assumed that this *Dynamic Safety Margin* is fully contained in the stability region while $\mathfrak{S}$ is defined to be around $\mathfrak{B}$. This margin is, like $\mathfrak{B}$, highly system specific, but can be set equal to $\mathfrak{B}$ in the case of restrictive systems.

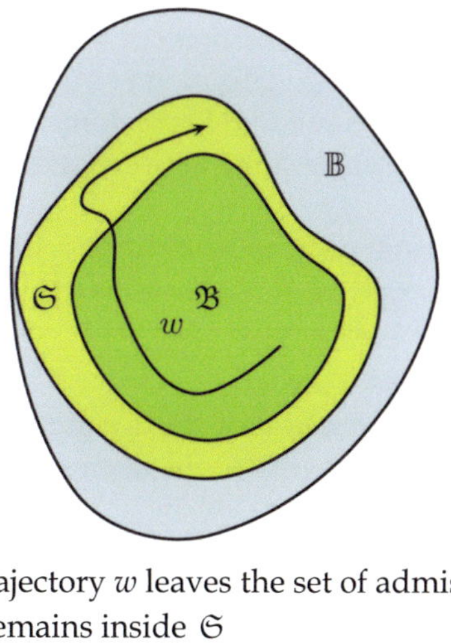


Figure 3. Safety: The system trajectory $w$ leaves the set of admissible trajectories $\mathfrak{B}$ but is still considered to be safe since it remains inside $\mathfrak{S}$

**Definition 4.6** *Let* $\Sigma = (\mathbb{T}, \mathbb{W}, \mathfrak{B})$, $\mathbb{T} = \mathbb{Z}$ *or* $\mathbb{R}$, *be a time-invariant dynamical system with a safe area* $\mathfrak{S} \supseteq \mathfrak{B}$. *The system is said to be* safe *if for all* $t \in \mathbb{T}$ *the system state* $w(t) \in \mathfrak{S}$.
This definition is consistent with the idea that a safe system is either operable or not operable but in a safe state.

### 4.3 Behaviour based dependability

Having defined the behaviour of a system and the mission, which corresponds to the service the system should deliver, the dependability of the system can be defined as:
**Definition 4.7** *A time-invariant dynamical system* $\Sigma = (\mathbb{T}, \mathbb{W}, \mathfrak{B})$ *with behaviours* $\mathbb{B}$ *and a mission* $w_m \in \mathfrak{B}$ *is said to be (gradually)* dependable *in the period* $T \in \mathbb{T}$ *if, for all* $t \in T$, *mission* $w_m$ *can be (gradually) accomplished.*

## 5. Behaviour based dependability measure

The basic idea behind the dependability measure proposed in the last section is to define the dependability based on the behaviour of the system. For this purpose, a desired behaviour, which was called mission $w_m(t)$, was defined for a system and the dependability measure was proposed to be depending on the total of deviation between the actual system behaviour $w(t)$ and the desired behaviour $w_m(t)$. In order to be able to actually measure the dependability this definition must, however, be more sophisticated.

### 5.1 Requirements for a dependability measure

Before proposing a function for measuring the dependability the characteristics this dependability function should posses are introduced. In the following, the function for the dependability will be called $\mathfrak{D}$.

- $\mathfrak{D}(t)$ should be a continuous time-dependent function
- $\mathfrak{D}(t)$ should be positive, strictly monotone decreasing
- $\mathfrak{D}(t)$ should be normalized between 0 and 1, where 1 means dependable and 0 means not dependable
- $\mathfrak{D}(t)$ should be a dimensionless quantity

The dependability must be measured during and after the mission, hence the dependability measure $\mathfrak{D}(t)$ must be a time dependant function.
The normalization and the non-dimensionalization is obvious in order to achieve a system and unit independent measure. The limitation to the domain between 0 and 1 was chosen so that dependability measure is comperable between different system and application domains.
$\mathfrak{D}(t)$ should be strictly monotonic decreasing since a system is less dependable, i.e. un-dependability is more likely to occur, the longer a system runs.

### 5.2 Definition of dependability measure

The system trajectory $w(t)$ is the evolution of the system state. The distance between this trajectory and the mission $w_m(t)$, together with the distance to the safety area $\mathfrak{S}$ will be the main idea of the measure for dependability.
After the system $\Sigma$ has completed its mission, the overall mission deviation $D_m$ of system and its mission $w_m$ is proposed as the sum of all deviations $\varepsilon^2(w(t),w_m(t))$. In the following,

the functional $\varepsilon^2(w(t),w_m(t))$ will be abbreviated as $\varepsilon^2(t)$. Thus, the overall mission deviation can be defined as:

$$D_m = \int_0^{t_m} max_\tau(\varepsilon(\tau)^2)d\tau - \int_0^{t_m} \varepsilon^2(\tau)d\tau \tag{1}$$

Where $\varepsilon^2(t)$ is an appropriate measure of the deviation between mission trajectory $w_m$ and system trajectory $w$ and consequently a combination of different distance measurements, including the distance to the safety area $\mathfrak{S}$ . The term $max_\tau$ $(\varepsilon\ (\tau\ )^2)$ represents the maximum deviation during this particular mission. Those distance measurements will be discussed in detail in the following.

More important than knowing the system dependability after completion of the mission is knowing the dependability during the mission. At time, $t$ the time dependent overall mission deviation $D(t)$ can be measured by means of

$$D(t) = \int_0^{t_m} max_\tau(\varepsilon^2(\tau))d\tau - \int_0^{t} \varepsilon^2(\tau)d\tau \tag{2}$$

Note that the integration limits for the second integral changed from (1) to (2).

In order to calculate $\mathfrak{D}$ $(t)$ during the mission an estimation for $max_\tau$ $(\varepsilon^2(\tau))$ must be used. This value depends on the distance function $\varepsilon^2(t)$ used and will be discussed together with the calculation of $\varepsilon^2(t)$ in the following.

Furthermore,

$$\int_0^{t_m} max_\tau(\varepsilon^2(\tau))d\tau$$

in (1) and (2) assures that the function for the time dependent overall deviation $D$ is a positive function.

The problem with this function for $D(t)$, is that, besides that it is unnormalized, $D(t)$ is equal to zero if there is no deviation between the desired trajectory $w_m(t)$ and the actual system trajectory $w(t)$. Hence, in this case, the dependability derived from this function would be zero.

### 5.3 Non-dimensionalization and normalization

Nondimensionalization is a technique for partial or full removal of units from a mathematical equation by a suitable substitution of variables. Normalization bounds the domain of a mathematical function to a given range of values.

Function $v$ with its codomain $[o_{min}..o_{max}]$ can be normalized to a function $v'$ with its co-domain $[n_{min}..n_{max}]$ by the following formula:

$$v' = \frac{v - o_{min}}{o_{max} - o_{min}} \cdot (n_{max} - n_{min}) + n_{min} \tag{3}$$

For the time dependent overall mission deviation (2) the value for $o_{min}$ is:

$$o_{min} = 0 \tag{4}$$

The dependability function, as stated in the introduction to this chapter, should have a co-domain of [0..1], consequently the values for $n_{min}$ and $n_{max}$ should be:

$$n_{min} = 0 \tag{5}$$

and

$$n_{max} = 1 \tag{6}$$

With these values the normalization function is reduced to:

$$v' = \frac{v}{o_{max}} \cdot (n_{max}) = \frac{v}{o_{max}} \tag{7}$$

The value $o_{max}$ for the unnormalized dependability $D$ can be set to

$$o_{max} = \int_0^{t_m} max_\tau(\varepsilon^2(\tau))d\tau \tag{8}$$

If at least one $\varepsilon^2(t) > 0$ for $t \in [0..t_m]$ the normalized dependability $\mathbb{D}(t)$ can be computed from (2) with (7) and (8) to:

$$\begin{aligned}\mathbb{D}(t) &= \frac{\int_0^{t_m} \max_\tau(\varepsilon^2(\tau))d\tau - \int_0^t \varepsilon^2(\tau)d\tau}{\int_0^{t_m} \max_\tau(\varepsilon^2(\tau))d\tau} \\ &= \frac{\int_0^{t_m} \max_\tau(\varepsilon^2(\tau))d\tau}{\int_0^{t_m} \max_\tau(\varepsilon^2(\tau))d\tau} - \frac{\int_0^t \varepsilon^2(\tau)d\tau}{\int_0^{t_m} \max_\tau(\varepsilon^2(\tau))d\tau} \\ &= 1 - \frac{\int_0^t \varepsilon^2(\tau)d\tau}{\int_0^{t_m} \max_\tau(\varepsilon^2(\tau))d\tau}\end{aligned} \tag{9}$$

Nevertheless, the problems with this function are:

1. It only exists if at least one $\varepsilon^2(t) > 0$ for $t \in [0..t_m]$. In other words, it only exists if at least a small deviation between the desired behaviour $w_m$ and the actual behaviour $w$ occurred.
2. It is subject to the calculation of $\varepsilon^2(t)$. Thereby $\max_\tau(\varepsilon^2(\tau))$ cannot be estimated in advance and dependability cannot be computed during the mission.

To finally overcome both problems, a system-independent way for computing $\varepsilon^2(t)$, which is additionally normalized between $[0 \ldots 1]$, is proposed.

Having this, $\max_\tau(\varepsilon^2(\tau))$ can be estimated equal to 1 and

$$\int_0^{t_m} \max_\tau(\varepsilon^2(\tau))d\tau \tag{10}$$

can be estimated to

$$\int_0^{t_m} \max_\tau(\varepsilon^2(\tau))d\tau = \int_0^{t_m} 1d\tau = \tau|_0^{t_m} = t_m$$

This finally leads to the desired system independent, normalized function $\mathfrak{D}(t)$ of dependability. $\mathfrak{D}$ can now be computed from (9) to:

$$\mathfrak{D}(t) = 1 - \frac{1}{t_m} \int_0^t \varepsilon^2(\tau) d\tau \tag{11}$$

If a systemindependent way to compute $\varepsilon^2(t)$ between [0 . . . 1] exists this function for the dependability posses all required properties stated at the beginning of this chapter.

### 5.4 Computing $\varepsilon^2(t)$

For computing the elements of $\varepsilon^2(t)$ it is not only important to address the distance between the system state and the mission trajectory but also to address the different dimensions of dependability such as reliability, availability, etc. For a behavioural definition of these attributes please refer to (Rüdiger et al., 2007a). Furthermore, the distance of the system state to the safe area $\mathfrak{S}$ also needs to be taken into account.

Thus, $\varepsilon^2(t)$ usually consists of different elements reflecting the different attributes of dependability for this special system. From (2) and (9) it follows that if $\varepsilon^2(t)$ is a combination of different measures $\varepsilon_1^2(t) \ldots \varepsilon_n^2(t)$, $\mathfrak{D}(t)$ is calculated

$$\mathfrak{D}(t) = 1 - \left( \frac{\int_0^t \varepsilon_1^2(\tau)d\tau + \cdots + \int_0^t \varepsilon_n^2(\tau)d\tau}{\int_0^{t_m} \max(\varepsilon_1^2)(\tau)d\tau + \cdots + \int_0^{t_m} \max(\varepsilon_n^2(\tau))d\tau} \right) \tag{12}$$

$$= \left( \frac{\int_0^t \sum_{i=1}^n \varepsilon_1^2(\tau)d\tau}{\int_0^{t_m} \sum_{i=1}^n \max(\varepsilon_1^2)(\tau)d\tau} \right) \tag{13}$$

setting again $\max(\varepsilon_i^2(t)) = 1$, for $i = 1 \ldots n$, this can be reduced to:

$$\mathfrak{D}(t) = 1 - \frac{1}{t_m} \left( \frac{1}{n} \int_0^t \sum_{i=1}^n \varepsilon_i^2(\tau)d\tau \right) \tag{14}$$

As stated in the previous section, $\varepsilon_i^2(t)$ must be normalized and between be [0 . . . 1]. The corresponding function of $\varepsilon(t)$ must be chosen in such a way that 0 means dependable, i.e. the system state $w(t)$ follows exactly the mission trajectory $w_m(t)$, and 1 means not dependable.

In order to compute the different $\varepsilon_i^2(t)$ a special distance measure is proposed derived from the euclidian distance measure between two points $x = (x_1 \ldots x_n)$ and $y = (y_1 \ldots y_n)$

$$d(x, y) = |x - y| = \sqrt{(x_1 - y_1)^2 + \cdots + (x_n - y_n)^2}. \tag{15}$$

This measure is, however, not normalized and not necessarily between 0 . . . 1. In order to achieve the remaining two points, too, the following distance measure is proposed derived from (15):

$$\varepsilon^2(w(t), w_m(t)) = 1 - e^{-\frac{|w(t) - w_m(t)|}{w_{dev}}} \tag{16}$$

In (16) $w_m(t)$ is the desired (mission) behaviour and $w(t)$ the actual behaviour of the system. The parameter $w_{dev}$ describes how severely a deviation from the mission trajectory influences the system's dependability. It must be chosen greater than zero and have the same dimension as $w(t)$. The lower $w_{dev}$ is chosen the more a deviation from the desired behaviour is rated (see Fig. 4). The proposed distance measure is therefore dimensionless and normalized between [0 and 1].

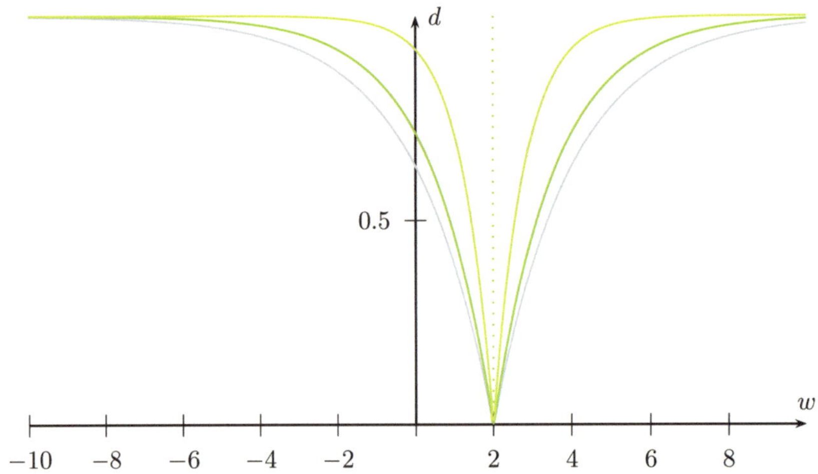


Figure 4. Example of the distance function to compute the different $\varepsilon_i(t)$ with $w_m = 2$ (dotted green line) and $w_{dev} = 1$ (blue), $w_{dev} = 0.8$ (green), and $w_{dev} = 0.4$ (light green)

As the euclidian distance measure, the proposed distance measure $\varepsilon^2(t)$ defines a metric over the space $\mathbb{W}$ since it satisfies all conditions for a metric which are:

1. $\varepsilon^2(x,x) = 0$, identical points have a distance of zero
2. $\varepsilon^2(x,y) = 0$ if and only if $x = y$, identity of indiscernible
3. $\varepsilon^2(x,y) = d(y, x)$, symmetry
4. $\varepsilon^2(x,y) \leq \varepsilon^2(x,z) + \varepsilon^2(z,y)$, triangle inequality

With the aid of this distance measure, the different attributes of dependability can be defined. For $\varepsilon_i^2(t)$ the corresponding euclidian distance measure $d_i(t)$ is used as a basis.

### 5.5 Mission deviation $\varepsilon_m^2(t)$

The mission deviation describes the normalized difference between the mission trajectory and the system state at time $t$. For this purpose the afore discussed distance measure is directly used with the euclidian distance $d_m$ between the mission trajectroy and the system state. When evaluating the dependability $\varepsilon_m^2(t)$ is used in most of the dependability measure. The mission deviation $\varepsilon_m^2(t)$ is defined as

$$\varepsilon_m^2(t) = 1 - e^{-\frac{|w(t) - w_m(t)|}{w_{dev}}} \tag{17}$$

Again, $w_m(t)$ is the desired mission trajectory and $w(t)$ is the actual behaviour of the system as described in (16). See Fig. 5 for examples of $d_m(t)$.

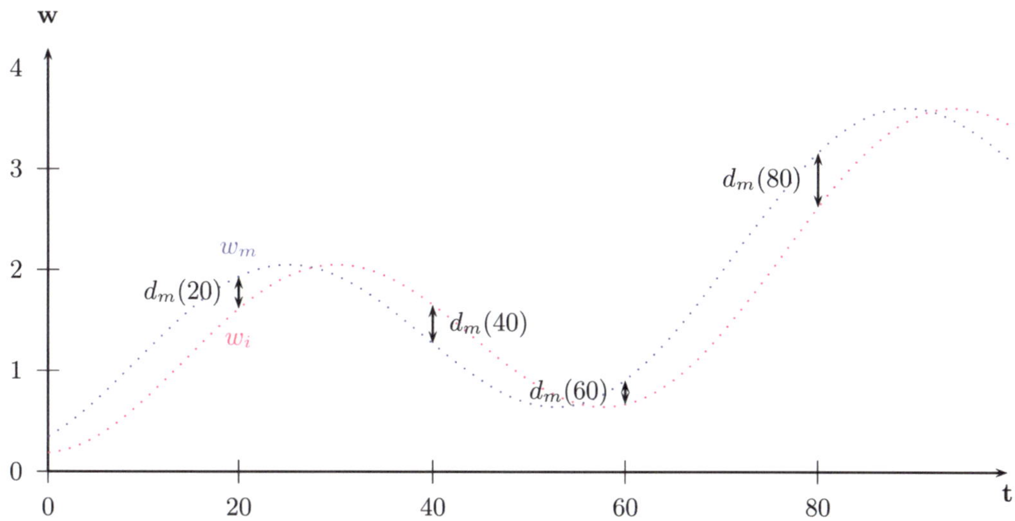


Figure 5. Mission trajectory $w_m(t)$ (blue) and system trajectory $w(t)$ (red) with examples for $d_m(t)$ at different timesteps.

### 5.6 Safety $\varepsilon_s^2(t)$

Beside the mission deviation $\varepsilon_m^2(t)$ is safety $\varepsilon_s^2(t)$ one of the most important elements of $\varepsilon^2(t)$. As proposed in Section 4.2.3 a safety area $\mathfrak{S}$ is introduced which when left will lead to catastrophic consequences. The minimum euclidian distance between a system trajectory $w(t)$ and the border of the safety area $\mathfrak{S}$ at time $t$ will be taken as a basis for the measure of $\varepsilon_s^2(t)$. This distance is called $d_S(w(t))$ and will be abbreviated as follows

$d_S(t)$ for the minimum distance between the actual system states $w(t)$ and the border of the safety area and

$d_{Sm}(t)$ for the minimum distance between the mission trajectory $w_m(t)$ and the border of the safety are at time $t$.

Obviously $\varepsilon_s^2(t)$ should be 1 when $d_S(t) = 0$, equivalent to the distance between the system state and the safety area being zero.

To be able to adequately cover cases where the mission trajectory $w_m(t)$ itself could be close to the border of the safety area $\mathfrak{S}$, not the absolute distance between the actual system trajectory and the border of the safety area $d_S(t)$ is taken but the relative distance between the minimum distance of the actual systemtrajectory and the safety area $d_S(t)$ and the minimum distance of the mission trajectory $w_m(t)$ to the border of the safety area $d_{Sm}$ is taken to compute $\varepsilon_s^2(t)$. Consequently, $\varepsilon_s^2(t)$ is proposed as:

$$\varepsilon_S^2(t) = 1 - e^{-abs\left(\frac{d_S(t)}{d_{Sm}(t)} - 1\right)} \tag{18}$$

Both, $d_S(t)$ and $d_{Sm}(t)$, are greater or equal to 0. The equation for $\varepsilon_s^2(t)$ is only defined for $d_{Sm}(t) \neq 0$. See Fig. 6 for examples for $d_S(t)$.

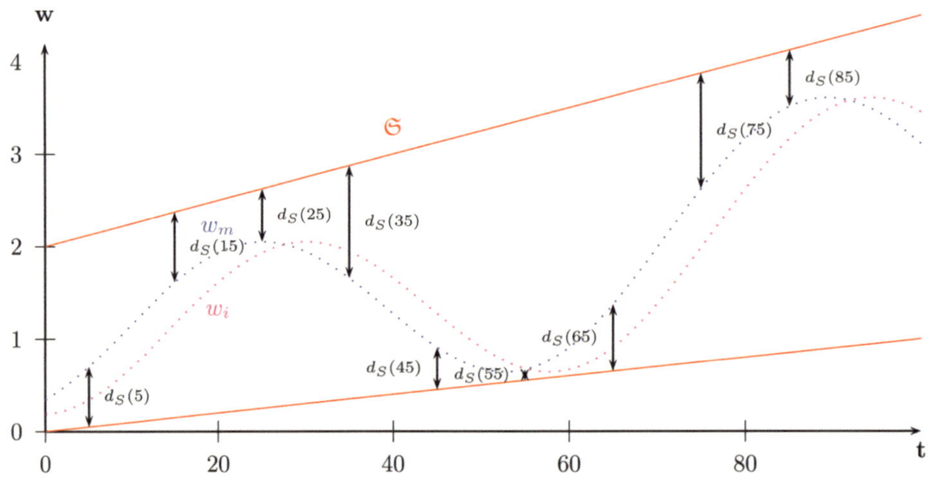


Figure 6. Mission trajectory $w_m(t)$ (blue) and system trajectory $w(t)$ (red) with examples for $d_{Sm}$ the distance between the mission trajectory $w_m(t)$ and the boarder of the safety area $\mathfrak{S}$ (read lines).

### 5.7 Timely mission accomplishment $\varepsilon_T^2(t)$

For a number of systems it is not only important that the system adequately follows the mission trajectory but that the system follows the mission trajectory at a given time. A good example for such systems is a heard-lung machine where it is not sufficient that the system gives the right pulses, they must be performed at given timesteps. Another important example, especially in the field of controlling autonomous mobile real-time systems, is the class of periodic behaviours, i.e. velocity control or collision avoidance. In the latter example, the exact time execution of a given behaviour is more important then the exact execution of the behaviour itself.

The calculation of $\varepsilon_T^2(t)$ is of course only possible if $w_m(t)$ is uniquely invertible. For periodic functions, often used on autonomous mobile systems, the uniquely invertible requirement of $w(t)$ can be simplified to a peacewise uniquely invertible requirement.

Let $w'_m(w) : \mathbb{T} \rightarrow \mathbb{W}^{\mathrm{T}}$ be the inverse function of $w_m(t)$ then $\varepsilon_T^2(t)$ is proposed as:

$$\varepsilon_T^2(t) = 1 - e^{-\frac{\left|t - w'_m(w)\right|}{t_{dev}}} \tag{19}$$

As in (16) and (17) the parameter $t_{dev}$ describes how severe a deviation from the mission trajectory influences the dependability of the system. See Fig 7 for an example of $\varepsilon_T^2(t)$

### 5.8 Reliability $\varepsilon_R^2(t)$

As stated in section 2, reliability $R\,|_t$ describes the probability according to which the system will operate correctly in a specified operating environment in an interval [0, $t$]. For $\varepsilon_R^2(t)$ this means that $1 - R\,|_t$ describes the probability that the system will fail in the interval [0...$t$]. Setting $t = t_m$ the latter probability can be directly used and thus $\varepsilon_R^2(t)$ is proposed as:

$$\varepsilon_R^2(t) = 1 - R|_t \tag{20}$$

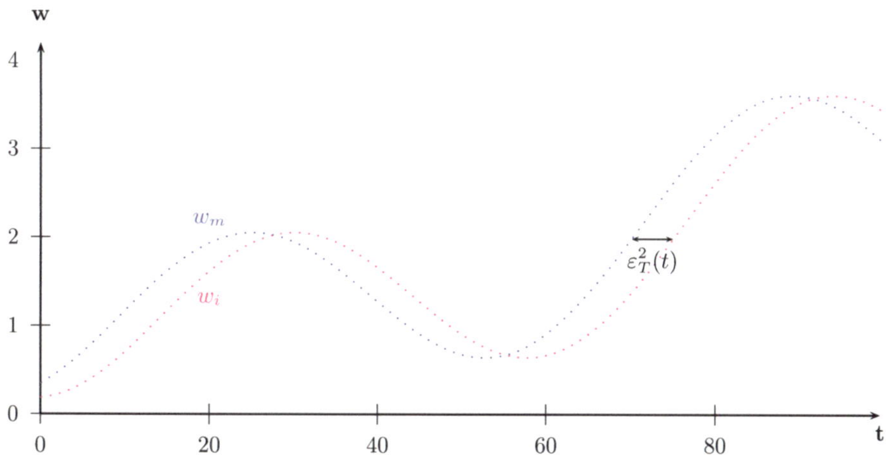


Figure 7. Mission trajectory $w_m(t)$ (blue) and system trajectory $w(t)$ (red) with examples for $d_T(t)$

### 5.9 Availability $\varepsilon_A^2(t)$

In contrast to reliability, availability is defined at a time instant $t$ while reliability is defined in a time interval. The availability $A|_t$ describes the probability that a system is operational at the instant of time $t$. As for the reliability, this means for $\varepsilon_A^2(t)$ that $1-A|_t$ describes the probability that the system is not operable at time instant $t$. This probability can be directly used when computing $\varepsilon_A^2(t)$. Thus $\varepsilon_A^2(t)$ is proposed as:

$$\varepsilon_A^2(t) = 1 - A|_t \tag{21}$$

This definition satisfies two statements about availability mentioned in section 2:

1. If a system cannot be repaired, its availability equals its reliability
2. The integral over the mission time of $\varepsilon_A^2(t)$ in the dependability function equal the average availability, also called interval or mission availability as introduced in section 2.

### 5.10 Additional $\varepsilon_X^2(t)$

According to the system and its mission, additional measures for $\varepsilon^2(t)$ might be needed to take into account further special requirements with respect to dependability.

As stated earlier, it is important that those $\varepsilon_X^2(t)$ are dimensionaless and are normalized between 0 and 1, where 0 means dependable and 1 means not dependable.

## 6. Examples for measuring the dependability

To present the adaptability of the dependability definition proposed above, the following two examples may serve as a demonstration.

### 6.1 Example 1: autonomous transport system

To clarify the behaviour based dependability measurement, an autonomous mobile system with only one position degree of freedom is used. The system is an autonomous transportation system build to autonomously reach different positions which could be, for example, stopping points on a track. For the dependability measurement only the position on the track is considered in the first example. The velocity and acceleration of the autonomous transportation system will be initially disregarded in this example.

### 6.1.1 Behaviour based system description

For the dependability measurement proposed in the last section, the system will be modelled as described in Section 3. Since the system only has one position degree of freedom it can only move forward and backward on the track, the signal space of the system is $\mathbb{W} = \mathbb{R}$. The time of interest for this system is $\mathbb{T} = \mathbb{R}^+$.

For the description of the behaviour $\mathfrak{B}$, the train model is needed. A simple train model with rolling friction derived from Newtons Law is used for that purpose. According to Newtons-Law, the sum of forces acting on an object is equal to the mass of that object, multiplied by its acceleration. The mass of the train is assumed to be $M$. The forces acting on the train are, on the one hand, the driving force $F_a$ and, on the other hand the friction force $F_r = \mu F_n$ ($\mu$ represents the coefficient of rolling friction, $F_n$ the force parallel to the planes normal). It is assumed that the train only moves in a plane, thus there is no inclination, etc. Consequently, the force parallel to the normal of the plane $F_n$ can be set equal to the force of gravity $F_n = F_g = Mg$, with $g$ being the acceleration due to gravity. A diagram of the system with the forces used in this model is shown in Fig. 8. The system can thus be described according to the following equations.

$$M\ddot{x} = F_a - F_r \tag{22}$$

$$M\ddot{x} = F_a - \mu g M \tag{23}$$

$$\ddot{x} = \frac{1}{M}\left(F_a - \mu g M\right) \tag{24}$$

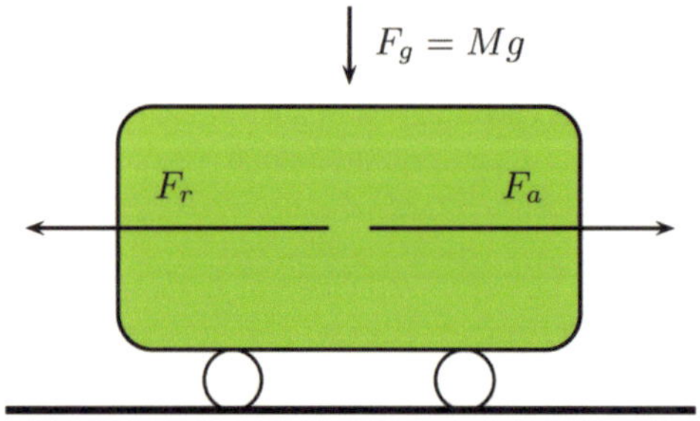


Figure 8. Example of an autonomous transportation system with the forces used to model the system. $F_a$ driving force, $F_r$ friction and $F_g$ gravitation force.

According to the behavioural based approach set forth in section 3, the autonomous mobile transportation system can be described as follows.

**Universe** $\mathbb{W} = \mathbb{R}$

**Time** $\mathbb{T} = \mathbb{R}^+$

**Behaviour** $\mathfrak{B} = \left\{x : \mathfrak{R}^+ \to \mathfrak{R} \middle| \ddot{x} = \frac{1}{M}\left(F_a - \mu g M\right)\right\}$

The corresponding Matlab Simulink Model is shown in Fig. 9. The position and the velocity of the system are controlled by simple PI-controllers (see Fig. 10 and 11). Of all possible

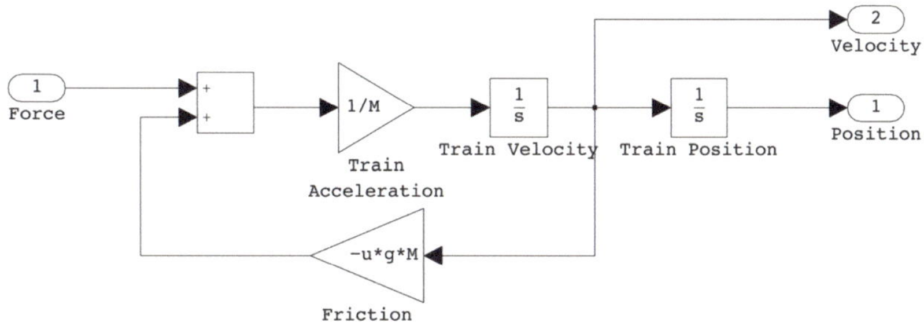


Figure 9. Matlan Simulink model of an autonomous transportation system. $M$ is the mass of the system, $\mu$ the friction coefficient and $g$ the acceleration due to gravity

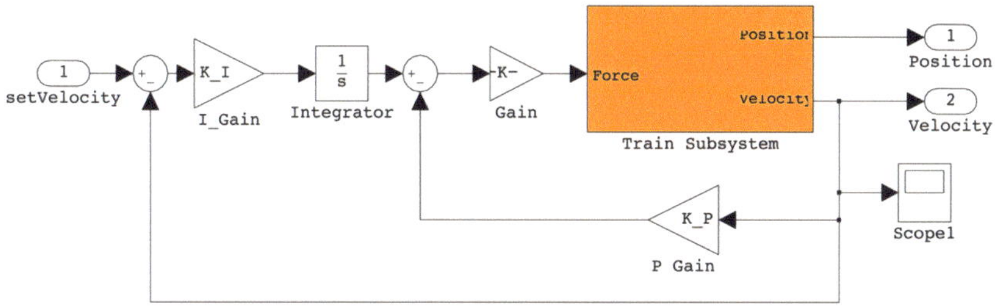


Figure 10. Velocity loop of an autonomous transportation system. The system velocity is controlled by a simple PI controller.

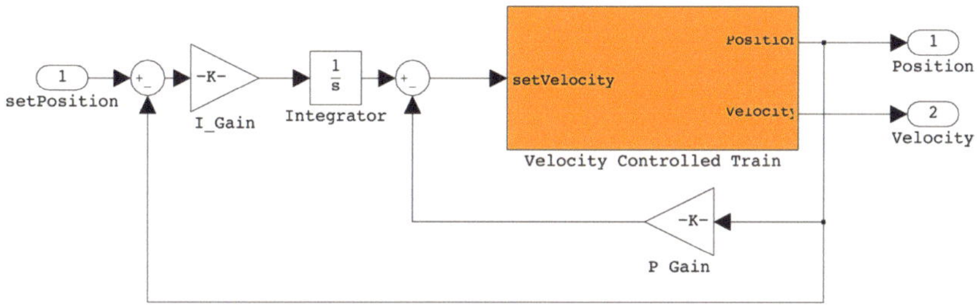


Figure 11. Position loop of an autonomous transportation system. The position of the system is controlled by a simple PI controller

system behaviours from the set $\mathfrak{B}$ only a subset $\mathbb{B} \subset \mathfrak{B}$ is available according to the mass and the maximum possible driving force of the system. In this example it is further assumed that the system is able to completely follow the given velocities and accelerations.

### 6.1.2 Behaviour based dependability measurement

The mission of the above modelled autonomous transportation system is to reach consecutively different positions on the track. The mission time in this example is set to 2400 time units.

The system should thus accomplish a desired behaviour $w_m(t)$ with its given behaviours $B \subset \mathfrak{B}$. The set of desired behaviours for this example is generated with a Matlab Simulink model. For this purpose, the signal builder block is used (see Fig 12) to define different desired positions on the track. The reference signal is fed to the real train system to simulate the actual behaviour (Model in Fig. 8) and also to the reference train system (Reference Model in Fig. 8) to generate the desired behaviour. With the aid of the generated behaviour

in the reference model, this will be taken as the desired behaviour $w_m(t)$ or mission of the autonomous transportation system and used for the computation of the system's dependability. This model shows an example of the different opportunities to measure the dependability of such systems.
At first, it is assumed that the position of the autonomous transportation system can be measured adequately. Consequently it is assumed that the measurement of the position itself does not produce additional errors.
Up till now only system internal errors or deviations were considered as deviations between the reference model and the real system. It is also possible that changes in the model or the environment, as implicitly considered in this case, may occur. Unexpected wearout of wheels, resulting from e.g. a smaller wheel radius can produce errors, and as such lead to a deviation from the desired behaviour, if the position of the train is only measured on the basis of the wheel rotations.

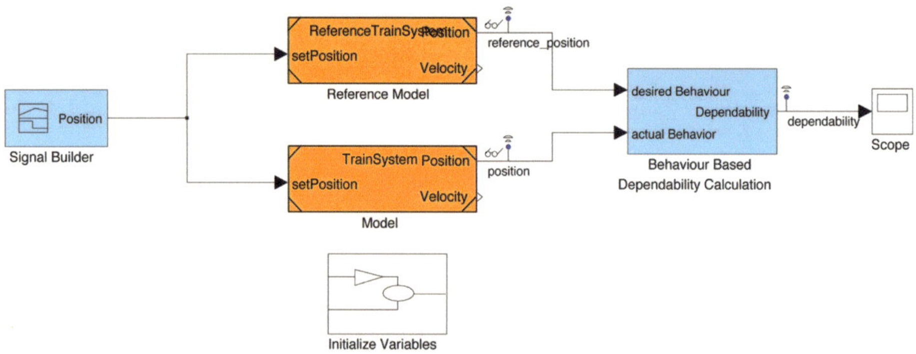


Figure 12. A Matlab signal builder block is used together with a reference and the real system in order to generate the actual and desired behaviour of the system.

When generating the desired behaviour in this example it is assumed that the system is functioning properly. Thus, the reference model reflects the system adequately. Noise in the sensors, for example, is not explicitly modelled. Of course, this could have been also introduced in the model for a better computation of the desired behaviour.
In the first example, two different simulations are carried out.

1. To simulate an additive error, a constant value is added to the position measurement. This error could be due to faulty initialization, slippage etc, but could also because of an error in the model of such autonomous transportation system.
2. To demonstrate as to what extend noise in sensors or measurement uncertainty affect the dependability of a system, noise is added to the measurement of the position.

The results of the two simulations are shown in Fig. 13. The dotted red line in each case represents the desired behaviour, thus the mission trajectory $w_m$. The actual system behaviour is shown as blue line. The measured dependability for this example is shown as a dashed green line.

### 6.2 Example 2: Small train

Since the autonomous transportation system is built for the transport of people and as such represents a safety critical system, system safety is also considered in the second example.
In the second example, besides the position of the system, the velocity is considered when calculating dependability. In addition to the above mentioned two simulations, two other scenarios were added for the computing of dependability.

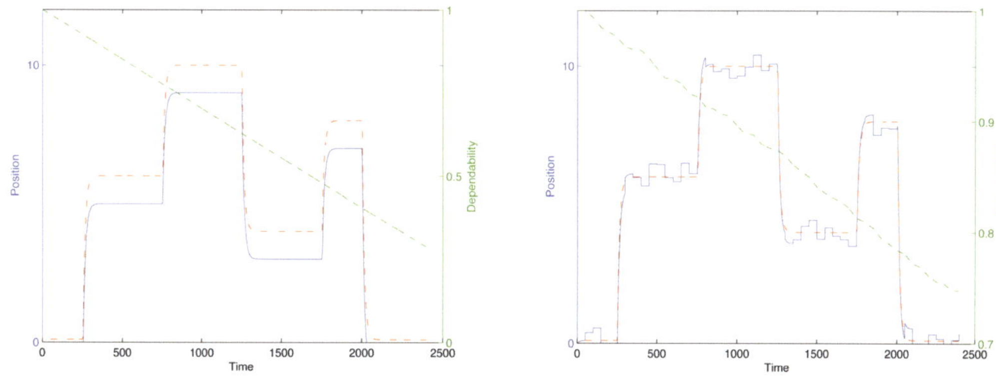


(a) Aboslut Value added to the position (b) Noise added to the position

Figure 13. Simulation Resutls for Example 1.

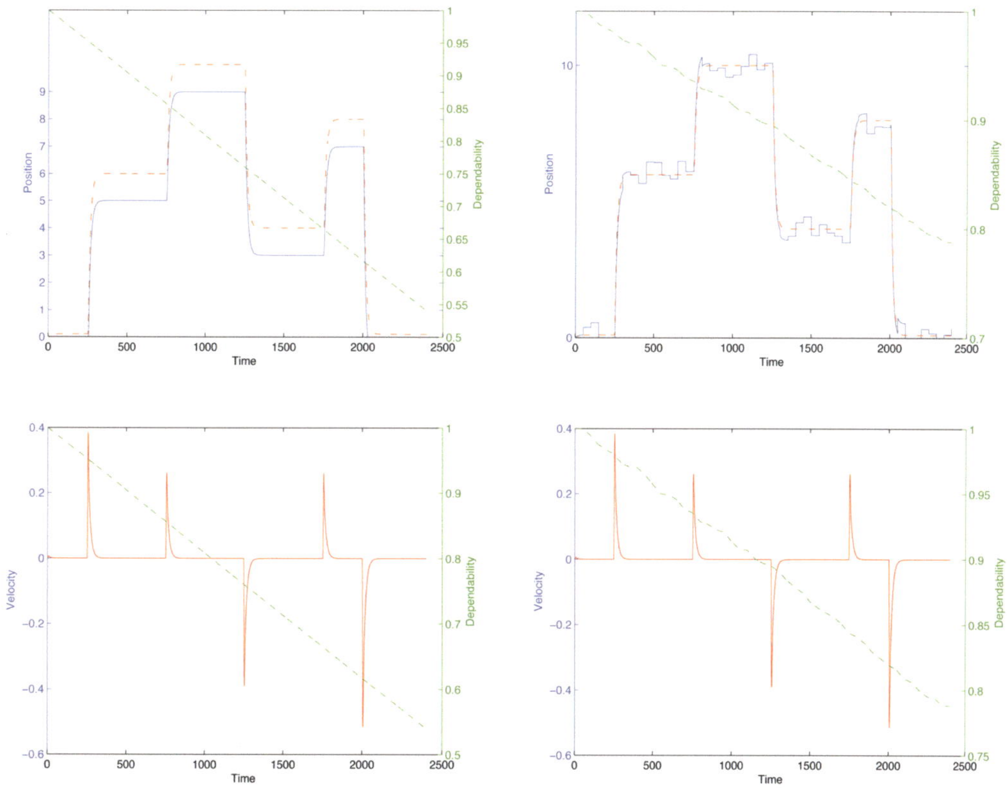


Figure 14. Simulation Results for example 2 with Position and Speed used for the dependability calculation

1. In order to enhance the dependability calculation, a desired and actual behaviour of the velocity was added. For the simulation of parameter errors, which are multiplicative, the velocity of the real system is multiplied by a constant value.
2. A safety area, as proposed, was added for the velocity. Consequently, the relative distance $\varepsilon_s^2(t)$ is also used when computing system's dependability.

For each of these two scenarios, again, both simulations allready used in the first examples where performed. The results of the individual four simulations are shown in Fig. 14 and 15. As in the last figure, the dotted red lines represents the desired behaviour for either the velocity or the position. The actual system behaviour in terms of velocity and position is shown as blue line. The measured dependability for the examples is shown as dashed green line.

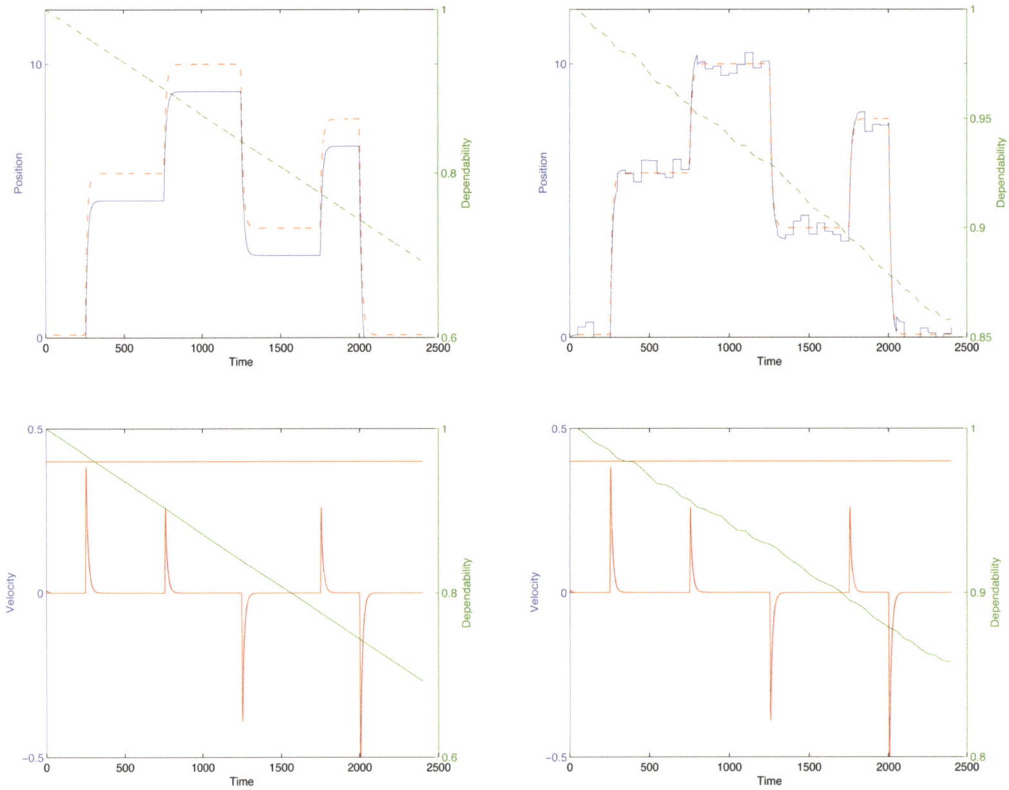


Figure 15. Simulation Results for example 2 with Position and Speed used for the dependability calculation. Additionally a safety area for the velocity is added.

## 7. Conclusion

There exist numerous non-formal definitions for dependability (see Carter, 1982; Laprie, 1992; Badreddin, 1999; Dubrova, 2006; Avizienis et al., 2004a just to name a few). When applying those non-formal definitions to a specific system the resulting dependability measure usually is only valid for this specific system and only in rare cases transferable to a family of equal systems. Small changes in the system or environment, however, render those measurements usually useless when it comes to measuring or even comparing the dependability of different systems.

Autonomous mobile robots are often described by their behaviour. This aspect was utilized in this chapter for the definition of dependability in a behavioural context in order to obtain an easy to apply and computable formula for the dependability of systems. Since this

formula for dependability is solely based on the behaviour and the mission of a system it can be easily compared with other systems having different missions.
The definition for dependability proposed in this chapter is straight forward, easily applicable and well suited for dependability comparison of different systems.

## 8. References


Avizienis, A., Laprie, J.-C., and Randell, B. (2004a). Dependability and its threats: A taxonomy.

Avizienis, A., Laprie, J.-C., Randell, B., and Landwehr, C. (2004b). Basic concepts and taxonomy of dependable and secure computing. *IEEE Trans. on Dependable and Secure Computing*, 1(1):11-33.

Badreddin, E. (1999). Safety and dependability of mechatronics systems. In *Lecture Notes*. ETH Zürich.

Badreddin, E. and Abdel-Geliel, M. (2004). Dynamic safety margin principle and application in control of safety critical systems. In *Proceedings of the 2004 IEEE International Conference on Control Applications, 2004.*, volume 1, pages 689-694Vol.1.

Brooks, R. A. (1986). A robust layered control systemfor a mobile robot. *IEEE Journal of Robotics and Automation*, 2(1):14-23.

Candea, G. (2003). The basics of dependability.

Carter, W. (1982). A time for reflection. In *Proc. 12th Int. Symp. on Fault Tolerant Computing (FTCS-12)*. FTCS-12) IEEE Computer Society Press Santa Monica.

Department of Defence, U. S. o. A. (1970). Military standard - definitions of terms for reliability and maintainability. Technical ReportMIL-STD-721C.

Dewsbury, G., Sommerville, I., Clarke, K., and Rouncefield, M. (2003). A dependability model for domestic systems. In *SAFECOMP*, pages 103-115.

Dubrova, E. (2006). Fault tolerant design: An introduction. Draft.

Filippini, R. and Bondavalli, A. (2004). Modeling and analysis of a scheduled maintenance system: a dspn approach.

Flammini, F. (2006). *Model-Based Dependability Evaluation of Complex Critical Control Systems*. PhD thesis, Universitá degli Studi di Napoli - Federico II.

Hermann, R.; Krener, A. (Oct 1977). Nonlinear controllability and observability. *Automatic Control, IEEE Transactions on*, 22(5):728-740.

IEC (1990). International electrotechnical vocabulary. chapter 191: Dependability and quality of service.

International Federation for Information Processing. Wg 10.4 on dependable computing and fault tolerance. http:// www.dependability.org/wg10.4/.

Laprie, J. C. (1992). Dependability. Basic Concepts and Terminology. Ed. Springer Verlag.

Randell, B. (2000). Turing Memorial Lecture: Facing up to faults. 43(2):95-106.

Rüdiger, J., Wagner, A., and Badreddin, E. (2007a). Behavior based definition of dependability for autonomous mobile systems. European Control Conference 2007. Kos, Greece.

Rüdiger, J., Wagner, A., and Badreddin, E. (2007b). Behavior based description of dependability - defining a minimum set of attributes for a behavioral description of dependability. In Zaytoon, J., Ferrier, J.-L., Andrade-Cetto, J., and Filipe, J., editors, *ICINCO-RA (2)*, pages 341-346. INSTICC Press.

Vesely, W. E., Goldberg, F. F., Roberts, N. H., and Haasl, D. F. (1981). *Fault Tree Handbook*. U. S. Nuclear Regulatory Commission, NUREG-0492, Washington DC.

Willems, J. (1991). Paradigms and puzzles in the theory of dynamical systems. *IEEE Transactions on Automatic Control*, 36(3):259–294.

# 23

# Model-free Subspace Based Dynamic Control of Mechanical Manipulators

Muhammad Saad Saleem and Ibrahim A. Sultan
*University of Ballarat*
*Australia*

## 1. Introduction

Realtime identification and dynamic control of mechanical manipulators is important in robotics especially in the presence of varying loading conditions and exogenous disturbances as it can affect the dynamic model of the system. Model free control promises to handle such problems and provides solution in an elegant framework. Model free control has been an active area of research for controlling plants which are difficult to model and time varying in nature.

The proposed framework takes the objective in operational space. Benefit of specifying objective in operational space along with direct adaptive control is self evident. In this framework, subspace algorithm is used for model identification. $\mathscr{H}_\infty$ is used for robust control of manipulator dynamics. Because of the seamless integration of identification and control modules, explicit values of dynamic parameters is not calculated. The model free control system is capable of explicitly incorporating uncertainties using $\mu$-synthesis. Uncertainty models can be calculated from experimental data using model unfalsification. The proposed control system employs a black box approach for dynamics of mechanical systems. The chapter also presents results from a simulation of a planar robot using MATLAB® and Simulink® from MathWorks Inc.

### 1.1 Notations

The rigid body model adopted in this chapter is given by

$$M(q)\ddot{q} + C(q,\dot{q}) + G(q) + \xi(q,\dot{q}) = u, \tag{1}$$

where $M(q)$ is the inertia tensor matrix, $C(q, \dot{q})$ is the Coriolis and centripetal forces, $G(q)$ is gravity, and $\xi(q, \dot{q})$ denotes unmodeled non-linearities. Joint variables, their velocities, and positions are donated by $q, \dot{q}, \ddot{q} \in \Re^n$. In case of revolute joint, $q$ is the angle while in prismatic joint, $q$ represents the distance. The torques generated by actuators are represented by $u \in \Re^n$. It is assumed that the mechanical manipulator is fully actuated, non-redundant and the Jacobian is known. If the position of endeffector is given by forward kinematics equation i.e. $x = f_{kinematics}(q)$. It can be differentiated by $\partial q$ to obtain

$$\dot{x} = J(q)\dot{q}, \ddot{x} = J(q)\ddot{q} + \dot{J}(q)\dot{q}, \tag{2}$$

### 1.2 Problem statement

The problem under discussion can be stated as

> Given a trajectory of end-effector in operational space $x, \dot{x}, \ddot{x} \in \Re^n$, joint variables $q$, $\dot{q}$, $\ddot{q} \in \Re^n$, and forward kinematics function $f_{kinematics}(\bullet)$. Find actuator torque $u \in \Re^n$ such that the end-effector smoothly follows the trajectory in the presence of gravity and exogenous disturbances and provides stable solution near singularities.

This problem is the amalgamation of inverse dynamics in which $M(q)$, $C(q, \dot{q})$, and $G(q, \dot{q})$ are known i.e. $u = f(q, \dot{q}, \ddot{q})$, dynamic parameter identification in which $M(q)$, $C(q, \dot{q})$, and $G(q, \dot{q})$ are calculated, and robust control to cater unmodeled non-linearities and disturbances in the system.

## 2. Control of articulated manipulators

Control of articulated manipulators can be divided into two main categories:

- Joint space control
- Operational space control

Joint space control is consisted of two subproblems. First, manipulator inverse kinematics is performed and then joint space control scheme is devised to allow the end effector to follow a reference input. The main computational burden in this scheme is incurred by the inverse kinematics procedure, which is normally performed by using different optimization techniques; particularly in redundant systems where there can be infinite solutions for a given task (Kim et al. (2003)). Many implementations of joint space control can be found in the literature (Laib (2000); Kelly (1997); Arimoto (1995); Kelly (1993); Wen et al. (1992); Tomei (1991); Takegaki & Arimoto (1981); Zhang et al. (2000)).

In many applications, the desired path of end effector is specified in the operational space (*e.g. Cartesian frame*). Operational space control, on the other hand, has also been used for constrained manipulator motions (Sapio & Khatib (2005)). These constraints can be because of gravity, or kinematically imposed. It can be seen in Figure 2 that inverse kinematics is embedded in the closed-loop control law but not explicitly performed as shown in Figure 1 (Sciavicco & Siciliano (2000)). Operational space control and task space control allude to the same concept (Xie (2003)).

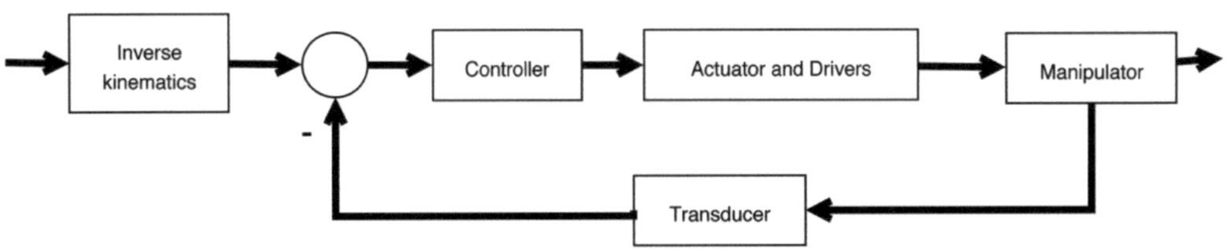


Figure 1. Joint space control

The proposed architecture controls the manipulator in joint space. The reason behind is the very fact that inverse kinematics is highly non-linear in nature. It is assumed the analytical

Jacobian $J_A$ is available, the manipulator is fully actuated and non-redundant. Numerical solutions to inverse kinematics are more complex and are manipulator specific (Khalil & Dombre (2004)). Numerical based inverse kinematics is out-of-scope of this chapter.

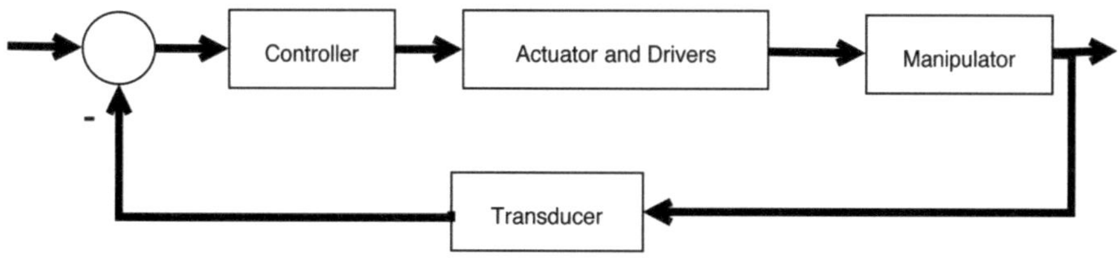


Figure 2. Operational space control

## 3. Model free control

There are four methods to use experimental data as shown in Table 3 (Woodley (2001)). Mainly, choice depends on application. For realtime systems which are fairly easy to model, indirect control is a better choice. The system then adapts itself and updates its model parameters according to the conditions gathered from the measured data. Normally an online model-based design is referred as indirect control. If a system is hard to model from first principles (e.g. Newton's laws of motion) or there are time varying nonlinearities then a direct adaptive control would suit the application. Examples of plants which are difficult to model are arc furnaces (Wilson (1997)) and helicopter rotors (Lohar (2000)). Biped robots on the other hand can be modeled but they exhibit time varying nonlinearities (Wolkotte (2003); Kim et al. (2004); Caballero et al. (2004)).

| | With Plant Model | Without Plant Model |
|---|---|---|
| Online | Indirect Adaptive | Direct Adaptive |
| Offline | Model Based Design | Direct Control Design |

Table 1. Four different techniques of control design from experimental data

### 3.1 General predictive control

Model free comes under the category of "general predictive control" (GPC). Model free implementations range from fuzzy and neural control (Boyd & Little (2000); Cheng (2004)) to crisp control techniques (Favoreel et al. (1999a); Woodley et al. (2001a)). However, crisp control is regarded as reliable and explicitly defines performance objective when compared with fuzzy control techniques (Athans (1999)).

In direct adaptive control techniques, an explicit model formation is not needed; this is why it is referred to as model free control. Plant input and output values are observed in realtime, and a controller is designed for the estimated plant model. Model free is actually a misnomer, as the data from the plant's input and output also represent some kind of plant information. In a model free control implementation, the system identification and the controller synthesis techniques are seamlessly integrated to reduce the computational burden, which makes it more suitable for realtime applications. Predictive control has been applied with $\mathscr{H}_\infty$ optimal predictors and $\mathscr{H}_2$ cost functions (Grimble (1998)), $\mathscr{H}_\infty$ optimal predictors and $\mathscr{H}_\infty$ control costs (Zhao & Bentsman (1999)), and mixed $\mathscr{H}_2/\mathscr{H}_\infty$ minimax predictors (Tse et al. (1993)). Subspace predictors have also been used for direct control with

quadratic (Favoreel et al. (1998, 1999b,a)) and $\mathscr{H}_\infty$ cost functions (Woodley et al. (2001b); Woodley (2001); Woodley et al. (2001a)). Other implementations include adaptive inverse control (Widrow & Walach (1994)), LMS[1] (Widrow & Stearns (1985)), FxLMS[2] and its alternatives (Sayyar-Rodsari et al. (1998)), identification and control based on the ($\mathscr{H}_\infty$) loop shaping (Date & Lanzon (2004)), and Lyapunov-based framework (Haddad et al. (2003); Hayakawa et al. (2004)).
$\mathscr{H}_\infty$ is famous amongst control engineers because of its ability to control MIMO[3] systems built on strong mathematical foundations.

## 3.2 System identification

System identification is used to build dynamical models from measured input-outpu data of a plant. There are many system identification techniques. The list starts with the classical prediction error (PE) and its variants; the auto regression with exogenous input (ARX), output error (OE), auto regression moving average with exogenous input (ARMAX), and Box Jenkins (BJ) (Norton (1986); Ljung (1999)). Subspace identification methods (SIM) have many advantages over classical system identification techniques (Overschee & Moor (1996)). Notables are;

- From plant's input and output data, a predictor is found. This bears similarity to the Kalman filter states, and transforms the analysis into a simple least square problem. As such, the whole architecture could be streamlined and in a user-friendly fashion.
- When implemented in direct adaptive control, the plant model does not require simplification. Simplification or model reduction can omit useful information. Instead, in subspace identification methods all the plant information is stored in a compact form of a subspace predictor.
- The output of subspace identification methods can be in the state space form which makes it easy to implement on a computer but its architecture has been exploited in different model free implementations as well (Woodley 2001b, Favoreel 1999a).

Wernholt used SIM to solve system identification problem for an ABB IRB 6600 robot (Wernholt (2004)). Hsu et al. used N4SID in style translation for human motion (Hsu et al. (2005)). These are some of the examples that show how SIMs are being used.

### 3.2.1 Reported problems in subspace identification methods

There are a few problems in subspace identification methods. Many of these problems have been discussed in recent literature and partial remedies have been suggested (Chou & Verhaegen (1997); Lin et al. (2004); Wang & Qin (2004); Chiuso & Picci (2005)). Some of these problems are:

- Biased estimate for closed loop data.
- Assumption of noise-free input.

The first problem can be solved by filtering the predicted data through a frequency weighted matrix. The second one is solved by using a robust control methodology, which would cater for disturbances and noise in the system.

---

[1] least mean square
[2] filtered-x least mean squares
[3] multiple-input multiple-ouput

### 3.3 Model unfalsification

For a true robust model free control, the system should be able to calculate an uncertainty model from the input and output data of plant. This can be done through model unfalsification. First an uncertainty model is unfalsified against the plant input and output data and then the uncertainty model ($\Delta$) is incorporated in the controller design. Model unfalsification *does not* get much appeal in practice because of its high computational burden (Woodley (2001)). There are many implementations available for model unfalsificaiton (Kosut & Anderson (1997); Agnoloni & Mosca (2003); Tsao et al. (2003); Wodoley et al. (1999); Safonov (2003); Tsao & Safonov (2001); Woodley et al. (1998); Cabral & Safonov (2004)). Wang et al. suggested a direct adaptive controller based on model unfalsification with the assumption that there would be a controller in the given set that would satisfy the control requirements for a particular plant (Wang et al. (2004)). The identification of uncertainty models using model unfalsification is out of scope of this chapter.

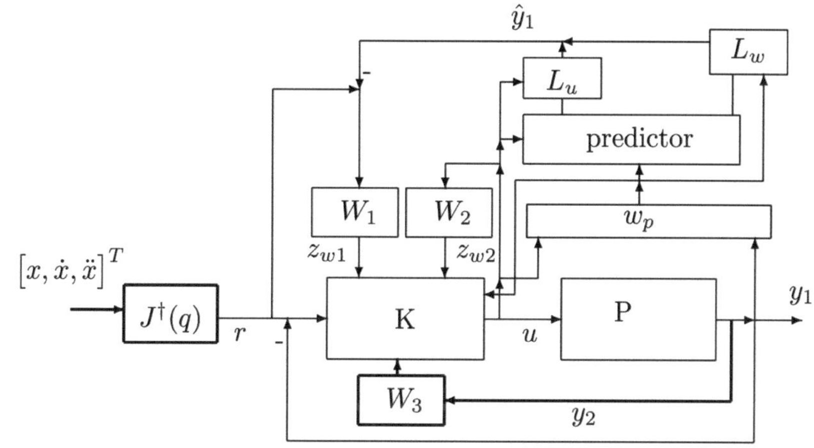


Figure 3. Model free subspace based $\mathscr{H}_\infty$ control. The plant transfer function $P$ is unknown and the controller transfer function $K$ is configured in realtime.

## 4. Model-free subspace based dynamic control of mechanical manipulators

The desired trajectory of end-effector is given by $[x, \dot{x}, \ddot{x}]^T$. If the initial position of end-effector $x_o$ and joint variables $q_o$ is known, then using equation (2), $q$, $\dot{q}$ and $\ddot{q}$ can be written as

$$\dot{q} = J^{-1}(q)\dot{x} \tag{3}$$

$$\ddot{q} = J{-}1(q)(\ddot{x} - \dot{J}(q)\dot{q}) \tag{4}$$

$$q_(t+1) = q_t + \dot{(q)}_t \Delta t \tag{5}$$

In case $J$ is not a square matrix, pseudo-inverse of $J$ i.e. $J^\dagger$ is used. Once the reference trajectory is in the joint space, model free control system which has been inspired by the work of Woodley et al. can be applied (Woodley et al. (2001b)). The cost function for this

framework minimizes the error in joint space $\left[\tilde{q},\dot{\tilde{q}},\ddot{\tilde{q}}\right]^T$, control effort $u$, and $\dot{q}$ for the maximum value of the input $q_r$. Here $\tilde{q} = q - q_r$ and $q_r$ is the reference trajectory in joint space. The reason behind minimizing $\dot{q}$ is that the system becomes unstable near singularities and it becomes important that near singularities system doesn't try to achieve extremely high velocities, which could make the system unstable.
The cost function to minimized can now be written as

$$L(\gamma) = z_{w_1}^T z_{w_1} + z_{w_2}^T z_{w_2} + z_{w_3}^T z_{w_3} - \gamma^2 r^T r, \tag{6}$$

where $\gamma$ is the performance objective, and $z_{w1}$, $z_{w1}$, $z_{w1}$ are weighted feedbacks of $\left[\tilde{q},\dot{\tilde{q}},\ddot{\tilde{q}}\right]^T$, $u$, and $\dot{q}$, respectively. For simplicity, lets suppose $y_1 = \left[\tilde{q},\dot{\tilde{q}},\ddot{\tilde{q}}\right]^T$, $y_2 = \dot{q}$, $e = \tilde{q}$, and $r = q_r$.
The weights are applied in frequency domain. The time domain equivalent of these weighted feedback signals can be written as

$$\begin{bmatrix} z_{w_1} \\ z_{w_2} \\ z_{w_3} \end{bmatrix} \triangleq \begin{bmatrix} H_1(r-y) + \Gamma_1(x_{w_1})_k \\ H_2(u) + \Gamma_2(x_{w_2})_k \\ H_3(y_2) + \Gamma_3(x_{w_3})_k \end{bmatrix}, \tag{7}$$

where $H_1$ and $H_2$ are lower triangular Toeplitz matrices developed from impulse responses (Markov parameters) of the discrete weighting filters, $W_1$ and $W_2$. These weights are normally assigned by the designer.

$$H_1 \triangleq \begin{bmatrix} D_{w1} & 0 & 0 & \cdots & 0 \\ C_{w1}B_{w1} & D_{w1} & 0 & \cdots & 0 \\ C_{w1}A_{w1}B_{w1} & C_{w1}B_{w1} & D_{w1} & \cdots & 0 \\ \vdots & \vdots & \vdots & \ddots & \vdots \\ C_{w1}A_{w1}^{i-2}B_{w1} & C_{w1}A_{w1}^{i-3}B_{w1} & C_{w1}A_{w1}^{i-4}B_{w1} & \cdots & D_{w1} \end{bmatrix} \tag{8}$$

$$H_2 \triangleq \begin{bmatrix} D_{w2} & 0 & 0 & \cdots & 0 \\ C_{w2}B_{w2} & D_{w2} & 0 & \cdots & 0 \\ C_{w2}A_{w2}B_{w2} & C_{w2}B_{w2} & D_{w2} & \cdots & 0 \\ \vdots & \vdots & \vdots & \ddots & \vdots \\ C_{w2}A_{w2}^{i-2}B_{w2} & C_{w2}A_{w2}^{i-3}B_{w2} & C_{w2}A_{w2}^{i-4}B_{w2} & \cdots & D_{w2} \end{bmatrix} \tag{9}$$

then

$$Q_1 \triangleq H_1^T H_1, Q_2 \triangleq H_2^T H_2$$

$\Gamma_1$ and $\Gamma_2$ are the extended observability matrices formed from the impulse responses of the weighting filters $W_1$ and $W_2$.

$$\Gamma_1 \triangleq \begin{bmatrix} C_{w1} \\ C_{w1}A_{w1} \\ \vdots \\ C_{w1}A_{w1}^{i-1} \end{bmatrix} \tag{10}$$

$$\Gamma_2 \triangleq \begin{bmatrix} C_{w2} \\ C_{w2}A_{w2} \\ \vdots \\ C_{w2}A_{w2}^{i-1} \end{bmatrix} \tag{11}$$

For simplicity, assume that $y = \left[\tilde{q}, \dot{\tilde{q}}, \ddot{\tilde{q}}\right]^T$ and $y_2 = y_{\{3\}}$[4].

For system identification, suppose a plant's input and output values at discrete times are given, respectively, by

$$\left( \begin{bmatrix} u_0 \\ u_1 \\ \vdots \\ u_{n-1} \end{bmatrix}, \begin{bmatrix} y_0 \\ y_1 \\ \vdots \\ y_{n-1} \end{bmatrix} \right)$$

where $u_i \in \Re^m$ and $y_i \in \Re^l$, where $m$ and $l$ are number of plant input and output signals respectively. The Hankel matrices for the past and future inputs are written as

$$U_p \triangleq \begin{bmatrix} u_0 & u_1 & \cdots & u_{j-1} \\ u_1 & u_2 & \cdots & u_j \\ \vdots & \vdots & \cdots & \vdots \\ u_{i-1} & u_i & \cdots & u_{i+j-2} \end{bmatrix} \in \mathbb{R}^{im \times j}$$

$$U_f \triangleq \begin{bmatrix} u_i & u_{i+1} & \cdots & u_{i+j-1} \\ u_{i+1} & u_{i+2} & \cdots & u_{i+j} \\ \vdots & \vdots & \cdots & \vdots \\ u_{2i-1} & u_{2i} & \cdots & u_{2i+j-2} \end{bmatrix} \in \mathbb{R}^{im \times j}$$

Similarly, the Hankel matrices for the past and future outputs can be written as $Y_p \in \Re^{il \times j}$ and $Y_f \in \Re^{il \times j}$ respectively. Hankel matrix for past outputs and inputs, $W_p$, could be defined as follows

$$W_p \triangleq \begin{bmatrix} U_p \\ Y_p \end{bmatrix}$$

The linear least squares predictor of $Y_f$ with given $W_p$ and $U_f$ can be written as Frobenius norm minimization as follows

$$\min_{L_w, L_u} \left\| Y_f - \begin{bmatrix} L_w & L_u \end{bmatrix} \begin{bmatrix} W_p \\ U_f \end{bmatrix} \right\|_F^2$$

where the subspace orthogonal projections, $L_w$ and $L_u$, are calculated as

$$\begin{bmatrix} L_w & L_u \end{bmatrix} = Y_f \begin{bmatrix} W_p \\ U_f \end{bmatrix}^{\mathrm{T}} \left( \begin{bmatrix} W_p \\ U_f \end{bmatrix} \begin{bmatrix} W_p \\ U_f \end{bmatrix}^{\mathrm{T}} \right)^{\dagger} \tag{12}$$

---

[4] every third element in the array

where † denotes pseudo-inverse. This solution assumes that the problem is overconstrained i.e., there are more independent equations than unknowns. If the problem is under-constrained, the pseudo-inverse cannot be computed. Future outputs can now be predicted from the past inputs, outputs, and future inputs.

$$\begin{bmatrix} \hat{y}_k \\ \vdots \\ \hat{y}_{k+i-1} \end{bmatrix} = L_w \begin{bmatrix} u_{k-i} \\ \vdots \\ u_{k-1} \\ y_{k-i} \\ \vdots \\ y_{k-1} \end{bmatrix} + L_u \begin{bmatrix} u_k \\ \vdots \\ u_{k+i-1} \end{bmatrix} \tag{13}$$

In order to calculate $L_w$ and $L_u$, matrix decomposition methods are used. Using $QR$ method, if

$$\begin{bmatrix} W_p \\ U_f \\ Y_f \end{bmatrix} = R^{\mathrm{T}} Q^{\mathrm{T}} = \begin{bmatrix} R_{11} & 0 & 0 \\ R_{21} & R_{22} & 0 \\ R_{31} & R_{32} & R_{33} \end{bmatrix} Q^{\mathrm{T}}$$

then

$$\begin{bmatrix} L_w & L_u \end{bmatrix} = \begin{bmatrix} R_{31} & R_{32} \end{bmatrix} \begin{bmatrix} R_{11} & 0 \\ R_{21} & R_{22} \end{bmatrix}^{\dagger} \tag{14}$$

Pseudo-inverse is normally calculated through singular value decomposition (SVD) but Woodley et al. presented another method which employs the Cholesky factorization instead of SVD (Woodley et al. (2001b)). This is computationally faster and requires less memory. Using the strictly causal estimate of $y_1$ and $y_2$ from equation (13), we get

$$\hat{y_1} = L_{w1}(w_{p1})_k + L_{u1}u \tag{15}$$

$$\hat{y_2} = L_{w2}(w_{p2})_k + L_{u2}u \tag{16}$$

Here $\hat{y}_1$ is the estimated value of the end effector position in the Cartesian coordinates and $\hat{y}_2$ is the estimated value of the joint angular velocities. From equations (7) to (16), we get

$$\begin{bmatrix} z_{w_1} \\ z_{w_2} \\ z_{w_3} \end{bmatrix} = \begin{bmatrix} H_1 & -H_1 L_{u1} & -H_1 L_{w1} & 0 & \Gamma_1 & 0 & 0 \\ 0 & H_2 & 0 & 0 & 0 & \Gamma_2 & 0 \\ 0 & H_3 L_{u2} & 0 & H_3 L_{w2} & 0 & 0 & \Gamma_3 \end{bmatrix} x \tag{17}$$

where

$$x \triangleq \begin{bmatrix} r \\ u \\ w_{p1} \\ w_{p2} \\ (x_{w1})_k \\ (x_{w2})_k \\ (x_{w3})_k \end{bmatrix} \tag{18}$$

Substituting equation (7) into (6) and (17) produces the objective

$$\min_u \sup_r x^T W x \leq 0 \tag{19}$$

where $W$ is given as follows;

$$W = \begin{bmatrix} Q_1 - \gamma^2 I & -Q_1 L_{u1} & -Q_1 L_{w1} & 0 & H_1^T \Gamma_1 & 0 & 0 \\ -L_{u1}^T Q_1 & L_{u1}^T Q_1 L_{u1} + Q_2 + L_{u2}^T Q_3 L_{u2} & L_{u1}^T Q_1 L_{w1} & L_{u2}^T Q_3 L_{w2} & -L_{u1}^T H_1^T \Gamma_1 & H_2^T \Gamma_2 & H_3^T \Gamma_3 \\ \vdots & \vdots & \vdots & \vdots & \vdots & \vdots & \vdots \\ \cdots & \cdots & \cdots & \cdots & \cdots & \cdots & \Gamma_3^T \Gamma_3 \end{bmatrix}$$

Differentiating (19) with respect to $[r\ u]^T$ and equating to zero produces the following;

$$\begin{bmatrix} Q_1 - \gamma^2 I & -Q_1 L_{u1} & -Q_1 L_{w1} & 0 & H_1^T \Gamma_1 & 0 & 0 \\ -L_{u1}^T Q_1 & L_{u1}^T Q_1 L_{u1} + Q_2 + L_{u2}^T Q_3 L_{u2} & L_{u1}^T Q_1 L_{w1} & L_{u2}^T Q_3 L_{w2} & -L_{u1}^T H_1^T \Gamma_1 & H_2^T \Gamma_2 & H_3^T \Gamma_3 \end{bmatrix} x = 0 \tag{20}$$

The linear system in (20) can be re-arranged as follows;

$$\begin{bmatrix} Q_1 - \gamma^2 I & -Q_1 L_{u1} \\ -L_{u1}^T Q_1 & L_{u1}^T Q_1 L_{u1} + Q_2 + L_{u2}^T Q_3 L_{u2} \end{bmatrix} \begin{bmatrix} r \\ u \end{bmatrix} \tag{21}$$

$$= \begin{bmatrix} Q_1 L_{w1} & 0 & -H_1^T \Gamma_1 & 0 & 0 \\ -L_{u1}^T Q_1 L_{w1} & -L_{u2}^T Q_3 L_{w2} & L_{u1}^T H_1^T \Gamma_1 & -H_2^T \Gamma_2 & -H_3^T \Gamma_3 \end{bmatrix} \begin{bmatrix} w_{p1} \\ w_{p2} \\ (x_{w1})_k \\ (x_{w2})_k \\ (x_{w3})_k \end{bmatrix}$$

Differentiating once again with respect to $[r\ u]^T$ suggests the following;

$$H \triangleq \frac{\partial^2 J(\gamma)}{\partial \begin{bmatrix} r \\ u \end{bmatrix}^2} = \begin{bmatrix} Q_1 - \gamma^2 I & -Q_1 L_{u1} \\ -L_{u1}^T Q_1 & L_{u1}^T Q_1 L_{u1} + Q_2 + L_{u2}^T Q_3 L_{u2} \end{bmatrix} \tag{22}$$

Schur decomposition offers the following definition;

$$\begin{bmatrix} A_1 & A_2 \\ A_2^T & A_3 \end{bmatrix} = \begin{bmatrix} I & A_2 A_2^{-1} \\ 0 & I \end{bmatrix} \begin{bmatrix} A_1 - A_2 A_3^{-1} A_2^T & 0 \\ 0 & A_3 \end{bmatrix} \begin{bmatrix} I & 0 \\ A_3^{-1} A_2^T & I \end{bmatrix} \tag{23}$$

Since $Q_1 = Q_1^T$ and $Q_1$ is positive definite, it could be concluded that $A_3 > 0$ which satisfies the saddle condition (Woodley (2001)). As $A_3 \in \Re^{im \times im}$ and $r \in \Re^{im}$, the condition for worst case input reference signal can be stipulated by the following inequality;

$$A_1 - A_2 A_3^{-1} A_2^T < 0 \tag{24}$$

Matching the definitions in (23) and (24) to the mathematical aspects of the model-free control introduced above, the following could be stated;

$$Q_1 - Q_1 L_{u1}(L_{u1}^T Q_1 L_{u1} + Q_2 + L_{u2}^T Q_3 L_{u2})^{-1} L_{u1}^T Q_1 < \gamma^2 I \tag{25}$$

which can be written as;

$$\gamma > \sqrt{\bar{\lambda}[Q_1 - Q_1 L_{u1}(L_{u1}^T Q_1 L_{u1} + Q_2 + L_{u2}^T Q_3 L_{u2})^{-1} L_{u1}^T Q_1]} \equiv \gamma_{min} \tag{26}$$

To calculate the optimum controller outputs, multiply (21) with [0 $I$].

$$u_{opt} = \begin{bmatrix} 0 & I \end{bmatrix} \begin{bmatrix} Q_1 - \gamma^2 I & -Q_1 L_{u1} \\ -L_{u1}^T Q_1 & L_{u1}^T Q_1 L_{u1} + Q_2 + L_{u2}^T Q_3 L_{u2} \end{bmatrix}^{-1} \tag{27}$$

$$\begin{bmatrix} Q_1 L_{w1} & 0 & -H_1^T \Gamma_1 & 0 & 0 \\ -L_{u1}^T Q_1 L_{w1} & -L_{u2}^T Q_3 L_{w2} & L_{u1}^T H_1^T \Gamma_1 & -H_2^T \Gamma_2 & -H_3^T \Gamma_3 \end{bmatrix} \begin{bmatrix} w_{p1} \\ w_{p2} \\ (x_{w1})_k \\ (x_{w2})_k \\ (x_{w3})_k \end{bmatrix}$$

### 4.0.1 Uncertainties

For a robust system, it is important that uncertainties are accounted for. Most uncertainties in a plant are hard to model. Figures (4) and (5) show the general layout of plant models with uncertainties in multiplicative and additive configurations, respectively. Woodley calculated $\gamma_{min}$ for different configurations of uncertainties in model-free control designs (Woodley (2001)). But the real challenge is to find the uncertainty block $\Delta$ through techniques like model unfalsification. A true robust system calculates $\Delta$ in realtime.

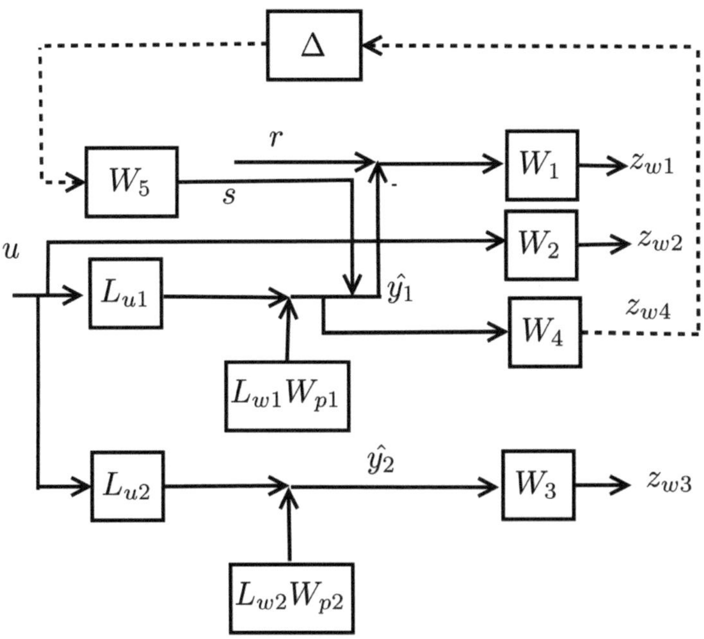


Figure 4. Plant with multiplicative uncertainties for robust $\mathcal{H}_\infty$ control design

## 5. Simulation

For a complete identification $13n$ parameters are to be identified (Khalil and Dombre (2004)). In model free framework, these parameters are not available explicitly. The prediction horizon should be two to three times the expected order of the system. From this rough estimate, prediction horizon as 30 is selected for a two joint planar manipulator. On one of the coordinate, a square signal and on the other one, a sinusoidal signal is given. The response of the end effector along with the performance objective $\gamma$ is given in the figure (6). One of the benefit of using subspace identification is the property of Hankel matrix that allows to concatenate the data from a previous session.

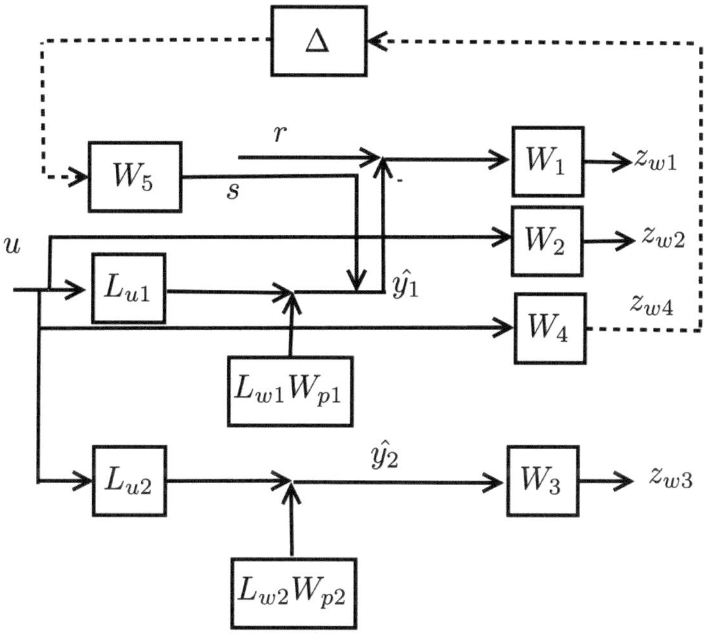


Figure 5. Plant with additive uncertainties for robust $\mathcal{H}_\infty$ control design

## 6. Conclusion

The proposed framework provides solution to inverse dynamics, parameter identification and robust control of mechanical manipulators in an elegant way. The fastest way to calculate pseudo-inverse is through Cholesky/SVD factorization (Golub & Loan (1996)). Its complexity is $O(ij + i^3)$, where $i$ is the prediction horizon and $j$ is number of prediction problems in a Hankel matrix. The computational burden becomes significant when a robot has large number of links and it has 6 degrees of freedom i.e. $13ln$, where $l$ is a arbitrary natural number from 1 to 5 as a safety factor in prediction and $n$ is the number of joints of manipulator. Calculations required to predict a predictor for a six joint robot with six degrees of freedom with safety margin of 2 is 3796416. This number is not so big for modern computers but increasing the number of joints will increase the complexity exponentially.

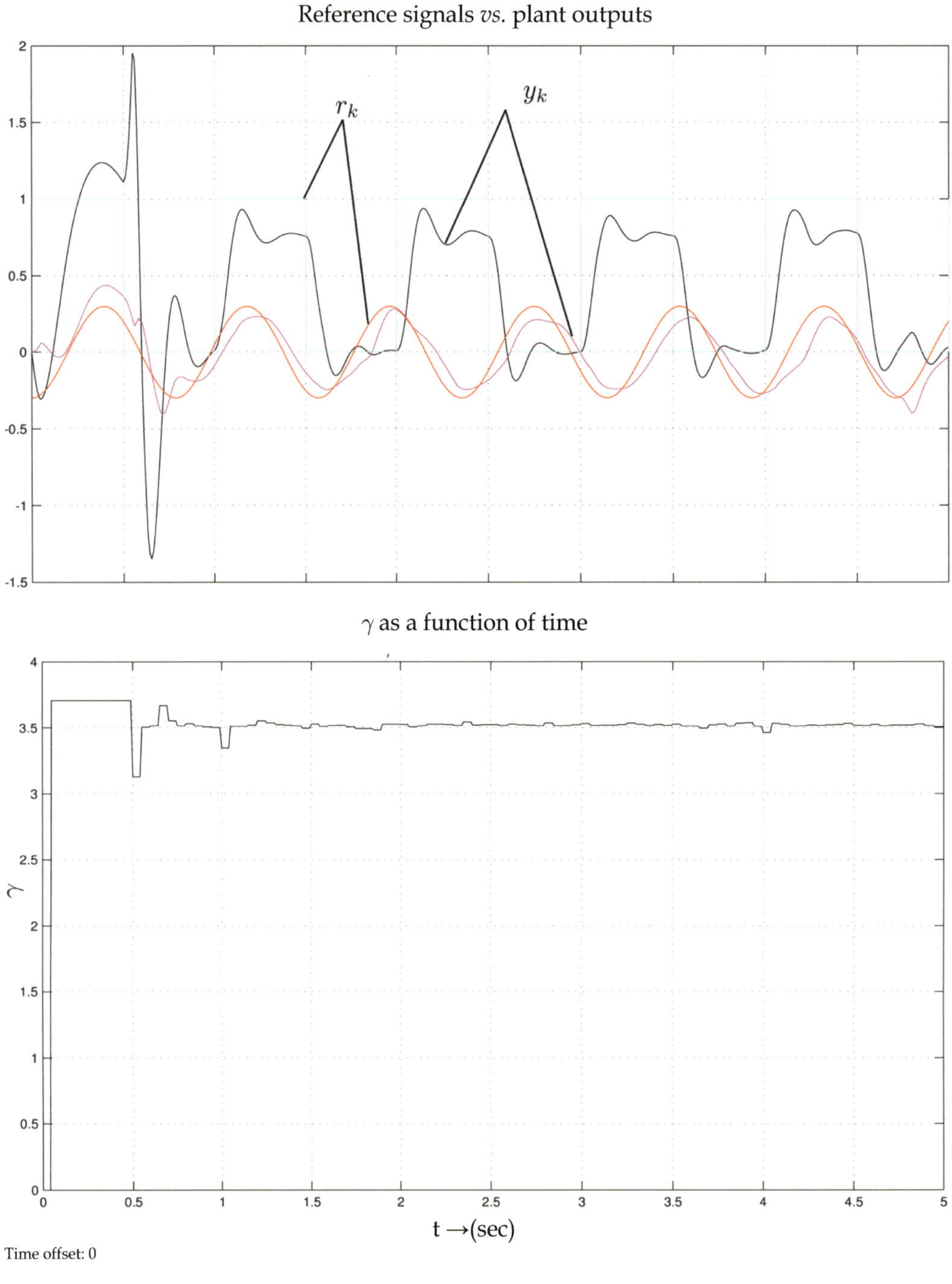


Figure 6. Response of a planar robot with two rotary joints. In this particular experiment, $W_p$ is initialized with null matrix. The performance is given by $\gamma$, which converges to a constant value when the input is consistent.

## 7. References


Agnoloni, T. and Mosca, E. (2003). Controller falsification based on multiple models. *International Journal of Adaptive Control and Signal Processing*, 17:163-177.

Arimoto, S. (1995). Fundamental problems of robot control: Parti, innovations in the realm of robot servo-loops. *Robotica*, 13:19-27.

Athans, M. (1999). Crisp control is always better than fuzzy feedback control. EUFIT t'99 debate with Prof. L.A. Zadeh, Aachen, Germany.

Boyd, J. E. and Little, J. J. (2000). Phase in model-free perception of gait. In *HUMO '00: Proceedings of the Workshop on Human Motion (HUMO'00)*, page 3, Washington, DC, USA. IEEE Computer Society.

Caballero, R., Armada, M. A., and Akinfiev, T. (2004). Robust cascade controller for nonlinearly actuated biped robots: experimental evaluation. *International Journal of Robotics Research*, 23(10/11):1075-1095.

Cabral, F. B. and Safonov, M. G. (2004). Unfalsified model reference adaptive control using the ellipsoid algorithm. *International Journal of Adaptive Control and Signal Processing*, 18(8):683-696.

Cheng, G. (2004). Model-free adaptive (mfa) control. *Computing & Control Engineering Journal*, 15(3):28-33.

Chiuso, A. and Picci, G. (2005). Consistency analysis of some closed-loop subspace identification methods. *Automatica*, 41(3):377-391.

Chou, C. T. and Verhaegen, M. (1997). Subspace algorithms for the identification of multivariable dynamic errors-in-variables models. *Automatica*, 33:1857-1869.

Date, P. and Lanzon, A. (2004). A combined iterative scheme for identification and control redesigns. *International Journal of Adaptive Control and Signal Processing*, 18(8):629-644.

Favoreel, W., Moor, B. D., Gevers, M., and Overschee, P. V. (1998). Model-free subspace-based LQG-design. Technical report, Katholieke Universiteit Leuven.

Favoreel, W., Moor, B. D., Gevers, M., and Overschee, P. V. (1999a). Closed loop model-free subspace-based LQG-design. In *Proceedings of the IEEE Mediterranean Conference on Control and Automation*, Haifa, Israel.

Favoreel, W., Moor, B. D., and Overschee, P. V. (1999b). Model-free subspace-based LQG-design. In *Proceedings of the American Control Conference*, pages 3372-3376.

Golub, G. H. and Loan, C. F. V. (1996). *Matrix Computations*. The Johns Hopkins University Press.

Grimble, M. J. (1998). Multi-step $\mathscr{H}_\infty$ generalized predictive control. *Dynamics and Control*, 8(4):303-339.

Haddad, W. M., Hayakawa, T., and and, J. M. B. (2003). Adaptive control for nonnegative and compartmental dynamical systems with applications to general anesthesia. *International Journal of Adaptive Control and Signal Processing*, 17(3): 209- 235.

Hayakawa, T., Haddad, W. M., and Leonessa, A. (2004). A Lyapunov-based adaptive control framework for discrete-time non-linear systems with exogenous disturbances. *International Journal of Control*, 77(3): 250-263.

Hsu, E., Pulli, K., and Popovi´c, J. (2005). Style translation for human motion. *ACM Transactions on Graphics (TOG)*, 24(3): 1082-1089.

Kelly, R. (1993). Comments on adaptive pd controller for robot manipulators. *IEEE Trans. Robot. Automat.*, 9: 117–119.

Kelly, R. (1997). Pd control with desired gravity compensation of robotic manipulators: A review. *Int. J. Robot. Res.*, 16(5): 660–672.

Khalil, W. and Dombre, E. (2004). *Modeling, Identification and Control of Robots*. Kogan Page Science.

Kim, D., Kim, N.-H., Seo, S.-J., and Park, G.-T. (2004). *Fuzzy Modeling of Zero Moment Point Trajectory for a Biped Walking Robot*. Lecture Notes in Computer Science. Springer-Verlag GmbH, 3214 edition.

Kim, J. O., Lee, B. R., Chung, C. H., Hwang, J., and Lee, W. (2003). *The Inductive Inverse Kinematics Algorithm to Manipulate the Posture of an Articulated Body*. Lecture Notes in Computer Science. Springer-Verlag GmbH, 2657 edition.

Kosut, R. L. and Anderson, B. D. O. (1997). Uncertainty model unfalsification. In *Proceedings of the 36th IEEE Conference on Decision and Control*, volume 1, pages 163–168, San Diego, CA.

Laib, A. (2000). Adaptive output regulation of robot manipulators under actuator constraints. *IEEE Trans. Robot. Automat.*, 16:29–35.

Lin, W., Qin, S. J., and Ljung, L. (2004). A framework for closed-loop subspace identification with innovation estimation. Technical Report 2004-07, Department of Chemical Engineering, The University of Texas at Austin, Austin, TX 78712, USA and Linköping University, SE-581 83 Linköping, Sweden.

Ljung, L. (1999). *System identification: theory for the user*. Prentice-Hall, Upper Saddle River, NJ, USA.

Lohar, F. A. (2000). $\mathcal{H}_\infty$ and $\mu$-synthesis for full control of helicopter in hover. In *38th Aerospace Sciences Meeting and Exhibit*, Reno, NV. American Institute of Aeronautics and Astronautics.

Norton, J. P. (1986). *Introduction to Identification*. Academic Press.

Overschee, P. V. and Moor, B. D. (1996). *Subspace Identificiation for Linear Systems*. Kluwer Academic Publishers.

Safonov, M. G. (2003). Recent advances in robust control theory. In *AIAA Guidance, Navigation and Control Conference and Exhibit*, volume 4, Austin, TX.

Sapio, V. D. and Khatib, O. (2005). Operational space control of multibody systems with explicit holonomic constraints. In *Proceedings of the 2005 IEEE International Conference on Robotics and Automation*.

Sayyar-Rodsari, B., How, J. P., Hassibi, B., and Carrier, A. (1998). An $\mathcal{H}_\infty$ optimal alternative to the fxlms algorithm. In *Proceedings of the American Control Conference*, pages 1116–1121.

Sciavicco, L. and Siciliano, B. (2000). *Modelling and Control of Robot Manipulators*. Springer, 2nd edition.

Takegaki, M. and Arimoto, S. (1981). A new feedback method for dynamic control of manipulators. *ASME J. Dyn. Syst., Meas., Control*, 102:119–125.

Tomei, P. (1991). Adaptive pd controller for robot manipulators. *IEEE Trans. Robot.* Automat., 7:565–570.

Tsao, T.-C., Brozenec, T., and Safonov, M. G. (2003). Unfalsified adaptive spacecraft attitude control. In *AIAA Guidance, Navigation, and Control Conference and Exhibit*, Austin, TX.

Tsao, T.-C. and Safonov, M. G. (2001). Unfalsified direct adaptive control of a twolink robot arm. *International Journal of Adaptive Control and Signal Processing*, 15(3):319–334.

Tse, J., Bentsman, J., and Miller, N. (1993). Properties of the self-tuning minimax predictive control (MPC). In *Proceedings of the 1993 American Control Conference*, pages 1721–1725.

Wang, J. and Qin, S. J. (2004). A new deterministic-stochastic subspace identification method using principal component analysis. Technical report, Department of Chemical Engineering, The University of Texas at Austin.

Wang, R., Stefanovic, M., and Safonov, M. G. (2004). Unfalsified direct adaptive control using multiple controllers. In *AIAA Guidance, Navigation, and Control Conference and Exhibit*, pages 1–16, RI, USA.

Wen, J., Kreutz-Delgado, K., and Bayard, D. (1992). Lyapunov function-based control laws for revolute robot arms. *IEEE Trans. Automat. Contr.*, 37:231–237.

Wernholt, E. (2004). *On Multivariable and Nonlinear Identification of Industrial Robots*. PhD thesis, Department of Electrical Engineering, Link¨oping University, SE-581 83 Linköping, Sweden.

Widrow, B. and Stearns, S. D. (1985). *Adaptive Signal Processing*. Prentice-Hall.

Widrow, B. and Walach, E. (1994). *Adaptive Inverse Control*. Prentice-Hall.

Wilson, E. (1997). Adaptive profile optimization for the electric arc furnace. In *Steel Technology International*, pages 140–145.

Wodoley, B., How, J., and Kosut, R. (1999). Direct unfalsified controller design - solution via convex optimization. In *Proceedings of the 1999 American Control Conference*, volume 5, pages 3302–3306.

Wolkotte, P. T. (2003). Modelling human locomotion. Technical report, Institute of Electronic Systems, Aalborg University.

Woodley, B., How, J., and Kosut, R. (2001a). Model free subspace based $\mathscr{H}_\infty$ control. In *Proceedings of the 2001 American Control Conference*, volume 4, pages 2712- 2717.

Woodley, B., Kosut, R., and How, J. (1998). Uncertainty model unfalsification with simulation. In *Proceedings of the 1998 American Control Conference*, volume 5, pages 2754–2755.

Woodley, B. R. (2001). *Model free subspace based $\mathscr{H}_\infty$ control*. PhD thesis, Department of Electrical Engineering, Stanford University.

Woodley, B. R., How, J. P., and Kosut, R. L. (2001b). Subspace based direct adaptive $\mathscr{H}_\infty$ control. *International Journal of Adaptive Control and Signal Processing*, 15(5):535–561.

Xie, M. (2003). Fundamentals of Robotics, volume 54 of Machine perception and artificial intelligence. World Scientific.

Zhang, Y., Tian, H., Wang, Q., and Qiang, W. (2000). Servo control in joint space of biped robot using nonlinear $\mathscr{H}_\infty$ strategy. In Jiang, D. and Wang, A., editors, *Proceedings of SPIE, International Conference on Sensors and Control Techniques (ICSC 2000)*, volume 4077, pages 386–391.

Zhao, H. and Bentsman, J. (1999). Multivariable $\mathscr{H}_\infty$ predictive control based on minimax predictor. In *Proceedings of the IEEE Conference on Decision and Control*, volume 4, pages 3699–3705.

# 24

# The Verification of Temporal KBS: SPARSE - A Case Study in Power Systems

Jorge Santos[1], Zita Vale[2], Carlos Serôdio[3] and Carlos Ramos[1]
[1]*Departamento de Engenharia Informática, Instituto Superior de Engenharia do Porto,*
[2]*Departamento de Engenharia Electrotécnica, Instituto Superior de Engenharia do Porto,*
[3]*Departamento de Engenharias, Universidade de Trás-os-Montes e Alto Douro,*
*Portugal*

## 1. Introduction

Although humans present a natural ability to deal with knowledge about time and events, the codification and use of such knowledge in information systems still pose many problems. Hence, the development of applications strongly based on temporal reasoning remains a hard and complex task. Furthermore, albeit the last significant developments in temporal reasoning and representation (TRR) area, there still is a considerable gap for its successful use in practical applications.

In this chapter we present VERITAS, a tool that focus time maintenance, that is one of the most important processes in the engineering of the time during the development of KBS.

The verification and validation (V&V) process is part of a wider process denominated knowledge maintenance (Menzies 1998), in which an enterprise systematically gathers, organizes, shares, and analyzes knowledge to accomplish its goals and mission. The V&V process states if the software requirements specifications have been correctly and completely fulfilled.

The methodologies proposed in software engineering have showed to be inadequate for Knowledge Based Systems (KBS) validation and verification, since KBS present some particular characteristics.

VERITAS is an automatic tool developed for KBS verification which is able to detect a large number of knowledge anomalies. It addresses many relevant aspects considered in real applications, like the usage of rule triggering selection mechanisms and temporal reasoning.

The rest of the chapter is structured as follows. Section 2 provides a short overview of the state-of-art of V&V and its most important concepts and techniques. After that, section 3 describes SPARSE, a KBS used to assist the Portuguese Transmission Control Centres operators in incident analysis and power restoration. Special attention is given to SPARSE's particular characteristics, introducing the problem of verifying real world applications. Section 4 presents VERITAS; special emphasis is given to the tool architecture and to the method used in anomaly detection. Finally, in section 5, achieved results are discussed and in section 6, we present some conclusions and ideas for future work.

## 2. Verification and validation of KBS

Many authors argued that the correct and efficient performance of any piece of software must be guaranteed through the verification and validation (V&V) process, and it becomes obvious that Knowledge Based Systems (KBS) should undergo the same evaluation process. Besides, it is known that knowledge maintenance is an essential issue for the success of the KBS since it assures the consistency of the knowledge base (KB) after each modification in order to avoid the assertion of knowledge inconsistencies. Unfortunately, the methodologies proposed in software engineering have showed to be inadequate for knowledge based systems validation and verification, since KBS present some particular characteristics (Gonzalez & Dankel 1993). Namely, the need for KBS to deal with uncertainty and incompleteness; the domains modelled normally do not underline physical models; it is not rare for KBS to have the ability to learn and improve the KB allowing a dynamical behaviour; in most domains of expertise, there is no concept of right results but only of acceptable ones.

Besides the facets of software certification and maintenance previously referred, the systematic use of formal V&V techniques is also a key for making end-users more confident about KBS, especially when critical applications are considered. In Control Centres domain the V&V process intends to assure the reliability of the installed applications, even under incident conditions.

The problem of Verification and Validation appears when there is a need to assure that some model (solution) correctly addresses the problem through the adequate techniques and methodology in order to provide the desired results. In the scope of this work, Validation and Verification will be referred as two complementary processes, both fundamental for KBS end-user acceptance.

Albeit there is no general agreement on the V&V terminology (Hoppe & Meseguer 1991), the following definitions will be used in the scope of this paper.

- **Validation** - Validation means building the right system (Boehm 1984). The purpose of validation is to assure that a KBS will provide solutions with similar (or higher if possible) confidence level as the one provided by domain experts. Validation is then based on tests, desirably in the real environment and under real circumstances. During these tests, the KBS is considered as a black box, meaning that only the input and the output are really considered important;
- **Verification** - Verification means building the system right (Boehm 1984). The purpose of verification is to assure that a KBS has been correctly designed and implemented and does not contain technical errors. During the verification process the interior of the KBS is examined in order to find any possible errors; this approach is also called crystal box.
- **Verification & Validation** - The Verification and Validation process allows determining if the requirements have been correctly and completely fulfilled in order to assure the system's reliability, safety, quality and efficiency. More synthetically, it can be said that the V&V process is to build the right system right (Preece 1998).

In the last decades, several techniques were proposed for validation and verification of Knowledge Based Systems, like inspection, formal proof, cross-reference verification or empirical tests (Preece 1998). The efficiency of these techniques strongly depends on the existence of test cases or on the degree of formalization used in the specifications. One of the most used techniques is static verification, which consists of sets of logical tests executed in order to detect possible knowledge anomalies.

- **Anomaly** - An anomaly is a symptom of one (or multiple) possible error(s). Notice that an anomaly does not necessarily denote an error (Preece & Shinghal 1994).

Rule bases are drawn as a result of a knowledge analysis/elicitation process, including, for example, interviews with experts or the study of documents such as codes of practice and legal texts, or analysis of typical sample cases. The rule base should reflect the nature of this process, meaning that if documentary sources are used, the rule base should reflect knowledge sources. Consequently, some anomalies are desirable and intentionally inserted in KB. For instance, redundancy on the documentary sources will lead to redundant KB.
Rule based systems are still the more often used representation in the development of KBS. The scientific community has deeply studied these systems. At the moment there is an assortment of V&V techniques that allow the detection of many anomalies in systems that use this kind of representation. Some well known V&V tools used different techniques to detect anomalies. The KB-Reducer (Ginsberg 1987) system represents rules in logical form, and then it computes for each hypothesis the corresponding labels, detecting the anomalies during the labelling process. Meaning that each literal in the rule LHS (Left Hand Side) is replaced by the set of conditions that allows to infer it. This process finishes when all formulas become grounded. The COVER (Preece, Bell & Suen 1992) works in a similar fashion using the ATMS (Assumption Truth Maintaining System) approach (Kleer 1986) and graph theory, allowing the detection of a large number of anomalies. The COVADIS (Rousset 1988) successfully explored the relation between input and output sets. The ESC (Cragun & Steudel 1987), RCP (Suwa, Scott & Shortliffe 1982) and Check (Nguyen et al. 1987) systems and more recently the PROLOGA (Vanthienen, Mues & Wets 1997) used decision table methods for verification purposes. This approach proved to be quite interesting, especially when the systems to be verified also used decision tables as representation support. These systems' major advantage is that it enables tracing the reasoning path quite clearly, while the major problem is the lack of solutions for verifying long reasoning inference chains. Some authors studied the applicability of Petri nets (Pipard 1989; Nazareth 1993) to represent the rule base and to detect the knowledge inconsistencies. More recently coloured Petri nets were used (Wu & Lee 1997). Although specific knowledge representations provide higher efficiency while used to perform some verification tests, arguably all of them could be successfully converted into production rules.

## 3. The case study: SPARSE

Control Centres (CC) are very important in the operation of electrical networks when receiving real-time information about network status. CC operators should take, usually in a short time, the most appropriate actions in order to reach the maximum network performance.
In case of incident conditions, a huge volume of information may arrive to these centres. The correct and efficient interpretation by a human operator becomes almost impossible. In order to solve this problem, some years ago, electrical utilities began to install intelligent applications in their control centres (Amelink, Forte & Guberman 1986; Kirschen & Wollenberg 1992). These applications are usually KBS and are mainly intended to provide operators with assistance, especially under critical situations.

### 3.1 Architecture and functioning

SPARSE (Vale et al. 1997) is a KBS used in the Portuguese Transmission Network (REN) for incident analysis and power restoration. In the beginning it started to be an expert system

(ES) and it was developed for the Portuguese Transmission Network (REN) Control Centres. The main goals of this ES were to assist Control Centre operators in incident analysis, allowing a faster power restoration. Later, the system evolved to a more complex architecture (Vale et al. 2002), which is normally referred as a Knowledge Based System (see Fig. 1).

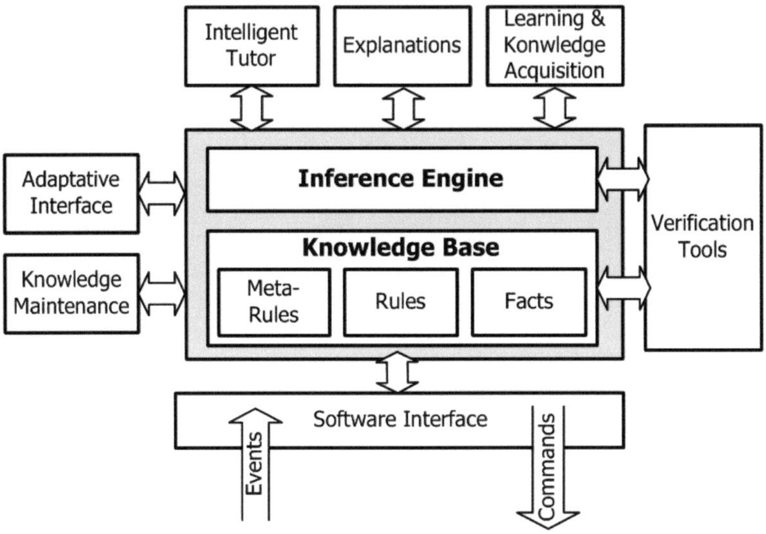


Fig. 1 - SPARSE Architecture

SPARSE includes many modules, namely for learning and automatic data acquisition (Duarte et al. 2001), adaptive tutoring (Faria et al. 2002) and automatic explanations (Malheiro et al. 1999). As it happens in the majority of KBSs, one of the most important SPARSE components is the knowledge base (KB) (see formula (1)):

$$KB = RB \cup FB \cup MRB \tag{1}$$

where:

- RB stands for rule base;
- FB stands for facts base;
- MRB stands for meta-rules base;

The rule base is a set of clauses with the following structure:

```
RULE ID: 'Description':
 [
  [C1 AND C2 AND C3]
  OR
  [C4 AND C5]
 ]
 ==>
 [A1,A2,A3].
```

The rule's Left Hand Side (LHS) is a set of conditions (C1 to C5 in this example) of the following types:

- A fact, representing domain events or status messages. Typically these facts are time-tagged;

- A temporal condition over facts;
- Previously asserted conclusions.

The rule's Right Hand Side (RHS) is a set of actions/conclusions to be taken (A1 to A3 in this example) and may be of one of the following types:

- Assertion of facts representing conclusions to be inserted in the knowledge base. A conclusion can be final (e.g., a diagnosis) or intermediate (e.g., a status fact concerning that later it will be used in other rule LHS);
- Retraction of facts (conclusions to be deleted from the knowledge base);
- Interaction with the user interface.

Let's consider the rule d3 as an example of a SPARSE rule:

```
rule d3 : 'Monophasic Triggering':
 [
  [
   msg(Dt1,Tm1,[In1,Pn1,[In2,NL]],'>>>TRIGGERING','01') at T1 and
   breaker(_,_,In1,Pn1,_,_, closed) and
   msg(Dt2,Tm2,[In1,Pn1,[In2,NL, _ ]],'BREAKER','00') at T2 and
   condition(abs_diff_less_or_equal(T2,T1,30))
  ]
  or
  [
   msg(Dt1,Tm1,[In1,Pn1,[In2,NL]],'>>>TRIGGERING','01') at T1 and
   msg(Dt2,Tm2,[In1,Pn1,[In2,NL]],'BREAKER','00') at T2  and
   condition(abs_diff_less_or_equal(T2,T1,30))
  ]
 ]
 ==>
 [
  assert(triggering(Dt1,Tm1,In1,Pn1,In2,NL,monophasic,not_identified,T2),T1),
  retract(breaker(_,_,In1,Pn1,_,_,closed),_,T2),
  assert(breaker(Dt2,Tm2,In1,Pn1,_,triggering,mov),T2),
  retract(msg(Dt1,Tm1,[In1,Pn1,[In2,NL]],'>>>TRIGGERING','01'),T1,T1),
  retract(msg(Dt2,Tm2,[In1,Pn1,[In2,NL | _ ]],'BREAKER','00'),T2,T2),
  retract(breaker_opened(In1,Pn1),_,T1),
  assert(breaker_opened(In1,Pn1),T1)
 ].
```

The meta-rule base is a set of triggers, used by the rule selection mechanism, with the following structure:

$$\text{trigger}(Fact, [(R_1, TB_1, TE_1), \ldots, (R_n, TB_n, TE_n)]) \qquad (2)$$

standing for:

- Fact - the arriving fact (external alarm or a previously inferred conclusion);
- ($R_i$, $TB_i$, $TE_i$) - the temporal window were the rule $R_i$ could by triggered. $TB_i$ is the delay time before rule triggering, used to wait for remaining facts needed to define an event, and the $TE_i$ is the maximum time for trying to trigger the rule $R_i$.

The inference process relies on the cycle depicted in Fig. 2. In the first step, SPARSE collects a message (represented as a fact in SPARSE scope) from SCADA[1], then the respective trigger is selected and some rules are scheduled. The scheduler selects the next rule to be tested (the inference engines try to prove its veracity). Notice that, when a rule succeeds, the conclusions (on the RHS) will be asserted and later processed in the same way as the SCADA messages.

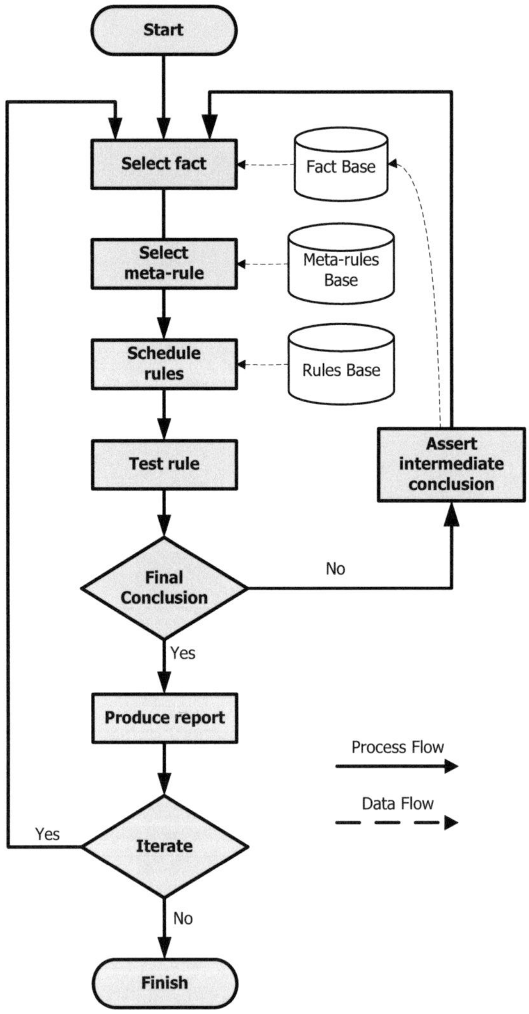


Fig. 2 - SPARSE main algorithm

Let's consider the following meta-rule that allows scheduling the rule d3:

trigger(msg(_,_,[Inst1,Painel1,_],'>>>TRIGGERING','01'), [(d1,30,50),(d2,31,51),(d3,52,52)] ).

---

1 Supervisory Control And Data Acquisition: this system collects messages from the mechanical/electrical devices installed in the network.

The use of rule selection mechanism allows configuring a heuristic approach with the following characteristics:

- **Problem space reduction** – since it assures that only a set o rules related to the Fact will be used and tested;
- **From specific to general rules** – the temporal windows allow defining an explicit order, so the system usually proceeds from the most specific to general.

In what concerns SPARSE, there were clearly two main reasons to start its verification. First, the SPARSE development team carried out a set of tests based on previously collected real cases and some simulated ones. Despite the importance of these tests for the final product acceptance, the major criticism that could be pointed out to this technique is that it only assures the correct performance of SPARSE under the tested scenarios.

Moreover, the tests performed during the validation phase, namely the field tests, were very expensive, since they required the assignment of substantial technical personnel and physical resources for their execution (e.g., transmission lines and coordination staff). Obviously, it would be unacceptable to perform those tests after each KB update. Under these circumstances, an automatic verification tool could offer an easy and inexpensive way of assuring knowledge quality maintenance, assuring the consistency and completeness of represented knowledge.

### 3.2 The verification problem

The verification problem based on the anomaly detection usually relies on the calculation of all possible inference chains that could be entailed during the reasoning process. Later, some logical tests are performed in order to detect if any constraints violation takes place.

SPARSE presents some features that make the verification work harder. These features demand the use of more complex techniques during anomaly detection and introduce significant changes in the number and type of anomalies to detect. The following ones are the most important:

- **Rule triggering selection mechanism** - In what concerns SPARSE, this mechanism was implemented using both meta-rules and the inference engine. As for verification work, this mechanism not only avoids some run-time errors (for instance circular chains) but also introduces another complexity axis to the verification. Thus, this mechanism constrains the existence of inference chains and also the order that they would be generated. For instance, during system execution, the inference engine could be able to assure that shortcuts (specialists rules) would be preferred over generic rules;
- **Temporal reasoning** - This issue received large attention from the scientific community in last two decades (surveys covering this issue can be found in (Gerevini 1997; Fisher, Gabbay & Vila 2005)). Although *time* is ubiquitous in society, and despite the natural ability that human beings show dealing with it, a widespread representation and usage in the artificial intelligence domain remains scarce due to many philosophical and technical obstacles. SPARSE is an *alarm processing application* and its major challenge is to reason about events. Therefore, it is necessary to deal with time intervals (e.g., temporal windows of validity), points (e.g., instantaneous events occurrence), alarms order, duration and the presence or/and absence of data (e.g., messages lost in the collection or/and transmission system);

- **Variables evaluation** - In order to obtain comprehensive and correct results during the verification process, the evaluation of the variables present in the rules is crucial, especially in what concerns temporal variables, i.e., the ones that represent temporal concepts. Notice that during anomaly detection (this type of verification is also called static verification) it is not possible to predict the exact value that a variable will have;
- **Knowledge versus Procedure** - Languages like Prolog provide powerful features for knowledge representation (in the declarative way) but they are also suited to describe procedures; so, sometimes knowledge engineers encode rule using procedural predicates. For instance, the following sentence in Prolog: `min(X,Y,Min)` calls a procedure that compares `X` and `Y` and instantiates `Min` with smaller value. Thus, it is not a (pure) knowledge item; in terms of verification it should be evaluated in order to obtain the `Min` value. It means that the verification method needs to consider not only the programming language syntax but also the meaning (semantic) in order to evaluate the functions. This step is particularly important for any variables that are updated during the inference process.

## 4. The verification tool: VERITAS

VERITAS is an automatic tool developed for KBS verification. This tool performs KB structural analysis allowing knowledge anomalies detection. Originally, VERITAS used a non temporal KB verification approach. Although it proved to be very efficient in other KBS verification - in an expert system for cardiac diseases diagnosis (Rocha 1990) and in other expert system otology diseases diagnosis and therapy (Sampaio 1996) -, in SPARSE case some important limitations were detected.

### 4.1 Main process

The VERITAS main process, depicted in Fig. 3, relies on a set of modules that assures the following competences: converting the original knowledge base; creating the internal knowledge base; computing the rule expansions and detecting anomalies; producing readable results. Regarding that, the user can interact with all stages of the verification process.

The conversion module translates the original knowledge base files into a set of new ones containing the same data but represented in an independent format, recognized by VERITAS. For this module functioning, a set of conversion rules is also needed, specifying the translation procedure. This module assures the independence of VERITAS to the format and syntax used in specification of the KB to be verified. The conversion operations largely depend on the original format, although the most common conversion operations are:

- If the LHS is a disjunctive form, a distinct rule is created for each conjunction contained by the LHS. The rule RHS remains the same;
- Original symbols, like logical operators, are replaced by others accordingly to the internal notation;
- The variables are replaced by internal symbols and a table is created in order to store these symbols.

During the conversion step, the rule d3 (previously presented) would be transformed in two d3-L1 and d3-L2, since the LHS contains two conjunctions. The tuple cvr/3 stores the intermediate rule d3-L1 presented after Fig.3.

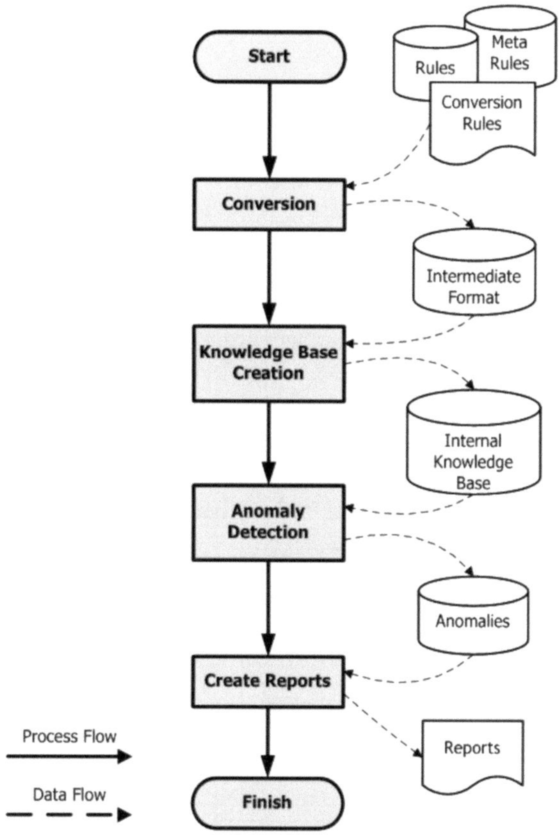


Fig. 3 - VERITAS main algorithm

```
cvr(
    d3-L1,
    Monophasic Triggering,
    [
    msg(#Dt1,#Tm1,[#In1,#Pn1,[#In2,#NL]],>>>TRIGGERING,01) at #T1,
    breaker(#_,#_,#In1,#Pn1,#_,#_, closed),
    msg(#Dt2,#Tm2,[#In1,#Pn1,[#In2,#NL,#_]],BREAKER,00) at #T2,
    abs_diff_less_or_equal(#T2,#T1,30)
    ],
    [
    assert(triggering(#Dt1,#Tm1,#In1,#Pn1,#In2,#NL,monophasic,not_identified,#T2),#T1),
    retract(breaker(#_,#_,#In1,#Pn1,#_,#_,fechado),#_,#T2),
    assert(breaker(#Dt2,#Tm2,#In1,#Pn1,#_,triggering,mov),#T2),
    retract(msg(#Dt1,#Tm1,[#In1,#Pn1,[#In2,#NL]],>>>TRIGGERING,01),#T1,#T1),
    retract(msg(#Dt2,#Tm2,[#In1,#Pn1,[#In2,#NL | #_]],BREAKER,00),#T2,#T2),
    retract(breaker_opened(#In1,#Pn1),#_,#T1),
    assert(breaker_opened(#In1,#Pn1),#T1)
]).
```

After the conversion step, the internal knowledge base is created (see section 4.2). This KB will store information about the following issues:

- The original rules and meta-rules, represented according to the structure that allows speeding up the expansion calculation;
- A set of restrictions over the knowledge domain modelled in KB regarding this set can be manually or semi-automatically created;
- A characterization over the data items extracted from the original KB.

In the following step, the anomaly detection (see section 4.3), the module responsible for this task will examine the KB in order to calculate every possible inference that could be entailed during KBS functioning, and then detect if any constraint (logic or semantic) is violated. Finally, the detected anomalies are reported in a suitable way for human analysis.

### 4.2 Knowledge base

The knowledge base schema was designed regarding the need to speed the expansion calculation since it is one of the most time consuming steps in the verification process. In Fig. 4 the concepts and relations contained in the schema are depicted.

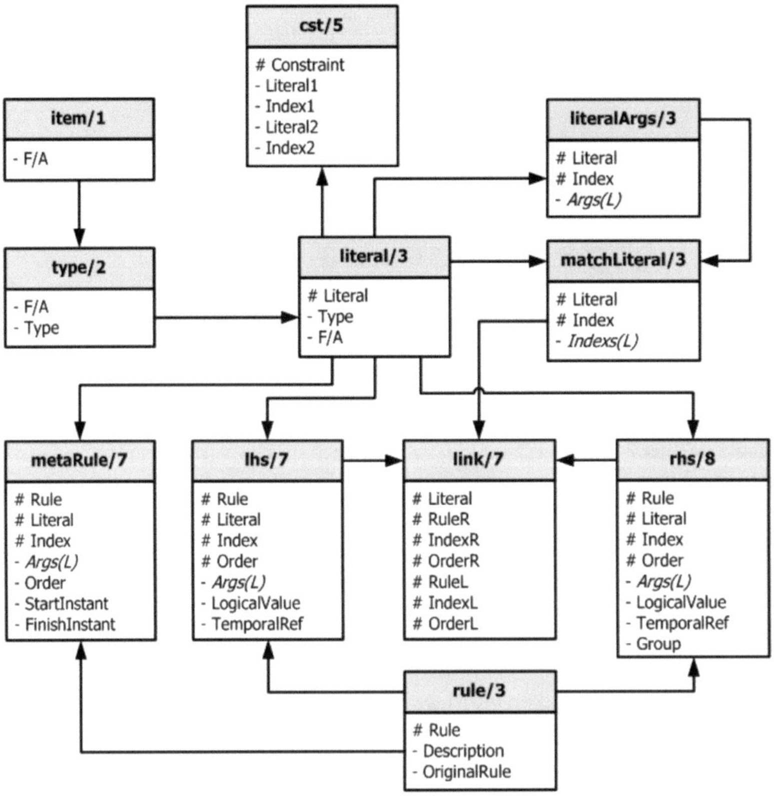


Fig. 4 - Knowledge Base Schema

This schema allows to: store the rules and meta-rules in an efficient way; classify the elements that compose such rules and meta-rules. Therefore, tuples item/1, type/2 and literal/3 allow to classify the elements (in VERITAS scope named literals) that compose both rule LHSs and RHSs. The tuples literalArgs/3 and matchLiteral/ represent, respectively: the arguments for each literal and the pairs of literals that could be matched during expansion calculation. The tuple rule/3 relates the old and new rule representations while lhs/7 and rhs/8 store the literals that compose each rule LHS and RHS, respectively. Finally, metaRule/7 stores the information related to original meta-rules.

$$\text{item}(F\ /\ A) \tag{3}$$

The tuple item/1 stores the functor and arity of each literal appearing in rules LHS and RHS contained in RB. For the rule d3-L1, the following items would be asserted:

item(abs_diff_less_or_equal/3).
item(breaker_opened/2).
item(triggering/9).
item(breaker/7).
item(msg/5).

$$\text{type}(F\ /\ A, Type) \tag{4}$$

The tuple type/2 allows the characterization of the items previously extracted and stored in item/1. The classification can be done manually or semi-automatically. The field Type can exhibit one of the following values:

- **interface** - an operation for user interface;
- **status** - state of a knowledge domain element (e.g., electrical devices);
- **process** - used to define eventualities with duration (non-instantaneous);
- **event** - used to define instantaneous eventualities;
- **time** - temporal operator used to reason about time;
- **operation** - used to define "procedural" operations like comparisons.

The following tuples type/2 would be created for the considered rule d3-L1:

type(abs_diff_less_or_equal/3,time).
type(breaker_opened/2,status).
type(triggering/9,status).
type(breaker/7,status).
type(msg/5,event).

$$\text{literal}(Literal, Type, F\ /\ A) \tag{5}$$

The tuple literal/3 stores the synthesis of type/2 and item/1. Besides, it allows labelling the literals with a key (Literal) for what they will be referred during the verification process. According to the example, the following instances of literal/3 would be created:

literal(tr1,time,abs_diff_less_or_equal/3).
literal(st5,status,breaker_opened/2).
literal(st9,status,triggering/9).
literal(st11,status,breaker/7).
literal(ev1,event,msg/5).

$$\text{literalArgs}(Literal, Index, Args) \tag{6}$$

During the expansions calculation, VERITAS replaces each literal X contained in a LHS rule by the literals that compose the LHS of the rule that allows inferring a literal Y if the literal Y matches X. Actually, this process simulates the inference engine using backward chaining. During the KBS usage the process of matching literals is straightway since the variables values are known; however, it doesn't happen in the verification process, so the number of expansions (possible inference chains) that the system needs to calculate grows exponentially. Additionally, during the expansions calculation each pair of literals needs to be checked often, so in order to avoid it, VERITAS computes *à priori* the pairs of literals that can be matched. Henceforth, the tuple literalArgs/3 stores the diverse occurrences of *similar* literals. Two literals are *similar* if they exhibit the same functor and arity but their respective lists of arguments are not *similar*. Two lists of arguments are similar if they have the same size and for each element in corresponding position one of the following situations happens:

- The argument is a free variable, meaning it can be matched with everything;
- The argument is a terminal (not a free variable) and in this case both arguments need to exhibit the same value.

Considering the literal st9 the following instances of literalArgs/3 would be created:

```
literalArgs(st9,1,[#Dt,#Tm,#In1,#Pn1,#In2,#NL,#Type,not_identified,#T2]).
literalArgs(st9,2,[#Dt1,#Tm1,#In1,#Pn1,#In2,#NL,monophasic,not_identified,#Tabr]).
literalArgs(st9,3,[#Dt2,#Tm2,#In1,#Pn1,#In2,#NL,triphasic,dtd,#Tabr2]).
literalArgs(st9,4,[#Dt2,#Tm2,#In1,#Pn1,#In2,#NL,monophasic,dmd,#Tabr2]).
literalArgs(st9,5,[#Dt1,#Tm1,#In1,#Pn1,#In2,#NL,triphasic,rel_rap_trif,#Tabr1]).
literalArgs(st9,6,[#Dt1,#Tm1,#In1,#Pn1,#In2,#NL,triphasic,dtr,#Tabr1]).
literalArgs(st9,7,[#Dt1,#Tm1,#In1,#Pn1,#In2,#NL,monophasic,rel_rap_mono,#Tabr1]).
literalArgs(st9,8,[#Dt1,#Tm1,#In1,#Pn1,#In2,#NL,monophasic,dmr,#Tabr1]).
literalArgs(st9,9,[#Dt1,#Tm1,#In1,#Pn1,#In2,#NL,triphasic,ds,#Tabr]).
literalArgs(st9,10,[#Dt1,#Tm1,#In1,#Pn1,#In2,#NL,#_, #TriggeringType,#Tabr]).
literalArgs(st9,11,[#Dt2,#Tm2,#In1,#Pn1,#In2,#NL,triphasic,not_identified,#Tabr]).
literalArgs(st9,12,[#Dt2,#Tm2,#In1,#Pn1,#In2,#NL,triphasic,close_defect,#Tabr]).
```

$$\text{matchLiteral}(\textit{Literal}, \textit{Index}, \textit{Indexes}) \tag{7}$$

The tuple matchLiteral/3 stores the possible matches between a literal defined by pair Literal/Index and a list of indexes. Notice that the process used to determine the possible matches between literals is similar to determining a Graph Transitive Closure where the pair Literal/Index is each graph node and the list Indexes is the set of adjacent edges. In the considered example the following instances of matchLiteral/3 would be created:

```
matchLiteral(st9,1,[1,2,10,11]).
matchLiteral(st9,2,[1,2,10]).
matchLiteral(st9,3,[3,10]).
matchLiteral(st9,4,[4,10]).
matchLiteral(st9,5,[5,10]).
matchLiteral(st9,6,[6,10]).
matchLiteral(st9,7,[7,10]).
matchLiteral(st9,8,[8,10]).
matchLiteral(st9,9,[9,10]).
matchLiteral(st9,10,[1,2,3,4,5,6,7,8,9,10,11,12]).
matchLiteral(st9,11,[1,10,11]).
matchLiteral(st9,12,[10,12]).
```

$$\mathrm{rule}(Rule, Description, OriginalRule) \qquad (8)$$

The tuple rule/3 allows relating the new rules used by VERITAS and the original ones. This tuple is quite useful in the user interaction since the user is more acquainted with the original rules. For the rule d3 the following instances of rule/3 would be created:

rule(d3-L1,Monophasic Triggering,d3).
rule(d3-L2,Monophasic Triggering,d3).

$$\mathrm{lhs}(Rule, Literal, Index, Position, TemporalArg, LogicalValue, Args) \qquad (9)$$

The tuple lhs/7 stores the set of conditions forming the rule LHS. Meaning, the occurrence of a literal (Literal/Index) in a rule and its position, temporal label (TemporalArg), logical value (since a condition can be a negation of the referred literal) and arguments. Considering the rule d3-L1, the following instances of lhs/7 would be asserted:

lhs(d3-L1,ev1,8,1,#T1,true,[#Dt1,#Tm1,[#In1,#Pn1,[#In2,#NL]],>>>TRIGGERING,01]).
lhs(d3-L1,st11,8,2,none,true,[#_,#_,#In1,#Pn1,#_,#_,fechado] ).
lhs(d3-L1,tr1,1,4,none,true,[#T2,#T1,30]).
lhs(d3-L1,ev1,7,3,#T2,true,[#Dt2,#Tm2,[#In1,#Pn1,[#In2,#NL,#_]],BREAKER,00]).

$$\mathrm{rhs}(Rule, Literal, Index, Position, TemporalArg, LogicalValue, Args, Type) \qquad (10)$$

The tuple rhs/8 stores the set of conclusions forming the rule RHS. This tuple has about the same structure as the lhs/7 but adds the field Type, which can exhibit the following values:

- **cf** - assertion of conclusion;
- **rf** - retraction of fact;
- **it** - user interface.

Considering the rule d3-L1, the following instances of rhs/8 would be asserted:

rhs(d3-L1,st9,1,1,#T1,true,cf,[#Dt1,#Tm1,#In1,#Pn1,#In2,#NL,monophasic,not_identified, #T2]).
rhs(d3-L1,st11,8,2,#T2,true,rf,[#_,#_,#In1,#Pn1,#_,#_,closed]).
rhs(d3-L1,st11,4,3,#T2,true,cf,[#Dt2,#Tm2,#In1,#Pn1,#_,triggering,mov]).
rhs(d3-L1,ev1,8,4,#T1,true,rf,[#Dt1,#Tm1,[#In1,#Pn1,[#In2,#NL]],>>>TRIGGERING,01]).
rhs(d3-L1,ev1,9,5,#T2,true,rf,[#Dt2,#Tm2,[#In1,#Pn1,[#In2,#NL | #_]],BREAKER,00]).
rhs(d3-L1,st5,1,6,#T1,true,rf,[#In1,#Pn1]).
rhs(d3-L1,st5,1,7,#T1,true,cf,[#In1,#Pn1]).

$$\mathrm{metaRule}(Rule, Literal, Index, Order, StartInstant, FinishInstant, Args) \qquad (11)$$

The tuple metaRule/7 stores the information the meta-rules used in the rule selection triggering mechanism. Hence, a rule is scheduled for the interval defined by StartInstant and FinishInstant if the literal defined by the pair Literal/Index is asserted in the KB. Concerning the tuple trigger/2 used by SPARSE (presented in section 3.1) and the rule d3, the following instances of metaRule/7 would be asserted:

metaRule(d3-L1,ev1,8,3,52,52,[#_,#_,[#Inst1,#Panel1,#_],>>>TRIGGERING,01]).
metaRule(d3-L2,ev1,8,4,52,52,[#_,#_,[#Inst1,#Panel1,#_],>>>TRIGGERING,01]).

$$\mathrm{link}(RuleL, Literal, IndexL, PositionL, RuleR, IndexR, PositionR) \qquad (12)$$

The tuple link/7 instantiates the tuple matchLiteral/3 for a specific knowledge base. While the matchLiteral/3 states which pairs of literals can be matched, the tuple link/7 stores which pairs of literals present in a KB can actually be matched during the expansions calculation. Concerning the literal st9 and the rule d3-L1, the following instances of link/7 would be created:

link(d3-L1,st9,2,1,d22,10,1).
link(d3-L1,st9,2,1,d21,10,2).
link(d3-L1,st9,2,1,d21,10,1).
link(d3-L1,st9,2,1,d5,2,1).
link(d3-L1,st9,2,1,i1,1,1).

$$\text{cst}(\mathit{Constraint}, \mathit{Literal1}, \mathit{Index1}, \mathit{Literal2}, \mathit{Index2}) \tag{13}$$

The tuple cst/5 allows storing the impermissible sets for a knowledge base. Each occurrence of this tuples states that a pair of literals is incompatible together.

### 4.3 Anomaly detection

The algorithm used for the anomaly detection, depicted in Fig. 5, works in the following way.

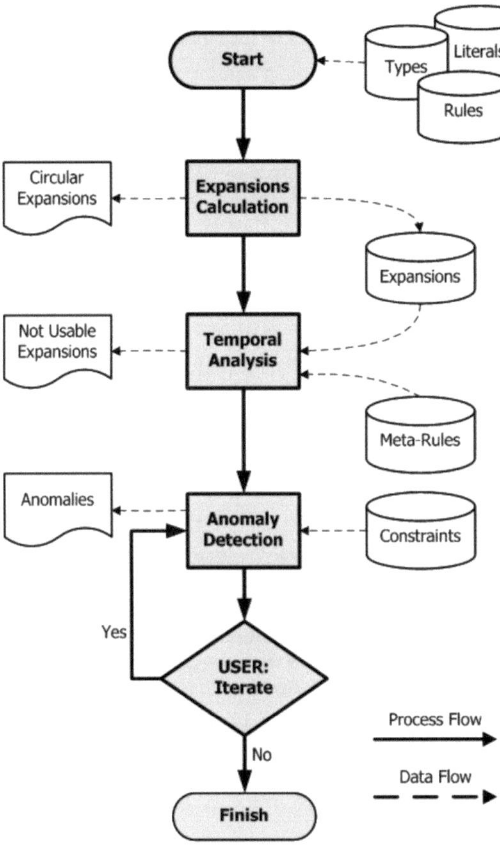


Fig. 5 - Anomaly Detection Algorithm

In the first step, the knowledge base is previously created (see section 4.2). After that, the expansions are calculated (see section 4.3.1); during this step the circular expansions are detected and labelled. In the next step, using both expansions and meta-rules, the temporal consistency of the expansions is evaluated (see section 4.3.2). Finally, specific algorithms are applied in order to detect the following anomalies: circularity (see section 4.3.3), ambivalence (see section 4.3.4) and redundancy (see section 4.3.5).

### 4.3.1 Expansions calculation

The expansions calculation process intends to thoroughly determine the inference chains that can be possibly drawn during the KBS functioning. Calculating an expansion consists in breadth-first search over a hypergraph, in which each hypernode is a set of literals and each transition represents a rule. The procedure calcExpansion works in the following way:

```
procedure calcExpansion(literal,expansion)
    input: literal
    output: expansion
    begin
    ;;Variables initialization
    rules = empty;
    expansion = empty;
    listE = pushBack(listE, (literal, rules));
    if (∃ rule(r,lhs,rhs) and (literal) ∈ rhs)) then
        (literal,rules) = popFront(listE);
        foreach(literal ∈ lhs) do
            listE = pushBack(listE,(literal,rules));
        end- foreach
        while listE ≠ empty do
            (literal,rules) = popFront(listE);
            if(∃ rule(r,lhs,rhs) and (literal ∈ rhs)) then
                ;;literal inferable;
                rules = pushBack(rules, r);
                if(r \in rules) then
                    ;;circular
                    expansion = pushBack(expansion, c(literal, rules));
                else
                    ;;inferable and not circular;
                    expansion = pushBack(expansion, e(literal, rules));
                    rules = pushBack(rules, r);
                    foreach literal ∈ lhs do
                        listE = pushBack(listE, (literal, rules));
                    end-foreach
                    expansion = updateLRulesExp(expansion, rules);
                end-if
            else
            ;;not inferable literal
            expansion = pushBack (expansion, f(literal));
        end-if
    end-while
end-procedure
```

First, some variables are initialized, namely: listE (the list of literals to be expanded); rules (list of rules that used in an expansion); expansion (list of the expanded literals). Hence, for each literal contained in the KB, a rule that allows its inference is selected and the literals contained in the LHS are stored in the list listE. This algorithm finishes when listE becomes empty, meaning that all literals were expanded. The elements contained in listE are iteratively popped and they can be one of the following types:

- **Not inferable** - this type of literals are ground facts or basic operations;
- **Inferable** - this type of literals can be inferred during KBS functioning. If a literal is part of a circular inference chain, the algorithm labels it and the expansion for this literal finishes; otherwise the algorithm calculates the set of literals needed to infer it and pushes this set into the listE.

Finally, the algorithm produces a list of the following kind of tuples:

- **f(literal)** - represents a literal not inferable;
- **e(literals,ruleList)** - represents a list of inferable literals and the rules used to infer them;
- **c(literals,ruleList)** - represents a list of inferable literals that configure a circular chain.

Let's consider the following set of rules:

```
rule(r1,[st1,ev1],[st2,st3])
rule(r2,[st3,ev3],[st6,ev5])
rule(r3,[ev3],[ev4])
rule(r4,[ev1,ev2],[ev4,st4])
rule(r5,[ev5,st5,ev4],[st7,st8])
```

After the use of the described algorithm over this set of rules, the following two expansions would be obtained:

```
f(ev3)
f(st5)
f(ev3)
f(ev1)
f(st1)
e([ev4],[r3])e([st2,st3],[r1])
e([st6,ev5],[r1,r2])
e([st7,st8],[r1,r2,r3,r5])
```

```
f(ev2)
f(ev1)
f(st5)
f(ev3)
f(ev1)
f(st1)
e([ev4,st4],[r4])
e([st2,st3],[r1])
e([st6,ev5],[r1,r2])
e([st7,st8],[r1,r2,r4,r5])
```

The dependencies between rules captured in the expansions can be graphically represented by the hypergraphs as depicted in the Fig. 6. This technique allows rule representation in a manner that clearly identifies complex dependencies across compound clauses in the rule base and there is a unique directed hypergraph representation for each set of rules (Ramaswamy & Sarkar 1997).

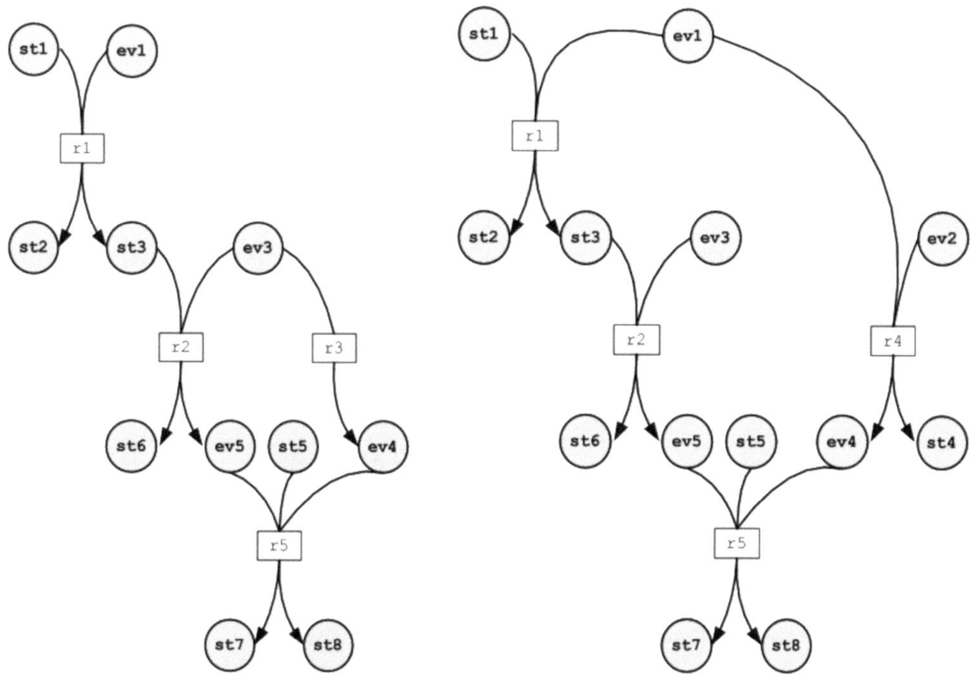


Fig. 6 - Hypergraphs for the literal st8

For sake of brevity, in the described algorithm some simplifications were considered, namely:

- **Local versus global** - as previously referred, in order to fasten the expansion calculation, the knowledge base includes the tuple link/7 that stores the pairs of literals able to mach together. Notice that link/7 assures only a local consistency (between the pair of literals), not global (along the entire expansion). Let's consider the following occurrences of the tuple literalArgs/3:

```
literalArgs(st11,5,[#Dt2,#Tm2,#In1,#Pn1,#_,relRapTrif,closed]).
literalArgs(st11,6,[#Dt2,#Tm2,#In1,#Pn1,#_,relRapMono,closed]).
literalArgs(st11,8,[#_,#_,#In1,#Pn1,#_,#Type,closed]).
```

  In this example, local consistency means that the pairs (st11/5, st11/8) and (st11/6, st11/8) can be matched. Global consistency means the chains (st11/5, st11/8, st11/5) and (st11/6, st11/8, st11/6) can be inferred. Furthermore, the chain (st11/5, st11/8, st11/6) isn't possible, since after the variable Type becomes instantiated with the relTrapTrif, it can't be re-instantiated with the value relTrapMono.
  In order to assure global consistency, the expansion calculation algorithm implements a table of symbols where the variables are stored and updated along with the expansion calculation;
- **Multiple conclusions per rule** - when a literal is expanded if a particular rule infers multiples conclusions that needs to be adequately stored. For instance, the rule r5 allows the inference of the literals st7 and st8, as depicted in the Fig. 6, obviously this situation is reflected in the tuples e/2 and c/2 contained in the expansion list;

- **Data representation** - in the described algorithm the rules are represented using a tuple rule(r,lhs,rhs) although in the algorithm implementation the tuples, previously described, are used (e.g., rule/3, lhs/7, rhs/8 or literal/3).

### 4.3.2 Temporal analysis

Concerning temporal analysis, the following issues where considered in the VERITAS implementation:

- **Distinct treatment for temporal variables** - this kind of variables are used for labelling literals, hence, during the internal knowledge base creation these variables where stored in lhs/7 and rhs/8 in the field TemporalRef. Later during the expansions calculation, they are indexed in a specific table of symbols with the following structure:

  $$(Var, BeginInst, EndInst)$$

  where each variable (Var) has a temporal interval of validity defined by its starting (BeginInst) and ending (EndInst) instants;
- **Capture and evaluation temporal operations** - the literals relating temporal operations contained in an expansion are captured in order to build a net representing their dependencies. This net is later evaluated aiming to assure temporal consistency over an entire expansion. Therefore, the following items are collected:
  - Literals for variables evaluation like: $t = t1 + 1$ and $t = \max(t1, t2)$;
  - Temporal relational operators like: $t1 \prec t2$ and $t \geq 30$;
  - Temporal operators used specifically in SPARSE like:
    `abs_Diff_Less_Or_Equal(t1,t2),` meaning $\|t2 - t1\| \leq t3$.

  Later the collected items are evaluated in order to:
  - Detect inconsistencies between related items like: $t1 \prec t2$ and $t1 \geq t2$;
  - Assert and/or update the table of temporal symbols, for instance, $t1 \prec t2$ and $t \geq \max(t1, t2)$ allow updating variable t with the value of t2.
- **Parametric temporal validity analysis** - the meta-rules stores temporal window of validity for a set of rules. The maximum validity for literals contained in an expansion is defined by the combination of the temporal validity intervals inherited from all rules used in the referred expansion. Additionally, this parametric evaluation is enriched with temporal operations described in the previous items. Regarding that, each literal usually has symbolic, starting and ending instants defined by knowledge base assert and retract operations.

### 4.3.3 Circularity detection

A knowledge base contains circularity if, and only if, it contains a set of rules, which allows an infinite loop during rule triggering. In order to accomplish modelling requirements, sometimes the knowledge engineer needs to specify rules that allow the definition of a circular inference chain. The algorithm used to compute expansions detects every circular chain existing in a rule base; although aiming to reduce the number of false anomalies, a heuristic was considered for circularity detection. This heuristic has two mandatory conditions:

- At least one literal of the type event needs to be present in the rule LHS. This implies the occurrence of an action requiring some rule triggering. Besides, the referred literal needs to be defined as the triggering fact in the metaRule/7 tuple;
- The culprit literals for circularity (equivalent literals) need to exhibit distinct temporal label. More precisely, the literal that appears in rule RHS needs to occur after the equivalent literal contained in the rule LHS.

### 4.3.4 Ambivalence detection

A knowledge base is ambivalent, if and only if, for a permissible set of conditions, it is possible to infer an impermissible set of hypotheses. Concerning the detection of ambivalence, VERITAS is capable of detecting two types: in a single expansion and in multiple expansions.

The detection of ambivalence in a single expansion is performed using an algorithm that works in the following way: for each pair literal1/index1 representing a conclusion (i.e., a literal which appears solely in the rules RHS) contained in the KB the following conditions are evaluated:

- The existence of an expansion that allows to infer the pair literal1/index1;
- The existence of an restriction relating the referred pair;
- The other literal contained in the restriction, literal2/index2, is contained in the expansion considered in the first point.

If all of these conditions are true, it means that two contradictory are contained in the same inference chain. In the last step the validity intervals defined for both literals, represented by literal1/index1 and literal2/index2, respectively, are evaluated, and if they intercept[2] each other then an anomaly is reported.

The detection of ambivalence in multiples expansions is performed using an algorithm that works in the following way: for each constraint contained in KB, if there are expansions supporting both literals contained in the restriction, represented by literal1/index1 and literal2/index2, respectively; finally the algorithm evaluates if the set of literals supporting literal1/index1 contains, or is contained by, the set of literals that support literal2/index2 and if so, an anomaly is reported. Notice that the notion of contain, or contained by, inherited from the set theory is refined with the condition of interception between corresponding literals as depicted in the Fig. 7. The set P (formed by the elements Pi, Pj, Pk and Pl) contains the set Q, since each element of Q exists simultaneously both in P and Q. The set R represents the temporal interception between P and Q.

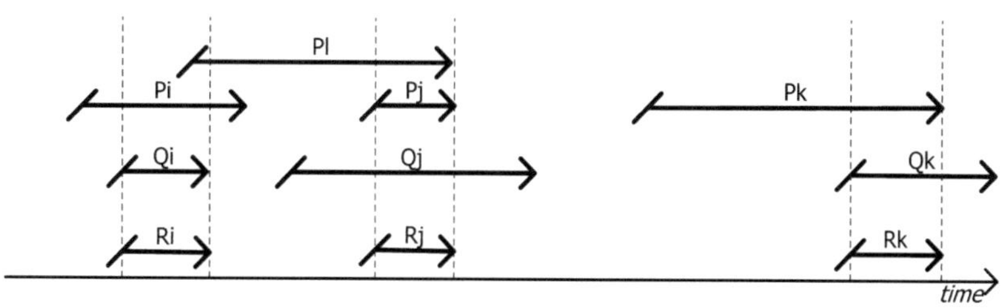


Fig. 7 - Set contains or contained by with temporal characterization

[2] Two temporal intervals intercept each other if they share at least an instant.

### 4.3.5 Redundancy detection

A knowledge base is redundant if, and only if, the set of final hypotheses is the same in the rule/literal presence or absence. Concerning the detection of the anomaly referred as redundancy, two distinct types where addressed: not usable rule and redundancy in groups of rules.

The detection of unused rules is performed using a rather simple algorithm that works in the following way: for each rule R contained in RB the following conditions need to be verified or an anomaly is reported:

- At least a meta-rule refers the R; if not, the rule wouldn't be ever called;
- The R rule LHS doesn't contain a pair of literals defined in any constraint, as long as an impermissible set of conditions would not be provided as input.

The detection of redundancy in groups of rules is performed using an algorithm that works in the following way: after all expansions calculation, all related expansions for each rule are checked, and if all the conclusions can be inferred by other expansions then the considered rule is redundant. Consequently, an anomaly is reported.

## 5. Results

In the development of VERITAS some well-known techniques were assembled with a set of new ones specifically designed for VERITAS. The set of techniques referred in literature includes: the calculation of the rule base expansions (Ginsberg 1987); detection of set of anomalies (Rousset 1988; Preece & Shinghal 1994); the use of graph theory for anomaly detection along with inference chains (Preece, Bell & Suen 1992); the use of logical and semantic constraints (Preece & Shinghal 1994; Zlatareva 1991) and directed hypergraphs for representing rule dependencies (Ramaswamy & Sarkar 1997). Therefore, VERITAS has the following characteristics:

- **Independence of the original rule grammar and syntax** - VERITAS includes a module that allows the conversion between original and verification representation formats;
- **Optimized rule base expansions calculation** – in order to fasten the expansions calculation two different techniques were considered: the information needed during the verification process was stored using a normalized data schema in which most important data issues where indexed; the matching pairs of literals were computed *à priori* and stored in the knowledge ensuring *local* consistency;
- **Variables and procedural instructions correctly addressed** – in order to process variables in an adequate way during knowledge base creation, the variables contained in the rule and meta-rule sets are extracted and later stored in the table of symbols. During the expansions calculation, the variables are evaluated and their values are updated in the table of symbols. The use of this mechanism allowed assuring global consistency through an expansion calculation and, consequently, reducing the number of computed expansions. The procedural instructions were considered during expansions calculation, and whenever it implies variables evaluation the table of symbols is updated accordingly;
- **Temporal aspects** – in the development of VERITAS a set of techniques and algorithms were considered in order to address the knowledge temporal reasoning representation issues, namely: definition of an anomaly classification temporal characterized, as well as the temporal characterization of logical and semantic restrictions; variables related

with time representation received a distinct treatment during expansion calculation; capture and evaluation of operations relating time, in order to evaluate the consistency of the net formed by these operations;

- **VERITAS testing** – the verification method supported by VERITAS was tested with SPARSE, an expert system for incident analysis and power restoration in power transmission networks. Regarding that, a previous version of VERITAS was tested with two expert systems for: cardiac diseases diagnosis (Rocha 1990) and otology diseases diagnosis and therapy (Sampaio 1996).

## 6. Conclusions and future work

This chapter focussed on some aspects of the practical use of KBS in Control Centres, namely, knowledge maintenance and its relation to the verification process.
The SPARSE, a KBS used in the Portuguese Transmission Network (REN) for incident analysis and power restoration was used as case study. Some of its characteristics that mostly constrained the development and use of verification tool were discussed, like: the use o rule selection triggering mechanism, temporal reasoning and variables evaluation, hence, the adopt solutions were described.
VERITAS is a verification tool that performs logical tests in order to detect knowledge anomalies as described.
The results obtained show that the use of verification tools increases the confidence of the end users and eases the process of maintaining a knowledge base. It also reduces the testing costs and the time needed to implement those tests.

## 7. References

Amelink, H., Forte, A. & Guberman, R., 1986. Dispatcher Alarm and Message Processing. *IEEE Transactions on Power Systems*, 1(3), 188-194.

Boehm, B.W., 1984. Verifying and validating software requirements and design specifications. *IEEE Software*, 1(1), 75-88.

Cragun, B. & Steudel, H., 1987. A decision table based processor for checking completeness and consistency in rule based expert systems. *International Journal of Man Machine Studies (UK)*, 26(5), 633-648.

Duarte, J. et al., 2001. TEMPUS: A Machine Learning Tool. In *International NAISO Congress on Information Science Innovations (ISI'2001)*. Dubai/U.A.E., pp. 834-840.

Faria, L. et al., 2002. Curriculum Planning to Control Center Operators Training. In *International Conference on Fuzzy Systems and Soft Calculational Intelligence in Management and Industrial Engineering*. Istanbul/Turkey, pp. 347-352.

Fisher, M., Gabbay, D. & Vila, L. eds., 2005. *Handbook of Temporal Reasoning in Artificial Intelligence*, Elsevier Science & Technology Books.

Gerevini, A., 1997. Reasoning about Time and Actions in Artificial Intelligence: Major Issues. In O.Stock, ed. *Spatial and Temporal Reasoning*. Kluwer Academic Públishers, pp. 43-70.

Ginsberg, A., 1987. A new aproach to checking knowledge bases for inconsistency and redundancy. *In procedings 3rd Annual Expert Systems in Government Conference*, 10-111.

Gonzalez, A. & Dankel, D., 1993. *The Engineering of Knowledge Based Systems - Theory and Practice*, Prentice Hall International Editions.

Hoppe, T. & Meseguer, P., 1991. On the terminology of VVT. *Proceedings of the European Workshop on the Verification and Validation of Knowledge Based Systems*, 3-13.

Kirschen, D. & Wollenberg, B., 1992. Intelligent Alarm Processing in Power Systems. *Proceedings of the IEEE*, 80(5), 663-672.

Kleer, J., 1986. An assumption-based TMS. *Artificial Intelligence \(Holland\)*, 2(28), 127-162.

Malheiro, N. et al., 1999. An Explanation Mechanism for a Real Time Expert System: A Client-Server Approach. In *International Conference on Intelligent Systems Application to Power Systems (ISAP'99)*. Rio de Janeiro/Brasil, pp. 32-36.

Menzies, T., 1998. Knowledge Maintenance: The State of the Art. *Engineering Review*.

Nazareth, D., 1993. Investigating the applicability of Petri Nets for Rule Based Systems Verification. *IEEE Transactions on Knowledge and Data Engineering*, 4(3), 402-415.

Nguyen, T. et al., 1987. Knowledge Based Verification. *AI Magazine*, 2(8), 69-75.

Pipard, E., 1989. Detecting Inconsistencies and Incompleteness in Rule Bases: the INDE System. In *Proceedings of the 8th International Workshop on Expert Systems and their Applications, 1989, Avignon,France;*. Nanterre, France: EC2, pp. 15-33.

Preece, A., 1998. Building the Right System Right. In *Proc.KAW'98 Eleventh Workshop on Knowledge Acquisition, Modeling and Management*.

Preece, A., Bell, R. & Suen, C., 1992. Verifying knowledge-based systems using the COVER tool. *Proccedings of 12th IFIP Congress*, 231-237.

Preece, A. & Shinghal, R., 1994. Foundation and Application of Knowledge Base Verification. *International Journal of Intelligent Systems*, 9(8), 683-702.

Ramaswamy, M. & Sarkar, S., 1997. Global Verification of Knowledge Based Systems via Local Verification of Partitions. In *Proceedings of the European Symposium on the Verification and Validation of Knowledge Based Systems*. Leuven,Belgium, pp. 145-154.

Rocha, J., 1990. Concepção e Implementação de um Sistema Pericial no Domínio da Cardiologia. Master Thesis, FEUP, Porto, Portugal.

Rousset, M., 1988. On the consistency of knowledge bases:the COVADIS system. In *Procedings of the European Conference on Artificial Intelligence (ECAI'88)*. Munchen, pp. 79-84.

Sampaio, I., 1996. Sistemas Periciais e Tutores Inteligentes em Medicina - Diagnóstico, terapia e apoio de Otologia. Master Thesis, FEUP, Porto, Portugal.

Suwa, M., Scott, A. & Shortliffe, E., 1982. An aproach to Verifying Completeness and Consistency in a rule based Expert System. *AI Magazine (EUA)*, 3(4), 16-21.

Vale, Z. et al., 1997. SPARSE: An Intelligent Alarm Processor and Operator Assistant. *IEEE Expert, Special Track on AI Applications in the Electric Power Industry*, 12(3), 86-93.

Vale, Z. et al., 2002. Real-Time Inference for Knowledge-Based Applications in Power System Control Centers. *Journal on Systems Analysis Modelling Simulation (SAMS), Taylor&Francis*, 42, 961-973.

Vanthienen, J., Mues, C. & Wets, G., 1997. Inter Tabular Verification in an Interactive Environment. *Proceedings of the European Symposium on the Verification and Validation of Knowledge Based Systems*, 155-165.

Wu, C. & Lee, S., 1997. Enhanced High-Level Petri Nets with Multiple Colors for Knowledge Verification/Validation of Rule-Based Expert Systems. *IEEE Transactions on Systems. Man, and Cybernetics*, 27(5), 760-773.

Zlatareva, N., 1991. Distributed Verification: A new formal approach for verifying knowledge based systems. In I. Liebowitz, ed. *Proceedings Expert systems world congress*. New York, USA: Pergamon Press, pp. 1021-1029.